Proceedings of the International Colloquium on Sports Science, Exercise, Engineering and Technology 2014 (ICoSSEET 2014)

Rahmat Adnan · Shariman Ismadi Ismail
Norasrudin Sulaiman
Editors

Proceedings of the International Colloquium on Sports Science, Exercise, Engineering and Technology 2014 (ICoSSEET 2014)

Springer

Editors
Rahmat Adnan
Shariman Ismadi Ismail
Norasrudin Sulaiman
Faculty of Sports Science and Recreation
Universiti Teknologi MARA
Shah Alam
Malaysia

ISBN 978-981-10-1372-0 ISBN 978-981-287-107-7 (eBook)
DOI 10.1007/978-981-287-107-7
Springer Singapore Heidelberg New York Dordrecht London

Printed on acid-free paper

Springer is part of Springer Science+Business Media (www.springer.com)

Foreword

On behalf of the Organizing Committee of the International Colloquium on Sports Science, Exercise, Engineering and Technology (ICoSSEET 2014), it gives me great pleasure to welcome all delegates to Penang, Malaysia, to our inaugural event. This event held from 7 to 9 April 2014 (ICoSSEET 2014) is sponsored by IEEE Malaysia. The purpose of the colloquium is to create a forum for sport scientists, engineers and practitioners throughout the world to present the latest finding, techniques and progress in Sports Science, Exercise Science, Sports Engineering and Sports Technology.

Although this is the first ever ICoSSEET event, we would like to establish a good quality control on the submitted manuscript. The acceptance rate for this colloquium is 70 % based on recommendation by a total of 150 reviewers worldwide who have selected submissions through a rigorous review process. On behalf of the organizing committee, we take this opportunity to thank reviewers for their excellent contribution towards ICoSSEET 2014.

Thank You!

Acknowledgments

We also thank all authors, session chairpersons and delegates for their strong support and contribution to ICoSSEET 2014. Special acknowledgement is reserved to Research Management Institute (RMI), UiTM and Assoc. Prof. Dr. Rugayah Hashim for their support. Not to forget as well, all committee members, colleagues and friends who worked hard to ensure the success of ICoSSEET 2014. I would like to encourage you to explore the beautiful sights of Penang Malaysia during your stay and do enjoy the colloquium.

Rahmat Adnan
Mustafar Kamal Hamzah

Contents

Part III Sports Physiology and Behaviour

Part VI Sports Industry and Management

Introduction

We are delighted to present the Book of Proceedings that is the outcome of an open call for papers related to the work delegates presented at ICoSSEET 2014 International Colloquium on Sports Science, Exercise, Engineering and Technology, Penang, Malaysia on 7–9 April 2014. The book covers topics on sports science and exercise, sports engineering and technology and sports industry. The quantitative and qualitative scientific researches in the above topics are in-line with current research interest.

Organizing Commitee

General Chair
Rahmat Adnan

General Co-Chair
Mustafar Kamal Hamzah

Technical Program Chair
Shariman Ismadi Ismail

Publication Chairs
Norasrudin Sulaiman

Technical Program Committee
Nagoor Meera Abdullah
Wahidah Tumijan
Adam Linoby
Hanifa Sariman
Mazlifah Omar

Technical Secretariat
Ebby Waqqash Mohamad Chan
Bashtiah Nahrul Khair
Nursyuhada Zainal Abidin
Silvia Augustine
Lucy Jawan
Mohammad Hisyam Izzuddin
Abdul Hadi Ruslan

Reviewers

Abdul Kadir Othman (Universiti Teknologi MARA, Malaysia)
Abdul Samad Abdul Rahman (Universiti Teknologi MARA, Malaysia)
Adam Linoby (Universiti Teknologi MARA, Malaysia)
Ahbul Zailani Begum Mohamed Ibrahim (Universiti Teknologi MARA, Malaysia)
Ahmad Naim Ismail (Ministry of Higher Education, Malaysia)
Ahmad Saiful Azlin Puteh Salin (Universiti Teknologi MARA, Malaysia)
Akash Singh (IBM, USA)
Alauyah Johari (Universiti Teknologi MARA, Malaysia)
Alex Morgan (Sunprogrammer Limited, UK)
Amelia Wong Azman (International Islamic University Malaysia, Malaysia)
Amin Mohajer (Iran Telecommunication Research Center (ITRC), Iran)
Amran Awang (Universiti Teknologi MARA, Malaysia)
Anthony Leow (Ministry of Education, Singapore)
Azhar Abdul Rahman (Universiti Teknologi MARA, Malaysia)
Baderisang Mohamed (Universiti Teknologi MARA, Malaysia)
Balbir Singh (University of Malaya, Malaysia)
Bashtiah Nahrul Khair (I-RENANG, Malaysia)
Boon Chong Ang (IEEE C&S, Malaysia)
Boon Hooi Lim (University Malaya, Malaysia)
Bossoufi Badre (University of Mohammed I, Oujda, Morocco)
Carla De Conti (E. Medea Research Institute Conegliano, Italy)
Carmel Nottle (University of South Australia, Australia)
Chandrela Das (Amity University, India)
Chengzong Pang (The Electric Reliability Council of Texas, USA)
Choo Kim Tan (Multimedia University, Malaysia)
Daniel Fernandes De Souza (Gold Coast City Council, Australia)
Dasimah Omar (Universiti Teknologi MARA, Malaysia)
Dinesh Sathyamoorthy (Science and Technology Research Institute for Defence (STRIDE), Malaysia)
Ebby Waqqash (Universiti Teknologi MARA, Malaysia)
Eduard Babulak (Sungkyunkwan University, Korea)
Eliana Werbin (Universidad Nacional de Cordoba, Argentina)
Emad Abdelrasoul (University of Calgary, Canada)
Erie Zuraidee Zulkifli (University Sains Malaysia, Malaysia)

Fadi Maayah (Edith Cowan University, Australia)
Fakhrul Hazman Yusoff (Universiti Teknologi MARA, Malaysia)
Ghazirah Mustapha (Universiti Teknologi MARA, Malaysia)
Giovanni B. Palmerini (Università di Roma La Sapienza, Italy)
Grzegorz Debita (Wroclaw University of Technology, Poland)
Gustavo Quintana-Carapia (Instituto Tecnologico de Toluca, Mexico)
Habibah Hashim (Universiti Teknologi MARA, Malaysia)
Hafizi Muhamad Ali (Yanbu University College, Saudi Arabia)
Haijun Pan (New Jersey Institute of Technology, USA)
Hamzah Asyrani Sulaiman (University Teknikal Malaysia Melaka, Malaysia)
Hanifa Sariman (Universiti Teknologi MARA, Malaysia)
Hasnul Faisal Hushin Amri (Universiti Teknologi MARA, Malaysia)
Ho Keat Leng (Nanyang Technological University, Singapore)
Hosni Hasan (Universiti Teknologi MARA, Malaysia)
Illa Kolani (Wireless Communications, Antenna and Propagation, BUPT, P.R. China)
Ina Suryani Ab Rahim (University Malaysia Perlis, Malaysia)
Ioannis Atsonios (Atypon Systems Inc, Greece)
Jaime Calvo-Gallego (University of Salamanca, Spain)
Jasni Sulong (University Sains Malaysia, Malaysia
Jeff Walkley (RMIT, Australia)
Jerome Rachele (Queensland University of Technology, Australia)
Jian-Jun Shu (Nanyang Technological University, Singapore)
Jinrui Huang (National Laboratory, USA)
John Mo (RMIT University, Australia)
Jonathon Headrick (Queensland University of Technology, Australia)
Josip Music (University of Split, Croatia)
Jouni Tervonen (University of Oulu, Finland)
Kalaivani Chellappan (University Kebangsaan Malaysia, Malaysia)
Kazuyuki Kojima (Saitama University, Japan)
Konstantinos Kalemis (National and Kapodistrian University of Athens, Greece)
Lucas Trambaiolli (Federal University of ABC, Brazil)
Mahenderan Appukutty (Universiti Teknologi MARA, Malaysia)
Mahendra Kumar Yadav (Sikkim Manipal University, India)
Manoj Sharma (BVCOE, India)
Mardiana Mazaulan (Universiti Teknologi MARA, Malaysia)
Matthias Vodel (Chemnitz University of Technology, Germany)
Mazlan Ismail (Universiti Teknologi MARA, Malaysia)
Mazlifah Omar (Universiti Teknologi MARA, Malaysia)
Mehdi Bahrami (University of California, USA)
Michael O'Grady (University College Dublin, Ireland)
Ming-Feng Wu (Taichung Veterans General Hospital, Taiwan)
Mohamad Fahmi Hussin (Universiti Teknologi MARA, Malaysia)
Mohamed Gowaid (Alexandria University, Egypt)
Mohammad Faiz Liew Abdullah (University Tun Hussein Onn, Malaysia)

Mohammad Nasir Abdullah (Universiti Teknologi MARA, Malaysia)
Mohd Azlan Mohd Ishak (Universiti Teknologi MARA, Malaysia)
Mohd Fakharul Zaman Raja Yahya (Universiti Teknologi MARA, Malaysia)
Mohd Helmy Abd Wahab (University Tun Hussein Onn Malaysia, Malaysia)
Mohd Rozi Ahmad (Universiti Teknologi MARA, Malaysia)
Mohd Shariman Shafie (Universiti Teknologi MARA, Malaysia)
Mohd Tahir Ismail (University Sains Malaysia, Malaysia)
Mohd. Zulkhairi Mohd. Azam (Universiti Teknologi MARA, Malaysia)
Muhamad Noor Mohamed (Universiti Teknologi MARA, Malaysia)
Muhamad Safiq Saiful Annur (Universiti Teknologi MARA, Malaysia)
Muhammad Mahadi Abdul Jamil (University Tun Hussein Onn Malaysia, Malaysia)
Muhammad Sufyan Mohamad Zaki (Universiti Teknologi MARA, Malaysia)
Muhammad Zulqarnain Mohd Nasir (Universiti Teknologi MARA, Malaysia)
Nagoor Meera Abdullah (Universiti Teknologi MARA, Malaysia)
Nasuha Lee Abdullah (University Sains Malaysia, Malaysia)
Nelfianty Mohd Rashid (University Pendidikan Sultan Idris, Malaysia)
Nor Azira Mohd Radzi (Universiti Teknologi MARA, Malaysia)
Nor Hazana Abdullah (University Tun Hussien Onn Malaysia, Malaysia)
Norasrudin Sulaiman (Universiti Teknologi MARA, Malaysia)
Norazah Mohd Suki (University Malaysia Sabah, Malaysia)
Norazan Mohamed Ramli (Universiti Teknologi MARA, Malaysia)
Norhasniza Mohd Hasan Abdullah (Universiti Teknologi MARA, Malaysia)
Norlizah Abdul Hamid (Universiti Teknologi MARA, Malaysia)
Nurhidayah Omar (University Malaysia Perlis, Malaysia)
Nurul Afiqah Bakar (Universiti Teknologi MARA, Malaysia)
Nurul Hidayah Amir (Universiti Teknologi MARA, Malaysia)
Ong Tah Fatt (Universiti Teknologi MARA, Malaysia)
Oscar Oviedo-Trespalacios (Universidad del Norte, Colombia)
Parminder Singh Reel (The Open University, UK)
Peter O'donoghue (Cardiff Metropolitan University, UK)
Phang Yook Ngor (Universiti Teknologi MARA, Malaysia)
Rachel Samuel (Universiti Teknologi MARA, Malaysia)
Rahizam Rahim (Universiti Teknologi MARA, Malaysia)
Rahmat Adnan (Universiti Teknologi MARA, Malaysia)
Raihana Sharir (Universiti Teknologi MARA, Malaysia)
Raja Mohammed Firhad Raja Azidin (Universiti Teknologi MARA, Malaysia)
Rajesh Duvvuru (National Institute of Technology Jamshedpur, India)
Rezian-na Muhammed Kassim (Universiti Teknologi MARA, Malaysia)
Rohana Sham (Universiti Teknologi MARA, Malaysia)
Rosdiyana Samad (University Malaysia Pahang, Malaysia)
Rozita Abdul Latif (Universiti Teknologi MARA, Malaysia)
Runer Marson (Brazilian Army Research Institute Physical Fitness, Brazil)
Safanah M. Raafat (University of Technology Baghdad, Iraq)
Santwana Mishra (Babasaheb Ambedkar Marathwada University, India)

Sareena Hanim Hamzah (University Malaya, Malaysia)
Sarel W. J. Bekker (Vaal University of Technology (VUT), South Africa)
Sarimah Ismail (Universiti Teknologi MARA, Malaysia)
Sarina Md. Yusof (Universiti Teknologi MARA, Malaysia)
Sasitharan Nagapan (Politeknik Sultan Salahuddin Abdul Aziz Shah (PSA), Malaysia)
Satoru Okamoto (Shimane University, Japan)
Shahrizat Shaik Mohamed (KDU College Penang, Malaysia)
Shamshul Anaz Kassim (Universiti Teknologi MARA, Malaysia)
Shariman Ismail (Universiti Teknologi MARA, Malaysia)
Shigeho Tanaka (National Institute of Health and Nutrition, Japan)
Siti Haryati Shaikh Ali (Universiti Teknologi MARA, Malaysia)
Siti Syairah Mohd Mutalip (Universiti Teknologi MARA, Malaysia)
Sofwan Naim (Universiti Teknologi MARA, Malaysia)
Suhaila Abd Muid (Universiti Teknologi MARA, Malaysia)
Suhana Mohamed (Universiti Teknologi MARA, Malaysia)
Syahrul Fahmy (TATI University College, Malaysia)
Tan Chee Hian (Universiti Teknologi MARA, Malaysia)
Tang Howe Eng (Universiti Teknologi MARA, Malaysia)
Tien Choon Toh (University Tunku Abdul Rahman, Malaysia)
Victor Selvarajah Selvanayagam (University Malaya, Malaysia)
Victoria Erosa (University of Applied Sciences, Hochshule, Bremen, Mexico)
Vincent Parnabas (Universiti Teknologi MARA, Malaysia)
Wahidah Tumijan (Universiti Teknologi MARA, Malaysia)
Wan Khairunizam Wan Ahmad(University Malaysia Perlis, Malaysia)
Wan Mohd Norsyam Wan Norman (Universiti Teknologi MARA, Malaysia)
Yasir Mohd Mustafah (International Islamic University Malaysia, Malaysia)
Zeashan Khan (NUST, Rawalpindi, Pakistan)
Zuraidah Ahmad (Universiti Teknologi MARA, Malaysia)

Part I
Sports Science and Technology Application

The Mechanical Aspects of Martial Arts: Total Time of Execution and Kinematics of *Kaedah A*

Ghazirah Mustapha, Muzammer Zakaria, Wan Ruzaini Wan Sulaiman and Jamaluddin Mahmud

Abstract The ability of Kaedah A as a technique to fend off an attack with bare hands is questionable since the estimated ordinary human minimal reaction time is 0.18 s while the offensive force can reach its target in less than 0.1 s. Therefore, this study aimed to analyze the effectiveness of Kaedah A based on its total execution time and to describe the kinematic characteristics of the hand movements during Kaedah A's execution. The experiment was carried out using the motion capture method. The Kinect sensor detects the hand motion, while the Virtual Sensei Lite directly processes the motion capture through a digitizing procedure to prepare the coordinate data for further analysis. The execution of Kaedah A was repeated five times by four experienced Seni Silat Cekak Malaysia (SSCM) practitioners to investigate its accuracy and repeatability. The obtained data have provided the input for the trajectory mapping procedure for initial and end point identifications. Time difference, Δt, between these two points has demonstrated that the total time of execution for Kaedah A is less than 0.1 s. Further analysis involved smoothing out the obtained coordinate data in order to generate the polynomial equations of the motions of the hand during Kaedah A's execution. Based on the velocity-time graph generated by the equation of motion,

---

This work is supported by Persatuan Seni Silat Cekak Malaysia, Universiti Teknologi MARA and Ministry of Education (MOE) Malaysia, Grant no. 600-RMI/FRGS5/3(76/2013).

G. Mustapha (✉) · J. Mahmud
Faculty of Mechanical Engineering, Universiti Teknologi MARA, Shah Alam, Malaysia
e-mail: ghazirah039@ppinang.uitm.edu.my

M. Zakaria
School of Electrical and Electronic Engineering, USM, Seri Iskandar, Malaysia

W. R. Wan Sulaiman
Advanced Material Research Centre, SIRIM Bhd, Kulim, Malaysia

R. Adnan et al. (eds.), *Proceedings of the International Colloquium on Sports Science, Exercise, Engineering and Technology 2014 (ICoSSEET 2014)*,
DOI: 10.1007/978-981-287-107-7_1,

it can be concluded that the execution of Kaedah A has the features of a ballistic movement. The findings have provided useful data for reliability prediction as well as enhancement of Kaedah A's execution.

Keywords Defensive · Ballistic · Martial art · Silat Cekak

1 Introduction

A fist or a punch is targeted to defeat the opponent at a controlled distance in a short time rate [1]. The time for a punch to reach its target is about 0.1 s [2]. The reactions upon receiving a punch are numerous. The opponent may evade, grab, fend off or just be stunned without doing anything. In addition, the estimated minimal theoretical time for a normal human reaction toward such attack is 0.18 s. This is the summation of the minimum time for each of these three phases: (i) realization of the stimulus (0.05 s), (ii) selecting an adequate reaction (0.05 s), and (iii) start movement (0.08 s) [2]. Based on this argument, the ability to defend oneself from such offensive attacks only with bare hands is almost impossible and is often associated with having supernatural skills.

Seni Silat Cekak Malaysia (SSCM) has a different perspective regarding this matter. *SSCM* is a genuine traditional Malay combat-oriented art and is different from other martial arts due to its focus on defensive aspects. *SSCM* consists of components of movements known as *Kaedah*, *Buah Asas*, *Buah Jatuh*, *Buah Potong,* and *Buah Serang*. Table 1 summarizes the definition of each terminology used to represent the techniques included in the *SSCM* syllabus. *SSCM* applies 99 % defending techniques and 1 % attacking technique.

SSCM practitioners are introduced to alternative methods to deal with various types of attacks. They do not evade when dealing with an attack. Instead, they will effectively fend off the attack and move forward. There are four fend off techniques in SSCM; *Kaedah A, Kaedah B, Kaedah C,* and *Kaedah D,* that serve as the first moves in every *Buah Asas* in SSCM. All of these *Kaedah* are able to repel attacks depending on the point of concentration of the attack.

Kaedah A is a simple and effective fending off move to be executed to repel an attack—such as a punch—within the thorax area, as indicated in Fig. 1. The execution begins by moving the hand to slap away the opponent's arm. Figure 2 shows the hand's (a) initial and (b) final position during the execution of *Kaedah A*.

However, there is currently a lack of biomechanical analysis for the execution of *Kaedah A*. The fundamental understandings on the mechanics of the human hands during *Kaedah A*'s application are important as they help to predict the reliability of the technique as well as increase its effectiveness.

According to a study in the psychology of human movements, movements with high-force and high-velocity characteristics are considered as being ballistic in nature [4]. Ballistic movements involve spontaneous propulsion of the limbs, a

Table 1 Description of terminology used in *Seni Silat Cekak Malaysia*

Terminology used	Technique description
Kaedah	Fend off an attack by changing the direction of incoming force
Buah Asas	Continuation of fend off technique before taking down the opponent
Buah Jatuh	Takedown technique
Buah Potong	Clinch techniques that teach the practitioner on how to escape even when in close proximity with the opponent
Buah Serang	Strike technique

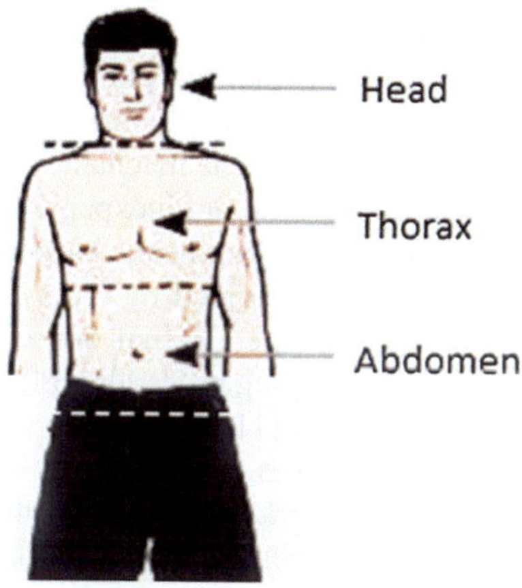

Fig. 1 The human body [3]

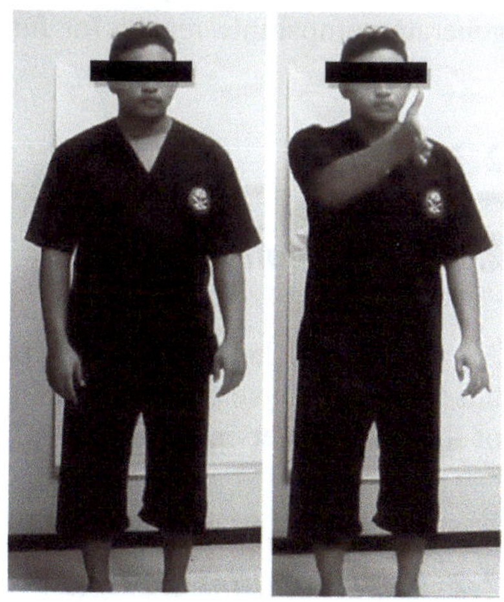

Fig. 2 *Kaedah A*'s execution: **a** initial position and **b** final position

class of point-to-point fast hand movements, characterized by straights paths and bell-shaped velocity [5]. Thus, punching and kicking may be characterized as possessing ballistic movements [6, 7]. This is due to the proficient movements, the

Table 2 Physical characterization of participants

Physical characterization	Mean value ± SD
Age	26.25 ± 5.6 years
Height	170.25 ± 2.5 cm
Weight	80.75 ± 6.5 kg

increase of speed, and the impulsive propulsion that may lead to a simple trajectory by the body parts involved [8]. This area of study indicates that the hand movements for these cases have a distinctive velocity profile. The bell-shaped velocity profile is observed in every ballistic movements [7]. As the central nervous system behaves accordingly depending on the type of movements, the attempt to classify a fend off motion seems to be a necessity.

The motion capture system is considered to be the most suitable apparatus to study human motions [9] since this method can provide the possible trajectories in the explorations of the mechanics in martial arts [10]. This study has employed the Virtual Sensei Lite, an inexpensive markerless motion analysis system developed by a mechanical engineer, Alesandro Timmi. During this protocol, the marking procedure is performed by the system itself, while the subject is performing the calibrating posture. Virtual Sensei Lite will directly process the motion capture through a digitizing procedure. Plus, it is capable of preparing a coordinate data for further analysis [11, 12].

The objectives of this study include the following: (i) to analyze the effectiveness of *Kaedah A* based on the total execution time and (ii) to describe the kinematic characteristics of the hand movements upon executing *Kaedah A*. Apart from predicting the performance of this particular technique, this study is an initiative taken to establish data gathering and processing methods in order to generate comparable results for future researches.

2 Method

2.1 Participants

A group of four healthy and experienced SSCM practitioners have participated in this study and have given their informed consents. The physical characterizations of the practitioners are shown in Table 2.

2.2 Experimental Design

This study is focused on the kinematic aspects of *Kaedah A*'s execution as performed by the practitioners. They were allotted 2-min warm-up sessions for them to prepare themselves and to help them become more comfortable with the experimental tasks. This technique was conducted with all participants starting from the initial position. This position is identified as standing upright and facing the sensor at a distance of 2.5 m. From this initial position, the practitioners would then execute *Kaedah A*. They were asked to perform a five-time repetition of *Kaedah A*, so that the nuisance factor can be reduced. In addition, to ascertain better results, the participants were required to wear tight outfits as loose clothing will interfere with the reading.

2.3 Apparatus

The movements were captured using a Kinect sensor, and the images were processed by the Virtual Sensei Lite software. The sensor is capable of capturing motion at a speed of 30 fps, and the Virtual Sensei Lite can replay movements of 100 fps. The Virtual Sensei Lite can track fifteen body landmarks trajectory. However, only the right hand serves as the source of data to be analyzed in this present study.

2.4 Data Visualization and Analysis

The average motion data for the x-axis was taken into consideration. The captured motion data were used in order to identify the initial and intersection point of the movements. The time for initial point, t_i, is when the hand starts its reaction, while the intersection time, t_f, is considered to be at the intersection position of the hand with the punch concentration point, X_c. The Xc point is determined using the practitioner's left shoulder, X_{LS}, and right shoulder, X_{RS}, coordinate data, as indicated in Eq. (1).

The punch concentration point was obtained using Eq. (1) below:

$$X_c = \left(\frac{X_{LS} + X_{RS}}{2}\right) + X_{LS} \tag{1}$$

The time interval between these two points (end point and intersection point) is considered as the total time for *Kaedah A*'s execution. Next, Eq. (2) was used to determine the kinematics descriptions.

$$\Delta t = t_f - t_i \tag{2}$$

Then, the obtained data concerning the coordinates of the hand movements were analyzed using the MATLAB software to determine the polynomial equation for the position of the hand. The general and derivative equations are as indicated by equations below [13]:

$$X(t) = a_0 + a_1 t + a_2 t^2 + a_3 t^3 + a_4 t^4 + a_5 t^5 \tag{3}$$

$$X^1(t) = a_1 + 2a_2 t + 3a_3 t^2 + 4a_4 t^3 + 5a_5 t^4 \tag{4}$$

$$X^2(t) = 2a_2 + 6a_3 t + 12a_4 t^2 + 20a_5 t^3 \tag{5}$$

From the obtained raw data, the best-fit spline curve fitting method was chosen in conjunction with the residual analysis using MATLAB as the data smoothing method. The equation is required to depict the motion of the hands in *Kaedah A*'s execution.

3 Results

Figure 3 shows the trajectory mapping for the repetition of hand movements relative to the x-axis. The data values were obtained from the motion capture system for *Kaedah A*'s left-hand execution. In order to reduce the nuisance factor, the movements were repeated five times by each practitioner.

Since the variance of the repeated data are not significant, the average value of each data was then calculated to observe the trajectory pattern for the execution of *Kaedah A*, as indicated in Fig. 4.

The trajectory mapping graph provides the information on the initial and intersection points. The initial point is identified when the hand starts moving, identified at $t_i = 0.02$ s for each practitioner, and the final point is at the intersection position of the hand with the punch concentration point. By applying Eq. (2), the total execution time of *Kaedah A* for each practitioner was less than 0.1 s. The average total execution time for *Kaedah A* is 0.06 s.

The equation of motion derived from the trajectory data was then differentiated with respect to time, in order to obtain the velocity profile of the hand motion. Figure 5 shows the velocity profile for *Kaedah A*'s execution for each practitioner. It can be observed that each practitioner has produced a bell-shaped velocity profile. The hand speed up at the initial phase and achieved the average maximum velocity, $V_{max\ ave}$ is 5.42 m/s.

Finally, by considering the second derivatives of Eq. (3), the acceleration-time graph in Fig. 6 was plotted to describe the motion phases of *Kaedah A*'s execution.

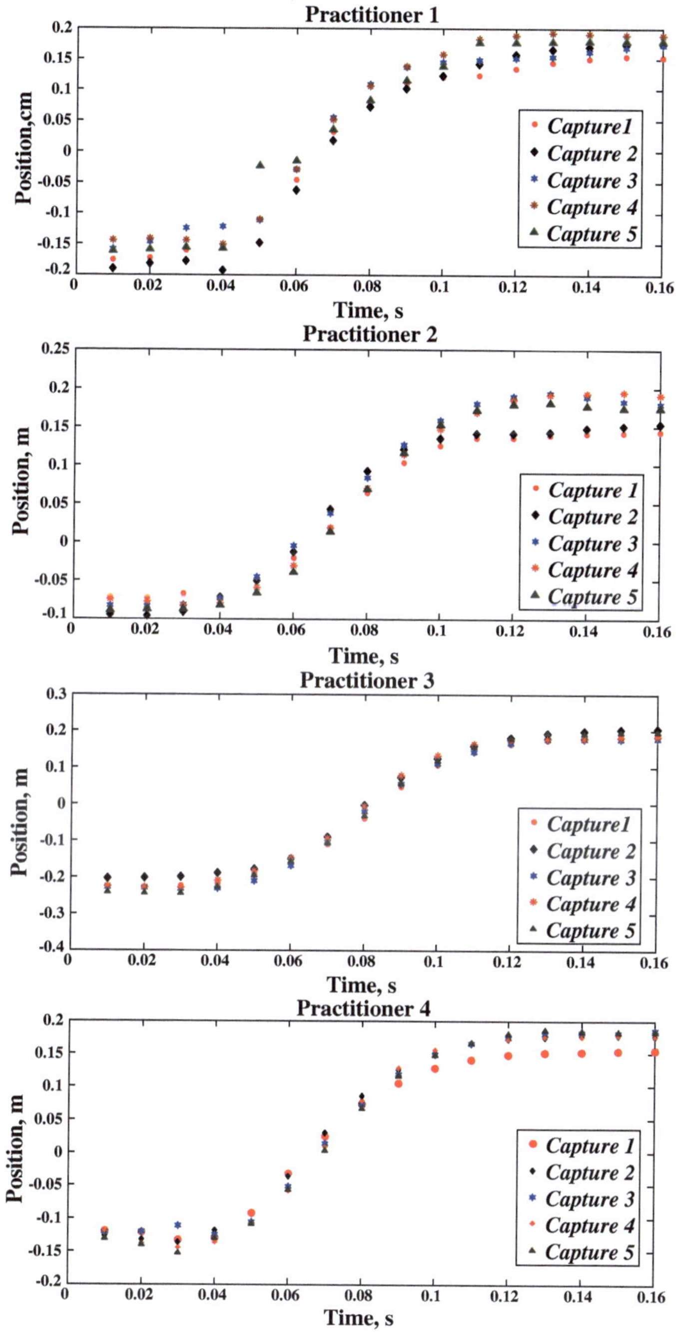

Fig. 3 Trajectory mapping of each practitioner's hand motion during *Kaedah A*'s execution

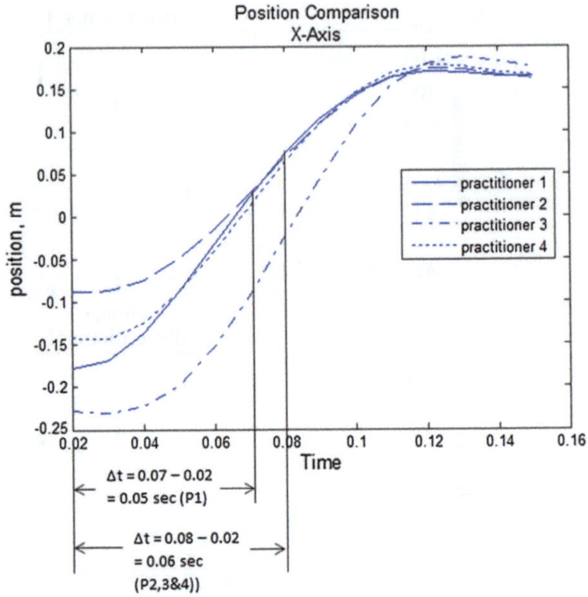

Fig. 4 The kinematic characteristics of *Kaedah A*'s execution—trajectory profile

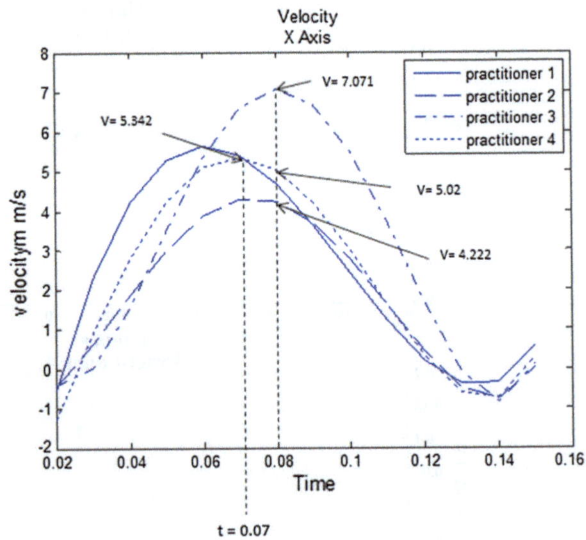

Fig. 5 The kinematic characteristics of *Kaedah A*'s execution—velocity profile

4 Discussion

The main objective of this study was to identify the total execution time of *Kaedah A* by SSCM practitioners. The results have demonstrated that the average total time to execute *Kaedah A* is less than 0.6 s. This indicates that *Kaedah A* meets the

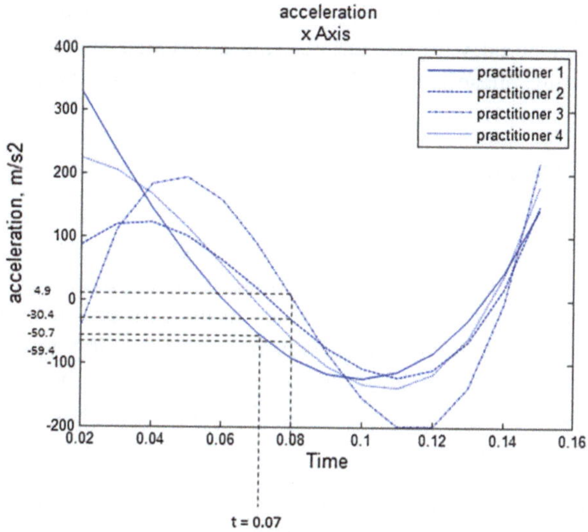

Fig. 6 The kinematic characteristics of *Kaedah A*'s execution—acceleration

main criteria for a technique to be established as an effective reaction toward a punch (the total execution time must be less than 0.1 s). In this study, *Kaedah A* was executed as a reaction toward an imaginary powerful punch. In comparison with the total time to react for a normal human (0.18 s), *Kaedah A* has eliminated 0.05 s which would have consisted of the time allocated for the second phase (thinking of the best reaction) and reduced the time to start moving due to the constant training.

The displacement-time and the velocity-time graphs have depicted that the hand movements during the execution of *Kaedah A* have the characteristics of a ballistic movement. In addition, the graphs obtained for the velocity profile and acceleration have trends that are similar to the ballistic movements for a palm strike as reported by Neto in 2008 [14]. The velocity profile presented a sudden deceleration due to the activation of antagonist muscles as a result of the sudden stop at the end of the motion [14].

According to the Bayesian model, the human ballistic movement is a sequence of movements between objects and the environment [8]. Therefore, it can be speculated that the actual performance will reduce the total time of *Kaedah A*'s application due to the influence of stimulating effects in real-time application. This finding may also influence other investigations regarding the effectiveness of *Kaedah A*.

Based on the graph of motion, the hand movements in *Kaedah A* may be divided into three phases. These phases are known as the initial phase, the phase of preparation, and the hitting phase. The initial phase is responsible for generating the period of activation of the muscles and forearm flexion for the following acceleration. The hitting phase is the point where *Kaedah A* starts repelling the attack/punch. From Fig. 4, it can be observed that the hand decelerated when it intersected with the target. This is because of tendency of sudden stop after the hitting phase, caused by the full extension of the arm.

5 Conclusion

The understanding of the kinematic characteristics of *Kaedah A* is important not only for improving the performance of the practitioners but also to demonstrate the applicability of this martial art in the process of motor development. It has been proven that the martial arts have helped develop defensive abilities in its members. For future work, all the geometric aspects may be used to study the kinetics of the practitioner and the attacker to determine their performance mechanically and biologically.

Acknowledgments The authors gratefully acknowledge the Ministry of Education Malaysia for sponsoring this research via Universiti Teknologi MARA (Grant no. 600-RMI/FRGS5/3(76/2013)) and Persatuan Seni Silat Cekak Malaysia for providing additional research fund and equipment to further facilitate this study.

References

1. Emmermacher, P. (2005). Acceleration course of fist push of Gyaku-zuki. In *International Symposium on Biomechanics Sports* (pp. 844–847).
2. Wąsik, J. (2009). *Chosen aspects of physics in martial arts.* Budo: Arch.
3. Foresman, P.S. (2012). Diagram of the upper torso and head of a human male.
4. Wallace, B., & Janz, J. (2009). Implications of motor unit activity on ballistic movement. *International Journal of Sports Science and Coaching, 4*(2), 285–292.
5. Yazdani, M., Gamble, G., Henderson, G., Hecht-Nielsen, R. (2012). A simple control policy for achieving minimum jerk trajectories. Neural Networks, 27, 74–80.
6. Uno, Y., Kawato, M., Suzuki, R. (1989). Formation and control of optimal trajectory in human multijoint arm movement. Biological Cybernetics, 61(2), 89–101.
7. Prasad, V., Kellokumpu, V., Davis, L. (2006). Ballistic hand movements: Articulated motion and deformable objects (pp. 153–164). New York: Springer.
8. Vitaladevuni, SNP. (2007). Human movement analysis: Ballistic dynamics, and edge continuity for pose estimation. USA: ProQuest.
9. Mahmud, J., Holt, C., Evans, S. (2010). An innovative application of a small-scale motion analysis technique to quantify human skin deformation in vivo. Journal of Biomechanics, 43(5), 1002–1006.
10. Azraai, N.Z., Bidin, M.O. (2013). Konsepsual Manifestasi Silat Titi Batang Melalui Teknologi Motion Capture in *The Regional Conference on Local Knowledge.*
11. Surer, E., Cereatti, A., Grosso, E., Croce, U. (2011). A markerless estimation of the ankle–foot complex 2D kinematics during stance, *Gait Posture.*
12. Alessandro, T. Virtual Sensei Lite. (2011). [Online]. Available: http://www.virtualsensei.it/lite/.
13. Hogan, N. (1984). An organizing principle for a class of voluntary movements. The Journal of Neuroscience, 11, 2745–2754.
14. Neto, O., Magini, M. (2008). Electromiographic and kinematic characteristics of Kung Fu Yau-Man palm strike. Journal of Electromyography and Kinesiology, 18(6), 1047–1052.

Determination of Tenpin Bowling Lane's Rolling Resistance Based on Kinetics and Kinematics Modeling

Shariman Ismadi Ismail, Rahmat Adnan and Norasrudin Sulaiman

Abstract Bowling alleys have significant influence on tenpin bowler's performance. The lane surface is one of the major factors in determining the player's results in a bowling tournament due to the ball–lane surface interaction. Understanding on how to adapt to different lane conditions will help bowlers to perform better. This study investigates rolling resistance coefficients of bowling ball–lane surface based on kinetics and kinematics modeling approach. Ball throwing phase was modeled where the bowler's throwing arm is defined as one rigid single connecting rod on a pivoted joint of the shoulder. Power generated from the throw is derived based on the work done by the arm during the throwing time interval. The amount of work done by the throwing arm is derived from the amount of torque generated from the throw during the throwing time interval. The rolling resistance coefficient of the thrown ball was calculated based on the power required to make the ball roll on the lane. Modeling results identify ball release velocity, ball velocity, and throwing arm angular velocity as parameters that are capable of being manipulated by the bowler in order to obtain desired throwing power. This, however, depends on the ball–lane rolling resistance coefficient and the ball's drag coefficient. Modeling information should enable bowlers to adjust their throwing mechanics to adapt with bowling lane conditions.

Keywords Tenpin bowling · Modeling · Rolling resistance · Kinematics and kinetics

S. I. Ismail (✉) · R. Adnan · N. Sulaiman
Sports Science Center of Studies, Faculty of Sports Science and Recreation,
Universiti Teknologi MARA, Shah Alam, Malaysia
e-mail: shariman_ismadi@salam.uitm.edu.my

R. Adnan et al. (eds.), *Proceedings of the International Colloquium on Sports Science, Exercise, Engineering and Technology 2014 (ICoSSEET 2014)*,
DOI: 10.1007/978-981-287-107-7_2,

1 Introduction

Tenpin bowling is a sport that combines speed, velocity, momentum, accuracy, and consistency. These elements relate with many other aspects, including the bowler's physical and mental strength, the ball used, lane conditions, and throwing technique.

Almost every time, lane conditions are the only aspects that bowlers have no control. The lane conditions are determined by the maintenance techniques performed by the bowling alley operators. Professional bowlers should be able to adjust their throwing technique when playing at a bowling alley that has different lane conditions compared to the one that the bowlers are already accustomed to, in terms of the lane surface conditions and also air circulation pattern near the bowling lane.

However, when one or two pin count differences determine winners and losers, these adjustment periods will be critical. Bowlers do not have empirical data and statistical data that show how the lane conditions on that particular day will affect their normal playing technique and plan. They will just adjust their throwing speed and technique to accommodate to the required adjustment. Information that explains about the lane conditions benefits bowlers in reducing the time of the adjustment period.

The physics of throwing can be understood either experimentally or numerically. Modeling has always been used at the beginning of an experimental based research. This is to provide guidelines for the researchers before embarking into the experimental phase. The modeling of throwing usually focuses on the related segmental body parts rather than the whole body movement model [1]. Limb-segment angles, joint angles, and velocities (linear and angular) are typically investigated to develop a throwing model [1, 2].

A previous study showed that typical angular velocity of the bowler's throwing hand is around 617 °/s [3]. This covers the arm's maximal hyperextension to flexion phase until the ball release time. In terms of ball release velocity, a previous study had indicated that for elite player the average ball release velocity was in the order of 28.4 km/h, whereas for semi-elite players, the average ball release velocity was 27.2 km/h [4]. Similar results were also found in another study [5]. It was also reported that the typical average ball velocity that travels toward the pin ranges between 22 and 26 km/h [6]. In terms of rolling resistance and drag force that associate the bowling ball with the bowling lane and bowling alley, there were no clear past studies that focused on tenpin bowling. However, it is understood that typical drag coefficient for sports ball with smooth surface that travels at relatively moderate to high velocity is below 0.3 [7, 8], and the coefficient of rolling resistance between two smooth surfaces with lubrication influence will be around 0.02–0.1 [9].

The present study focuses on identifying a simple model of bowling throw. After identifying this model, the rolling resistance between the ball and the lane will be estimated. Information regarding the lane's rolling resistance is important to help bowlers to understand better about lane conditions.

2 Method

2.1 Mathematical Modeling

The modeling of the throwing phase is performed as if the arm is one rigid single connecting rod on a pivoted joint of the shoulder. Power generated from the throw is derived based on the work done by the arm during the throwing time interval. The amount of work done by the throwing arm is derived from the amount of torque generated from the throw during the throwing time interval. This is shown in the equation below.

Power requirement to throw the ball:

$$P = \frac{W}{t}, \tag{1}$$

where W is the work done and t is the time used to perform the task. The work done during the throwing phase is defined by the following equation.

Work done during the throwing phase:

$$W = \int \tau \mathrm{d}\theta = \frac{1}{2}mv^2 + \frac{1}{2}I\omega^2 \tag{2}$$

Therefore, the power required to throw the ball is:

$$P = \frac{mv^2 + I\omega^2}{2t} \tag{3}$$

where t is the torque, θ is the angle, m is ball mass, v is ball release velocity, I is moment inertia of the arm, w is angular velocity of the throwing arm, and t is time used to perform the task. Here, the amount of torque generated in performing the work is defined as the amount of linear and rotational kinetic energy of the movement.

As the arm is modeled as a single connecting rod on a pivoted joint at shoulder (Fig. 1), the moment of inertia of the throwing arm can be modeled as shown in the equation below:

$$I = \frac{1}{12}ML^2 + Mr^2 \tag{4}$$

Here, M is arm mass, L is the arm length, and r is the arm distance between arm's center of mass to the shoulder pivot joint. Here, we consider that the arm's mass is equally distributed.

The power generated by the throwing arm is also equal to the force that is required to roll the ball on the lane and the drag force generated during the ball moving on the lane. This is illustrated in the following equation.

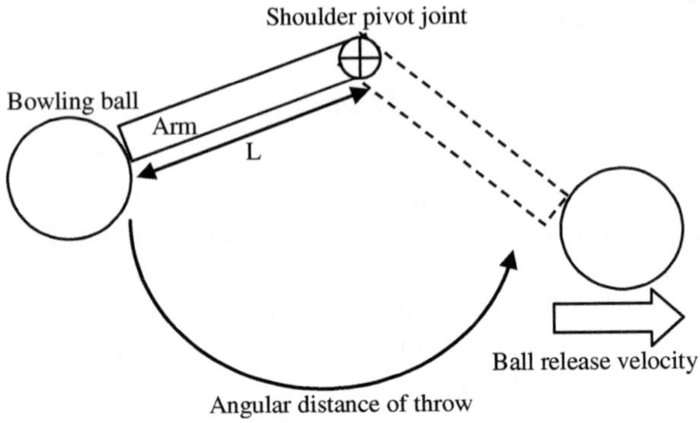

Fig. 1 Modeling of throwing arm

Power required for moving the ball on the lane:

$$P = \left(C_r mg + \frac{1}{2} \rho_a C_d A V^2 \right) V \tag{5}$$

where C_r is the rolling resistance coefficient between ball and lane, m is ball mass, g is gravity acceleration, ρ_a is air density surrounding the lane, C_d is drag coefficient between ball and air during ball movement on lane, A is ball cross-sectional area, and V is the ball velocity.

If we consider that the power that is required to throw the ball is equal to the power that is required to move the ball on the lane, the following equation will represent the rolling resistance coefficient of the ball–lane:

$$C_r = \frac{mv^2 + I\omega^2 - \rho_a C_d A V^3 t}{2Vmgt} \tag{6}$$

2.2 *Modeling Parameters*

The parameters utilized in this study to calculate the model are presented in Table 1. All of the chosen values for the modeling parameters are based on typical and average quantities that were found or were used in previous studies [1, 3–9].

Based on the modeling parameters shown in Table 1, the amount of work and power required to throw the ball can be calculated. This will help to solve the next stage of the simulation model.

3 Results and Discussion

Figure 2 shows the modeling results of bowling ball–lane rolling resistance coefficient, C_r, and the bowling ball average ball speed. Based on average ball speed that travels toward the pin as reported from a previous study [6], it was

Table 1 Modeling parameters

No.	Modeling parameters	Value
1	Bowler's height	180 cm
2	Bowler's weight	80 kg
3	Bowler's arm mass	10.5 kg
4	Bowler's arm moment of inertia	1.97 kg m^2
5	Bowler's arm angular velocity	11 rad/s
6	Ball release velocity	28.4 km/h
7	Ball throwing period	2 s
8	Bowler's arm length	75 cm
9	Bowling ball mass	7 kg
10	Bowling ball diameter	21.6 cm
11	Bowling ball cross-sectional area	0.04 m^2
12	Air density at 20 °C	1.2 kg/m^3
13	Drag coefficient of bowling ball	0.1–0.3
14	Rolling resistance coefficient between ball and lane	0.02–0.1
15	Gravity acceleration	9.81 m/s^2

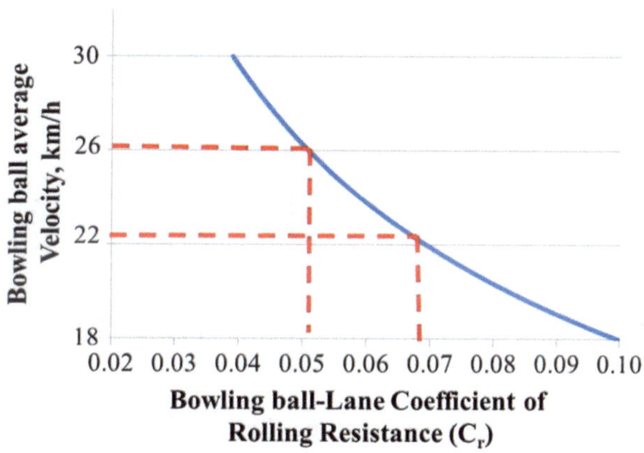

Fig. 2 Ball average velocity and rolling resistance coefficient (C_r)

indicated that the typical average ball speed was between 22 to 26 km/h. We can identify from the modeling results in Fig. 2 that the possible value of C_r is between 0.05 and 0.07. In this simulation, it was also found that the potential power requirement to throw the ball based on this model was 0.34 KWh.

Based on the results in Fig. 2, a relationship between the required power to throw the ball and the ball average velocity toward the pin was modeled. Figures 3, 4, and 5 represent the simulation model with drag coefficient of bowling ball C_d at 0.1, 0.2, and 0.3, respectively. In each of this simulation, the C_r was set between the ranges of 0.02–0.1. This should cover more than ±25 % of the estimated C_r based on the early modeling results.

Simulation results shown in Figs. 3, 4, and 5 indicated that C_r does not influence much the ball that travels below 10 m/s on average. It can also be seen that at

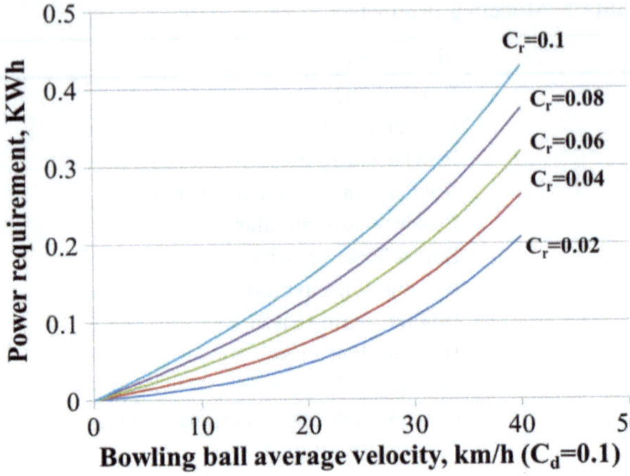

Fig. 3 Ball average velocity–power requirement ($C_d = 0.1$)

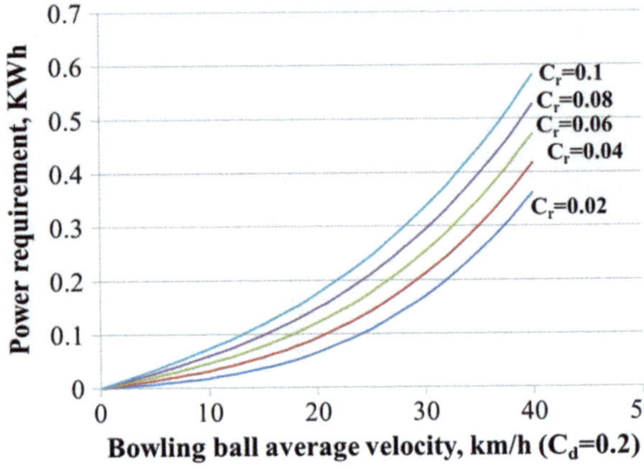

Fig. 4 Ball average velocity–power requirement ($C_d = 0.2$)

10 km/h average velocity and below, there is little influence of C_d on the power requirement. The increasing amount of C_r and C_d will proportionally increase the power requirement during throwing the ball. This is because more energy is needed to overcome the resistance generated by the increasing amount of drag and rolling resistance. Based on the average ball velocity at 22 km/h, the power required to throw the ball is in the range of 0.05–0.23 KWh.

In simulating the power required to throw a bowling ball, the ball release velocity and the arm angular velocity are utilized in the model. Therefore, in real world situations, if there are possible way to measure the ball release velocity, arm angular velocity and other basic information such as the bowler's anthropometry data, bowling ball information, and environmental aspects like the room temperature and air density, we can identify the amount of rolling resistance of the bowling lane. Based on this information, we can systematically identify the suitable average ball velocity as well as its throwing power requirement.

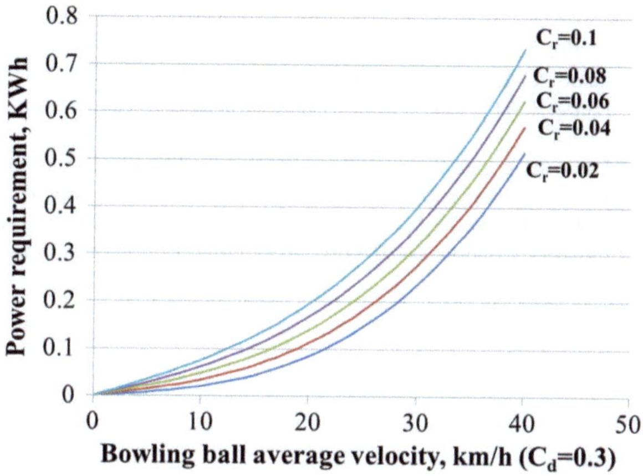

Fig. 5 Ball average velocity–power requirement ($C_d = 0.3$)

This real-time information will help bowlers to adjust their throwing effort in order to adapt to different bowling lane condition, especially if they are not familiar with the bowling lane surface.

4 Conclusion

This study proposed a simple model to understand the throwing ball phase of tenpin bowling. In developing this model, important aspects such as power requirements, lane's rolling resistance coefficient, and ball's drag coefficient were estimated. An understanding of the information helps bowlers to adapt faster during playing at unfamiliar bowling alleys. By reducing the adjustment time to adapt to the lane's condition, improvement on total pin score can be achieved. However, further experimental work should be carried out to validate the model. Furthermore, the arm swing model during ball throw need to be further improved.

References

1. Cross, R. (2004). Physics of overarm throwing. *American Journal of Physics, 72*(3), 305–312.
2. Winter, D. (2009). *Biomechanics and motor control of human movement* (4th ed.). New York: Wiley.
3. Chu, D. P. K., Zhang, B. M., & Mau, K. (2002). Tenpin bowling technique on elite players. In *Proceedings of the 20th International Conference on Biomechanics of Sports, Caseres, Spain* (pp. 123–125).
4. Razman, R., Wan Abas, W. A. B., Abu Osman, N. A., & Cheong, J. P. G. (2011). Temporal characteristics of the final delivery phase and its relation to tenpin bowling performance. In *IFMBE Proceedings Volume 35, 5th Kuala Lumpur International Conference on Biomedical Engineering* (pp. 222–224).

5. Wilson, B., Slowinski, J., & Jalil, A. P. (2010). Contribution to tenpin bowling ball speed. *ISN Buletin, 3*, 21–30.
6. Martin, C. L. (1984). Generating bowling ball speed. *Athletic Journal, 54*, 38–39.
7. Mehta, R. D. (1985). Aerodynamics of sports ball. *Annual Review of Fluid Mechanics, 17*, 151–189.
8. Mehta, R. D., & Pallis, J. M. (2001). Sports ball aerodynamics: Effects of velocity, spin and surface roughness. *Materials and Science in Sports*, pp. 186–197.
9. Bartlett, R. (2007). *Introduction to sports biomechanics: Analysing human movement patterns* (2nd ed.). London, New York: Routledge Taylor & Francis Group.

Reflection Rate Index of Markers for Motion Capture Application

Shariman Ismadi Ismail, Rahmat Adnan and Norasrudin Sulaiman

Abstract Motion capture application continues to grow in many sectors of the industry, including medical science, sports science, animation, robotics, and many more. In order to capture any specific movement to be analyzed, markers are utilized to identify joint movement. Passive reflective markers are usually utilized in motion capture systems equipped with infrared cameras. Active reflective markers are usually utilized in motion capture systems with digital video recorders. These markers usually transmit or diffuse specular light distribution to be captured by the camera system. In this study, the reflection rate index (RRI) of 4 different types of markers was measured (M1, M2, M3, and M4). Based on the RRI value, the type and level of reflection and light distribution from each marker can be identified. Results indicated that M1 and M2 had RRI values above 1, which means that these markers produced diffuse reflection, whereas M3 and M4 had RRI values below 1, which means that these markers produced specular reflection. Based on this study, we can categorize each markers light distribution or reflection rate based on the calculated RRI. This is helpful to researcher for deciding what type of marker that needs to be utilized in each respective research area.

Keywords Motion capture · Motion analysis · Markers · Biomechanics · Bioengineering

1 Introduction

Motion capture systems have been around since the nineteen century [1] and have become increasingly important in today's research associated with motion analysis. Many industries conduct motion analysis in their working area, which include

S. I. Ismail (✉) · R. Adnan · N. Sulaiman
Sports Science Center of Studies, Faculty of Sports Science and Recreation,
Universiti Teknologi MARA, Shah Alam, Malaysia
e-mail: shariman_ismadi@salam.uitm.edu.my

R. Adnan et al. (eds.), *Proceedings of the International Colloquium on Sports Science, Exercise, Engineering and Technology 2014 (ICoSSEET 2014)*,
DOI: 10.1007/978-981-287-107-7_3,

sports science, health and clinical science, automotive industries as well as entertainment and gaming industries. Current state-of-the-art motion capture systems require markers to identify movement performed by objects in motion that are being tracked. These markers come in many sizes according to the requirements of the industries. The markers that are used in motion capture systems could be either active marker, where the markers have lighting capabilities with either conventional lamps or light-emitting diodes (LED) or it could also be passive markers, where the markers are built from light-reflecting materials and the motion capture systems require lighting during the motion capture. However, no lighting effect is required for an active marker system.

Typical active markers in motion capture systems will usually utilize a standard video-recording camera, while passive markers in motion capture systems typically utilize infrared cameras for video recording. However, passive marker systems can also use a standard video-recording camera, but require lighting to create a source of reflection for the markers. Both active and passive marker systems are suitable for either 2-dimensional or 3-dimensional motion analysis as long as the motion analysis system can interpret the coordinates of the tracked objects that have been identified by the markers used in the motion capture system.

There is a perception that active markers generate better movement tracking when compared to passive markers. Although almost all current motion capture systems could easily provide the movement information, a previous study [2] indicated that much more improvement is needed in order to produce a real-time motion capture and motion monitoring system. In certain situations, marker recognition can be difficult due to the marker positioning. It was also found [2] that passive markers sometimes could not recognize two markers sitting very close with each other. It is believed that reflection created by the markers affected this recognition capability. Different types of reflective markers interact differently with different light sources. Different specifications of light distribution for active markers also create different states of recognition.

Passive marker motion capture systems are preferred compared to active markers due to the fact that it is easy to use. Although there is evidence that active marker motion capture systems are becoming more user friendly [3], nonetheless, it is still not hassle free. More time is utilized at the setup stage when using active marker motion capture systems. Marker-less motion capture systems are also becoming more popular [4, 5], but there is no real indication as yet that there is a marker-less motion capture system that is as accurate and reliable as marker motion capture systems.

In terms of the passive marker, a reflective surface influences the reflection capability, where previous studies [6, 7] had indicated that there was evidence that a smooth surface creates a longer light reflection distance, whereas a less smooth surface, such as orange peel-like surface, creates a sharper reflection of the light. Based on these findings, markers with different purposes are being designed according to application requirements. More information regarding marker classifications and specifications for different scenarios, situations, and conditions are

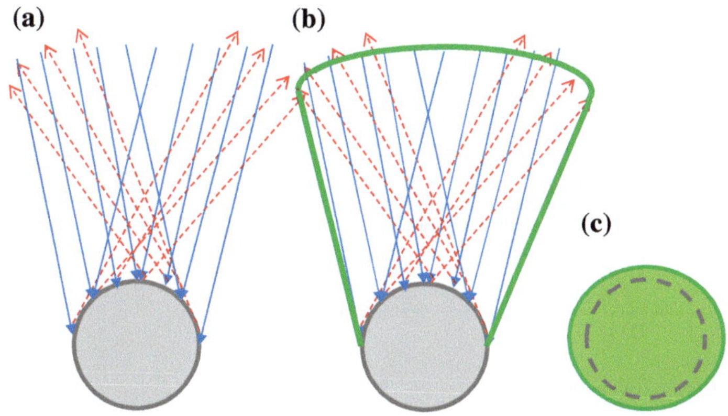

Fig. 1 **a** Incoming and reflection light, **b** range of light distribution beam, and **c** glowing lighting effects

needed in order to help researchers utilize the most suitable markers for their studies and applications.

The purpose of this study was to establish a simple method to indicate the rate of light reflection created by markers that are typically utilized in motion capture systems.

2 Method

2.1 Understanding of Light Emission, Reflection and Distribution

Light travels in a straight line. A light-emitting object produces its on lighting and distributes them. LED is one example of a light-emitting object. Materials without its own light-emitting capabilities, but capable of reflecting incoming light are known as reflective materials. The beam of the light distribution from either a light-emitting object or a light-reflected object will either be a diffused or specular distribution beam. A diffused distribution represents glowing lighting effects surrounding the object, while specular distribution lighting do not have glowing lighting effects surrounding the object's surface. Figure 1 illustrates the concept of light reflection and light distribution viewed from the top view ('a' and 'b') and front view ('c').

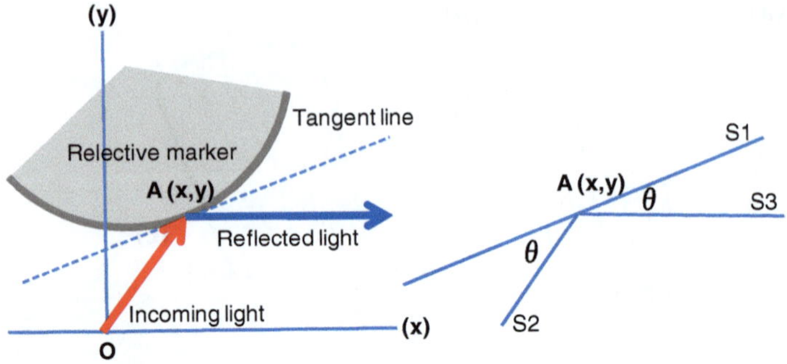

Fig. 2 Light reflection model

2.2 Light Reflection Model

Light reflection is influenced by the lighting source and also the surface conditions of the reflective object. Figure 2 illustrates a light reflection model on a passive marker.

The model shown in Fig. 2 considers that when the reflective marker surface is evenly smooth and equal, the angle created between the incoming light and tangent line on point A (reflection point) and the reflected light and tangent line on point A to be equal. Based on these assumptions, we would obtain the following relationship:

$$\tan\theta = \frac{S1 - S2}{1 + S1S2} \quad (1)$$

Here, $S1$, $S2$, and $S3$ are the slopes of each of the respective sources.

Under real situations, the angle will not be the same due to the marker surface conditions and incoming light properties. This is why we would obtain reflection light with either diffused of specular reflection. In order for the reflection light distribution to be diffused reflection, the angle of the reflection light will have to be a smaller angle when compared to the angle of the incoming light. Inversely, in order for the reflection light distribution to be a specular reflection, the angle of the reflection light will have to be a larger angle when compared to the angle of the incoming light.

Calculating these angles on a spherical shaped object is complicated, but we are interested to understand about the reflection rate of a marker. Therefore, this study proposes an alternative method to categorize light reflection or distribution rate of a marker.

Table 1 Marker description

Type	Symbol	Description	Size (diameter) (mm)
Active marker	M1	Bulb type	11
Active marker	M2	LED type	10
Passive marker	M3	Reflective markers conventional	12
Passive marker	M4	Reflective markers prototype	16

2.3 Marker Description

In this study, four different types of markers were investigated (Table 1). All of these markers are commonly used in motion capture and motion analysis application systems, except for M4 that is a prototype and is not yet available commercially.

Each of the markers light reflection and distribution images were captured using a still image camera (Casio ZX-100) in a dark room. The camera was positioned approximately 5 cm in front of the markers, while it was in the light-reflecting and distributing condition. All dimensional calculations were obtained using the Kinovea Software (v 0.8.15). Each of the marker images was taken 10 times, and the average value was calculated.

The following equation was utilized in order to categorize the rate of light reflection and distribution of the markers. This was named as the rate of reflection index (RRI):

$$\mathrm{RRI} = \frac{\text{Total glowing light cross-section area}}{\text{Total cross-section area of marker}} \tag{2}$$

RRI is the ratio of the cross-sectional area of reflected or distributed light with respect to the markers total cross-sectional area. Figure 3 illustrates the flow chart for this study.

3 Results and Discussion

Results of the RRI are presented in Table 2. Based on the results, it was evident that passive markers have lower RRI when compared with the active markers. Figure 4 illustrates the representative still images of each of the markers.

Based on the results of the still images, it was clear that active markers generate diffuse light distribution, whereas passive markers generate specular light reflection.

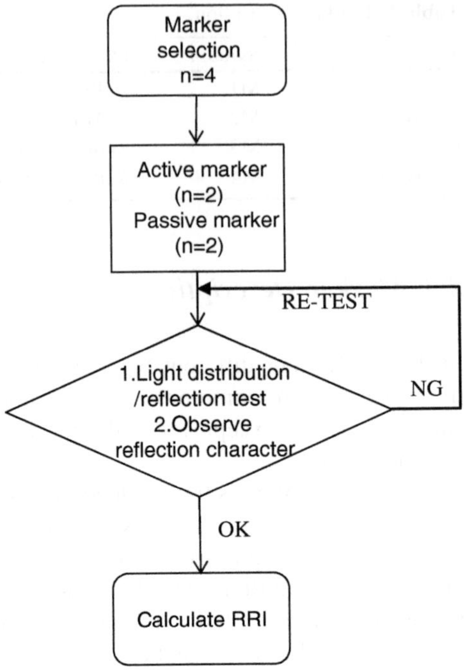

Fig. 3 Flow chart of study

Table 2 RRI of markers

Marker	Light distribution/reflection	RRI
M1	Diffuse	1.38
M2	Diffuse	2.45
M3	Specular	0.95
M4	Specular	0.78

It is also acknowledged that this study found that markers with RRI of below 1 generate specular light distribution or reflection, whereas RRI of above 1 generates diffuse light distribution or reflection.

The value of RRI would enable us to identify the reflection and light distribution rate based on the marker size. It is also useful for researchers to select the type of marker to be used in their respective research area. This is important because certain motion capture studies would require the markers to be more glowing, or more diffusive rather than specular light reflection or distribution. There would also be motion capture studies that prefer specular lighting reflection. However, this is associated with many parameters, including research location, type of video recorder, lighting effects, and also the type of the motion itself.

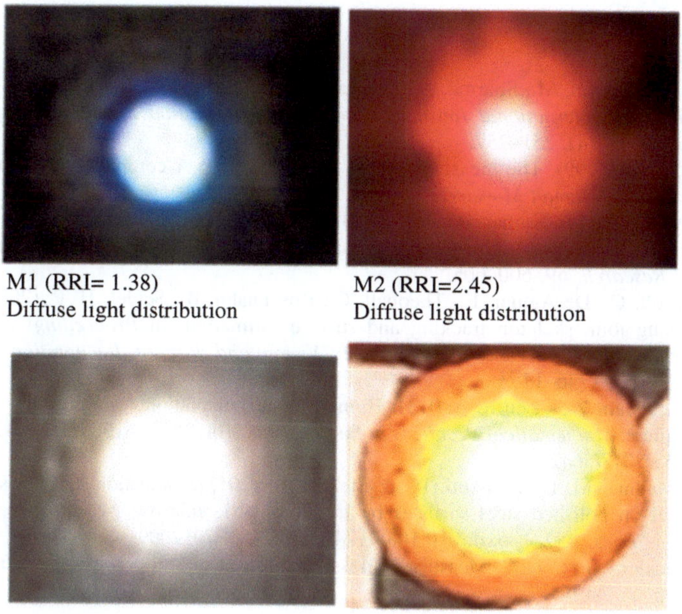

M1 (RRI= 1.38)
Diffuse light distribution

M2 (RRI=2.45)
Diffuse light distribution

M3 (RRI=0.95)
Specular light reflection

M4 (RRI=0.78)
Specular light reflection

Fig. 4 Light distribution and reflection of markers

4 Conclusion

A simple method, of determining the RRI, for categorizing types of light reflection and distribution of markers that are typically utilized in motion capture systems is proposed. By identifying the RRI of the markers, the types of light distribution or reflections are known. Based on this information, a better understanding of the markers is established. Thus, these results will enable researchers and practitioners of motion capture and motion analysis systems to select the markers more correctly for their work.

Acknowledgments Thanks are due to the Faculty of Sports Science and Recreation, Universiti Teknologi MARA, Malaysia, for the financial support of this study.

References

1. Sharma, A., Agarwal, M., Sharma, A., Dhuria, P. (2013). Motion capture process, techniques and applications. *International Journal on Recent and Innovation Trends in Computing and Communication, 1*(4), 251–257.
2. Richards, J. G. (1999). The measurement of human motion: A comparison of commercially available system. *Human Movement Science, 18*, 589–602.
3. Kumar, N., Kunju, N., Kumar, A., & Sohi, B. S. (2010). Active marker based kinematic and spatio-temporal gait measurement system using LabView Vision. *Journal of Scientific and Industrial Research, 69*, 600–605.
4. Gall, J., Stoll, C., De Aguiar, E., Theobalt, C., Rosenhahn, B., Seidel, H. P. (2009). Motion capture using joint skeleton tracking and surface estimation. In *Proceedings of the IEEE Computer Society Conference on Computer Vision and Pattern Recognition Workshops (CVPR '09)* (pp. 1746–1753).
5. Zhang, D., Miao, Z., Chen, S. (2013). Human model adaptation for multiview markerless motion capture. *Mathematical Problems in Engineering*, Vol. 2013, Article ID 5642147 pages. doi:10.1155/2013/564214.
6. Miranda-Medina, M. L., Somkuti, P., & Steiger, B. (2013). Detection and classification of orange peel on polished steel surfaces by interferometric microscopy. *Journal of Physics: Conference Series, 450*, 1–6. doi:10.1088/1742-6596/450/1/012009.
7. Kigle-Boeckler, G. (1996). Measuring gloss and reflection properties of surfaces. *Tappi Journal*, pp. 194–198.

Efficacy of Handgrip Strength in Predicting Total Body Strength Among High Performance Athletes

Lucy Jawan, Rahmat Adnan, Norasrudin Sulaiman and Shariman Ismadi Ismail

Abstract The purpose of this study was to determine the efficacy of handgrip strength in predicting total body strength among high-performance male athletes. A total of 100 male athletes (mean ± SD), with a mean age of 20.92 years old (±2.53) and BMI 24.77 kg/m^2 (±4.70), were willing to participate in this study. All subjects were considered as high-performance athletes as they represented their university, state, or country in sports and were reported as being healthy and fit through the interview sessions held before selecting them into the program. Subjects were required to perform indirect 1RM test (lat pull-down, incline bench press, leg press, and leg extension), handgrip strength test, and the vertical jump test. The handgrip strength test was evaluated using the digital handgrip dynamometer (Takei A5401) to test the handgrip strength test. Meanwhile, the Vertec vertical jump equipment was used to determine the leg power. The data were analyzed using the Statistical Package for the Social Science (SPSS) version 19.0 with the significance level set at $p \leq 0.05$. This study found that the handgrip strength test showed no significant correlation between all the 1RM tests, including the bench press (0.057), lat pull-down (0.304), the leg press (0.113), leg extension (0.015), and strength test. Likewise, the leg power test also showed no significant correlation (0.119) with the handgrip strength test. Hence, this study showed that the handgrip was not a significant measure of total body strength of high-performance athletes.

Keywords Component · 1RM test · Total body strength · Handgrip strength · High-performance athletes

---

L. Jawan
Strength Training and Conditioning Laboratory, Universiti Teknologi MARA, Shah Alam, Malaysia

R. Adnan (✉) · N. Sulaiman · S. I. Ismail
Faculty of Sports Science and Recreation, Universiti Teknologi MARA, Shah Alam, Malaysia
e-mail: rahmatadnan@salam.uitm.edu.my

R. Adnan et al. (eds.), *Proceedings of the International Colloquium on Sports Science, Exercise, Engineering and Technology 2014 (ICoSSEET 2014)*,
DOI: 10.1007/978-981-287-107-7_4,

1 Introduction

In general, high-performance athletes have high total body strength compared to other athletes and the general population. Strength is an essential function of the human body, which can manifest itself in various ways, depending on individual conditions and objectives used to perform different actions or exercises. The term "strength" can be employed to identify the force or torque developed by a muscle during a particular joint movement [1]. Handgrip strength is an important measure of general health and is regarded as one of the most reliable clinical methods for estimating strength [2, 3] and determines the handedness of an individual, an important field of population variation study. It is often used as an indicator of overall physical strength and is affected by a number of factors including age, gender, and body size [4, Elissa, Title of paper if known, unpublished].

Besides handgrip strength test, the one repetition maximum (1RM) testing is another way to predict specific muscular strength in athletes. The 1RM test is a method widely used for the determination of the intensity of strength training [5]. Besides, the 1RM test is considered the gold standard for assessing muscle strength in non-laboratory situations [6]. It is defined as the maximal weight that can be lifted once with proper lifting technique. It requires comparatively easy and inexpensive non-laboratory equipment. Due to the wide use of 1RM testing, the 1RM measurement is very important. Mostly, exercises such as the bench press and squat have been shown to be reliable measurements for 1RM testing [6]. The 1RM test goal is to mobilize the greatest possible resistance via a specified range of movements (ROM), without additional feedback on the rate of force development (RFD), or time consumed producing the movement [1]. Handgrip strength measurements are reliable, safe, easy, and fast to perform [7]. Besides, handgrip strength is suitable for prediction of total body strength in healthy young adolescents [8]. Most of the previous studies on handgrip strength were carried out on the general population; however, there was only limited studies carried out among high-performance athletes particularly in Malaysia.

Therefore, the purpose of this study was to estimate the efficacy of handgrip strength in predicting total body strength among high-performance athletes.

2 Methodology

2.1 Sample

One hundred ($N = 100$) male athletes at high-performance level from two categories of sports (individual sports and team sports) were selected to perform the indirect 1RM strength test, handgrip strength test, and the vertical jump test.

Fig. 1 Research framework

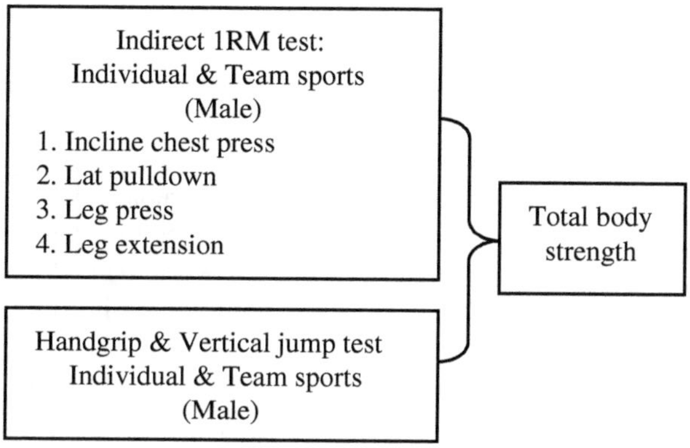

2.2 Research Design

This study used correlations in order to find the relationship between total body strength with handgrip strength among high-performance athletes.

Besides, this study was also designed to find the relationship and differences in the dominant versus non-dominant hand for the handgrip strength between the two types of sports (individual and team sports) (Fig. 1).

2.3 Test Instrument

The Faculty of Sport Science and Recreation Gymnasium UiTM was used as the venue for the data collection.

2.3.1 Handgrip and Vertical Jump Test

Takei A5401 (Japan) digital handgrip dynamometer was used to measure the handgrip strength among the high-performance athletes. The Vertec device was used for measuring vertical jump height in order to obtain muscular power of the participants.

2.3.2 Indirect 1RM Test

The Olympic bar was used on an incline bench press (45° angle) test to identify the specific frontal upper body strength. The Cybex dual axis pull-down machines were used to measure the back posterior muscles that comprises of Latissimus dorsi, rhomboid, and posterior deltoid. Leg extension machine was used in the leg extension test to measure quadriceps muscles alone, which were the most dominant muscles in the leg.

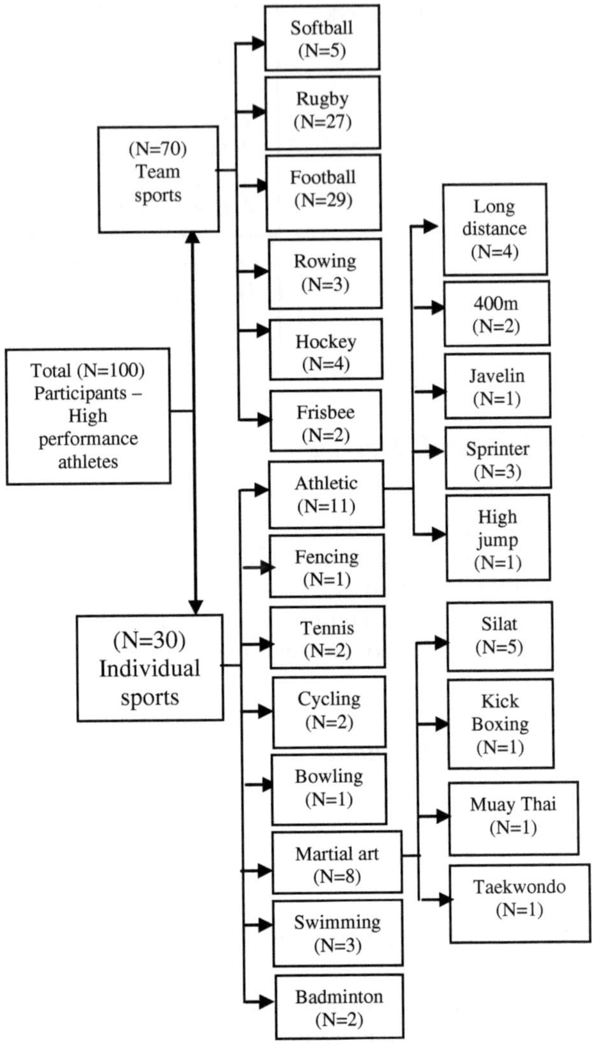

Fig. 2 Sampling procedure

2.4 Data Collection

This study began with the indirect 1RM strength test (incline bench press, lat pull-down, leg extension, and leg press).

After 72 h, all of the participants went through the second phase of the test, which began with the handgrip strength test and followed by the vertical jump test [8]. A rest period was given for muscle recovery after participants underwent the indirect 1RM tests in order to get the best result for total body strength (Fig. 2).

2.5 Procedure

Participants were required to sign the informed consent form. Then, the basic demographic (age, gender) and anthropometric (height, weight, body mass index) data were obtained.

Next, the participants were allowed to perform individual warm-up sessions by running on a treadmill for 5 min at moderate intensity and for a brief familiarization with the equipment. The indirect 1RM test (lat pull-down, bench press, leg extension, and leg press) and vertical jump test were conducted by experienced sports science practical students (Fig. 3).

2.5.1 1RM Strength Testing

Participants warmed up prior to testing by running for 5 min on a treadmill. After a 1-min rest period, participants were familiarized with each of the resistance machines by performing 8–10 repetitions of a light load (~50 % of predicted 1RM). After 1 min of rest, participants performed a load (~80 % of estimated 1RM) through the full range of motion. After each successful performance, the weight was increased until a failed attempt occurred. One-minute rests were given between each attempt, and the 1RM was attained within 5 attempts and 5-min rest separated each test [6]. In order to facilitate the recovery and reduce the effect of fatigue, exercises were alternated between the upper and lower body [6]. The orders of the test were as follow: incline chess press, leg press, dual axis pull-down, and leg extension. After determining the 1RM value and allowing 72 h of rest, the maximum repetitions to failure for eight different sub-maximal percentages was assessed (30, 40, 50, 60, 70, 80, 90, and 95 % of 1RM) [1].

2.5.2 Handgrip Strength Testing

To perform this test, the participants have to squeeze a *Takei* dynamometer as strongly as possible with a hand, which is held in a rotationally neutral position. The width of the grip is individually adjusted; once the handgrip dynamometer was adjusted to the size of the player's hands, each subject performed three attempts trying to reach the peak force [9]. The test is executed in the standing position, with the arm extended downward along the body for 90° and 180° both for the left and right hand [10]. Thirty-seconds to one-minute time intervals were maintained between each of the handgrip strength testing [11]. The results were recorded in kilogram (kg).

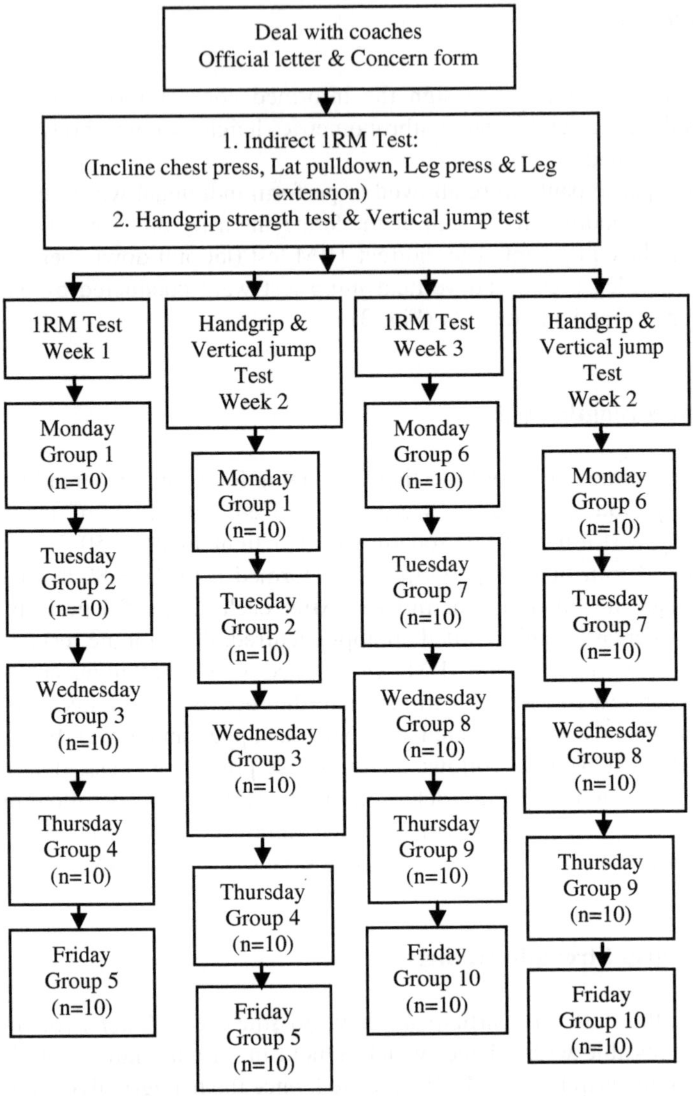

Fig. 3 Procedure chart

2.5.3 Vertical Jump Testing

The athletes were required to perform adequate warm-up and stretching prior to testing. The athletes are required to stand below the Vertec device, without moving the feet, bend the knees, and jump as high as possible to displace the markers. The athletes were given three trials. The best of the three trials were recorded to be analyzed.

Table 1 Demographic profile of subjects

Variable	Male ($N = 100$)		
	Mean ± SD	Min	Max
Age in years	20.92 ± 2.53	18	30
BMI	24.77 ± 4.70	18	43

2.6 Data Analysis

Data were analyzed using the Statistical Package for Social Science (SPSS 19.0) software. Descriptive statistics were used to analyze the data to seek the means and standard deviations of participant's age and BMI besides, finding the range of the difference between the dominant and non-dominant hand.

It was observed that the data for the individual and team sports met the assumptions for parametric statistics, in which there were two categories (individual and team sports), with independent observations and a normal population ($N > 30$) based on the central limit theorem (CLT). Thus, the independent-sample t test was used to compare the score of handgrip strength tests between individual and team sports. Hence, the median and standard deviations were used to determine which result showed the highest.

The Pearson's correlation was used to determine the relationships between the handgrip strength with 1RM strength tests and the vertical jump test. The values were between +1 and −1 inclusive.

3 Data Analysis and Results

The mean age in years for participants was 20.92 (±2.53), while the youngest was 18 years old and the oldest was 30 years old (Table 1). The mean value for BMI was 24.77 (±4.70), while the lowest value for BMI was 18 and the highest was 43.

There were weak positive correlations between both dominant extension and dominant 90° elbow flexion with indirect 1RM strength (lat pull-down and leg press) and the vertical jump (Table 2). However, there was a weak negative correlation between handgrip strength with indirect 1RM test for leg extension. Therefore, athletes who had high scores for indirect 1RM strength does not really tend to have high scores for the handgrip strength test, except for leg extension.

The range strength for dominant 90° elbow flexion was 41.8 kg and for elbow extension was 40.5. Hence, there is no significant different between elbow position. The differences of range strength between the dominant and non-dominant side for both 90° elbow flexion and elbow extension handgrip strength test among high-performance athletes were 8.2 and 3.1 kg, respectively, (Table 3). Range reflects the distance from the minimum to the maximum score.

Table 2 Pearson's correlations between the arm extension handgrip strength test, the indirect repetition maximum test, and the lower body power test

Indirect 1RM test	Dominant extension	Dominant 90° elbow flexion
Incline bench press	0.057	0.109
Lat pull-down	0.304**	0.345**
Leg press	0.113	0.065
Leg extension	−0.015	0.018
Vertical jump	0.119	0.154

**Correlation is significant near to ±1

Table 3 Descriptive statistics of differences range strength between the dominant and non-dominant side for the 90° elbow flexion and elbow extension handgrip strength test

Variables		*N*	Range (kg)	Min (kg)	Max (kg)
90° elbow flexion	Dominant	100	41.8	24.6	66.4
	Non-dominant	100	33.6	25.1	58.7
	Different	100	8.2	0.5	7.7
Extension	Dominant	100	40.5	30.1	70.6
	Non-dominant	100	37.4	25.5	62.9
	Different	100	3.1	4.6	7.7

Table 4 Independent-sample t test for 90° elbow flexion handgrip test and elbow extension handgrip between individual sports and team sports

Variables		*n*	Mean ± SD	Sig (2-tailed)
Extension dominant	Individual sports	30	48.26 ± 8.08	0.354
	Team sports	70	46.59 ± 8.43	
Extension non-dominant	Individual sports	30	43.60 ± 7.35	0.673
	Team sports	70	38.74 ± 6.75	
90° elbow flexion dominant	Individual sports	30	44.57 ± 8.05	0.096
	Team sports	70	41.67 ± 7.70	
90° elbow flexion non-dominant	Individual sports	30	40.50 ± 7.20	0.237
	Team sports	70	38.74 ± 6.75	

The mean of handgrip strength test of dominant elbow extension showed highest results in both individual and team sports (Table 4). For individual sports, the mean for dominant elbow extension was 48.26 (±8.08) and for the dominant 90° elbow flexion, it was 44.57 (±8.05). Meanwhile, the mean in team sports for the dominant elbow extension was 46.59 (±8.43) and for the 90° elbow flexion, it was 41.67 (±7.70). Hence, this showed that individual sports gave higher results on handgrip strength compared to team sports.

4 Discussion

Based on the results of this study, there was no significant relationship in handgrip and indirect 1RM strength tests among high-performance athletes. All indirect strength tests as selected were represent major muscles group. From all strength tests, there was only lat pull-down exercise that shown very low relationship between two elbow positions. However, this relationship was very weak to accept its finding and this answering all high-performance athletes may have their specific muscle strength based on their selected sports. However, handgrip strength test is suitable for prediction of total body strength in healthy young adolescents [8]. Besides, the handgrip strength test is one of the most reliable clinical methods for estimating strength [12].

This study found there was no significant different between elbow position. This finding was supported by previous study done by Carrasco et al. [9] found that there was small different grip strength between dominant and non-dominant hand. However, this study contra with Rahmat et al. [13] finding as elbow extension position shown with superior strength compare to flexion position.

There were significant differences in strength between the dominant and non-dominant hand among high-performance athletes. The maximum strength of hand was higher in the dominant versus the non-dominant hand, both for left-handed and for right-handed high-performance athletes. Similar results were reported in a previous study on table tennis players [9].

The findings of the present study also found that there was no significant relationship in the handgrip strength between individual and team sports. Individual sport is believed to give higher results of handgrip strength, which is considered as total body strength, compared to team sports. This is because different types of sports have different levels of strength and power demands [14]. The velocities that elicited maximal power in the lower extremities were lower than in the upper extremities [14].

However, this study group was limited to healthy male high-performance athletes (age range 18–30 years). One handicap was the lack of a female study group. However, according to Ertem et al. [15], the increase in grip strength was similar in both genders and appeared to be independent of sex hormones.

5 Conclusion

This study found handgrip strength test is not sensitive to predict muscle strength among high-performance athlete. Thus, handgrip strength is not recommended tool in measuring muscular strength in high-performance athlete.

Acknowledgments The researcher would like to express their utmost gratitude to RMI—UiTM RAGS for funding this study. We would also like to thank the Faculty of Sports Science and Recreation Shah Alam campus for giving this opportunity to complete this research project within the time frame.

References

1. Naclerio, F. J., Jiménez, A., Alvar, B. A., & Peterson, M. D. (2009). Assessing strength and power in resistance training. *Journal of Human Sport and Exercise, 4*(2), 100–113.
2. Visnapuu, M., & Jürimäe, T. (2007). Handgrip strength and hand dimensions in young handball and basketball players. *Journal of Strength and Conditioning Research, 21*(3), 923–929.
3. Maxwell, J. C. (1892). *A treatise on electricity and magnetism* (3rd ed., Vol. 2, pp. 68–73). Oxford: Clarendon.
4. Koley, S., & Kaur, P. S. (2011). Correlations of handgrip strength with selected hand-arm-anthropometric variables in Indian inter-university female volleyball players. *Asian Journal of Sports Medicine, 2*(4), 220–226.
5. Cadore, E., et al. (2012). Prediction of one repetition maximum load by total and lean body mass in trained and untrained men. *Medicina Sportiva, 16*(3), 111–117.
6. Seo, D., Kim, E., Fahs, C. A., Rossow, L., Young, K., & Ferguson, S. L. (2012). Reliability of the one-repetition maximum test based on muscle group and gender.
7. Leyk, D., Gorges, W., Ridder, D., Wunderlich, M., Rüther, T., Sievert, A., et al. (2007). Hand-grip strength of young men, women and highly trained female athletes. *European Journal of Applied Physiology, 99*(4), 415–421.
8. Rahmat, A., Elkhalifah M, Zulkifli A. K, Norasrudin S, Harris K. K, Shariman I, et al. (2012). Efficacy of the handgrip strength test amongst healthy young adolescents.
9. Carrasco, L., Pradas, F., Floría., P., Martínez, A., Herrero, R., Jurado J. A. G., et al. (2010). Grip strength in young top-level table tennis players. *International Table Tennis Federation Sports Science Congress Conference Proceedings* (vol. 6, pp. 64–66).
10. Cular, D., Miletic D, & Miletic, A. (2010). Influence of dominant and non-dominant body side on specific performance in taekwondo.
11. Sharma, A., Tripathi, V., & Koley, S. (2012). Correlations of anthropometric characteristics with physical fitness tests in Indian professional hockey players. *Journal of Human Sport and Exercise, 7*(3), 698–705.
12. Torvo, M., & Torvo, M. (2007). Handgrip strength and hand dimensions in young handball and basketball players.
13. Rahmat, A., Elkhalifah, M., Zulkifli, A. K., Harris, K. K., Norasrudin, S., Shariman, I. I., Nagoor, M., & Rahizam, M. (2013). Efficacy of the handgrip strength test amongst healthy young adolescent. *Proceeding IEEE Symposium on Business, Engineering, and Industrial Applications (ISBEIA 2013)*. Un publish research, Kuching.
14. Izquierdo, M., et al. (2002). Effects of long-term training specificity on maximal strength and power of the upper and lower extremities in athletes from different sports. *European Journal of Applied Physiology, 87*(3), 264–271.
15. Ertem, K., Inan, M., Yologlu, S., Elmali, N., Harma, A., Sahin, S., et al. (2003). Effects of dominance, body mass index and age on grip and pinch strength. *Isokinetics and Exercise Science, 11*(4), 219–223.

Emotional Intelligence and Sports Performance Among Malaysian Ethnics

Vincent Parnabas, Nagoor Meera Abdullah, Mohd Rahizam Abd Rahim, Mohamad Nizam Mohamed Shapie and Julinamary Parnabas

Abstract Athlete must recognize one's emotions, as well as their opponents and teammates emotions, in order to perform well in sports. For athletes, higher emotional intelligence has been linked to higher performance in sports. Previous studies showed that only scant research done on emotional intelligence in sports compared emotional intelligence among the athletes. It was also evident that comparing emotional intelligence on different ethnic, whom come from different culture, was very limited. The aim of this research was to compare the level of emotional intelligence among different ethnic of athletes and their sports performance. It sought to correlate the relationship between emotional intelligence and sports performance of athletes from different ethnics (Malay, Chinese and Indians). Emotional intelligence scale (EIS) and Sport Performance Questionnaire were used to collect the data. The sample consisted of 217 athletes, who competed in Majlis Sukan University Malaysia (MASUM), also known as Malaysian University Sports Council. The result of this research showed that overall Chinese athlete scored higher in emotional intelligence and Indians, the lowest. The result also showed that the existence of a strong positive relationship between emotional intelligence and sports performance. Sports psychologist and coaches can use the findings to apply coping strategies to enhance emotional intelligence among Indian and Malay athletes.

Keywords Emotional intelligence · Sports performance · Ethnics

V. Parnabas (✉) · N. M. Abdullah · M. R. Abd Rahim · M. N. Mohamed Shapie
Faculty of Sport Science and Recreation, Universiti Teknologi MARA,
40450, Shah Alam, Selangor, Malaysia
e-mail: vincent@salam.uitm.edu.my

J. Parnabas
Institut Pendidikan Guru, Kampus Darulaman, Jitra, Kedah, Malaysia

R. Adnan et al. (eds.), *Proceedings of the International Colloquium on Sports Science, Exercise, Engineering and Technology 2014 (ICoSSEET 2014)*,
DOI: 10.1007/978-981-287-107-7_5, © Springer Science+Business Media Singapore 2014

1 Introduction

Emotional intelligence (EI) is defined as 'the ability to employ, monitor, perceive, and manage emotions within one and in others' [1]. Furthermore, according to [2], emotional intelligence is 'the ability to understand, recognize, and effectively use our feelings, which make it easier to manage the people around us and ourselves'. This definition consists of five components such as motivating ourselves, managing them, knowing our emotions (self-awareness), recognizing emotions in others (empathy) and handling relationships.

In light of these definitions in sport, emotional intelligence can be defined that an athlete being able to motivate himself/herself, to control his/her feelings (towards his/her teammate, the opponents and supporters, referee, the managers, managers of the opposing team and other similar factors), to direct controlled feelings in a positive way, to establish positive and effective communication, to develop communication skills, to gain control over stress levels and utilize this control [3]. In other words, an emotional intelligent athlete is skilled in four areas, such as understanding, identifying, using and regulating emotions [1].

In the sports context, emotional intelligence is important for athletes. For athletes, higher emotional intelligence has been linked to higher performance in sports. Research done by Perlini and Halverson [4] showed that emotional intelligence of hockey players was higher and this contributes to their success in the super league. Athlete must recognize one's emotions, as well as the opponents' and teammates emotions, in order to perform well in sports [5]. Therefore, emotional intelligence plays an important role in influence one's ability to succeed in coping with environmental demands, challenges and pressures [6].

Many studies showed that emotional intelligence resembles success even though not many studies done on these athletes. The study done by Bar-On [6] and Swart [7] showed that those students who scored higher in emotional intelligence are those students who are good in academic. Both studies showed that students, who were able to manage their emotions, were more success to solve problems related to personal and interpersonal, which was considered as an important element to success in studies.

The main problem regarding this study is when previous studies showed that only scant research was conducted on emotional intelligence in sports compared to emotional intelligence among the athletes [8–11]. Furthermore, comparing emotional intelligence on different ethnics, who come from different culture is very limited. Culture may be defined as the changing patterns of learning behaviour and the products of learning behaviour (including attitudes, values, knowledge and material objects) that are shared by and transmitted among the members of the society; hence, it is an ongoing social heritage [12].

Research of [13] on Whites, Blacks and Hispanics showed that female Hispanics scored the highest in EI and Female-Whites scored the lowest. Similarly, research of [14], comparing emotional intelligence on Whites, African Americans, Latino-Hispanics and Asian Americans, shows a very contradictory results.

Whitman et al. [15] had done an interesting study using emotional intelligence test on 910 job applicants. The purpose of testing emotional intelligence on job applicant was to select the best candidate, since EI determined success in job. The result showed that the level of the emotional intelligence was highest on the Whites, followed by Hispanics and Blacks.

In Malaysia, Malay is the largest ethnic (50.4 %) population, followed by Chinese (23.7 %), indigenous (11 %), Indians (7.1 %) and others (7.8 %, especially from East Malaysia) [16]. Even though, all of them staying in the same country but each of the ethnic followed their own culture. Sometimes, it is hard to believe that they remain in the same country, but their way of living differs dramatically from the others. Since many studies showed that there is a very strong connection between culture and emotional intelligence [17], we are not sure how far culture of each ethnic influences their emotional intelligence.

2 Literature Review

Emotional quotient (EQ) is one of the methods to measure a person's ability to be successful in their life. According to the theories of brain function, a high EQ means of someone's self-aware, self-confident and able to navigate through trying emotional times. EQ is often directly to the degree of success in job and personal relationships. A few studies had been proved that a person with better emotional intelligence typically has more success in personal and interpersonal interaction compared with people who have less effective ways of dealing with emotions [18, 19].

The concept of emotional intelligence has been put to use in various psychological research settings in recent years and has been the main focus of many different theories throughout the literature [20–22].

Besides that, several researchers in the past introduced and developed the concept of emotional intelligence as a type of intelligence that includes abilities, competencies and skills and that enables a person to identify emotions, be able to solve emotional issues, as well as to manage and control own emotions [6, 9, 23]. Furthermore, it is commonly acknowledged that an athlete's emotions play an integral part in their performances [24]. We often hear that an athlete or team is 'playing on emotion' or that on any given day, feelings such as aggression may be up or down during the competition. However, measuring the emotion that underlies these observable behaviours or accessing them when needed by the athlete is less simple than seeing them in action during the performance stage [1].

Based on the health setting, researches suggest that participants with higher emotional intelligence reported better psychological well-being, social interactions, physical health and self-actualization [18, 19, 25, 26].

3 Aim of the Study

The aim of this research was to identify the level of emotional intelligence among Malaysian athletes of different ethnics. It aimed at comparing the level of emotional intelligence among different ethnic of athletes and their sports performance. It sought to correlate the relationship between emotional intelligence and sports performance of athletes from different ethnics (Malay, Chinese and Indians).

4 Method

This survey was conducted using a set of questionnaire namely, emotional intelligence scale (EIS) [27] and Sport Performance Questionnaire to collect the data. The 33 items of EIS assesses six factors: appraisal of own emotions, appraisal of others' emotions, regulation, social skills, utilization of emotions and optimism. These six factors represent a person's emotional competence, which is very important in the performance and interaction with others. The EIS is a self-report measure of emotion that provides an estimation of the level of emotional intelligence.

Appraisal of own emotions is the ability of self-awareness and expressing the feelings in appropriate way. It also includes the feelings when making decisions and solving problems. Appraisal of other emotional assess is the ability to appraise an emotional states experienced by team athletes, coaches and audience. Regulation of their own emotions assesses the management strategies that may be used to alter an emotional stability. Social skills refer to the ability to change the emotions of other people by general strategies such as showing empathy towards others, cooperation, satisfying relationships and helping others. Utilizations of emotions concerned with the adaption and awareness of the influence of emotions on a range of different performance outcomes. Optimism is the positive beliefs and look at the brighter side regarding the future towards general outcomes (e.g. improved happiness, good things to happen, doing well).

Besides that, Sports Performance Questionnaire, which contains 15 items evaluated athletes' performance based on their confident, satisfaction, enjoyment and concentration level.

The sample consisted of 217 athletes, who voluntary participated in this study. The sample was drawn from athletes who competed in Majlis Sukan University Malaysia (MASUM) or Malaysian University Sports Council.

Table 1 Respondents' profile

Emotional intelligence	Malay	Chinese	Indian
Appraisal of own emotions	3.2212	3.3451	3.1718
Appraisal of others' emotions	2.7181	3.2217	2.8731
Regulation	3.6781	3.1721	2.5461
Social skills	2.4739	2.5811	2.7614
Utilization of emotion	3.4732	3.7418	2.7811
Optimism	3.7895	3.5541	3.0008

5 Result

5.1 Respondents' Profile

Frequency, percentage, mean and standard deviation are presented in Table 1, which shows the overall results of the respondents' profile. The profile of the respondents described the ethnics, gender, age, universities and type of sports taken part. Based on the ethnic, the majority of the athletes belong to Malays ($n = 105$) since they are the majority in Malaysia. The second largest ethnic in Malaysia is Chinese. There are 65 Chinese athletes. Indians are the lowest ethnic in Malaysia. There are 47 Indian athletes.

The mean age for overall respondents was 21.71 years old (Table 1). The age of male varied from 18 to 27 years, where the mean age was 22.07 years old. The age of females ranged from the minimum of 18 to the maximum of 25 years old. The mean age for female respondents was 22.01 years old.

The variable 'types of sports' are gathered through the studies. This variable is categorized into seven sports. The highest respondents involved football (20.28 %), followed by basketball (18.89 %), swimming (15.67 %), silat (14.28 %), track and field (12.90 %), hockey (9.68 %) and takraw (8.30 %).

Based on universities (Table 1), the majority of the respondents were from Universiti Teknologi MARA (UiTM), followed by University Kebangsaan Malaysia (UKM), University Utara Malaysia (UUM), University Pendidikan Sultan Idris (UPSI), University Teknikal Melaka, Universiti Teknologi Malaysia (UTM) and University Perlis Malaysia (UNIMAP).

5.2 Mean of Emotional Intelligence

The results (Table 2) showed that Malay athletes were higher in regulation ($x = 3.678$) and optimism ($x = 3.789$). While Chinese athletes were higher in appraisal of own emotions ($x = 3.345$), appraisal of others' emotions ($x = 3.345$) and utilization of emotion ($x = 3.345$). Furthermore, Indian athletes were higher in social skills ($x = 2.761$).

Table 2 Mean of emotional intelligence based on the categories of athletes ethnic

Variables	Frequency	Percentage	Mean	SD
Ethnics				
Malay	105	48.39		
Chinese	65	29.95		
Indian	47	21.66		
Gender				
Male	75	34.56		
Female	142	65.44		
Age				
Male			22.07	2.18
Female			22.01	1.89
Overall			21.71	1.53
Types of sport				
Football	44	20.28		
Basketball	41	18.89		
Swimming	34	15.67		
Silat	31	14.28		
Track and Field	28	12.9		
Hockey	21	9.68		
Takraw	18	8.3		
Universities				
UiTM	55	25.34		
UKM	35	16.13		
UUM	30	13.82		
UPSI	28	12.91		
USM	27	12.44		
UTM	23	10.61		
UNIMAP	19	8.75		

5.3 Level of Emotional Intelligence

One-way ANOVA showed high significant differences of levels of emotional intelligence among categories of ethnic athletes, $F(2, 217) = 15.001$, $p < 0.01$ (Table 3). Ethnics of athletes represented those who belong to Malay, Chinese and Indian race.

Post Hoc Tukey's test (Table 4) showed that the level of emotional intelligence of Indian athletes was lower than Malay ($p < 0.05$) and Chinese athletes ($p < 0.05$). Furthermore, the level of emotional intelligence among Malay athletes was higher than Indian athletes ($p < 0.05$) and less than Chinese athletes ($p < 0.05$). In addition, the level of emotional intelligence of Chinese athletes was higher than Malay ($p < 0.05$) and Indian athletes ($p < 0.05$).

Table 3 Level of emotional intelligence based on categories of athletes' ethnic

Categories according to ethnics	Mean	F value	p value
Malay	20.7789	15.001[a]	0.000
Chinese	23.4321		
Indian	18.8129		

[a] $p < 0.01$

Table 4 Post hoc Tukey: level of emotional intelligence among categories of athletes' ethnic

Categories according to ethnic	Malay	Chinese	Indian	N
Malay		[a](2.221)	[a](3.312)	05
Chinese			[a](2.619)	65
Indian				47

[a] $p < 0.05$

Table 5 The relationship between the emotional intelligence and sports performance among athletes

Subject	Sports performance
Emotional intelligence	0.018[a] (0.000)

[a] $p < 0.01$

5.4 *Correlation of Emotional Intelligence and Sport Performance*

The correlation coefficient of 0.018 was noted between the emotional intelligence and sports performance in the evaluation of 217 athletes, which is significantly ($P < 0.01$). In other words, the positive relationship existing between these variables is statistically significant (Table 5).

6 Discussion

6.1 *Mean of Emotional Intelligence*

The mean of elements of emotional intelligence showed that Chinese athletes were higher in appraisal of their own emotions, appraisal of others' emotions and utilization of emotion.

Malay athletes were higher in regulation and optimism, while Indian athletes were higher in social skills. A few researches indicated it is normal that different ethnics showed different levels of certain elements of emotional intelligence because of the cultural differences. Research of [28] showed that African Americans, aboriginals and Asians scored lower in regulation of emotion, compared to

European Americans. We can be concluded that the differences in the element of emotional intelligence related with the cultural difference. Hence, one can use EIS to predict the cultural difference between ethnics.

6.2 Level of Emotional Intelligence

Overall the result showed that the athletes in the Chinese categories exhibit higher level of emotional intelligence than Malay category athletes, whereas Indian athletes showed the lowest levels of emotional intelligence. In Malaysia, no research has been done involving these three categories of ethnics, so the findings of the current research could not be compared with that of previous researches. Anyway, culture is the main factor that determined the level of emotional intelligence among different ethnics. According to [29], people differ from one another in their abilities to interpret and the use emotional information in each culture. Therefore, the level and element of emotional intelligence may differ from culture to culture. A few comparison researches done on emotional intelligence between North Americans and East Asians showed that the East Asians groups obtained lower EI scores compared to Northern Americans because of practicing the contradictory cultural values and beliefs [30].

According to [31], behaviour, which is in one culture considered as clever or intelligent called to be regard as stupid or nonsense in another culture. For example, item 21, 'I have control over my emotions'. In Malay culture, one believes that everyone should control over their emotions, but in Indian culture, it is encouraged to express their emotions rather than controlling it. For example, during funeral, Malays are forbidden to cry and should accept the God's will to take one's life. But, in Indian and Chinese culture people are encourage to cry, to express their feelings of sadness in losing their love ones. A few questions in EIS [27] related with non-verbal cues, for example item no. 5, 'I find it hard to understand the non-verbal messages of other people', item 'I am aware of the non-verbal messages I send to others', and item 25, 'I am aware of the non-verbal messages that other people send'. If we compare the usage of non-verbal cues in the daily behaviour, Malays are using it more compared to other ethnics. For Indians, non-verbal message sometimes can be considered as lack of manners. For instance, to answer 'yes' or 'no' is considered as normal in Malay culture by shaking the head but it is considered as lack of manners in Indian culture. In Indian culture, one should open his mouth to say 'yes' or 'no'. Therefore, most probably, this cultural behaviour may influence their answers.

Furthermore, item no. 28, 'When I am faced with a challenge, I give up because I believe that I will fail', we can find that Malays and Indians have the tendency to give up very easily compared to Chinese. Many Malays believed that when you face a lot of challenges in attempting a goal, it shows that God does not accept it or in other words, one does not have the 'rezeki' or blessings. Therefore, in Malaysia, Chinese is considered successful in business compared to Malays and Indians.

6.3 *Correlation of Emotional Intelligence and Sport Performance*

The result showed the existence of positive relationship between emotional intelligence and sports performance. It is not surprising since emotional intelligence strongly linked with job performance, quality of relationships, and well-being [32, 33]. Therefore, emotional intelligence of athletes enhances their sports performance. The higher the emotional intelligence, the higher their performance in sports is.

In sports, it is necessary to be sensitive with non-verbal cues among team members. Success in team sports such as football and hockey depends how far team members can communicate with each other using non-verbal cues. Therefore, a few items like 'I find it hard to understand the non-verbal messages of other people', 'I am aware of the non-verbal messages I send to others', 'By looking at their facial expressions, I recognize the emotions people are experiencing', 'I am aware of the non-verbal messages other people send' show that the more knowledgeable athletes on non-verbal cues, the more success in sports is.

Besides that, a few other items in the EIS have a strong connection with sports performance. For example, 'When I am faced with a challenge, I give up because I believe I will fail'. Athletes who are easily giving up and have a very low self-confidence, its impossible for them to succeed in sports.

Furthermore, controlling the emotions is very important for an athlete to succeed in sport. Therefore, a few items that related to control emotion are considered important for success in sport, for example 'I have control over my emotions'. For success in sports, it is very important to control emotion. There are many challenges in sports like the nasty behaviour of the audience, the irritated behaviour of coach, the provoking behaviour of opponents and so on. If athletes fail to control their anger, anxiety, worry and shame, it is impossible to succeed in sports.

7 Conclusion

The result of this research showed that overall Chinese athletes scored higher in emotional intelligence and Indians, the lowest. The result also showed the existence of exist a strong positive relationship between emotional intelligence and sports performance. Sport psychologists and coaches can use the findings to apply coping strategies to enhance emotional intelligence among athletes of Indians and Malays.

Future research should explore the reasons of Indians athletes scored lower in emotional intelligence. Future research also should attempt to test the emotional intelligence based on different level skills of athletes, for example elite or national, state, district, university and school levels. Research needed to be done to find out the influence of socio-economic factors on emotional intelligence of athletes.

References

1. Salovey, P., & Mayer, J. D. (1990). Emotional intelligence. *Imagination Cognition and Personality, 9*, 185–211.
2. Goleman, D. (1995). *Emotional intelligence*. New York: Bantam Books.
3. Adilogullari. Relationship between organizational commitment and emotional intelligence in professional soccer players. PhD thesis, Marmara University Institute of Health Sciences, Istanbul, 2011.
4. Perlini, A. H., & Halverson, T. R. (2006). Emotional intelligence in the National Hockey League. *Canadian Journal of Behavioural Science, 38*(2), 109–119.
5. Zizzi, S. J., Deaner, H. R., & Hirshhorn, D. K. (2003). The relationship between emotional intelligence and performance among college baseball players. *Journal of Applied Sport Psychology, 15*, 262–269.
6. Bar-On, R. (2006). The Bar-On model of emotional-social intelligence (ESI). *Psicothema 18*, 13–25.
7. Swart, A. (1996). The relationship between well-being and academic performance. *Unpublished master's thesis*, University of Pretoria, South Africa.
8. Meyar, B., & Fletcher, T. (2007). Emotional intelligence: A theoretical overview and Implications for research and professional practice in sport psychology. *Journal of Applied Sport Psychology, 19*, 1–15.
9. Meyer, B. B., & Zizzi, S. (2007). Emotional intelligence in sport: Conceptual, methodological, and applied issues. In A. M. Lane (Ed.), *Mood and Human Performance: Conceptual, Measurement, and Applied Issues* (pp. 131–154). Hauppauge, NY: Nova Science.
10. Saklofske, D. H. (2007). Personality, emotional intelligence and exercise. *Journal Health Psychology, 12*(6), 937–948.
11. Petrides, K. V., & Furnham, A. (2000). Gender differences in measured and self estimated trait emotional intelligence. *Sex Roles, 42*(5/6), 449–461.
12. Soyinka, W. (1991). Culture, memory and development. *Proceedings of an International Conference held at The World bank*, Washington D.C.
13. Van Rooy, D. L., Alonso, A., & Viswesvaran, C. (2005). Group differences in emotional intelligence test scores: Theoretical and practical implications. *Personality and Individual Differences, 38*, 689–700.
14. Roberts, R. D., Zeidner, M., & Matthews, G. (2007). Emotional intelligence: Knowns and unknowns. In G. Matthews, M. Zeidner, & R. D. Roberts (Eds.), *The Science of Emotional Intelligence: Knowns and Unknowns* (pp. 419–474). New York: Oxford University Press.
15. Whitman, D. S., Van Rooy, D. L., Viswesvaran, C., Burns, T., & Kraus, T. (2006). Group differences in job applicant scores on an emotional intelligence test. *Paper presented at the Society of Industrial and Organizational Psychology*, Dallas, TX.
16. Malaysia Demographics Profile 2012. (http://www.indexmundi.com/malaysia/demographics_profile.html).
17. Roberts, R., Zeidner, M., & Matthews, G. M. (2001). Does emotional intelligence meet traditional standards for an intelligence? Some new data and conclusions. *Emotions, 1*, 196–231.
18. Extremera, N., & Fernandez-Berrocal, P. (2002). Relation of perceived emotional intelligence and health related quality of life of middle-aged women. *Psychological Reports, 91*, 47–59.
19. Slaski, M., & Cartwright, S. (2002). Health, performance and emotional intelligence: an exploratory study of retail managers. *Stress and Health, 18*, 63–68.
20. Brackett, M. A., & Mayer, J. D. (2003). Convergent, discriminant, and incremental validity of competing measures of emotional intelligence. *Personality and Social Psychology Bulletin, 29*, 1147–1158.
21. Mayer, J. D., Caruso, D., & Salovey, P. (1999). Emotional intelligence meets traditional standards for an intelligence. *Intelligence, 27*, 267–298.

22. Davies, M., Stankov, L., & Roberts, R. D. (1998). Emotional intelligence: In search of an elusive construct. *Journal of Personality and Social Psychology, 75*, 989–1015.
23. Goleman, D. (2004). Emotional intelligence—why it can matter more than IQ. In D. Goleman (Ed.), *Emotional Intelligence and Working with Emotional Intelligence (Omnibus)*. London: Bloomsbury Publishing.
24. Lane, A. M., Sewell, D. F., Terry, P. C., Bartram, D., Nesti, M. S. (1999) Confirmatory factor analysis of the Competitive State Anxiety Inventory-2. *Journal of Sports Sciences, 17*(6), 505–512,.
25. R. Bar-On. (2005). Emotional intelligence and subjective wellbeing. Manuscript submitted for publication.
26. Bar-On, R. (2001). Emotional intelligence and self-actualization. In Joseph Ciarrochi, Joe Forgas, & John D. Mayer (Eds.), *Emotional Intelligence in Everyday Life: A Scientific Inquiry*. New York: Psychology Press.
27. Schutte, N. S., Malouff, J. M., Hall, L. E., Haggerty, D. J., Cooper, J. T., Golden, C. J., & Dornheim, L. (1998). Development and validation of a measure of emotional intelligence. *Personality and Individual Differences, 25*, 167–177, 1998.
28. Neisser, U., Boodoo, G., Bouchard, T. J., Boykin, A. W., Brody, N., Ceci, S. J., Halpern, D. F., Loehlin, J. C., Perloff, R., Sternberg, R. J., & Urgina, S. (1996). Intelligence: Knowns and unknowns. *American Psychologist, 51*, 77–101, 1996.
29. Grewal, D., & Salovey, P. (2005). Feeling smart: The science of emotional intelligence. *American Scientist, 93*, 330–339.
30. Moon, S. (2011). East meets West: The cultural-relativity of emotional intelligence. A thesis submitted in conformity with the requirements for the degree of Doctor of Philosophy Joseph L. Rotman School of Management University of Toronto.
31. Sternberg, R. J. (2004). Culture and Intelligence. *American Psychologist, 59*(5), 325–338.
32. Brackett, A., Rivers, S. E., Hoffman, S., Lerner, N., & Salovey, P. (2006) Relating emotional abilities to social functioning: A comparison of self-report and performance measures of emotional intelligence. *Journal of Personality and Social Psychology, 91*, 780–795.
33. Cote, M. S., & Miners, C. T. H. (2006) Emotional intelligence, cognitive intelligence, and job performance. *Administrative Science Quarterly, 51*, 1–28.

Pulmonary Function Profiling Among Young Athletes of SUKMA Terengganu

Norlizah Abdul Hamid, Suzanayantie Salleh, Nagoor Meera Abdullah, Sarimah Ismail, Mohamad Nizam Mohamad Shapie and Rozita Abdul Latif

Abstract Studies on pulmonary function usually revolved in clinical research of chronic obstructive pulmonary disease (COPD). The current study was designed to set a profile and compare the pulmonary function among young athletes competing in the Sukan Malaysia (SUKMA) 2011. The parameters measured include forced vital capacity (FVC), one-second forced expiratory volume (FEV_1), and FEV_1/FVC ratio (FEV_1/FVC%). A total of 75 (males: 40, female: 35) young athletes (9–22 years old) participated in this study. The athletes were randomly selected according to their sports' involvement (i.e., swimming, cycling, and hockey). Pulmonary function was measured using COSMED Pony FX Desktop Spirometer. Descriptive and a one-way analysis of variance (ANOVA) were used to analyze the data. Cyclists showed to have significantly higher FVC and FEV_1 compared to swimmers in both genders. Male cyclists also showed significantly higher in FEV_1 than male hockey players. Female hockey players showed significantly higher FVC compared to swimmers. However, FEV_1/FVC% showed no significant differences among cyclists, swimmers, and hockey players. In general, results indicated that cyclists had better pulmonary functions compared to the other two sports.

Keywords Component · Pulmonary disease · Endurance athletes · Forced vital capacity · Pulmonary function

1 Introduction

Pulmonary system plays an important role in the process of pulmonary ventilation. Therefore, cardiorespiratory fitness is one of the dominating factors of exercise performance. Competitive swimmers require high aerobic capacity to support and

N. Abdul Hamid (✉) · S. Salleh · N. M. Abdullah · S. Ismail · M. N. Mohamad Shapie · R. Abdul Latif
Faculty of Sport Science and Recreation, Universiti Teknologi MARA, Shah Alam, Selangor, Malaysia
e-mail: norli144@salam.uitm.edu.my

R. Adnan et al. (eds.), *Proceedings of the International Colloquium on Sports Science, Exercise, Engineering and Technology 2014 (ICoSSEET 2014)*,
DOI: 10.1007/978-981-287-107-7_6,

sustained performance in a severe exercise [1]. Cycling and hockey are sports that require the athletes to have enough cardiorespiratory fitness to perform in prolonged time. In addition, swimming also needs the same aspect as breathing is important for this sport. Swimming, cycling, and hockey involve the use of aerobic and anaerobic energy system. Most of the training programs were designed to develop and improve aerobic and anaerobic conditioning. Aerobic training improves cardiopulmonary system that trains lungs to supply oxygen efficiently. It enhances blood flow and overall oxygen supply throughout the body. This system positively affects the ability of the lungs to take oxygen and match with hemoglobin receptors in the blood during gas diffusion. It helps athletes to perform better. Besides improving cardiovascular functioning, aerobic exercise also influences lung capacity and at the same time improves lung function too. As cardiovascular system is important in sport performance, so does the pulmonary system since the heart, lungs, and circulatory systems typically work together to deliver oxygen to all parts of the body. Moreover, the most efficient way to improve lung capacity is by engaging in aerobic exercises. However, other high intensity exercises such as anaerobic exercise may also help improve lung capacity as long as the intensity put on it is appropriate.

The lung function performance also depends on several factors such as age, gender, height, and weight [2]. Pulmonary function test (PFT) is a standard test used to measure lung capacity that includes lung volumes, airway functions, and gas exchanges. It is generally used in the clinical setting to assess pulmonary disease. In sport performance, it becomes a basic test to evaluate athletes' respiratory efficiency and obligatory in assessing sport fitness [2]. So, lung function test is not only to indicate the risk of pulmonary disease but also to determine lung efficiency among athletes. Endurance training such as running and cycling involves the performance of high repetitions and low resistance exercise over a period of time. This type of training is attained to increase aerobic capacity and is important in hockey as it is a fast game and athletes need to endure more than one hour. According to [3], field hockey requires high level of aerobic fitness and it is a prerequisite for a superior anaerobic performance during sustained intermittent activities. In Malaysia, there are a number of studies on pulmonary functions carried out on healthy adults, sedentary people, factory workers, and pulmonary patients [4–10]. In 2009, [4] have studied and formulated prediction equation for ventilatory function among 869 participants of different ethnic groups in Kuching, Malaysia. Pulmonary functions were studied according to various age groups including both male and female among the Malays, Chinese, Dayak, and Indians.

Latest research was done by Gnanou et al. [5] to establish the prediction equation of lung function in healthy non-smoking Malaysian. The study aimed to develop predictive equations for male and female in Malaysian population age ranged from 22 to 92 years old. Since lung function is associated with the anthropometric characteristics, the values of FVC and FEV_1 regularly showed significant lower values when compared to other references from European and American populations. A total of 5,708 participants were involved in this study but only 1,483 (386 male and 1,097 females) who matched the descriptions were selected. Results showed that male FVC was 2.7 L, female 2.1 L. FEV_1 showed 2.4 L in male and 1.8 L in female. From the study, they have formulated twelve prediction formulae of FVC and FEV_1. Result of the study showed Malay smokers in the age group of 22–44 and 45–60 years old had higher values of FVC and FEV_1 than non-smokers. For FEV_1/FVC ratio, the value was lower than non-smoker. Study by Shashikala and Sarath [11] found that regular exercise has influence on the cardiovascular function and pulmonary function. The effect of a 12-week exercise and sports activities on PFT was studied. Results showed that PFT was higher after exercise training program.

Hence, physical activity improved some aspect of lung function and had significant improvement in aerobic fitness (VO_2max) [12]. The present study was intended to profile and compare the pulmonary function among young Malaysian athletes of different sports.

2 Methodology

2.1 Sample

A total of 75 athletes currently training in preparation for SUKMA tournament the following year were randomly selected for this study. They were well-trained athletes with no history of pulmonary disease and all of them were non-smokers. Sports involved in this study were swimming, cycling, and hockey of both genders.

2.2 Instrumentation

Spirometry test was conducted to analyze and evaluate the lung function of the subjects using COSMED Pony FX Desktop Spirometer, where values of FVC, FEV_1, and FEV_1/FVC% were obtained. Permission from the Terengganu Sports Council was granted prior to the study and subjects' information sheet explaining the details of the test were also given. Informed consent was acquired through the sports' managers since most of the athletes involved in this study were under age.

2.3 Test administration

The participants were first briefed on the procedures of the test. Informed consents were obtained. All participants were given time to familiarize with the correct breathing technique of the test. The best of three trials were taken as the scores for analysis.

The test was performed in the standing position. The participants first breathed normally and quietly through a mouthpiece that is connected to the spirometer. The amount and the rate of air breathed in and out over a period of time were examined. A nose clip was used throughout the test to ensure optimum recording of the exhaled air volume.

For measurement of the lung volumes, maximal and forceful exhalation was performed following maximal inhalation. During the test, participants were given verbal instructions to "take a deep breath in" and then "blow it all out hard". Acceptable value was to reach a one-second FVC plateau and to record 6 s or more during exhalation. All data were saved and printed out at the end of the test.

2.4 Analysis of data

The data were analyzed by using the statistical package for social sciences (SPSS) software and tabulated. In this study, descriptive and one-way ANOVA were used to compare the significant difference between groups. Descriptive analysis was used to describe the physiological characteristics of the subjects. Descriptive data were expressed as mean and standard deviation (SD). Independent *t* test used to examine significant difference on lung function performance between male and female athletes. p value <0.05 indicated a statistically significant difference.

3 Results

The mean age, weight, and height for male and the female athletes are presented in Table 1. Mean and standard deviation (±SD) of age for male cyclists were 16.60 ± 1.55 years old, swimmers 13.09 ± 2.77 years old, and hockey players were 13.57 ± 0.52 years old. Mean weight for cyclists were 59.25 ± 6.51 kg, swimmers recorded mean weight of 45.40 ± 11.59 kg, and mean weight for hockey players were 46.74 ± 9.65 kg. Mean for height, male cyclists recorded at 163.01 ± 8.43 cm, swimmers were 155.46 ± 21.65 cm, and hockey players were 159.36 ± 16.16. Mean age and standard deviation (±SD) for female cyclists were 17.60 ± 1.52 years old, swimmers were 10.80 ± 1.64 years old, and hockey players were 16.32 ± 2.90 years old. Mean and standard deviation (M ± SD) for weight in female cyclists were 53.64 ± 5.77 kg, in swimmers 38.90 ± 8.58 kg, and in hockey players were 51.08 ± 7.19 kg. For height, the mean and standard deviation in cyclists were 155.80 ± 1.92 cm, while swimmers and hockey players were 143.80 ± 11.45 cm and 155.88 ± 4.37 cm, respectively.

Analysis of mean and standard deviation (M ± SD) of lung function performance among different sports has been provided in Table 2. In male athletes, FVC value in cyclists was 3.45 ± 0.51 L, swimmers was 2.61 ± 0.91 L, and hockey players was 2.93 ± 0.53 L. FEV_1 for cyclists was 3.13 ± 0.45 L, swimmers recorded 2.44 ± 0.85 L, and hockey players showed 2.54 ± 0.59 L. For FEV_1/FVC%, male cyclists were 90.87 ± 4.60 %, swimmers were 93.95 ± 4.35 %, and hockey players were 86.83 ± 13.02 %. Analysis of mean and standard deviation

Table 1 Mean and standard deviation of age, weight, and height of cyclists, swimmers, and hockey players

		Cycling		Swimming		Hockey	
		M	F	M	F	M	F
Age (years)	M	16.60	17.60	13.09	10.80	13.57	16.32
	SD	1.55	1.52	2.77	1.64	0.52	2.90
Weight (kg)	M	59.25	53.64	45.40	38.90	46.74	51.08
	SD	6.51	5.77	11.59	8.58	9.65	7.19
Height (cm)	M	163.01	155.80	155.46	143.80	159.36	155.88
	SD	8.43	1.92	21.65	11.45	16.16	4.37

Table 2 Mean and standard deviation of FVC, FEV_1, and FEV_1/FVC% of cyclists, swimmers, and hockey players

		Cycling		Swimming		Hockey	
		M	F	M	F	M	F
FVC (L)	M	3.45	2.78	2.61	2.12	2.93	2.53
	SD	0.51	0.10	0.91	0.18	0.53	0.33
FEV_1 (L)	M	3.13	2.53	2.44	1.98	2.54	2.25
	SD	0.45	0.07	0.85	0.07	0.59	0.27
FEV_1/ FVC%	M	90.87	89.42	93.95	91.48	86.81	90.43
	SD	4.60	5.09	4.35	6.90	13.02	5.17

(M ± SD) of lung function performance in female athletes among different sports shows FVC value in female cyclists was 2.78 ± 0.10 L, swimmers 2.12 ± 0.18 L, and hockey players 2.53 ± 0.33 L. In FEV_1, cyclists scored 2.53 ± 0.07 L, swimmers were 1.98 ± 0.07 L, and hockey players were 2.25 ± 0.27 L. In female cyclists, the FEV_1/FVC% value was 89.42 ± 5.09 %, while swimmers and hockey players scored 91.48 ± 6.90 % and 90.43 ± 5.17 %, respectively.

The descriptive data between genders were compared by using independent *t* test as shown in Table 3. Mean and standard deviation (M ± SD) of FVC in male was 3.04 ± 0.72 L and was significantly higher than female 2.5 ± 0.34 L ($p = 0.000$). In FEV_1, male also recorded significantly higher value of 2.74 ± 0.69 L compared to female of 2.25 ± 0.28 L ($p = 0.000$). However, for FEV_1/FVC%, female showed slightly higher value of 90.35 % than male 90.30 %.

One-way ANOVA in Table 4 shows that *p* values of FVC and FEV_1 were 0.007 and 0.014, respectively. This shows that there were significant differences in FVC and FEV_1 among the sports. However, no significant difference was found for FEV_1/FVC%.

For the females, the one-way ANOVA as in Table 5 shows that *p* values of FVC and FEV_1 were 0.05 and 0.06, respectively. This indicates that there were significant differences in FVC and FEV_1 among the sports. However, no significant difference was found for FEV_1/FVC%.

Table 3 Mean (M) and standard deviation (SD) of pulmonary function between genders

	Male		Female	
	M	SD	M	SD
FVC (L)	3.04	0.72	2.50	0.34
FEV_1 (L)	2.74	0.69	2.25	0.28
FEV_1/FVC%	90.30	8.79	90.35	0.40

Table 4 One-way ANOVA of pulmonary function performance for male

		Sum of squares	df	Mean square	F	Sig.
FVC (L)	Between groups	4.785	2	0.2393	6.349	0.007[a]
FEV_1 (L)	Between groups	3.834	2	1.917	6.316	0.014[a]
FEV_1/ FVC%	Between groups	320.424	2	160.212	2.203	0.125

[a] Sig. at $p < 0.05$

Table 5 One-way ANOVA of pulmonary function performance for female

		Sum of squares	df	Mean square	F	Sig.
FVC (L)	Between groups	1.094	2	0.547	6.349	0.005[a]
FEV1 (L)	Between groups	0.762	2	0.381	6.316	0.06[a]
FEV_1/ FVC%	Between groups	10.669	2	5.334	0.182	0.834

[a] Sig. at $p < 0.05$

4 Discussion

This study was conducted to compare pulmonary function performance among athletes of different sports. Spirometry test was used to evaluate and analyze pulmonary function performance of the athletes. Participants were 19 swimmers, 21 cyclists, and 35 hockey players. Based on the data analysis, the following findings were made: Cyclists had significantly higher FVC and FEV_1 compared to swimmers in both genders. Female swimmers had significantly lower FVC compared to hockey players. Male cyclists had significantly higher FEV_1 compared to hockey players. There was no significant difference between male cyclists and male hockey players in FVC. There was no significant difference between female swimmers and female hockey players in FEV_1. There was no significant difference in FEV_1/FVC% among swimmers cyclists and hockey players in both genders.

There were no previous literatures that study the difference of pulmonary function between the selected sports as conducted in this study. The present study demonstrated that there were significant differences among cyclists, swimmers, and hockey players in FVC and FEV_1. However, there was no significant difference in FEV_1/FVC% among the athletes.

There were significantly higher values of FVC and FEV_1 in both male and female cyclists compared to swimmers. However, there was no significant difference observed between male cyclists with male hockey players in FVC. In FEV_1, there was no significant difference between female swimmers and female hockey players. There was no significant difference in FEV_1/FVC%, but swimmers showed slightly higher value than cyclists and hockey players.

Athletes who were trained for many years have higher values of pulmonary function compare to athletes who were new in training exposure. Besides, the intensity of training imposed could also influence the lung capacity. Increase in intensity of exercise contributes to the increase in lung function. As supported from [13], it had been proposed that intensity of physical activity was associated with better lung function. Most previous studies agreed that among other sports, swimmers have shown better in pulmonary function [14, 15]. However, the present study demonstrated that cyclists had the highest value and swimmers had lowest value in both FVC and $FEV_{1.}$

Hockey players have higher value of FVC and FEV_1 compared to swimmers but lower value in FEV_1/FVC%. This is supported by previous study by Ghosh et al. [16] who had found that hockey players have higher value in FVC but lower in FEV_1/FVC% compared to swimmers. It can be concluded that even hockey players had higher FVC, although swimmers could perform a more forceful expiration in the first second of exhalation. This may be contributed by swimmers having generally train in the water where naturally more pressure is placed on the lungs.

Study by Bandyopadhyay [17] showed similar findings whereby pulmonary function performance was better in the athletes when compared to the young healthy Malaysian in Kelantan. Result in the current study showed FVC and FEV_1 in male cyclists were almost equivalent with male healthy Malaysian population. However, female cyclists showed higher FVC and FEV_1 compared to female Malaysian population. In general, swimmers demonstrated to have higher FEV_1/FVC% compared to the Malaysian population. Young athletes in this study showed better pulmonary function performance even though their age range was younger than subjects in previous Malaysian population study.

5 Conclusion

The present study shows that both male and female cyclists had better lung function performance compared to swimmers and hockey players. However, swimmers had higher FEV_1/FVC% even though they were younger compared to the other two sports. It is recommended that coaches and the management of the State Sports Council to enforce regular assessment on the current athletes' fitness performance to evaluate the effectiveness of the training programs.

References

1. McKay, E. E., Braund, R. W., Chalmers, R. J., & Williams, C. S. (1983). Physical work capacity and lung function in competitive swimmers. *British Journal of Sports Medicine, 17*(1), 27–33.
2. Ostrowski, S., & Barud, W. (2006). Factors influencing lung function: are the predicted value for spirometry reliable enough? *Journal of Physiology and Phamarcoogy, 57*(4), 263–271.
3. Adams, G. M., & Beam, W. C. (2007). *Exercise physiology laboratory manual* (5th ed.). New York: McGraw-Hill.
4. Djojodibroto, R. D., Pratibha, G., Kamaluddin, B., Manjit, S. S., Sumitabha, G., Kumar, A. D., et al. (2009). Predicted equations for ventilatory function among kuching (Sarawak, Malaysia) population. *Medical Journal of Malaysia, 64*(4), 275–279.
5. Gnanou, J., Caszo, B., WanHaniza, W. M., Hapizah, N., Khalid, Y., & Tengku, I. (2011). Prediction equations for lung function in healthy, life time never-smoking Malaysian population. *Southeast Asian Journal, 42*(4), 965–976.
6. Ismail, Y., Azmi, N. N., & Zulkarnain, Y. (1993). Lung function in Malay children. *Medical Journal of Malaysia, 48*(2), 171–174.
7. Razlan, M., Naing, L., Zulkifli, A., & Yassin, K. (2000). Respiratory health of rice miller in Kelantan Malaysia. *Southeast Asian Journal, 31*(3), 575–578.
8. Singh, R., Singh, H. J., & Sirisinghe, R. G. (1992). Forced vital capacities in Malaysia females. *Japanese Journal of Fisiology, 42*, 407–414.
9. Singh, R., Singh, H. J., & Sirisinghe, R. G. (1993). Spirometric studies in Malaysians—between 13 and 69 years of age. *Medical Journal of Malaysia, 48*(2), 175–184.
10. Singh, R., Singh, H. J., & Sirisinghe, R. G. (1995). Cardiopulmonary capacities of Malaysia female. *Singapore Medicine Journal, 36*, 169–172.
11. Shashikala, L., & Sarath, R. (2011). Effect of exercise on pulmonary function test. *Indian Journal of Fundamental and Applied Life Science, 1*(3), 230–231.
12. Hulke, S. M., & Phatak, M. S. (2011). Effect of endurance training on lung function: a longitudinal study. *International Journal of Biological and Medical Research,* 2(1), 443–446.
13. Holmen, T. L., Connor, E. B., Clausen, J., Holmen, J., & Bjermer, L. (2002). Physical exercise, sports, and lung function in smoking versus nonsmoking adolescents. *Europen Respiratory Journal, 19*, 8–15.
14. Doherty, M., & Dimitriou, L. (1997). Comparison of lung volume in Greek swimmers, land based athletes, and sedentary controls using allometric scaling. *British Journal of Sports Medicine, 31*(4), 337–341.
15. Kippelen, P., Caillaud, C., Robert, E., Connes, P., Godard, P., & Prefaut, C. (2005). Effect of endurance training on lung function: A one year study. *British Journal of Sports Medicine, 39*, 617–621.
16. Ghosh, A. K., Ahuja, A., & Khanna, G. L. (1985). Pulmonary capacities of different groups of sportsmen in India. *British Journal of Sports Medicine, 19*(4), 232–234.
17. Bandyopadhyay, A. (2011). Pulmonary function studies in young healthy Malaysian of Kelantan, Malaysia. *Indian Journal of Medical Respiratory, 134*, 653–657.

The Effects of Aging on Body Composition, Leg Power and Balance Performance Among Malaysian Women

Sarina Md. Yusof, Zaiton Zakaria, Aminuddin Abd Hamid Karim, Suhana Aiman and Zulkifli Abdul Kadir

Abstract The study was conducted to ascertain the age at which muscle power and balance begins to decline, and to establish a baseline pattern of the changes of body composition. A total of 96 Malaysian women age ranged 20–65 years old participated in this study. The subjects were assigned into four age groups, representing a decade each. Fat mass and fat-free mass was measured using Bioelectrical Impedance. Vertical jump test was used to measure leg power. Balance was also performed as functional ability test. One-way ANOVA test detected differences between age groups. Fat-free mass, leg power and balance decline with age. The analysis also revealed that leg power exhibit a significant drop from the fifth decade group (50 years and above). Significant difference is found in leg power in the fifth decade. Leg power decreases from the 5th decade onwards. There is also a significant drop in the ability to balance from the 5th decade. The study is unable to determine at which particular age group body composition significantly increases as it exhibit a steady trend.

Keywords Aging · Body composition · Power · Functional ability

---

S. Md. Yusof (✉) · S. Aiman · Z. Abdul Kadir
Faculty of Sports Science and Recreation, Universiti Teknologi MARA, Shah Alam, Malaysia
e-mail: sarin864@salam.uitm.edu.my

Z. Zakaria
Physiology Department, Universiti Kebangsaan Malaysia, Bangi, Malaysia

A. Abd Hamid Karim
Physiology Department, National Defence University of Malaysia, Kuala Lumpur, Malaysia

R. Adnan et al. (eds.), *Proceedings of the International Colloquium on Sports Science, Exercise, Engineering and Technology 2014 (ICoSSEET 2014)*,
DOI: 10.1007/978-981-287-107-7_7,

1 Introduction

As an individual experience advancing age, difficulties in performing basic daily activities will be manifested. The common fitness declines that occur with aging include changes in body composition with increased body fat and decreased muscle mass [1], diminished cardiorespiratory capacity [2] and muscle atrophy [3]. Muscle power is defined as the product of muscle force and the velocity of muscle shortening [4].

Aging results in decrement in muscle power faster than the decrement in endurance in both men and women. A much faster and earlier decline in power and strength were found in both men and women. Aging implicated on the muscle power and strength, at which power was shown to decline at a faster rate than strength [5–8].

Reduced muscle power has been identified as one of the important factors, which may jeopardize maintenance of mobility and independency for older individuals [9], mortality [10] and preventing falls. Lower limb muscle power is strongly related to mobility performance [11]. Lower-extremity weakness is a well-recognized contributing factor to the incidence of falls. Ability to balance is another contributing factor. Daily activities involve several physical and physiological demands, it is expected that the older the adults will require some amount of these physiological aspects of muscle performance to sustain their abilities to complete functional tasks. These declines can lead to poor worker productivity since older workers would not be able to work at their optimal ability. The aims of the study were to ascertain the age at which muscle power and balance begins to decline, and to establish a baseline pattern of the changes of body composition.

2 Methodology

2.1 Participants

A total of 96 Malaysian female volunteers' age ranged 20–66 years old engaged in this cross-sectional design study. Participants comprised of staff, students and guests who visited the Faculty of Sport Science and Recreation Gym at Universiti Teknologi MARA Shah Alam during women's week promotion. Inclusion criteria were women aged 20 and above, healthy, reported no known neuromuscular diseases or musculoskeletal problems specific to the ankle, knee or hip joints or cardiovascular pathology affecting their functional ability, no participation in regular activity, nonsmoker and have been weight stable for the past 3 months prior to the study. Exclusion criteria included pregnant women, persons with chronic wasting disease, diagnosed with heart and respiratory problems, resting systolic and diastolic pressure greater than 160 mm Hg and 100 mm Hg, respectively. Each participant completed and passed a physical readiness

Table 1 Participants characteristics according to age decade

Age group	n	Age (years)	Height (cm)	Weight (kg)	BMI (kg/m^2)
20–29	26	24.85 ± 2.20	1.57 ± 0.06[a]	57.26 ± 12.71[b]	23.06 ± 4.80[d]
30–39	26	34.85 ± 2.85	1.56 ± 0.05	60.41 ± 11.03[c]	24.69 ± 4.05[e]
40–49	24	44.63 ± 2.52	1.54 ± 0.06	62.13 ± 9.53	26.09 ± 4.08[f]
>50	20	54.55 ± 3.33	1.53 ± 0.05[a]	69.25 ± 11.65[b,c]	29.60 ± 4.72[d,e,f]

$p < 0.05$
Mean ± SD, *N* Number of participants
[a] Significantly different between 20–29 and >50
[b] Significantly different between 20–29 and >50
[c] Significantly different between 30–39 and >50
[d] Significantly different between 20–29 and >50
[e] Significantly different between 30–39 and >50
[f] Significantly different between 40–49 and >50

questionnaire (PAR-Q) as a screening tool and signed an informed consent document after a complete explanation of the objectives and the procedures of the study was given. Participants' characteristics are shown in Table 1. This study was approved by the UKM Research and Ethical Committee for human subjects in accordance with the principles of the 1964 Declaration of Helsinki.

2.2 Anthropometry and Body Composition

Anthropometric measurements were made with the participants in minimal clothing without socks and shoes in stretched stature. Height was measured with a stadiometer to the nearest 0.1 cm. Body mass was measured with a digital scale (Tanita) to the nearest 0.1 kg. Body mass index (BMI) was expressed as weight in kg/height in m^2. Bioelectric impedance analysis (Maltron Bioscan 916 Analyser) was used to measure fat-free mass and fat mass and was performed early in the morning after an overnight fast [12]. A tetra polar device (Human IM; Dieto-system, Milan, Italy) with an applied current (50 kHz, 0.7 mA) with whole-body resistance was used. Electrode was each placed on the right hand and wrist, and ankle and feet of the subjects lying in supine position with all limbs abducted away from the body.

2.3 Leg Power

Two trials of maximal countermovement vertical jump (CMVJ) with a 1-min rest period in between trials were given. To perform CMVJ, participant stood upright with both arms fully extended overhead and then performed a rapid descent before executing the jump [13] and touch the highest possible vane. To achieve the greatest possible height from this test, participants were required to extend the hip and knee joints and plantar flexed the ankle explosively. The vertical jump was measured by using a Vertec. Vertical jump score is measured by the difference of the standing height and the jumping height.

2.4 Balance Performance Test

Participants were to balance by standing on the dominant leg with eyes closed. Timing started when the participant raised the free leg off the ground and stopped when balance was lost or the free leg touched the floor again.

2.5 Statistical Analysis

Data is presented as means ± standard deviation (SD). To compare between age groups, participants were divided into four age decades. One-way ANOVA was utilized to compare the physical characteristics (height and body weight), body composition (fat-free mass and BMI), power and balance between groups. Tukey post hoc test was performed to determine the difference between groups. In order to determine the strength and directionality of the relationship among variables and age groups, Pearson's product moment correlation was further utilized. Significant level was set at 0.05 ($p < 0.05$). The SPSS software statistical program version 16 was used to analyze the data.

3 Results

3.1 Participants' Physical Characteristics

Participants' physical characteristics according to age group are presented in Table 1. The ANOVA showed significant differences in height and weight among groups. The fifth decade group was significantly shorter than those in the 20–29 years group ($p < 0.05$) and heavier than those in the 20–29 years and 30–39 years group ($p < 0.05$). The 40–49 years group is not significantly different in height and weight than all other groups. Significant difference was noted in BMI among the groups ($p < 0.05$). Fifth decade group had significantly higher BMI compared to all the other 3 groups, that is the 20–29, 30–39 years and 40–49 years group ($p < 0.05$). Besides the fifth decade groups, all the other groups are not significantly different from one another.

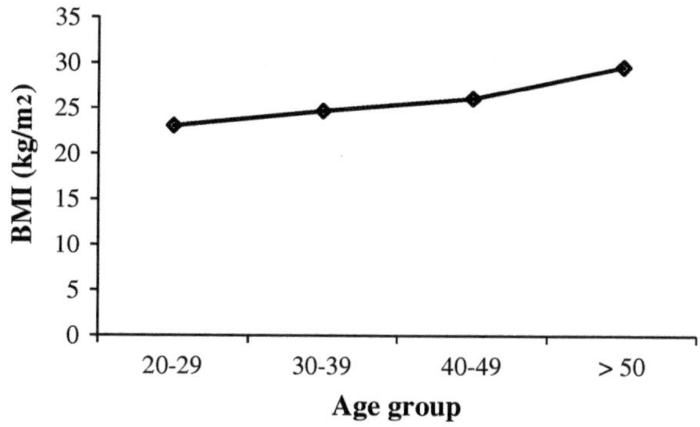

Fig. 1 Mean plot of BMI by age group

The mean plot of BMI by age group (Fig. 1) shows an increase in BMI from the fifth decade onwards. A significant correlation was found between FFM and leg power in 40–49 years and >50 years age groups ($r = 0.51$, and $r = 0.49$, $p < 0.05$, respectively).

3.2 Body Composition, Power and Balance

The ANOVA showed significant differences in fat-free mass percentage among the groups. The fifth decade group is significantly higher compared to the 20–29 years and 30–39 years age group ($p < 0.05$). The 40–49 years group is significantly higher with the 20–29 years group ($p < 0.05$) (Table 2).

The ANOVA showed significant differences in fat-free mass percentage among the groups. The fifth decade group is significantly lower with the 20–29 years and 30–39 years age group ($p < 0.05$). The 40–49 years group is significantly different with the 20–29 years group ($p < 0.05$).

3.3 Power

There is a significant difference among the groups ($p < 0.05$). The fifth decade group is significantly lower from all the other groups ($p < 0.05$). The 40–49 years group is significantly different with the 20–29 years ($p < 0.05$). The mean plot of CMJ by age group shows remarkable decline from the fifth decade onwards (Fig. 2).

3.4 Balance

Balance is significantly different between the groups ($p < 0.05$). The Tukey post hoc analysis indicated significant differences between the fifth decade and all other

Table 2 Mean ± SD of fat-free mass, fat mass, power and balance of different age groups

Age group	FFM (%)	FM (%)	Vertical jump (cm)	Balance (s)
20–29	73.97 ± 9.28[a, c]	26.03 ± 9.27	26.3 ± 5.73[d,g]	19.19 ± 1.88[h]
30–39	69.63 ± 8.26[b]	30.37 ± 8.26	24.15 ± 3.56[e]	18.06 ± 3.55[i]
40–49	64.72 ± 7.70[c]	35.28 ± 7.67	22.21 ± 4.36[f,g]	18.51 ± 7.87[j]
>50	58.46 ± 6.45[a, b]	41.54 ± 6.45	18.80 ± 4.04[d,e,f]	5.80 ± 3.50[h,i,j]

$p < 0.05$
[a] Significantly different between 20–29 and >50
[b] Significantly different between 30–39 and >50
[c] Significantly different between 20–29 and 40–49
[d] Significantly different between 20–29 and >50
[e] Significantly different between 30–39 and >50
[f] Significantly different between 40–49 and >50
[g] Significantly different between 20–29 and 40–49
[h] Significantly different between 20–29 and >50
[i] Significantly different between 30–39 and >50
[j] Significantly different between 40–49 and >50

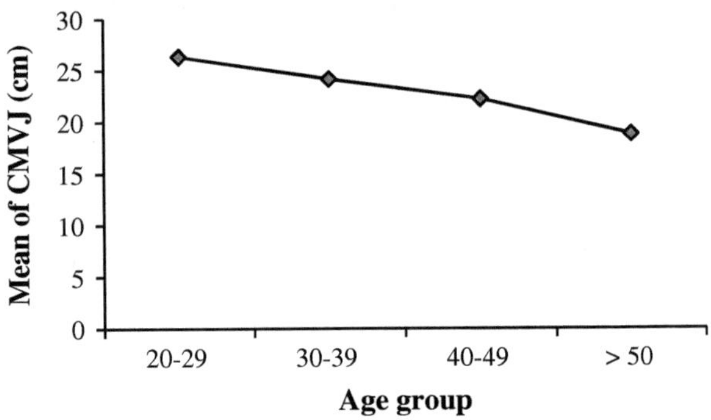

Fig. 2 Mean plot of CMVJ by age group

groups ($p < 0.05$). Besides the fifth decade group, all the other groups are not significantly different from one another. The mean plot of balance by age group again shows remarkable decline in balance from the fifth decade onwards (Fig. 3).

4 Discussion

As a result of advancing age, the findings may give us an idea that physical performance deterioration was not due to any disease or disability. This study is a self selected group. From this study, FFM is found to exhibit a steady decline following age. A remarkable significant loss in power and balance started to manifest in the fifth decade. Significant correlation was observed between percentage of FFM and leg power in 40–49 years and >50 years age group. However,

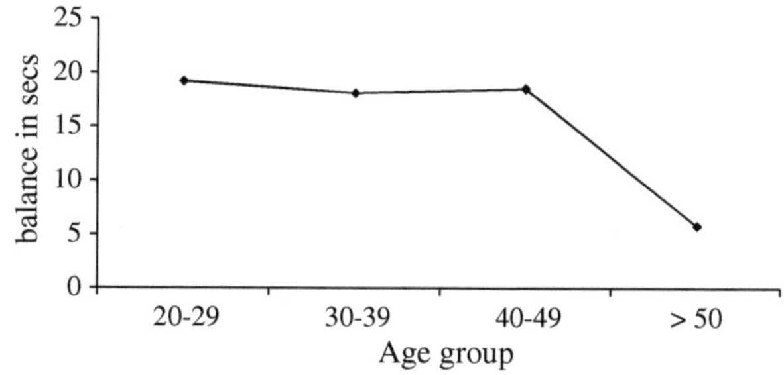

Fig. 3 Mean plot of balance by age group

it is not clear, whether the changes of body composition and the loss were attributed to sedentary lifestyle or due to age-related influences. However, [12] reported fat-free mass was positively associated with physical activity level. Reference [14] postulated less involvements in physical activity is a contributing factor for decrement in muscle protein synthesis and mass in the aged. Limitation of this study was that the lifestyle aspect that is the physical activity level was not measured.

When muscle mass diminished due to aging, it is replaced by fat, while the remaining muscle would be infiltrated with fat [15]. Reference [15] has shown that high body fat in women is associated with poor lower-extremity performance, while muscle area is accounted for men. High fat mass was also associated with low muscle quality in the lower limbs [16]. Often increment in body fat is linked to metabolic syndrome and loss of lean mass contributes to disability and frailty [17]. Reduction in FFM is the results of losses in skeletal muscle mass [18]. Reference [19] suggested that appendicular skeletal muscle mass change in older subjects was greater than the FFM change. Muscle mass with age decline greater than non-muscle lean (organ) mass [19]. Loss of muscle mass due to advancing also known as sarcopenia is prevalent in the elderly and is associated with reduced mobility and quality of life and increased morbidity and mortality [20–23].

Our results confirm those of previous reports suggesting that decline in power is age-related. Reference [24] suggested that reduction in fast twitch muscle fiber area with advancing age will influenced force production. Fast twitch fibers are characterized by larger size, higher force capacity and lower endurance capacity. Atrophy in fast twitch fibers among elders are also associated with a specific fiber type decline in satellite cells content which accompanies the loss of skeletal muscle mass with aging [25]. There was a tendency of delay in muscle fiber conduction velocity due to aging [26]. Reference [27] suggested that improvement in power in the lower limb may increase strength which also required in performing daily activities.

The findings of this study as with others [28, 29] portray reduction in balance as a result of aging. References [28, 30] indicated that eyes closure will further impaired balance. Decline in balance ability would increase the risk of falling [31]. Reference [31] suggested that lower stability was contributed by higher fat mass, experience by heavier people which leads to higher fall risk.

Measures to maintain muscle mass, power and strength should be taken as it has functional implications for older adults. Reference [32] suggested that a varied range of intensities should be applied to increase lower-extremity power and strength and maintained in order to reduce functional limitations and disability in later life. Exercise such as strength training programs has been proven to increase muscle mass [12], power and balance in older men and women [3, 33, 34] and mobility [35]. Therefore, health promotion and awareness of the effect of regular exercise in advancing age should be conducted widely to reduce disability in old age.

5 Conclusion

Age was found to have a significant effect on body composition changes, reduction in power and ability to balance. Body composition, power and balance demonstrated significant changes at the fifth decade. Earlier involvement in regular exercise may help to delay the effect of aging.

Acknowledgments The authors are grateful to Tan Kok Siang for the fruitful discussion, and to all participants in the project. This study is supported by UKM Medical Faculty post graduate grant.

References

1. Gallagher, D., Visser, M., De Meersman, R. E., Sepulveda, D., Baumgartner, R. N., Pierson, R. N., et al. (1997). Appendicular skeletal muscle mass: Effects of age, gender, and ethnicity. *Journal of Applied Physiology, 83*(1), 229–239.
2. Oxenham, H., & Sharpe, N. (2003). Cardiovascular aging and heart failure. *The European Journal of Heart Failure, 5*, 427–434.
3. Bross, R., Storer, T., Bhasin, S. (1999) Review. *Aging and Muscle Loss, 10*(5), 194–198.
4. Kraemer, W. J., & Newton, R. U. (2000) Training for muscular power. *Physical Medicine and Rehabilitation Clinics of North America, 11*(2), 341–368.
5. Metter, E. J., Conwit, R., Tobin, J., & Fozard, J. L. (1997). Age-associated loss of power and strength in the upper extremities in women and men. *Journals of Gerontology. Series A, Biological Sciences and Medical Sciences, 52*, B267–B276.
6. Martin, J. C., Farrar, R. P., Wagner, B. M., & Spirduso, W. W. (2000). Maximal power across the lifespan. *Journals of Gerontology. Series A, Biological Sciences and Medical Sciences, 55*, M311–M316.
7. Bean, J. F., Kiely, D. K., Herman, S., Leveille, S. G., Mizer, K., Frontera, W. R., et al. (2002). The relationship between leg power and physical performance in mobility-limited older people. *Journal of the American Geriatrics Society, 50*, 461–467.
8. Bean, J. F., Leveille, S. G., Kiely, D. K., Bandinelli, S., Guralnik, J. M., & Ferrucci, L. (2003). A comparison of leg power and leg strength within the InCHIANTI study: Which influences mobility more? *Journals of Gerontology. Series A, Biological Sciences and Medical Sciences, 58A*, 728–733.

9. Foldvari, M., Clark, M., Laviolette, L. C., Bernstein, M. A., Kaliton, D., Castaneda, C., et al. (2000). Association of muscle power with functional status in community-dwelling elderly women. *Journals of Gerontology. Series A, Biological Sciences and Medical Sciences, 55*(4), M192–M199.
10. Metter, E. J., Talbot, L. A., Schrager, M., & Conwit, R. A. (2004). Arm-cranking muscle power and arm isometric muscle strength are independent predictors of all-cause mortality in men. *Journal of Applied Physiology, 96*, 814–821.
11. Herman, S., Kiely, D. K., Leveille, S., O'Neill, E., Cyberey, S., & Bean, J. F. (2005). Upper and lower limb muscle power relationships in mobility-limited older adults. *Journal of Gerontology Series A: Biological Sciencse and Medical Sciences, 60*(4), 476–480.
12. Westerterp, K. R., & Meijer, E. P. (2001). Physical activity and parameters of aging. *The Journals of Gerontology, 56A*, 7–12.
13. Harman, E., Rosenstein, M. T., Frykman, P. N., & Rosenstein, R. M. (1990). The effects of arms and countermovement on vertical jumping. *Med Sci Sports Exerc, 22*, 825–833.
14. Tessari, P. (2000). Changes in protein, carbohydrate, and fat metabolism with aging: Possible role of insulin. *Nutrition Reviews, 58*(1), 11–19.
15. Visser, M., Kritchevsky, S. B., Goodpaster, B. H., Newman, A. B., Nevitt, M., Stamm, E., et al. (2002). Leg muscle mass and composition in relation to lower extremity performance in men and women aged 70 to 79: The health, aging and body composition study. *Journal of the American Geriatrics Society, 50*, 897–904.
16. Newman, A. B., Haggerty, C. L., & Goodpaster, B. (2003). Strength and muscle quality in a well-functioning cohort for older adults: The health, aging and body composition study. *Journal of American Geriatric Society, 51*, 323–330.
17. Harris, T. B. (2002). Invited commentary: Body composition in studies of aging: New Opportunities to better understand Health Risks Associated with Weight. *American Journal of Epidemiology, 56*(2), 122–124.
18. Evans, W. J. (1998). Exercise and nutritional needs of elderly people: Effects on muscle and bone. *Gerontology, 15*, 15–24.
19. Kyle, U. G., Genton, L., Hans, D., Karsegard, L., & Slosman, D. O., Pichard, C. (2001). Age-related differences in fat-free mass, skeletal muscle, body cell mass and fat mass between 18 and 94 years. *European journal of clinical nutrition, 55*(8), 663–672.
20. Kehayias, J. J., Fiatarone, M. A., Zhuang, H., & Roubenoff, R. (1977). Total body potassium and body fat: Relevance to aging. *American Journal of Clinical Nutrition, 66*, 904–910.
21. Baumgartner, R. N., Koehler, K. M., Gallagher, D., Romero, L., Heymsfeld, S. B., & Ross, R. R. (1998). Epidemiology of sarcopenia among the elderly in New Mexico. *American journal of epidemiology, 147*, 755–763.
22. Roubenoff, R. (2000). Sarcopenia and its implication for the elderly. *European Journal of Clinical Nutrition, S2*, 128–131.
23. Zhu, S., Heo, M., Plankey, M., Faith, M. S., & Allison, D. B. (2003). Associations of body mass index and anthropometric indicators of fat mass and fat free mass with all-cause mortality among women in the first and second National Health and Nutrition Examination Surveys follow-up studies. *Annals of Epidemiology, 13*, 286–293.
24. Hortobagyi, T., Zheng, D., Weidner, M., Lambert, N. J., Westbrook, S., & Houmard, J. A. (1995). The influence of aging on muscle strength and muscle fiber characteristics with special reference to eccentric strength. *Journals of Gerontology. Series A, Biological Sciences and Medical Sciences, 50*(6), B399–B406.
25. Verdijk, L. B., Koopman, R., Schaart, G., Meijer, K., Savelberg, H. H, van Loon, L. J. (2007). Satellite cell content is specifically reduced in type II skeletal muscle fibers in the elderly. *American Journal of Physiology-Endocrinology and Metabolism, 292*, E151–E157.
26. Mase, K., Kamimura, H., Imura, S., & Kitagawa, K. (2006). Effect of age and gender on muscle function—analysis by muscle fiber conduction velocity. *Journal of Physical Therapy Science, 18*(1), 81–87.

27. Yamauchi, J., & Ichii, N. (2007). Relations between force-velocity characteristics of the knee-hip extension movement and vertical jump performance. *Journal of Strength Conditioning Research, 21*(3), 703–709.
28. Massimo, C. (2001). Effect of aging on human postural control and the interaction with attention (unpublished). PhD thesis.
29. Lee, H. J., & Chou, L. S. (2007). Balance control during stair negotiation in older adults. *Journal of Biomechanics, 40*(11), 2530–2536.
30. Katayama, Y., Senda, M., Hamada, M., Kataoka, M., Shintani, M., & Inoue, H. (2004). Lower extremity muscle power did not appear to be the dominant factor in maintaining balance in these young subjects. *Acta Medica Okayama, 58*(4), 189–195.
31. Winters, K. M., & Snow, C. M. (2000). The body composition predicts bone mineral density and balance in premenopausal. *Women Journal of Women's Health and Gender-Based Medicine*, *9*(8), 865–872.
32. Puthoff, M. L., & Newman, D. H. (2007). Relationships among impairments in lower-extremity strength and power, functional limitations, and disability in older adults. *Physical Theraphy, 87*(10), 1334–1347.
33. Faber, M. J., Bosscher, R. J., Chin, A. P. M. J., & van Wieringen, P. C. (2006). Effects of exercise programs on falls and mobility in frail and pre-fail older adults: a multicentre randomized controlled trial. *Arch Phys Medi Rehabil, 87*, 885–896.
34. Hourigan, S. R., Nitz, J. C., Brauer, S. G., O'Neill, S., Wong, J., & Richardson, C. A. (2008). Positive effects of exercise on falls and fracture risk in osteopenic women. *Osteoporosis International, 19*(7), 1077–1086.
35. Fatouros, I. G., Kambas, A., Katrabasas, I., Nokolaidis, K., Chatzinikolaou, A., Leonstini, D., et al. (2005). Strength training and detraining effects on muscular strength, anaerobic power, and mobility of inactive older men are intensity dependent. *British Journal of Sports Medicine, 39*, 776–780.

A Study on Factors Associated with Physical Fitness Status Among Emergency Response Team of Oil and Gas Company in Peninsular Malaysia

Zulkifli M. Yunus, Ahmad Faizal Zuli, Norasrudin Sulaiman, Rahmat Adnan and Shariman Ismadi Ismail

Abstract A cross-sectional study was conducted to determine factors associated with physical fitness status among emergency response team (ERT) personnel of an Oil and Gas Company (Company P) in Peninsular Malaysia. The study involved 381 respondents who were selected through universal sampling. The respondents comprised of workers from two companies located in Kertih, Terengganu and Gebeng, Pahang. Data collection was through self-administered questionnaire, anthropometric [height, weight, and waist circumference (WC)] measurement, body fat percentage measurement, hemoglobin and cholesterol blood sampling, and National Physical Fitness Test assessment. Sociodemographic data, smoking habit, physical parameters, hemoglobin level, fasting glucose level, and fasting cholesterol level were studied in relation to their association with participant's physical fitness. This study found that 101 (31.8 %) of ERT members were fit and 217 (68.2 %) were unfit. Age, body mass index (BMI), WC, body fat percentage, hemoglobin level, and HDL cholesterol showed significant relationship with physical fitness status. Logistic regression analysis, however, revealed only age ($\beta = 0.135$, $p = 0.001$, $p < 0.05$) was found to have a clear and significant association with physical fitness status.

Keywords Health-related fitness · Physical fitness status · Emergency response team

Z. M. Yunus (✉)
Group HSE Division PETRONAS, Kuala Lumpur, Malaysia
e-mail: zulkifli_yunus@petronas.com.my; zulcjk@gmail.com

A. F. Zuli
PETRONAS Gas Berhad, Kuala Lumpur, Malaysia

N. Sulaiman · R. Adnan · S. I. Ismail
FSR Performance Analysis Laboratory, Universiti Teknologi MARA, Shah Alam, Malaysia

R. Adnan et al. (eds.), *Proceedings of the International Colloquium on Sports Science, Exercise, Engineering and Technology 2014 (ICoSSEET 2014)*,
DOI: 10.1007/978-981-287-107-7_8,

1 Introduction

Physical fitness can be described as an integrated elements of all human body functions (skeletomuscular, cardiorespiratory, hematocirculatory, psychoneurological, endocrine-metabolic, etc.) involved in the performance of daily physical activity and/or physical exercise [1].

Specifically illustrating the importance of physical fitness in relation to the capacity of responding to emergency, Clarke [2] in his physical education text book gave more accurate definition. His definition considers physical fitness as the ability to carry out daily tasks with vigor and alertness, without undue fatigue, and with ample energy for leisure-time pursuits, and to meet unforeseen emergencies. However, in terms of practicality of assessment, nowadays, the most common and objectively measurable physical fitness components are being grouped into two types: health related and skill related. Health-related physical fitness includes cardiorespiratory fitness, muscular strength, muscular endurance, flexibility, and body composition. Skill-related physical fitness includes agility, balance, coordination, power, speed, and reaction time.

Cardiorespiratory fitness, also called cardiovascular fitness or maximal aerobic power, is the overall capacity of the cardiovascular and respiratory systems and the ability to carry out prolonged strenuous exercise. The maximal oxygen consumption (VO_2max) attained during a graded maximal exercise to voluntary exhaustion has long since been considered by the World Health Organization (WHO) as the single best indicator of cardiorespiratory fitness, and this reference standard is still being used as an international guide [3]. The VO_2max can be estimated using maximal or submaximal tests, by direct or indirect methods. The most commonly used tests are walking/running tests followed by cycling and step tests. In epidemiological studies involving young people, the most common test for assessing cardiorespiratory fitness has been the 20 m shuttle run test, or adaptations/modifications of this test. The VO_2max can then be estimated from the score by a specific formula conversion.

Muscular fitness is the capacity to carry out work against a resistance. Since the maximum force that can be generated depends on several factors (for example, the size and number of muscles involved, the proportion of muscle fibers called into action, the coordination of the muscle groups, etc.), the handgrip test is one of the most used tests for assessing muscular fitness in epidemiological studies. In adults, handgrip strength has been reported to be a strong predictor of morbidity and life expectancy [1].

The study was conducted to support and compliment company's fitness to work program. Evaluating employees' fitness to perform their work in the respective working environment is the main aim of fitness to work program. Availability of scientific evidence which can lead to determination of factors associated or even influence physical fitness will definitely help researcher to design intervention plan/fitness enhancement program.

Fitness to work program refers to the planning, implementation, and monitoring of a systematic approach in the evaluation of employees' personal health in relation to their work and work environment. Its main objective is to ensure employees have adequate access to health assessment facilities and are protected from unacceptable health risks arising from their work and work environment.

Fitness to Work assessment are universal practice mostly by international companies especially in oil and gas, petrochemical, and aviation industries. Stringent health, safety and environment requirement being enforced to ensure substandard workers' fitness status will not impose safety risk to workers' own health and the company as a whole.

Company (P) already sets a certain standard whereby certain minimum fitness requirement needs to be followed prior to certification of fitness to work by relevant authorities. Health assessments for fitness to work are objective assessments of the health of employees in relation to their jobs and job environment. Consideration should be given to the potential impacts at a person's health on his job as well as the impact of the job on his health. Additionally, a person's health should also be viewed with regard to its potential impacts on the health and safety of others in the workplace as well its impacts on a company's business and image.

2 Method

2.1 Sampling

All 318 subjects completed health parameters checkup, blood sampling and Ujian Kecergasan Jasmani Kebangsaan (UKJK) fitness assessment are involve in this study. The 318 subjects comply with minimum sample size with the additional of 20 % respondents to cater for drop out.

2.2 Instrument

2.2.1 Questionnaires

The questionnaire consists of five parts:

1. Sociodemographic particular of employee which include age, sex, ethnicity, marital status, and department.
2. Previous employment and occupational health history.
3. Family history of medical illness.
4. Medical history declaration.

5. Smoking status.

The questionnaires were distributed to the respondents' at the beginning of program during registration. The workers were given a briefing on the questionnaire and were asked to fill in a consent form for health parameter checkup and participate in the study. The questionnaires were answered under the observation of the researcher. At the end of the session, each of the questionnaires was checked by the researcher to ensure that all questions had been answered. Questionnaires answering session lasted approximately 15 min.

2.2.2 Anthropometric Data Measurement

The respondents were instructed to remove their shoes, articles in their pockets, and exterior clothing other than a light shirt and pants. Their body weights and body fat percentage were taken using digital weighing and fat analysis machine, and the readings were printed and recorded accordingly. Readings obtained were rounded to the nearest kilogram and percentage. The digital weighing and fat analysis machine were within its validity of calibration as per manufacturer specification. Height measurements were taken using specially designed tape attached to the wall, and the readings obtained were rounded to the nearest centimeter. The recorded weight and height of each respondent were used to compute their body mass index (BMI) in kg/m^2.

WC measurements were done following WHO recommended methods observing the importance of standard anatomical locations.

(a) Subject stands with feet 25–30 cm apart, weight evenly distributed.
(b) Waist measurement is taken midway between the inferior margin of the last rib and the crest of the ilium in a horizontal plane.
(c) The measurer sits by the side of the subject and fits the tape snugly but not compressing soft tissues.
(d) Circumference is measured to the nearest 0.1 cm.

2.2.3 Blood Sampling

The respondents were instructed to fast for 8 h prior to blood sampling. Venous blood samplings were done by competent laboratory technician from private accredited laboratory. Sample was transported to laboratory and analyzed within the same day following sampling. Blood parameters included in the analysis process were full blood count, fasting glucose, and fasting lipid profile.

2.2.4 National Physical Fitness Test "Ujian Kecergasan Jasmani Kebangsaan"

UKJK physical fitness testing was designed for evaluation of holistic body fitness which gave consideration on cardiovascular illness risk, cardiovascular fitness, muscle strength, muscle endurance, and muscle flexibility. Each UKJK test component was adopted from established test supported by scientific studies/research. All the test norms are based on Malaysian population except for BMI which using WHO norm. The norms are based on age and gender. In terms of score calculation, total individual points in UKJK module assessment range from 5 to 25.

Specific for this research, physical fitness status referred to status of fitness, i.e., either fit or unfit. Status of fit is given to ERT personnel upon achievement of total points of ≥ 18, whereas range of points from 5 to 17 is taken as unfit. Requirement to achieve Bronze Award category for UKJK is a total score at least 15 points, which is a recommended minimal points for ordinary person should have for a healthy living. One step higher and a chosen demarcation line between fit or unfit is Silver Award with a requirement score at least 18 points. Such requirement was decided in view of ERT tasks, which are physically demanding.

2.3 Standardization of Data Collection Methods

For the purpose of standardization, data collections were done under controlled process and procedure:

(a) Questionnaires were distributed and administered by the researcher throughout the data collection period.
(b) Blood sampling was done by competent laboratory technician.
(c) Blood results were processed by accredited Malaysian laboratory.
(d) UKJK assessments were done by certified trainer by Ministry of Sport and Youth, Malaysia.

2.4 Ethical Considerations

The permission to carry out this study was obtained from the Medical and Ethics Research Committee of University Kebangsaan Malaysia Medical Centre. This research was conducted to support and compliment Company (P) fitness to work program. All respondents were given a briefing on the conduct of the program and research procedure. All respondents were required to give written permission using the dedicated form.

2.5 Data Analysis

Data analysis was performed using statistical package for social sciences (SPSS) version 17.0. Both descriptive statistical and analytical methods were used. Data which have been collected were coded and cleaned prior to analysis. The level of significance was set at 0.05.

3 Result

3.1 Baseline Characteristics

All ERT members are male with Operator-technician as job title. Three of ERT members are Indian, and the rest are Malays. Three hundred and eighteen (318) subjects (age [mean ± SD], 36.26 ± 6.99 years old; BMI, 26.45 ± 4.06 kg/m^2; systolic blood pressure, 128.94 ± 12.97 mm Hg; diastolic blood pressure, 80.66 ± 11.08 mm Hg; body fat percentage, 25.14 ± 6.54 %; hemoglobin level, 15.44 ± 0.98 g/dL; fasting blood sugar level, 5.91 ± 3.17 mmol/L; total cholesterol (TC) level, 6.35 ± 3.46 mmol/L; HDL-C level, 1.26 ± 0.27 mmol/L; UKJK score, 15.96 ± 3.10) are willing to participate in this study. The analysis is shown in Table 1.

3.2 Factors Associated with Physical Fitness Status

Based on physical fitness status which was decided by researcher based on adaptation from UKJK module, 101 (31.8 %) were found fit and 217 (68.2 %) were unfit (Table 2).

Independent t test was performed to determine the association. Result shows that physical fitness status positively associated with age and high density lipoprotein cholesterol (HDL-C). Body mass index (BMI), body fat percentage, WC, hemoglobin, and TC were inversely associated with physical fitness. The rest, i.e., systolic blood pressure, diastolic blood pressure, and fasting blood sugar, shows no significant association (Table 3).

Chi-square test was done to determine the association between smoking status and physical fitness status. Result shows no significant association ($p = 0.168$, $p > 0.05$). Refer Table 4.

Table 1 Baseline characteristic of respondent

Characteristic	*N*	Mean	Standard deviation
Age (years)	318	36.26	6.99
BMI (kg/m^2)	318	26.45	4.06
Systolic blood pressure (mm/Hg)	318	128.94	12.97
Diastolic blood pressure (mm/Hg)	318	80.66	11.08
Body fat percentage (%)	318	25.14	6.54
Waist circumference (cm)	318	88.78	11.23
Hemoglobin level (g/dL)	318	15.44	0.98
Fasting blood sugar level (mmol/L)	318	5.91	3.17
Total cholesterol level (mmol/L)	318	6.35	3.46
HDL-C level (mmol/L)	318	1.26	0.27
UKJK score	318	15.96	3.10

Table 2 Physical fitness status of ERT members

Physical fitness status	Frequency (*n*)	Percent (%)
Fit	101	31.8
Unfit	217	68.2
Total	318	100

Table 3 Factors associated with physical fitness status

Variable	Mean ± SD		*t* statistic (d*f*)	*p* value
	Fit	Unfit		
Age (years)	39.06 ± 6.70	34.96 ± 6.76	5.04 (316.00)	0.001
BMI (kg/m^2)	24.75 ± 3.14	27.63 ± 4.20	−5.40 (254.46)	<0.001
Systolic blood pressure (mm/Hg)	128.86 ± 13.03	128.98 ± 12.97	−0.07 (316.00)	0.941
Diastolic blood pressure (mm/Hg)	79.81 ± 9.11	81.06 ± 11.89	−0.94 (316.00)	0.351
Body fat percentage (%)	22.25 ± 5.71	26.58 ± 6.40	−1.03 (248.10)	<0.001
Waist circumference (cm)	84.61 ± 9.90	90.67 ± 11.30	−4.58 (316.00)	<0.001
Hemoglobin level (g/dL)	15.26 ± 1.00	15.53 ± 0.95	−2.31 (186.14)	0.021
Fasting blood sugar level (mmol/L)	5.62 ± 1.45	6.04 ± 3.70	−1.11 (316.00)	0.266
Total cholesterol level (mmol/L)	6.16 ± 1.10	6.40 ± 3.70	−0.73 (273.64)	0.464
HDL-C level (mmol/L)	1.32 ± 0.30	1.23 ± 0.25	2.54 (168.58)	0.012

Table 4 Association between smoking status and physical fitness status

Variable	Physical fitness status *N* (%)		X^2 value (d*f*)	*p* value
	Fit	Unfit		
Smoking Status			3.573 (2)	0.168
Non-smoker	39 (32.5)	81 (67.5)		
Ex-smoker	21 (42.0)	29 (58.0)		
Current smoker	41 (27.7)	107 (72.3)		

4 Discussion

This research is able to illustrate a positive correlation between physical fitness status and age. Physiological changes upon aging are well known from ancient medieval civilization, and most Western-orientated experiments and studies were done as early as 1930s. Specific to physical fitness, Cullumbine and Bibile [4] studied various aspects of functional fitness in 7,000 Ceylonese subjects and have observed the effects of age, sex, physique, and muscular development on performance. Subjects range from 10 years school pupils to groups of adults with variety of occupations. It was found that speed, strength, and stamina increase with age to reach a maximum in early adult years and in each the highest levels were reached by those of normal or tough body build. Heavy exercise was best performed by the slim subjects. Female subjects gave poorer assessments than the males.

Currently, almost all internationally accepted physical fitness assessment norms were set based on gender and age group. Performances rating for each test component were graded based on gender and age, i.e., passing point threshold for fitness level are lower for female and those with increasing age (UKJK Tester Manual 2011). This worldwide fitness norms pattern was formulated in such a way to compensate with physiological changes of aging. Ironically, those belong to senior age group and have a good fitness level such as for athlete, they easily can achieve high performance mark.

Specific to this research, in general, older age group ERT members seem to have better physical fitness status compared with the younger one. They might have good fitness level due to higher involvement in sports or any regular physical activity. Another factor that is difficult to be controlled by researcher is willingness of participants to push self to maximal best effort. If presumptively many of younger ERT members are not cooperative in doing their fitness test, this explains the correlation. Further study and analysis need to be done to clarify the possible explanation of the positive correlation. Further details such as physical exercise frequency and intensity of each ERT member need to be analyzed.

This research is able to illustrate an inverse correlation between physical fitness status and three major body parameters which relate to risk of developing lifestyle illness which are BMI, waist circumference (WC), and body fat percentage. BMI, WC, and body fat percentage are body parameters, which are known to be related to each other. BMI and WC perform similarly as indicators of body fatness and are more closely related to each other than with percentage body fat. These variables may be an inaccurate measure of percentage body fat for an individual, but they correspond fairly well overall with percentage body fat within sex-age groups and distinguish categories of percentage body fat. Therefore, it is not surprising that these body parameters correlate in the same pattern with physical fitness.

This research is able to illustrate a negative correlation between physical fitness status and hemoglobin level. In terms of general population, most study efforts were done among elderly population with main aim to evaluate correlation between anemia and fitness for daily living. Healthy elderly population that anemia (defined as hemoglobin level less than 13 g/dL for men) was associated with greater mean decline in physical performance both for cardiovascular fitness and muscle endurance. Similar pattern of correlation between hemoglobin level and skeletal muscle properties especially muscle strength was identified by Cesari [5] through a prospective population-based study of older people. Study concludes that low hemoglobin levels are associated with decreases in muscular strength

Scenario is different for group of people with high physical fitness level such as athlete or army recruits whereby higher hemoglobin level is not necessarily always associated with higher physical performance. "Sports Anemia" is a phenomenon of false or pseudo-anemia mostly observed in athletes. They had a normal ferritin level but due to regular aerobic exercise, physiologically their baseline plasma volume are expended thus diluting red blood cell caused a low hemoglobin concentration [6]. Low hematocrit resulting in lower viscosity has been shown to have a positive effect on peripheral perfusion, including to muscles, thus causing better oxygen delivery and consumption.

Novack [7] studied the relationship between hemoglobin concentration and physical fitness among new infantry recruits in defense forces. Participants were divided to groups with hemoglobin value lower than 12 g/dl, intermediate hemoglobin values (between 12 and 14 g/dl) and greater than 14 g/dl. Participants with hemoglobin values in the 12–14 g/dl range achieved significantly better results than those with hemoglobin levels lower than 12 g/dl and greater than 14 g/dl. The low physical fitness of the participants with hemoglobin values below 12 g/dl can be explained by existence of iron deficiency anemia, resulting in lower oxygen transporting capacity confirmed by their low ferritin values. On the other hand, ferritin values in intermediate group were normal and not different from the group with normal hemoglobin. Thus, values of hemoglobin between 12 and 14 g/dl very likely do represent dilution anemia.

Specifically for Company (P) ERT population, though majority of them are non-athlete individual, phenomena mimicking "sports anemia" were seen in this ERT group. In a capacity of researcher's experience as Occupational Health Doctor in-charge of Plant in-house clinic for six years, it was observed that job demand as Operator-technician roughly consist of 20 % light task and 80 % manual handling or field task. In few occasion per year, they were involved with more strenuous task during periodic plant maintenance period. The frequency of ERT physical exercise training was usually done once or twice over three month period, and major disaster drill mostly done once per year. Due to reasonably high involvement with physical activities throughout the year, their routine workplace task already provides them with relatively adequate aerobic exercise.

This research shows the TC inverse correlation with physical fitness and "good cholesterol" HDL-C in opposite, positively correlate with physical fitness. Few studies describe cholesterol fraction as part of metabolic profile with high LDL-C contribute to favorable profile and TC in reverse. Analysis upon large pool of data from Spanish AVENA (Food and Assessment of the Nutritional Status of Adolescents) study indicates that high levels of cardiorespiratory fitness are associated with a more favorable metabolic profile [8]. Still in aspect of cardiovascular fitness, Garcia et al. [9] analyzed same pool of data from AVENA study found that physical fitness, but not physical activity, is related to lipid and metabolic cardiovascular risk. Higher aerobic capacity in males and greater muscle strength in females were associated with lower lipid and metabolic risk factors for cardiovascular disease.

In terms of clinical use, until local data are available, Malaysian Clinical Practice Guideline 2011 recommends adopting the latest (2008) Framingham Cardiovascular risk score with TC and HDL-C as two parameters to be considered upon calculation of 10 years CV risk (Management of Dyslipidaemia, Malaysian CPG 2011). Similarity in correlation seen in cardiovascular risk and physical fitness status in relation to TC and HDL-C level does in some degree illustrate the importance of physical fitness as a predictor or marker of health status. This research failed to support a correlation between physical fitness status and blood pressure level.

This research failed to support a correlation between physical fitness status and blood pressure level. As mentioned in literature review, most available study did not directly design to determine direct relationship between physical fitness status and blood pressure level. Majority of researcher were interested to evaluate correlation between physical activity frequency and intensity with physical fitness status but in same time capturing an improvement in lifestyle illness risk parameters including blood pressure. Correlation varies from study to study.

Dunn et al. [10] designed a randomized clinical trial to compare the 24-month intervention effects of a lifestyle physical activity program with traditional structured exercise (fitness center orientated) on improving physical activity, cardiorespiratory fitness, and cardiovascular disease risk factors. Both the lifestyle and structured activity groups had significant and comparable improvements in physical activity and cardiorespiratory fitness, reductions in systolic, and diastolic blood pressure. In conclusions, a lifestyle physical activity intervention is as effective as a structured exercise program in improving physical activity, cardiorespiratory fitness, and blood pressure.

A meta-analysis study [11] revealed contradicting result whereby for children and adolescent, study suggest that exercise does not reduce resting systolic and diastolic blood pressure. As mentioned in Malaysian Clinical Practice Guideline 2008, as part of non-pharmacological treatment, the effect of at least 6 months of exercise on BP reduction among patients with hypertension is very modest. Despite that, CPG still recommends an aerobic-type exercise which is more effective than exercise that involves resistance training (e.g., weight lifting).

Mean systolic blood pressure for ERT members was 128.94 ± 12.17 mm Hg, diastolic was 80.66 ± 11.08 mm Hg which was slightly lower than criteria of diagnosing hypertension, i.e., 130/90 mm Hg. More prominent negative correlation might be observed if many of respondent have a higher blood pressure level. Furthermore, this research does not exclude those who already diagnosed with hypertension but current blood pressure level has likely been controlled by medication. This research failed to support a correlation between physical fitness status and fasting blood sugar level. Correlation between physical fitness and glucose level mostly illustrated by researcher as clinical correlation of development of impaired fasting glucose (IFG) and type 2 diabetes. Lee [12] conclude that poor cardiorespiratory fitness was associated with increased risk of IFG and type 2 diabetes. Mean fasting blood sugar for ERT members was 5.91 ± 3.17 mmol/L, which was relatively lower than diagnostic range of fasting sugar glucose (IFG) and type 2 diabetes. More prominent negative correlation might be observed if many of respondents have a higher sugar level. Furthermore, this research does not exclude those who already diagnosed with type 2 diabetes whereby current sugar level might be controlled by medication.

This research failed to support a correlation between physical fitness status and smoking status. Majority of available research on smoking and physical activity provides strong evidence of smoking's negative impact and physical activity's positive impact on long-term health. Conway and Cronan [13] examined associations of exercise activity, smoking behavior, and physical fitness among 3,045 Navy personnel. Exercise and smoking behaviors were measured using a lifestyle survey. Result showed that smoking was associated with lower exercise levels and lower physical endurance both cardiorespiratory and muscular. Study findings suggest that smokers will have lower physical endurance than non-smokers, even after controlling for exercise activity and event among relatively young, fit individuals. A longitudinal study of 1393 healthy middle-aged Norwegian men for 7 years was on long-term effects of smoking on physical fitness and lung function. Physical fitness was substantially lower in persistent smokers than in persistent non-smokers, and decline in physical fitness and lung function was considerably greater among smokers than among non-smokers and could not be explained by differences in age and physical activity.

Razalee et al. [14] conducted a cross-sectional study on 59 Royal Malaysian Navy recruits aged 18–25 years basically to identify association between aerobic physical fitness and smoking, finally surprised that no significant correlation was found. Another study on group of young trainees from the US Army which undergo Basic Combat Training, with a mean age close to 22 years, it was found that there were no difference in VO_2max between smokers and non-smokers [15].

Looking back to the design of this research, in order to establish proper correlation, consideration should be given on definition of smoking. Clear demarcation or social smoker, persistent smoker and even the number of stick per day which may affect the physical performance. Conway and Cronan [13] had done an extra steps by controlling physical activity and age in order to ensure respondent have almost similar baseline fitness. Samples of study should be higher because negative correlation of effect is more significantly seen in older age group

5 Conclusion

In general, physical fitness should be considered as a useful health marker due to correlation with lifestyle illness risk factors, i.e., age, BMI, WC, body fat percentage, TC, and HDL-C.

Physical fitness status for ERT serving Company (P) is below satisfactory. Management should plan a prompt mitigation plan to address the issue. Physical fitness enhancement program should be design with recommendation such as structured regular physical exercise sessions and healthy diet awareness workshop. Body composition, blood parameters, and UKJK performance derived from this research serve as a baseline data and repeat assessment should be done at least after 6 months intervention program.

References

1. Ortega, F. B., Ruiz, J. R., Castillo, M. J., & Sjo, M. (2007). Cardiorespiratory fitness and sedentary activities are associated with adiposity in adolescents. *International Journal of Obesity, 32*, 1–11.
2. Clarke, H. H. (1967). *Application of measurement to health and physical education*. Englewood Cliffs: Prentice-Hall Inc.
3. Shephard, R. J., Allen, C., Benade, A. J., Davies, C. T., di Prampero, P. E., & Hedman, R. (1968). The maximum oxygen intake. An international reference standard of cardiorespiratory fitness. *Bull World Health Organ, 38*, 757–764.
4. Cullumbine, H., & Bibile, S. W. (1950). Influence of age, sex, physique and muscular development on physical fitness. *Journal of Applied Physiology, 2*(9), 488–511.
5. Cesari, M. (2004). Hemoglobin levels and skeletal muscle: Results from the In CHIANTI study. *Journals of Gerontology. Series A, Biological Sciences and Medical Sciences, 59*(3), 249–254.
6. Eichner, E. R. (1992). Sports anemia, iron supplements, and blood doping. *Medicine and Science in Sports and Exercise, 24*(9 Suppl), 315–318.
7. Novack, V. (2007). The prevalence of low hemoglobin values among new infantry recruits and nonlinear relationship between hemoglobin concentration and physical fitness. *American Journal of Hematology, 82*, 128–133.
8. Mesa, J. L., Ruiz, J. R., Ortega, F. B., Warnberg, J., Gonzalez-Lamuno, D., & Moreno, L. A. (2006). Aerobic physical fitness in relation to blood lipids and fasting glycaemia in adolescents: Influence of weight status. *Nutrition, Metabolism and Cardiovascular Diseases, 16*, 285–293.
9. Garcia-Artero, E., Ortega, F. B., Ruiz, J. R., Mesa, J. L., Delgado, M., & Gonzalez-Gross, M. (2007). Lipid and metabolic profiles in adolescents are affected more by physical fitness than physical activity (AVENA study). *Revista Espanola de Cardiologia, 60*, 581–588.
10. Dunn, A. L., Marcus, B. H., Kampert, J. B., Garcia, M. E., Kohl, H. W, I. I. I., & Blair, S. N. (1999). Comparison of lifestyle and structured interventions to increase physical activity and cardiorespiratory fitness: A randomized trial. *JAMA, 281*(4), 327–334.
11. Kelley, G. A., & Kelley, K. S. (2003). Exercise and resting blood pressure in children and adolescent: A meta-analysis. *Pediatric Exercise Science, 15*, 83–97.
12. Lee, D. (2009). Associations of cardiorespiratory fitness and obesity with risks of impaired fasting glucose and type 2 diabetes in men. *Diabetes Care, 32*(2), 257–262.
13. Conway, T. L., & Cronan, T. A. (1992). Long term effects of smoking on physical fitness and lung function: A longitudinal study of 1393 middle aged Norwegian men for seven years. *BMJ, 311*, 718–721.
14. Razalee, S., Poh, B. K., & Ismail, M. N. (2012). Relationship between body composition. *Smoking and Physical Fitness of Malaysian Armed Forces Naval Trainees, 1*, 548. doi: 10.4172/scientificreports.548.
15. Daniels, W. L., Patton, J. F., Vogel, J. A., Jones, B. H., & Zoltick, J. M. (1984). Aerobic fitness and smoking. *Medicine and Science in Sports and Exercise, 16*, 195–196.

Performance Indices of Two Different Repeated Ability Tests Based on Playing Positions

Annisaa Basar, Muhammad Sufyan Mohamad Zaki, Sarina Md. Yusof, Suhana Aiman and Adam Linoby

Abstract Soccer is characterized as complex sport which requires high aerobic, anaerobic capacity, and tactical skills. The purpose of this study was to examine the performance indices of repeated sprints ability (RSA) and repeated dribbling ability (RDA) tests among Malaysian University soccer players. A total of fifty-two soccer players were participated (age 21.73 ± 1.82 years, height 171.31 ± 5.27 cm, weight 63.78 ± 7.30 kg, BMI 21.69 ± 2.29 kg m^{-2}) to perform all 4 tests: repeated sprint ability (7 × 34.2 m), RDA (7 × 34.2 m), anaerobic power (vertical jump), and aerobic power (20-m multistage shuttle run test). Significant level was set at $p < 0.05$ according to the player's position in the teams. There were significant differences in VO_{2max}, total time, best time, and mean time of both repeated ability tests. Peak blood lactate of RDA test also showed the significant difference between playing positions. Anaerobic interrelated to aerobic power with greater VO_{2max}, which would speed up recovery from explosive power activities in soccer.

Keywords Aerobic power · Leg strength · Fitness testing · Repeated ability · Soccer

This work was supported in part by the Fundamental Research Grant Scheme (FRGS), Ministry of Higher Education of Malaysia FRGS/1/2011/SKK/UITM/03/32.

A. Basar (✉) · M. S. Mohamad Zaki · S. Md. Yusof · S. Aiman · A. Linoby
Faculty of Sports Science and Recreation, Universiti Teknologi MARA,
Shah Alam, Malaysia
e-mail: annisaa_basar@yahoo.com

A. Basar · M. S. Mohamad Zaki · S. Md. Yusof · S. Aiman · A. Linoby
Faculty of Sports Science and Recreation, Universiti Teknologi MARA,
Jengka, Pahang, Malaysia

R. Adnan et al. (eds.), *Proceedings of the International Colloquium on Sports Science, Exercise, Engineering and Technology 2014 (ICoSSEET 2014)*,
DOI: 10.1007/978-981-287-107-7_9,

1 Introduction

Soccer is an intermittent sport which requires high aerobic [1] and anaerobic capacity [2]. Intermittent sports require players' skills incorporate with technical, tactical, and physical fitness to succeed [3]. Studies have shown the anaerobic component in soccer involves sprints, tackles, and jumps in one game. Therefore, sprinting could be the most important skill that determines the success of the game [4]. In order to perform high speed of running, the maximum anaerobic power requires powerful and strength of leg in competitive games [5, 6]. The ability to produce fastest sprint in shortest time separated by recovery known as repeated sprint ability (RSA) [7]. Repeated dribbling ability (RDA) is determined by the ability to dribble at highest speed. Dribbling or running with ball requires the player to control the ball's direction while maintaining high speed of running [8]. Running with the ball is a good skill that provides advantages over opponents in ball possession.

Duration of 90 min of soccer game is important for recovery from any anaerobic activities and to maintain player's performance. The efficiency of cardiovascular system to supply oxygen to all working muscle indicates player having the greatest VO_{2max} level and could optimally perform exercise. The literature reported professional soccer players reached between 55 and 67 ml kg^{-1} min^{-1} [1, 7].

Using match analysis technology, players performed almost 30 % of running and 60 % of sprinting in soccer [9]. Due to anaerobic activities, blood lactate increases and affects blood flow to tissue. The accumulation of the blood lactate will lead fatigue and decrease player's performance. Higher capacity of oxygen consumption tends to delay the concentration of blood lactate.

Studies have shown that there is decline in amount of sprinting [10] as well as decrement in high-intensity efforts in second half of a soccer game [9, 11, 12]. Distance covered by elite players during match play has been reported in range between 10,000 and 14,000 m in mean based on playing position [13, 14]. Midfielders showed the greatest distance covered during the play match with approximately 12,000 m [15]. Strikers performed the highest number of sprints, but the total of distance covered decreased in the second half [15]. There was significant difference of activities such as sprinting, jumping, tackling, and other performance according to their playing position [16]. No significant difference was found between the total of ball possession in a match play with the different playing positions in the team [16].

Therefore, the purpose of this study was to examine the performance indices of two different of RSA and RDA tests among Malaysian University soccer players.

2 Methods

2.1 Participants

Fifty-two, well-trained soccer players (age 21.73 ± 1.82 years, height 171.31 ± 5.27 cm, weight 63.78 ± 7.30 kg, BMI 21.69 2.29 kg m^{-1}) selected participants were enrolled by university students. All participants were members of top three teams from the first division of Malaysian Higher Education Institution Soccer league, 2013. These teams were consistently ranked among the top 3 teams in the league for last 2 years. The players trained 5 days every week (~90 min per session) with the university team for at least 4 months and active in sports within 1 year.

During the session, participants were competed every weekend. This study was performed in the middle of the soccer season, when the players were assumed to be in their best physical fitness. Player who was free from any injuries was selected for this study. Exclusion criteria consisted of history of any cardiovascular, metabolic disorders, musculoskeletal, and neurological problem. Subjects answered a demographic or injury history questionnaire which was used to obtain background information from each participant.

All the procedures were approved by the Institutions Ethic and Research Committee (Ref: 600-RMI 5/1/6) of Universiti Teknologi MARA, Malaysia. The testing procedures were explained to participants, and a written informed consent was obtained before testing started. All participants were fully familiar and clearly understood with the procedures used in this testing. They were informed that they could withdraw from the testing at any time without penalty.

2.2 Procedures

Participants started the tests varying between 1,630 and 1,900 in order to maintain their performance similar to the routine training and to avoid any circadian variability. On the RSA and RDA testing day, the reading of temperature and wind velocity was recorded to ensure both factors were controlled. The temperature was about 33 °C and wind velocity was 1.2 mps. All tests were conducted in two sessions after at least 48 h which last performed match or vigorous exercise to prevent unnecessary fatigue accumulation. First session was comprised of anaerobic power test (vertical jump) and aerobic power test (20-m multistage shuttle run test). Second session was comprised of RSA and RDA tests. All tests were conducted outside on a grass training field, using regular soccer shoes to replicate playing condition.

Before testing, a standardized warm-up consisted of jogging and stretching for 20 min was conducted.

2.3 Vertical Jump Test

Vertec is testing device, where steel frame construction has horizontal vanes and would be rotated out of the way by the hand to indicate the height reached. For starting, participants had their standing reach measured, and then, they pointed of the arm fully extended upward of the dominant arm against the Vertec apparatus. The Vertec standing reach height marker was adjusted to the tip of their dominant hand's middle finger. Participants performed the test with feet flat on the ground and about shoulder width apart, ready position to perform. Then, they started the projectile motion, and participants were on the upward phase of the jump where they extended their dominant hand into the air in attempt to displace the Vertec measuring vanes. The displacement of the vanes was used to measure a participant's vertical jump height, per 1/2 inch increments; the jump height was the difference between standing height and jumping height.

All power output measurements were based on the highest point deduction to the lowest height. The Sayer's equation was used to calculate peak anaerobic power output in the vertical jump was (60.7 × height (cm) squat jump + (45.3 × body mass (kg) – 2,055 [17].

2.4 20 m Multistage Shuttle Run Test

The 20-m multistage shuttle run test was conducted according to established procedures. When the participants were ready, the CD was started and the participant commenced the test. Participants had to follow the procedures and performed their maximal effort. If the participant arrived at the end of a shuttle before the beep, the participant had to wait for the beep and then resume running. If the participants failed to reach the end of the shuttle before the beep, they should be allowed two or three further shuttles to attempt to regain the required pace before being withdrawn. Researcher recorded the level and number of shuttles completed at that level either when the participant withdraws voluntarily from the test failed to be within 3 m of the end lines on two consecutive tones [18].

2.5 Repeated Sprint and Repeated Dribbling Ability Tests

The RSA and RDA tests were conducted after 48 h from the first session. Blood sample was taken at rest as baseline prior to warm up. Before RSA test, each subject was allowed to perform two sprints, and for RDA test, participant allowed to perform two repetitive dribbling skills along the test course for familiarization purposes. Although in the absence of any learning effects from this type of multiple sprint protocol [19], all subjects completed a familiarization trial of both protocols.

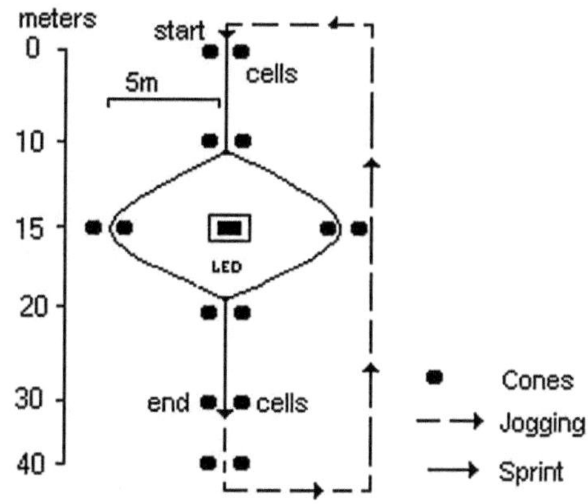

Fig. 1 Diagram of Bangso test protocol [21], adapted by Aguiar et al. [20]

Testing was started with RSA followed by RDA test. Same participant would perform the next testing after 1-h rest interval. The protocol comprised of seven maximal 34.2 m sprints [4, 20]. Participant sprint along a distance of 10 m changes direction and sprint between an obstacle along 14.2 m and sprint along 10 m as showed on Fig. 1. Electric timing gates were used to measure participant's performance. Following each sprint, there was a period of active recovery (25 s to cover a distance of 40 m), which consisted of jogging to return to the initial point. Recovery was taken using stopwatch in order to ensure subject return to initial point of course between 23rd and 24th s. Verbal feedback was given by assistants at 5, 10, 15, 20, and 25 s of recovery [4, 20]. Similar protocols were used to perform RDA test except participant while dribbling a ball.

2.6 Peak Blood Lactate and Lactate Removal Rate

Blood sample was collected from fingertip at the end of RSA and RDA tests (zero time recovery) and at 1st, 3rd, 5th, 7th, 9th, and 12th minutes of recovery [13]. The participant's finger was washed with water and dried, then it was disinfected by alcohol-contained cotton, and using single use of disposable lancet device, participant's finger was pricked. Laboratory assistant would swab alcohol-contained cotton again at the participant's finger to ensure the hygiene of blood sample. The blood sample was collected into lactate strip and analyzed using an automated lactate analyzer (Accutrend). After 1 min, the analyzed blood sample was recorded.

2.7 Statistical Analyses

Descriptive statistics were reported as mean ± SD for all measures with different playing position in team. A one-way ANOVA with Tukey's post hoc test was used to examine possible differences between four playing position: goalkeepers,

Table 1 Anthropometric measures, aerobic and leg power ($N = 52$)

Measures	Mean ± SD
Age (years)	21.73 ± 1.82
Height (cm)	171.31 ± 5.27
Weight (kg)	63.78 ± 7.30
BMI (kg m^{-2})	21.69 ± 2.29
Leg power (W)	3,473.93 ± 3,473.93
VO_{2max} (ml min^{-1} kg^{-1})	47.91 ± 6.378

strikers, midfielders, and defenders. All data were tested for assumptions of normality, homogeneity of variance and covariance matrices. The results were not violated the assumptions. The alpha level of statistical significance was set at $p < 0.05$.

3 Results

The anthropometric characteristics of the players, aerobic, and leg strength power are summarized in Table 1. The aerobic and leg power by playing positions are summarized in Table 2. The results of the performance indices of two different repeated ability tests by playing position are summarized in Tables 3 and 4.

4 Discussion

The purpose of this study was to examine the performance indices of repeated sprints ability, RDA, anaerobic and aerobic power by playing positions among Malaysian University soccer players. The significant difference was showed between maximal oxygen consumption and positions in Table 2. This would be expected that the midfielders (50.67 ± 5.65 ml min^{-1} kg^{-1}) would have the highest VO_{2max} and slightly difference between strikers and defenders. Moreover, VO_{2max} was the main factor would maintain performance players [21]. An increase in VO_{2max} blood transport oxygen efficiency will improved athlete's performance [22]. Limitation of oxygen delivery to muscle due to decrease of oxygen supply during the test activities [23].

There was no significant difference of leg power of different positions. The same result from previous studies reported no significant difference of leg power among players [24]. However, the highest leg power (3,779.56 ± 404.46 W) was found among goalkeeper compared to other positions. The probable reason is that goalkeepers had the highest mean may due to body structure as goalkeeper and their difference of training role.

Table 2 Aerobic and leg power by playing positions

Characteristics	Position				
	GK ($n = 4$)	ST ($n = 11$)	MID ($n = 14$)	DF ($n = 23$)	F value $p < 0.05$
Leg power (W)	3,779.56 ± 404.46	3,395.16 ± 430.99	3,324.42 ± 338.89	3,549.93 ± 490.30	1.55
VO_{2max} (ml min^{-1} kg^{-1})	38.99 ± 6.49	48.47 ± 5.50	50.67 ± 5.65	47.53 ± 6.03	4.20*

Values are (mean ± SD), * significance value set at $p < 0.05$

Table 3 Performance indices of RSA (mean ± SD) by playing positions

Indices	Position				
	GK	ST	MID	DF	*F* value $p < 0.05$
Total time (s)	46.50 ± 2.34	38.97 ± 1.88	40.55 ± 4.76	42.21 ± 3.21	5.41*
Fastest time (s)	6.31 ± 0.26	5.34 ± 0.21	5.58 ± 0.65	5.70 ± 0.47	4.21*
Mean time (s)	6.64 ± 0.34	5.57 ± 0.27	5.79 ± 0.68	6.03 ± 0.46	5.43*
Performance decrement (%)	5.26 ± 1.86	4.31 ± 2.02	4.23 ± 1.73	5.30 ± 2.37	1.04
Peak blood lactate (mmol L^{-1})	13.80 ± 1.10	13.21 ± 1.71	11.81 ± 2.23	12.62 ± 1.80	1.75
Removal rate (mmol L^{-1} min^{1})	0.39 ± 0.28	0.37 ± 0.15	0.34 ± 0.21	0.34 ± 0.19	0.11

* Significance value set at $p < 0.05$

Table 4 Performance indices of RDA (mean ± SD) by playing positions

Indices	Position				
	GK	ST	MID	DF	*F* value $p < 0.05$
Total time (s)	67.97 ± 6.77	57.52 ± 4.64	58.62 ± 7.88	62.91 ± 4.88	4.66*
Fastest time (s)	8.86 ± 0.67	7.39 ± 0.64	7.76 ± 1.23	8.28 ± 0.75	4.15*
Mean time (s)	9.35 ± 0.76	8.21 ± 0.66	8.37 ± 1.13	8.87 ± 0.76	2.82*
Performance decrement (%)	9.56 ± 3.21	10.42 ± 4.19	7.83 ± 4.45	8.72 ± 3.91	0.88
Peak blood lactate (mmol L^{-1})	13.95 ± 2.26	11.99 ± 1.91	10.08 ± 1.90	11.06 ± 3.13	2.81*
Removal rate (mmol L^{-1} min^{1})	0.35 ± 0.30	0.48 ± 0.23	0.37 ± 0.11	0.47 ± 0.24	1.05

* Significance value set at $p < 0.05$

The probable reason is that goalkeepers had the highest mean may due to body structure as goalkeeper and their difference in training intensity compared to other position. Although the mean of fastest time was showed slightly difference, the statistical analysis confirmed strikers' (5.34 ± 0.21 s) sprint was the fastest. The fastest time would influence mean time of players [10]. Thus, the mean time was significant difference among strikers. Sprinting, jumping, and tackling were used to test anaerobic power in soccer [25].

The RDA would explain players' skill of controlling the ball possession as well as maintaining high running speed [8]. Strikers showed to have good dribbling skill among positions. They reported to have lowest mean in total time, fastest time, and mean time. An interesting observation of the fastest time was performed by player during the 6th and 7th in repeated sprinting and dribbling ability. Similar result from 5th to the 7th sprint players met to speed up their running pace [26].

The peak blood lactate was significant difference among goalkeepers (13.95 ± 2.26 mmol L^{-1}) followed by strikers (11.99 ± 1.91 mmol L^{-1}) and defenders (11.06 ± 3.13 mmol L^{-1}). The extended performed of anaerobic activities of sprinting or dribbling during match play may cause the elevation of blood lactate [27]. Performance decrement was indicator of fatigue; however, the rest intervals were induces oxygen uptake.

5 Conclusion

In improving aerobic power, implementing specific soccer training would be beneficial. The identical total work, RSA, and RDA with rest intervals demonstrate different physiological implications. In this context, players' positions present the best method to identify and propose specific soccer training. Both tests attribute different fitness and address abilities of strength and condition among players. The aerobic power that addressed midfielders was the highest VO_{2max} with greatest distance covered. The tests replicate the playing game of tests where it stimulates optimal performance. The fastest time indicates running as quickly over the opponents for scoring goals. Therefore, dribbling and sprinting are interrelated skills in soccer.

References

1. Aziz, A. R., Tan, F. H. Y., & Teh, K. C. (2005). A pilot study comparing two field tests with the treadmill run test in soccer players. *Journal of Science and Medicine in Sport, 4*, 105–112.
2. Sporis, G., Ruzic, L., & Leko, G. (2008). The anaerobic endurance of elite soccer players improved after a high-intensity training intervention in the 8 week conditioning program. *Journal of Strength and Conditioning Research, 22*(2), 559–566.
3. Kunduracioglu, B., Guner, R., Ulkar, B., & Erdogan, A. (2007). Can heart rate values obtained from laboratory and field lactate tests be used interchangeably to prescribe exercise intensity for soccer players? *Advances in Therapy, 24*(4), 890–902.
4. Abrantes, C., Macas, V., & Sampaio, J. (2004). Variation in football players' sprint test performance across different ages and levels of competition. *Journal of Sport Medicine, 3*, 44–49.
5. Amusa, L. O., & Toriola, A. L. (2003). Leg power and physical performance measures of top national track athletes. *Journal of Exercise Science and Fitness, 1*(1), 61–67.
6. Meysa, C., Rouholah, F., & Shahsavar, (2012). Relationship between speed, agility and anaerobic power of 14–16 years elite soccer players. *Journal of Applied and Basic Sciences, 3*(2), 427–432.
7. Bishop, D., Girard, O., & Mendez-Villanueva, A. (2011). Repeated-sprint ability—part II recommendations for training. *Sports Medicine, 41*(9), 741–756.
8. Lemmink, K. A. P. M., Elferink-Gemser, M. T., & Visscher, C. (2004). Evaluation of the reliability of two field hockey specific sprint and dribbles tests in young field hockey players. *Journal of Sport Medicine, 38*, 138–142.

9. Mohr, M., Krustrup, P., & Bangsbo, J. (2003). Match performance of high standard soccer players with special reference to development of fatigue. *Journal of Sport Science, 21*, 519–528.
10. Chaouachi, A., Manzi, V., Wong, D. P., Chaalali, A., Laurencelle, L., Chamari, K., et al. (2010). Intermittent endurance and repeated sprint ability in soccer players. *Journal of Strength and Conditioning Research, 24*(10), 2663–2669.
11. Barros, R. M. L., Misuta, M. S., Menezes, R. P., Figueroa, P. J., Moura, F. A., Cunha, S. A., et al. (2007). Analysis of the distances covered by first division Brazilian soccer players obtained with an automatic tracking method. *Journal of Sport Science, 6*, 222–242.
12. Mohr, M., Krustrup, P., Anderson, H., Kirkendal, D., & Bangsbo, J. (2008). Match activities of elite women soccer player at different performance levels. *Journal of Strength and Conditioning Research, 22*, 341–349.
13. Da Silva, C. R., Bloomfield, J., & Marins, B. J. C. (2008). A review of stature, body and maximal oxygen uptake profiles of U17, U20 and first division players in Brazilian soccer. *Journal of Sport Science, 7*, 309–319.
14. Di Salvo, V., Baron, R., Tschan, H., Calderon, M. F. J., Bachl, N., & Pigozzi, F. (2007). Performance characteristics according to playing position in elite soccer. *International Journal of Sports Medicine, 28*, 222–322.
15. Dellal, A., Wong, D. P., Moalla, W., & Chamari, K. (2010). Physical and technical activity of soccer players in the French first league—with special reference to their playing position. *International Sport Medicine Journal, 11*(2), 278–290.
16. Bloomfield, J., Polman, R., & O'Donoghue, P. (2007). Physical demands of different positions in FA premier league soccer. *Journal of Sports Science and Medicine, 6*, 63–70.
17. Sayers, S., Harackiewicz, D., Harman, E., Fryman, P., & Rosenstein, M. (1999). Cross-validation of three power equations. *Medicine and Science in Sports and Exercise, 31*(4), 572–577.
18. Paliczka, V. J., Nichols, A. K., & Boreham, C. A. G. (1987). A multi-stage shuttle run as a predictor of running performance and maximal oxygen uptake in adults. *British Journal of Sports Medicine, 21*(4), 163–165.
19. Glaister, M., Howats, G., Lockey, R. A., Abraham, C., Goodwin, J., & Mcinnes, G. (2007). Familiarization and reliability of multiple sprint running performance indices. *Journal of Strength and Conditioning Research, 21*(3), 857–859.
20. Aguiar, M., Abrantes, C., Macas, V., Leite, N., Sampaio, J., & Ibanez, S. (2008). Effects of intermittent or continuous training on speed, jump, and repeated sprint ability in semi-professional soccer players. *The Open Sports Sciences Journal, 1*, 15–19.
21. Bangsbo, J. (1994). *Fitness training for football: A scientific approach.*
22. Svensson, M., & Drust, B. (2005). Testing soccer players. *Journal of Sport Science, 23*(6), 601–618.
23. Bassett, D. R, Jr, & Howley, E. T. (2000). Limiting factors for maximum oxygen uptake and determinants of endurance performance. *Medicine and Science in Sports and Exercise, 32*, 70–84.
24. Swapan, K. D., Nabanita, K., & Parthasarthi, D. (2010). Anthropometric, motor ability and physiological profiles of Indian national club footballers: A comparative study. *South African Journal for Research in Sport, Physical Education and Recreation, 32*(1), 43–56.
25. Odetoyinbo, K., & Ramsbottom, R. (1995). Aerobic and anaerobic field testing of soccer players. *Journal of Sports Sciences, 13*, 506.
26. Spencer, M., Lawrence, S., Rechichi, C., Bishop, D., Dawson, B., & Goodman, C. (2004). Time-motion analysis of elite field hockey with special reference to repeated sprint ability. *Journal of Sport Science, 9*, 843–850.
27. Sahlin, K., & Henriksson, J. (1984). Buffer capacity and lactate accumulation in skeletal muscle of trained and untrained men. *Acta Physiologica Scandinavica, 122*(3), 331–339.

Adjustable Crank: A Comparison Between Wireless Motion Sensor and Motion Capture Analysis Camera for Crank Kinematic Measurement

Fezri Aziz, Ahmad Faizal Salleh, Sukhairi Sudin, Wan Mohd Radzi Rusli, Norazian Abdul Razak, Mohd Asyraf Faris Abdol Aziz, Fathinul Syahir Ahmad Saad and Ali Yeon Md Shakaff

Abstract This paper is focused on the development of wireless measured kinematics specifically for cycling. The aims of this study are to create sensory system with portability, reliability and based on the wireless system. The adjustable crank is novel type of prototype crank that design to maximize the minimum torque at bottom dead center (BDC) and top dead center (TDC) where the crank can be set for $+10°$ maximum with addition of 5° back and forth from inertial 0° TDC point. The system will measure the power output during cycling to evaluate the cyclist performance. In order to measure power output, the angle displacement (kinematic) and force measurement (kinetic) are needed. The inertial measurement unit (IMU) combination of accelerometers and gyrometers was used to measure the angle and angular velocity of the crank. The system has been validated using a visual system to compare the output provided by IMU. The RMS error value between motion capture camera and IMU for crank angle was $0.480 \pm 0.325°$. The RMS error value for normalized angular velocity was 0.743 ± 0.911 %. The wireless-based system will aid to reduce the wiring complexity and user-friendly portable measuring system. The wireless communication using Zigbee protocol with two Xbee devices point-to-point will be used to transfer the information to the computer controlled system. The enhancement of this system can be used for coaches for cycling monitoring system to improve cyclist coordination, strategy, and technique.

Keywords IMU · Crank · Kinetic · Kinematic and wireless

F. Aziz (✉) · A. F. Salleh · W. M. R. Rusli · N. A. Razak · M. A. F. Abdol Aziz
School of Mechatronics (Biomedical Electronics Engineering), Biomechanics Laboratory, University Malaysia Perlis (UniMAP), Arau, Malaysia
e-mail: fezriaziz89@gmail.com

S. Sudin · F. S. Ahmad Saad · A. Y. Md Shakaff
Center of Excellence Advanced Sensor and Technology, University Malaysia Perlis (UniMAP), Kampus Ulu Pauh, 02600 Arau, Perlis, Malaysia

R. Adnan et al. (eds.), *Proceedings of the International Colloquium on Sports Science, Exercise, Engineering and Technology 2014 (ICoSSEET 2014)*,
DOI: 10.1007/978-981-287-107-7_10,

1 Introduction

Pedaling is a circular motion with a repetitive pattern of force application. The force application, however, is not constant throughout the pedal–crank revolution stressed by Coyle et al. [1]. Ericson and Nisell showed that the tangential force exertion is maximal when the position of the crank is at 90° crank angle [2]. Moreover, in circular motion, there are two dead points bottom dead center (BDC) which at bottom most 180° and top dead center (TDC) which at 0° top most of the circular motion. They are called dead points because throughout the circular motion of cycling, the points produce minimum tangential force. The tangential force that acts parallel to the crank arm length produces minimum torque. Any optimization of this crank cycle would necessarily lead to higher net torque and power output (assuming an equivalent cadence). Therefore, many designs of independent crank arms (crank no longer fixed at 180°), non-circular chain rings, and varying lengths of crank have been manufactured. The main goal for this design is to improve the transfer of human power into cycling performance.

Gregor and Rugg studied the activity pattern of muscle recruitment at 85 RPM against moderate load in ten competitive male cyclists riding at their own personal comfort level [3]. They indicated that the crank cycle can be categorized into three phases: (i) the pushing phase which the foot pushed forward at the TDC, (ii) the down stroke/propulsive/power phase, and (iii) the up stroke/pulling/recovery phase. Pushing or power phase of the crank cycle is within 10° before and after 10° TDC at 0° (Fig. 1).

The rotor crank system is an independent crank arm which has been introduced since 2002. Manufacturers' claim rotor crank system eliminates the dead points where torque production is minimal, and this system has been approved by the union cyclist international (UCI) for use in competitive events. The rotor cranks system allowed varying crank angle of maximum crank arm length between 116°, 122°, 128°, and 134°, increment of 3° of each level where 0° is at TDC stated by Rodríguez-Marroyo et al. [4]. Santalla et al. researched the effect of the rotor crank system using carbon dioxide uptake (VCO_2) cycling performance in a group of healthy non-cyclists to calculate delta efficiency during the 3-min work stage [5]. In contrast, Lucia et al. [6] assessed the effect of the rotor crank system on a group of well-trained cyclists also using 3-min work stages; they suggested that rotor crank system improved contralateral cooperation of the legs.

Contralateral cooperation is achieved in trained cyclists who have already learned the appropriate technique. Using rotor crank system, contralateral cooperation is minimized for normal recreational cyclist. In more recent studies, Simon et al. investigated rotor crank system on laboratory cycling performance for 40.23-km time trial [7]. The main finding was that the rotor system had no measurable impact on the time taken to complete a 40.23-km time trial. Jose et al. tested the effect rotor pedaling system on aerobic and anaerobic cycling performance. They found that the rotor pedaling system improves anaerobic (heart rate, power output, and speed) but not aerobic (VO_2). However, cycling with adjustment of the crank at maximize ±10°

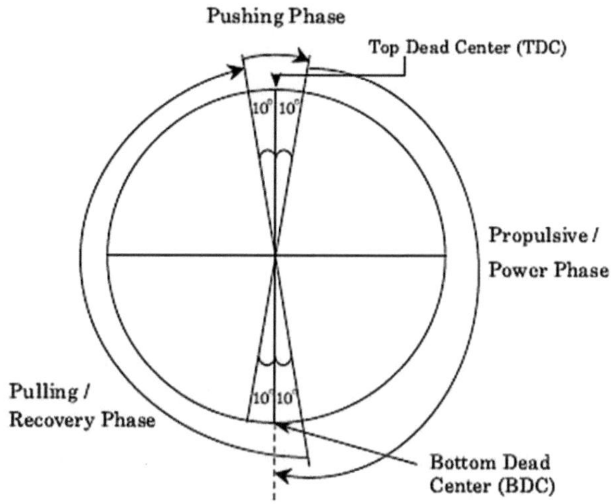

Fig. 1 The three phases of the crank cycle during the cycling action [3]

back and forth angle from dead point centers on power output of the cyclist has not yet been studied. The ergonomics of this crank angle adjustment would affect the performance cyclist.

IMU is commonly used in system for balancing, orientation, and navigation. Due to combination of accelerometers and gyrometers, the system gives more accurate reading than standalone single accelerometers and gyrometers. Hence, the IMU is applied for joint kinematics angle detection [8] in sports and rehabilitation. In sports, IMU is applied in many kinds of sport fields such as swimming, cycling, and kayak. However, due to gravitational problems in its micromechanical systems (MEMS) chip, the accelerometers and gyrometers produce noises. The accelerometer's noise is produced by the influence of gravitational. The gyrometer's noise produced suffers from temperature-dependent drift. Due to the noises constraint, the researchers used the Kalman filtering to produce more accurate data estimation. Despite for accurate data estimation, the IMU is used in adjustable crank to provide instantaneous angle displacement and normalized angular velocity (two axis) for the power output measurement system calculation.

Nowadays, wireless communication has been revolutionized by more advanced technology to justify problems in reducing energy consumption, easy portability, and enhancing reliability. The wireless communication protocol systems are merged into radio mobile networks (GSM and GPRS), internet protocols (IP), and wireless protocols (Bluetooth and Zigbee). In this context, Zigbee wireless communication protocol is selected. Zigbee wireless communication is selected to consume low battery power and reducing cost. Xbee module is based on the IEEE 802.15.4/Zigbee wireless personal area network (WPAN). The Xbee module is integrated with microcontrollers for monitoring temperature [9], wireless sensor network [10], and human movement detection [11]. Hence, in this study, the Xbee module shield is applied and stack on top of the Arduino Mega 2,560. The Arduino Mega microcontroller is powered by battery and the data transmit wirelessly. The data consist of Kalman filter angle displacement and two-axis angular velocity to computer controlled system. The data received are evaluated and correlate with motion capture analysis.

2 Methodology

2.1 Overall Overview

The complete system deployment is shown in Fig. 2; the challenges faced in this study are to design, develop, and implement for the end user. The system structure consist of fuse integrated system from IMU 5 DOF unit sensor, two-axis forces (normal and anterior–posterior axis) which connected to Arduino Mega microcontroller analog pins. The board was placed on the adjustable crank. The wireless transmission is communicating using Xbee module. The two-axis model for force sensor is designed by Daniele [12].

The pedal is called strain gauges instrumented (SGI) clipless pedal. The SGI clipless pedal made from aluminum material. The pedal force is measured by the strain gauges attached properly to aluminum core. The deformation of aluminum material will convert the potential energy to the electrical energy. The aluminum core of the pedal is attached with shoe and cleat (clipless) for closed loop chain kinematic chain.

The clipless pedal allowed professional and amateur cyclists to reach maximum mechanical efficiency which generated 11 % greater mechanical efficiency and used only 9 % more work to do so [13]. The relationship between strain and deformation is correlated by the modulus of elasticity coefficient (70.0 N/mm^2). The aluminum can withstand force until 1,000 N before it reaches the ultimate tensile strength (UTS).

The integrated system of SGI clipless pedal and IMU combining accelerometers and gyrometers provide instantaneous forces, angle displacement, and angular velocity for power output measurement. Power output can be calculated using Eq. (1).

$$\text{Power} = \text{Force} * \text{Angular Velocity} \tag{1}$$

The forces in normal axis and anterior–posterior axis are normalized as well as the angular velocity in x-axis and y-axis before inserting into the power output equation. The angle displacement, angular velocity, and forces data are transferred wirelessly via Xbee module to computer controlled system using MATLAB 2012a for further calculation analysis. Xbee module series 2 capable to transmit 6–10 bit ADC up to 40 m range in indoor and have 2 mW (+3 dbm) transmit power. The benchmarking comparison wireless data transmission and high-speed motion capture camera as a reference will be discussed.

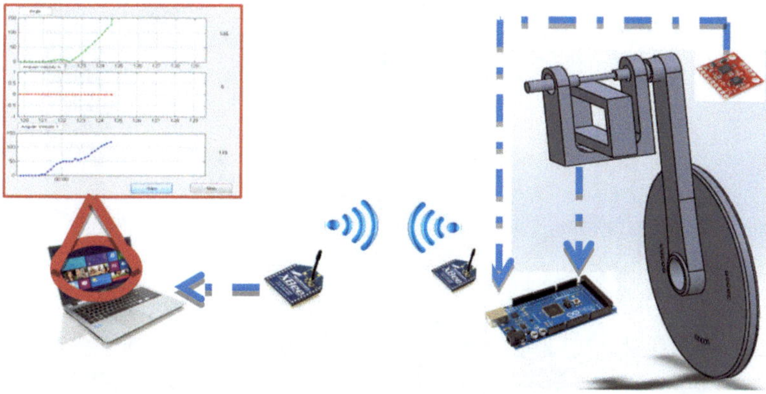

Fig. 2 The overall adjustable crank system

2.2 Design and Development

The adjustable crank prototype is designed with Solidworks 2013 software. The crank can be set for fixed angle +10° and −10° maximum with the 5° back and forth (Figs. 2 and 3).

The idea was to eliminate the minimum torque produce at both dead centers (BDC and TDC) in enhancing cycling performance. The range of angle is based on the pushing phase range.

The first plate is attached with the second plate, and the second plate was fixed at the shaft. The first plate can be rotated as the angle needed and screws the remaining drill holes. The mass load of the crank was supported by the crank shaft.

The SGI pedal was combined with 3 component parts which are pedal shaft, square loadcell core, and stirrup '*U*' shape. The cleat of the cycling shoe is attached directly to square loadcell core. The cleat is screwed in between the pedal shaft without touching the shaft. The strain gauges attached along the square loadcell core to convert the pushing/pulling energy to electrical resistance energy. The electrical resistance was due to the deformation of the aluminum square loadcell core. The combination of the strain gauges can provide two-axis pedal force in two directions of forces in normal force axis and anterior–posterior axis (Fig. 4). The strain gauges were arranged with Wheatstone Full-Bridge circuit to produce stable signal. The signal was amplified using AD623 differential amplifier circuit.

3 Kinematics Measurement

The IMU 5-degree of freedom IMU (Sparkfun Electronics, SEN-11072) was used to calculate angle displacement and angular velocity. The IMU incorporates the triple-axis accelerometers (ADXL 335) and dual-axis gyrometers (IDG 500) and

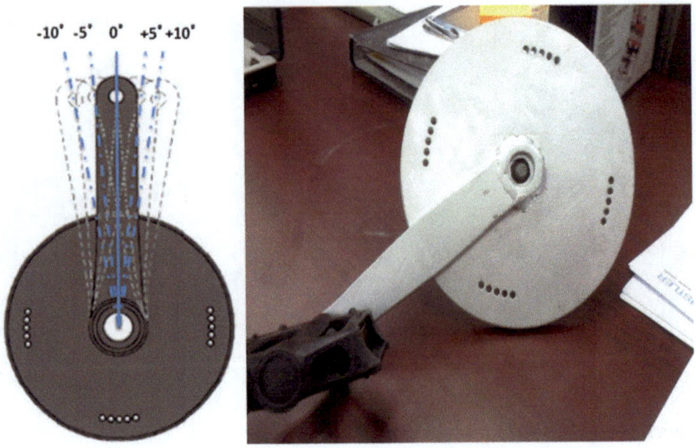

Fig. 3 The crank design

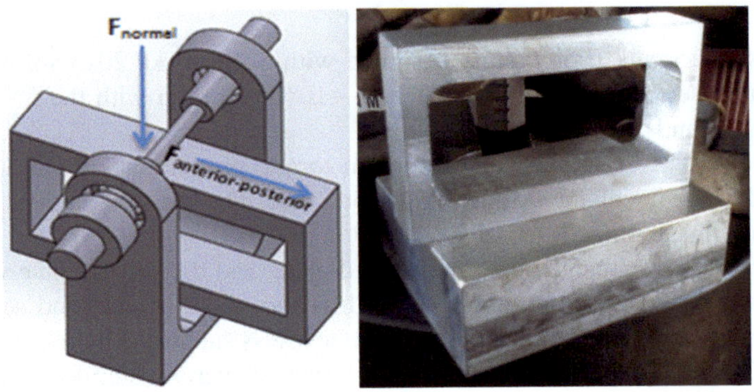

Fig. 4 The SGI pedal design

size of 2.54×10^{-2} m^2 placed on the crank. However, both ICs have sensitivity values that are contained in datasheet. The IDG 500 and ADXL 335 have sensitivity value of 2 mv/deg/s and 1 g = 330 mV (±10 %), respectively. The IMU was placed on crank of the bicycle symmetry to the pedal. The compact IMU circuit board was mounted on strip board and layered with non-conductive material to avoid short circuit. The IMU accelerometers and gyrometers were fused together by the Kalman filter algorithm to provide angle displacement estimation (Fig. 5).

The gyrometers measure the angular velocity that refers to time in second. The gyrometers that vary with time need to convert the analog–digital values to actual values (2).

$$\text{GyroRate} = \frac{(\text{GyroAdc} - \text{GyroZero})}{\text{Sensitivity}} \tag{2}$$

Fig. 5 Block diagram for IMU

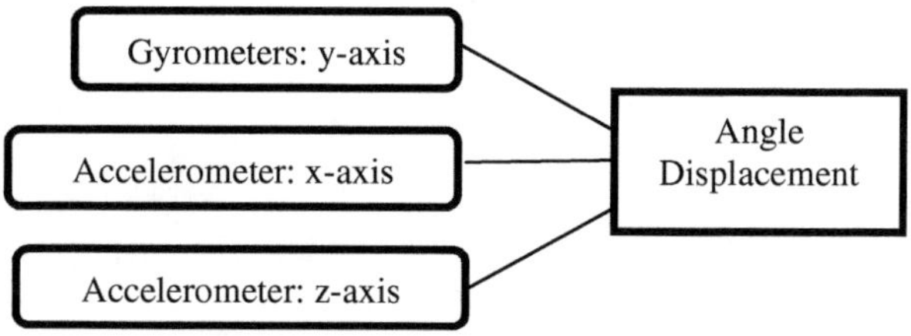

To measure the angle displacement, the angular rate needs to be sampled continuously over time (3).

$$\text{Angle} = \frac{\text{GyroRate}}{\Delta \text{time}} \tag{3}$$

The accelerometer measured the acceleration. The algorithm is based on 'C' language programming which then uploaded the program into the Arduino Mega 2,560 microcontroller. The data recorded instantaneous angle for full circle 360° and back to 0° when the crank completed the circle (Fig. 6).

The gyrometers in angular rate of x-axis and y-axis were needed for power output measurement. However, the angular velocity from gyrometers is affected by noise.

Fig. 7 shows the markers placement setup to calculate the kinematic parameters. The parameters measured angle displacement and angular velocity. Marker placement at the body of the bicycle is perpendicular (90°) with the second marker at the center crank shaft. The relative angle is measured by placing the third marker at the crank as same position with the IMU sensor placement. The Qualisys Oqus 100 series camera in the biomechanics laboratory of UniMAP was used to capture the motion.

3.1 Graphical User Interface

The first column in the Fig. 8 is the graph plotted the Kalman filtered angle displacement. Next, the second column is the graph plotted the angular velocity in x-axis, and the third column is the graph for angular velocity in y-axis. The graphs displayed were current angle for the crank movement.

3.2 Closed Loop Kinematic Model

In this study, the closed loop kinematic model is used for kinematic measurement as in the Fig. 9. A and B are the cyclist's thigh. B and C are the cyclist's shank. C and D are the cyclist's foot attached by clipless pedal at the crank. Lastly, O and D are the bicycle's crank.

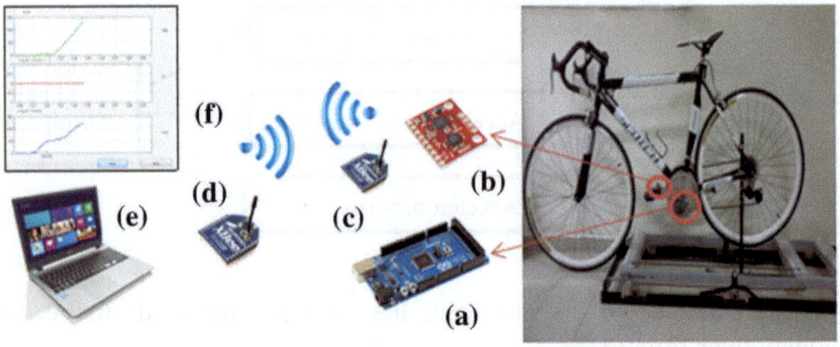

Fig. 6 The connection of transmitter module for **a** Arduino MEGA 2,560 battery powered and from **b** IMU to transmit data independently by Xbee shield **c** position on *top* of the Arduino **d** the connection of receiver module by USB powered cable setting point–point by X-CTU software and the data are transferred to **e** end user displayed in **f** matlab

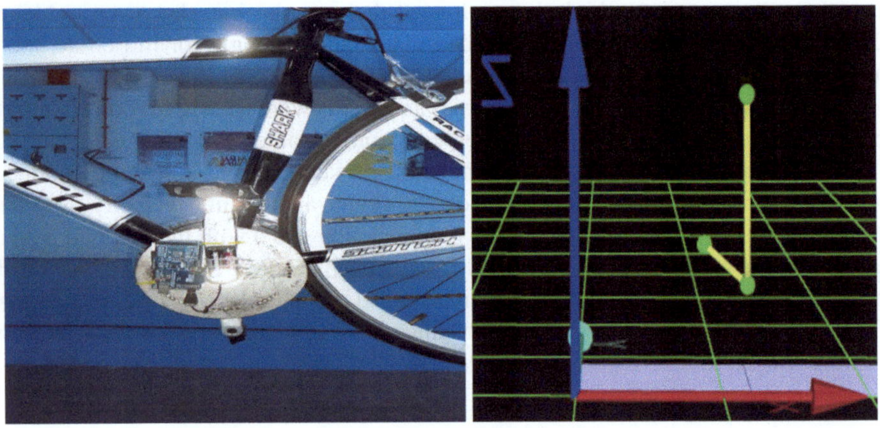

Fig. 7 Motion capture analysis camera setup for kinematics measurement

$$\theta(t) = \int \omega(t)\mathrm{d}t + \theta_{\mathrm{inertial}} \tag{4}$$

The equation that is used in this study is in the Eq. (4). $\theta(t)$ is the angle. $\omega(t)$ is angular velocity, and θ(inertial) is the point starting at 0° angle. The angle is calibrated using goniometer.

4 Benchmarking Measurement

The motion capture camera was used to convert the markers position into measurable kinematic parameters. To this scope, the angle displacement and angular velocity are measured. Five (Qualisys Oqus 100 series) motion capture cameras

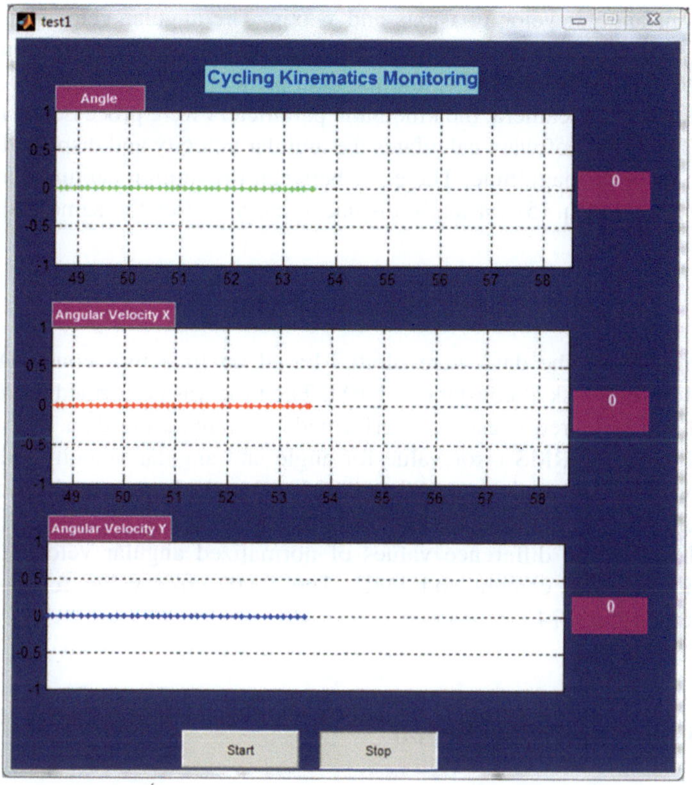

Fig. 8 Graphical user interface (*GUI*) using MATLAB software

Fig. 9 Closed loop kinematic chain

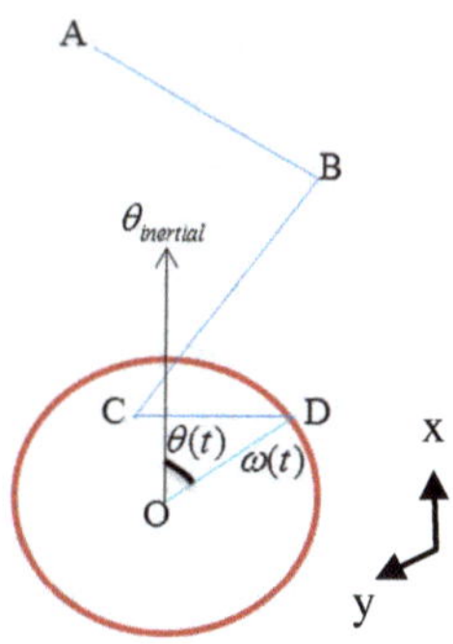

measured in 1 k/s (1,000 Hz) sampling rate. The data were processed using QTM software.

The IMU was placed on the crank symmetrically to the crank marker placement. The IMU converts the position into voltage and connected to analog read Arduino Mega 2,560 microcontroller. The data collected were Kalman filtered angle displacement angular velocity x-axis and y-axis. The angular velocity is divided into two components x-axis and y-axis. The angular velocities were normalized by the average angular velocity [14] and the data present in percentage (%).

In motion capture camera, the kinematic parameters were processed using QTM software. The QTM software calculates the angular velocity and displays the graph in angular velocity magnitude. The error between the motion capture camera and IMU is defined in Eq. (5) for angle displacement and uses the same equation for the angular velocity.

$$\sigma = \theta_{\text{camera}} - \theta_{\text{IMU}} \tag{5}$$

For comparison, the data were plotted based on time in seconds. A subject performed cycling task at constant 30 RPM. The task was performed in three trials and in 30 s. The paired t test was performed to evaluate the data. The result in Table 1 showed the RMS error value for angle and angular velocity values. The RMS error value for angle is $0.480 \pm 0.325°$. $P < 0.05$ was considered statistically different. Fig. 10 shows the plot for three trials averaged value.

Fig. 11 shows the difference values of normalized angular velocity between both motion capture camera and IMU. The RMS error value for normalized angular velocity is 0.743 (0.911) %.

5 Discussion and Conclusion

The RMS error value was $0.480 \pm 0.325°$. The significant different value for the angle was 0.146° which indicated motion capture camera, and IMU for angle was not significantly different. The error bar graph in Fig. 12 shows the compare means between motion capture camera and IMU.

Subsequently, the RMS error value normalized angular velocity was 0.743 ± 0.911 %. The significant different value for the normalized angular velocity was 0.217 % which indicated motion capture camera, and IMU for angular velocity was not significantly different. Nonetheless, the normalized error bar graph for angular velocity was plotted (Fig. 13). The angular velocity between motion capture camera and IMU was not significantly different.

The correlation coefficient for the angle was shown in the Fig. 10. The angle of motion capture camera is increased, while the angle of IMU is increased. Equivalent to angle, the normalized angular velocity correlation coefficient was shown in the Fig. 11. The normalized angular velocity of motion capture camera is increased, while the normalized angular velocity of IMU is increased.

However, there is some constraint in this study. The bicycle needs to mount at fix position in order to obtain the data. The axis was not varying over time. In this study, the axis was calibrated assuming that the bicycle was vertically to ground along the cycling task. If the constraint can be overcome, the IMU can be used for outdoor cycling such as in the track cycling.

Table 1 Comparison of IMU and motion capture camera (mean ± SD)

Parameter	Mean (SD)	Sig. (2-tailed)*
Angle	0.480 (0.325)	0.146
Angular velocity	0.743 (0.911)	0.217

* $P < 0.05$

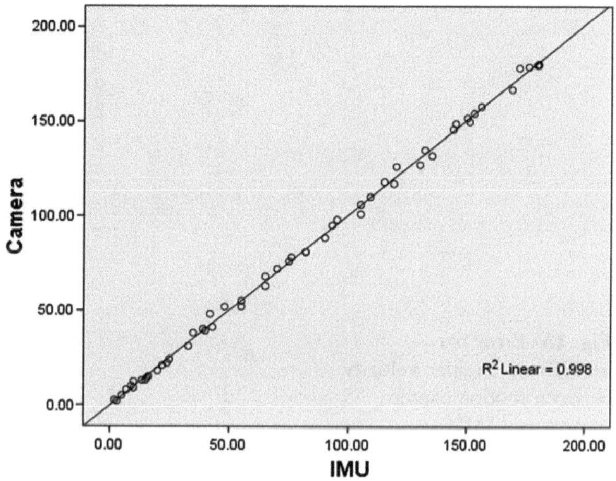

Fig. 10 Comparison of angle plot between motion capture camera and IMU

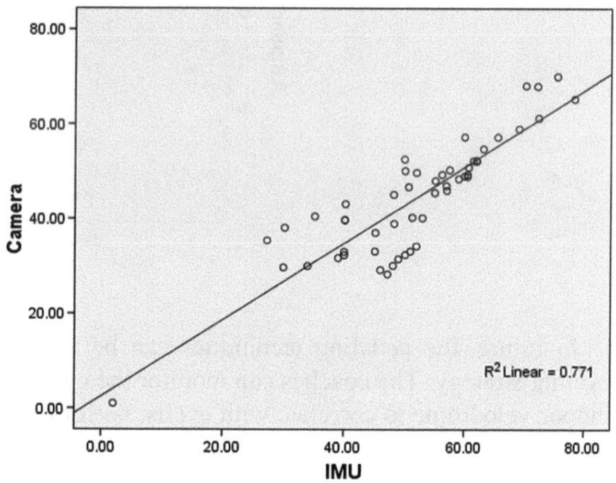

Fig. 11 Comparison of normalized angular velocity plot between motion capture camera and IMU

The adjustable crank is a device for monitoring the cycling power output at different crank angle wirelessly. The effect of adjusting crank angle back a forth to eliminate minimum torque is still unknown. In theory, the additional torque produced at the pushing phase may aid the cyclist to enhance the performance cyclist. The device has the flexibility to measure the cycling performance, biomechanical, and rehabilitation.

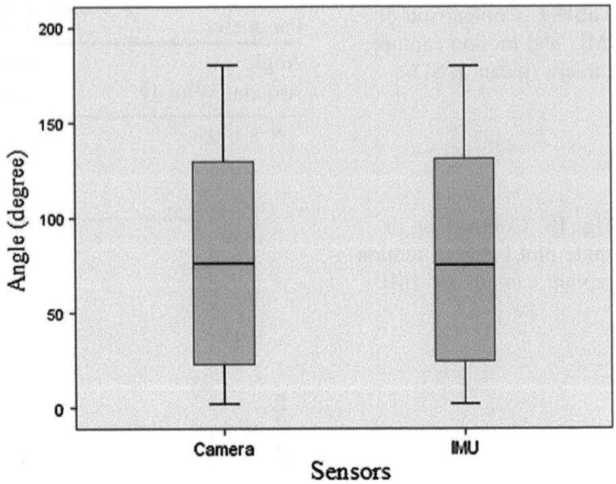

Fig. 12 Error bar angle between motion capture camera and IMU

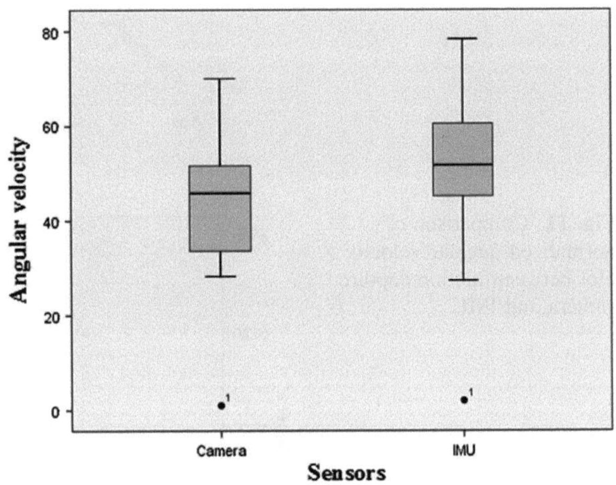

Fig. 13 Error bar normalized angular velocity between motion capture camera and IMU

In future, the pedaling techniques can be monitored by coaches to improve cycling strategy. The coaches can monitor the cyclist power output performance in indoor velodrome to correlate with cyclist performance. In rehabilitation, therapist may monitor pathology and non-pathology patients observing the force pattern or pedal coordination as the bicycle was the safest tool in patient rehabilitation.

Acknowledgments This research was performed in biomechanics laboratory and collaboration with Center of Excellence Advanced Sensor and Technology (CEASTech), University Malaysia Perlis.

References

1. Coyle, E., Feltner, M. E., Kautz, S., Hamilton, M., Montain, S., Baylor, A., et al. (1991). Physiological and biomechanical factors associated with elite endurance cycling performance. *Medicine and Science in Sports and Exercise, 23*, 93–107.
2. Ericson, M. O., Nisell, R., & Németh, G. (1988). Joint motions of the lower limb during ergometer cycling. *Journal of Orthopaedic and Sports Physical Therapy, 9*, 273–278.
3. Gregor, R. J., Broker, J. P., & Ryan, M. M. (1991). 4 The biomechanics of cycling. *Exercise and Sport Sciences Reviews, 19*, 127–170.
4. Rodríguez-Marroyo, J. A., García-López, J., Chamari, K., Córdova, A., Hue, O., & Villa, J. G. (2009). The rotor pedaling system improves anaerobic but not aerobic cycling performance in professional cyclists. *European Journal of Applied Physiology, 106*, 87–94.
5. Jobson, S. A., Hopker, J., Galbraith, A., Coleman, D. A., & Nevill, A. M. (2009). Effect of the rotor crank system on cycling performance. *Journal of Sports Science and Medicine, 8*, 463.
6. Lucia, A., Balmer, J., Davison, R., Perez, M., Santalla, A., & Smith, P. (2004). Effects of the rotor pedalling system on the performance of trained cyclists during incremental and constant-load cycle-ergometer tests. *International Journal of Sports Medicine, 25*, 479–485.
7. Jobson, S. A., Hopker, J. G., Korff, T., & Passfield, L. (2012). Gross efficiency and cycling performance: A brief review. *Journal of Science and Cycling, 1*, 3–8.
8. Bakhshi, S., Mahoor, M. H., & Davidson, B. S. (2011). Development of a body joint angle measurement system using IMU sensors. In *Engineering in Medicine and Biology Society, EMBC, 2011 Annual International Conference of the IEEE* (pp. 6923–6926).
9. Boonsawat, V., Ekchamanonta, J., Bumrungkhet, K., & Kittipiyakul, S. (2010). XBee wireless sensor networks for temperature monitoring. In *The Second Conference on Application Research and Development (ECTI-CARD 2010)*. Chon Buri, Thailand.
10. Faludi, R. (2010). Building wireless sensor networks: With ZigBee, XBee, Arduino, and processing. O'Reilly Media, Inc.
11. Chen, C., & Pomalaza-Ráez, C. (2009). Monitoring human movements at home using wearable wireless sensors. In *Proceedings of the Third International Symposium on Medical Information and Communication Technology*.
12. Bibbo, D. (2007). New techniques for the pedalling performance assessment in cycling.
13. Fonda, B., & Sarabon, N. (2010). Biomechanics of cycling. *Sport Science Review, 19*, 187–210.
14. Kitawaki, T., Tokuyasu, T., & Oka, H. (2011). *Evaluation of bicycle pedaling skills using a bicycle crank angular velocity irregularity index* (vol. 111, pp. 323–329).

The Initial Design of Learning Outcomes in the Sport Training Application

Noornasirah Nasri, Yulita Hanum P. Iskandar, Lester Gilbert, Gary B. Wills, Wan Asim Wan Adnan, Nordin Zakaria, Dayang Rohaya Awang Rambli and Helmi Md Rais

Abstract Teaching and learning activities should be designed and developed based on a pedagogical approach. These activities occur within a particular context and are designed to achieve intended learning outcomes through a series of tools and resources. This paper shows an initial design of learning outcomes that will be specifically designed for reusability to support automation and computer-assisted discovery for sport training application.

Keywords Learning outcomes · Motor skill · Sport training

1 Introduction

Skills lie at the heart of athletes' performances. Athletes develop their skill through the regular practice of training. Training involves continual practice of the motion and is typically composed of repetitions of movements [1].

N. Nasri (✉) · Y. H. P. Iskandar
Graduate School of Business, USM, Penang, Malaysia
e-mail: nasirahnasri@gmail.com

L. Gilbert · G. B. Wills
Electronics and Software Systems, University of Southampton, Southampton, UK

W. A. Wan Adnan
School of Health Sciences Health Campus, University Sains Malaysia, 16150 Kubang Kerian, Kelantan, Malaysia

N. Zakaria · D. R. Awang Rambli · H. Md Rais
Faculty of Science and Information Technology, University Teknology PETRONAS, 31750 Bandar Seri Iskandar, Perak, Malaysia

R. Adnan et al. (eds.), *Proceedings of the International Colloquium on Sports Science, Exercise, Engineering and Technology 2014 (ICoSSEET 2014)*,
DOI: 10.1007/978-981-287-107-7_11,

The coach assists the athletes to enhance their skill by determining the training objectives or intended outcomes during the period of instruction [2]. The coach determines the training materials and techniques or tactics to be used in the coaching activities to achieve particular learning outcomes. The procedures usually integrate conditions for performing the skill, providing training with feedback, and reflecting the athletes for a given type of learning outcome. Behaviorists recognized these examples as rules and practice with feedback.

Planned, coordinated, and progressive coaching are needed for the athlete to develop successfully toward the intended outcomes [3]. Systematic coaching activities that particularly focus on task analysis are derived from the behaviorist perspective. A behaviorist approach to learning provides simple and clear coaching activities. Task analysis involves a detail analysis of complex skills by detailing each muscle, nerve, and tendon involved in a given motion to generate an accurate technique and tactic analysis that is congruent with the learning outcomes [4]. This analysis generates precise and detailed instruction that allows the coach to conduct the coaching activities pertaining to the athletes' achievement on the intended learning outcomes and thus allows the athletes to effectively develop their skills and abilities [4].

Thus an effective planning in sport training allows the congruity between techniques and tactics to be taught (represented in learning outcomes), and supports the assessment of learning outcomes and the teaching and learning activities used to foster their achievement.

2 Learning Outcomes in Motor Skill Domain

Motor skills, although often not specifically targeted by educational objectives in higher education, are components of a distinct type of learning outcome. This motor skill which concerned with the area of muscle development and coordination is essential for teaching and learning of human performance. Cognitive domain typically involves with declarative, procedural, or conditional knowledge. Learning outcomes in the motor skill domain, however, involve precise, smooth, continuous, and accurately timed performances, characteristically associated with surgical training, pilot training, and sport training.

Figure 1 shows several taxonomies of learning outcomes in the motor skill domain [5–7]. Generally, these taxonomies describe a hierarchical model of skill development from simple observation to mastery level.

However, several limitations of current learning outcomes are as follows:

- cannot be applied without having certain, and sometimes quite expensive as well as systemic conditions in place [8],
- lack of adequate evaluation and instructional resources that could adapt to the multitude of student interests in creating context-based learning [9], and

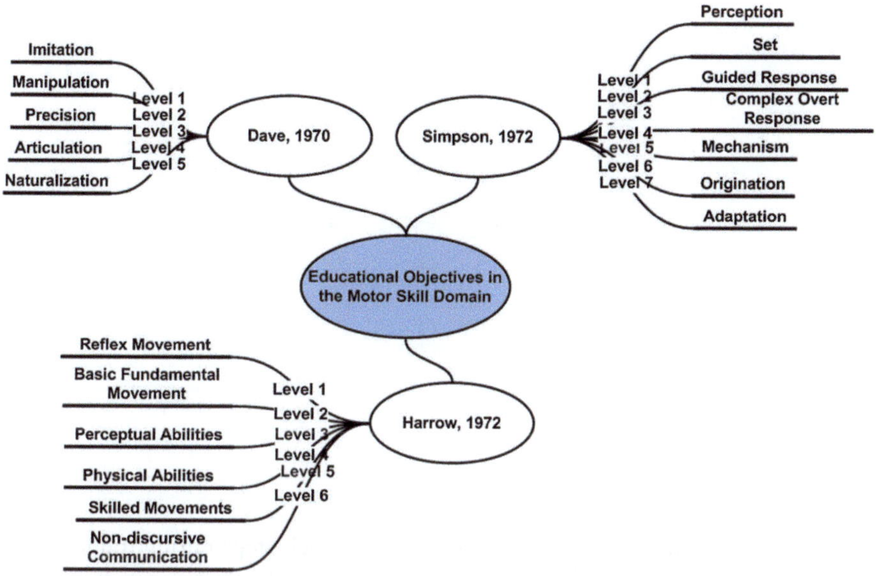

Fig. 1 Current learning outcomes

- additional research will be needed to assess the impact of individual readiness and the role of trust, which was not a review for enabling the learning experience [10].

3 Proposed Model of Learning Outcomes

A rigor literature leads to the findings of learning outcomes model that are relevant to this study. All the 6 identified learning outcomes are illustrated in Fig. 2, these categories namely sport pedagogy, teaching and learning, sport training and coaching, e-learning, computer-based learning, and sport competence.

Sport pedagogy is about the creation of athlete's learning outcomes, literacy, and enthusiasm. Teaching and learning on the other hand provides guidance, service, and substance in the progress as well as dissemination of the research-based teaching and learning resources. Meanwhile, sport training and coaching concerns improve technique, performance, and expertise in a particular area. While e-learning comprise all forms of electronically supported learning and teaching or pedagogy, sport competence refers to skills that athletes could understand and appreciate until the end of the program.

Considering their reliability and frequency, 19 variables that best-suit learning outcome components in the motor skill domain were selected. And these variables are as follows: Intended learning outcome is the evidence learner understand,

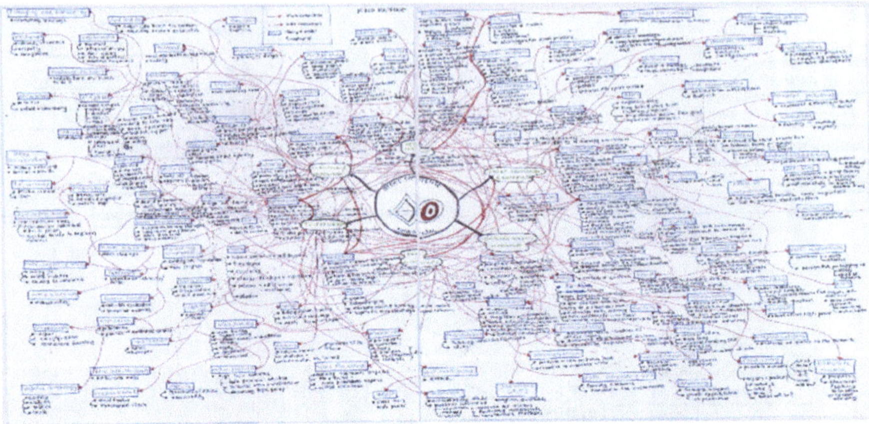

Fig. 2 Proposed model of learning outcomes

know, and be able to do until the end, situation of each steps during training, tool such as equipment or machine as well as a physical object that can be used to achieve a goal [11] proficiency level is that portion of the current situation, physical the body activity [12], completion time is the period of task performs, motivation refers to the desire to do something, feedback reaction to the performance [13] capability to do something well [8], confident is that feeling be able to perform the task [9], accuracy as an ability to assess the result to the true value, reliability ability of a system to achieve and maintain its functions, validity such as a condition to being acceptable [10], environment refers to conditions of the activity surrounding [14], the design of the plan or program, process to bringing about a result, skill to make something, interaction with the system or coaches [12], and role to delegate the working conditions [10]. Therefore, to ensure an internal consistency between items for each dimension remained, the reliability analysis was done. This is to ensure that the answers given by each respondent (individual) are the same.

4 Pilot Study

Adopting from literature, 212 questions were developed, featuring 19 independent variables. And a pilot study involving all level of expert archers was carried out using Cronbach Alpha to test reliability of the said instruments. The study undertook a pilot study with respondents consisted of all levels of expertise archers. The study completed the pilot study in order to achieve the reliability of the instrument. To assess the reliability, Cronbach's Alpha is used. Table 1 shows the overall results of this reliability test.

Table 1 Reliability test

Factor	Dimension	Items	Cronbach's alpha value	Status
1	IOL	18	0.735	Accepted
2	Situation	15	0.795	Accepted
3	Tool	15	0.925	Accepted
4	Proficiency level	15	0.832	Accepted
5	Completion time	15	0.865	Accepted
6	Interaction	15	0.903	Accepted
7	Motivation	15	0.940	Accepted
8	Feedback	15	0.927	Accepted
9	Capability	15	0.919	Accepted
10	Physical	8	0.915	Accepted
11	Confident	8	0.887	Accepted
12	Accuracy	6	0.576	Rejected
13	Reliability	7	0.469	Rejected
14	Validity	8	0.636	Rejected
15	Environment	6	0.418	Rejected
16	Design	8	0.687	Rejected
17	Process	8	0.700	Accepted
18	Subject matter	4	0.699	Rejected
19	Role	4	−0.078	Rejected

5 Results

Table 1 shows the Cronbach Alpha result of all items tested. And the overall value used is 0.985 (Pallant [15]). This result suggested that there were 7 dimensions to be rejected in this test as they showed values less than 0.7.

Corrected item—total correlation analysis performed as the suggested items were related to the study. This was necessary to refine the dimensions needed for the next experiment. The pivotal result of this test is summarized in Table 2. Result from the second test corroborates the importance of the said 7 items, and therefore, they accepted, not rejected for the study. Eighty-two dimensions were deleted from the first 212 entirety leaving only 130 dimensions that positively correlated in the study.

Table 2 shows that 7 dimensions are acceptable when corrected item total correlation was used. Eighty-two of the items were deleted to make 130 items of 212 items total. From it, we can observe how well the items in a group are indeed correct after unwanted item deleted. There are 7 dimensions used to measure aspects of accuracy, reliability, validity, environment, design, interaction, and role. From reliability testing inspection to ensure that each item was measuring the same characteristics that have done. Next, a scale to measure aspects accessibility have shown there is no internal consistency based on Cronbach's Alpha coefficient that was 0.576 (accuracy), 0.469 (reliability), 0.636 (validity), 0.418 (environment), 0.687 (design), 0.699 (subject matter), and −0.078 (role). Thus, the 197

Table 2 Cronbach alpha if item deleted

Factor	Dimension	Cronbach's alpha value	Item deleted	Status
1	IOL	0.863	8	Accepted
2	Situation	0.912	8	Accepted
3	Tool	0.925	4	Accepted
4	Proficiency level	0.832	8	Accepted
5	Completion time	0.865	6	Accepted
6	Interaction	0.903	4	Accepted
7	Motivation	0.940	1	Accepted
8	Feedback	0.927	3	Accepted
9	Capability	0.919	1	Accepted
10	Physical	0.915	3	Accepted
11	Confident	0.887	No item deleted	Accepted
12	Accuracy	0.862	3	Accepted
13	Reliability	0.857	5	Accepted
14	Validity	1.000	5	Accepted
15	Environment	0.889	4	Accepted
16	Design	1.000	6	Accepted
17	Process	0.818	5	Accepted
18	Subject matter	0.700	No item deleted	Accepted
19	Role	0.706	2	Accepted

items that have a coefficient Alpha Cronbach with values above 0.07 removed to provide a Cronbach Alpha coefficient for better. As a result, the Cronbach Alpha coefficient increased to 0.7 and above. Table 2 shows the value of alpha for each item in the dimensions.

While there were 12 dimensions used to measure aspects of ILO, situation, tool, proficiency level, completion time, interaction, feedback, capability, physical activity, confident and process from reliability testing inspection to ensure that each item was measuring the same characteristics that have done. Next, a scale to measure aspects accessibility found that there is a good internal consistency with Cronbach's Alpha coefficient that was 0.735 (ILO), 0.795 (situation), 0.925 (tool), 0.832 (proficiency level), 0.865 (completion time), 0.940 (interaction), 0.927 (feedback), 0.919 (capability), 0.915 (physical activity), 0.887 (confident), and 0.700 (process). Table 2 shows the value of Alpha for each item in the dimension.

6 Discussion

The study affirms that the selected 19 variables are useful to measure athletes' learning outcomes in sport competence model. Results attest as such since the items were proven to account for 77.8 % reliable as well as good fit to the data. This brought to our attention that the act of developing a learning outcome model

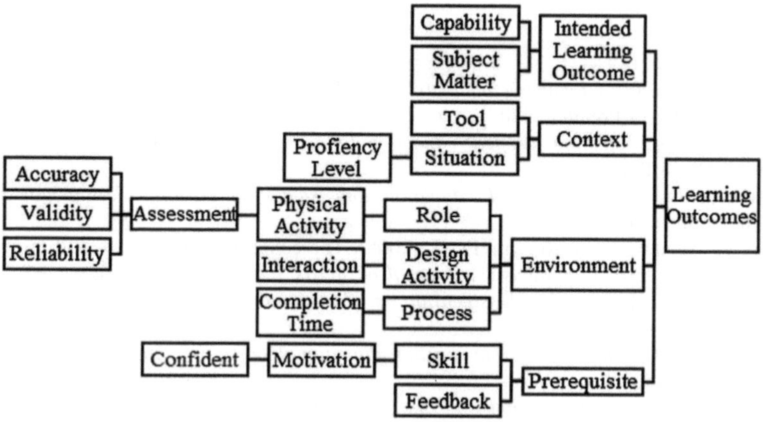

Fig. 3 Refine model of learning outcome

is indispensable. Having selected all the important categories from the results, a refine model of learning outcome in the motor skill domain is presented in Fig. 3.

It is highly recommended that the learning outcomes in the motor skill domain is used preliminarily to measure athlete's competence learning outcomes relating to motor skills. With overall score above 0.7, the study strongly views the following 19 competencies as significantly important to athletes concerning learning outcomes. Moreover, each model with a mean score above 0.7 could be regarded as being of considerably more influence by participants; 19 competencies accepted from the list. The list acceptable are as follows: (1) intended learning outcome, (2) situation, (3) tool, (4) proficiency level, (5) completion time, (6) motivation, (7) feedback, (8) capability, (9) physical, (10) confident, (11) accuracy, (12) reliability, (13) validity, (14) environment, (15) design, (16) process, (17) subject matter, (18) interaction, (19), and role.

7 Conclusion

The paper argues for the design of learning outcomes in the motor skill domain for sport training application. Future work will focus on expressing the learning outcomes as a series of UML models for the purposes of higher engineering education and training in sport applications. Thus, this learning outcome will achieve better reasoning and classification expression with regard to knowledge management and sharing.

Acknowledgments The authors would like to thank University Sains Malaysia and Ministry of Higher Education, Malaysia for funding this research under the "Fundamental Research Grant Scheme".

References

1. Gredler, M. E. (2001). *Learning and instruction: Theory into practice* (4th ed.). Upper Saddle River, NJ: Merrill Prentice Hall.
2. Rink, J. (1985). *Teaching physical education for learning*. St. Louis, MO: Times Mirror/ Mosbt College Publishing.
3. Siedentop, D. (1996). *Physical education and education reform*. Champaign: Human Kinetics Publisher.
4. Irwin, G., Hanton, S., & Kerwin, D. G. (2005). The conceptual process of skill progression development in artistic gymnastics. *Journal of Sports Sciences, 23*(10), 1089–1099.
5. Simpson, E. J. (1966). *The classification of educational objectives, psychomotor domain*. Urbana: University of Illinois.
6. Dave, R. (1970). *Developing and writing educational objectives*. In R. Armstrong (Ed.). Tucson: Educational Innovators Press.
7. Harrow, A. (1972). *A taxonomy of the psychomotor domain: a guide for developing behavioral objectives*. New York: McKay.
8. Wills, G., Gilbert, L., & Recio, A. (2012). *Towards a framework for games and simulations in STEM subject assessments*.
9. Van Bruggen, J., et al. (2004). Latent semantic analysis as a tool for learner positioning in learning networks for lifelong learning. *British Journal of Educational Technology, 35*(6), 729–738.
10. Huang, C.-H., et al. (2010). A sports e-learning platform: Teaching and learning by using multimedia contents. In *3rd IEEE International Conference on Ubi-media Computing (U-Media)*, IEEE.
11. Krathwohl, D. R. (2002). A revision of bloom's taxonomy: An overview. *Theory into Practice, 41*(4), 212–218.
12. Wang, C. (2001). Young people's motivational profiles in physical activity: A cluster analysis. *Journal of Sport and Exercise Psychology, 23*, 1–22.
13. Koohang, A., et al. (2009). E-learning and constructivism: From theory to application. *Interdisciplinary Journal of E-Learning and Learning Objects, 5*(1), 91–109.
14. Le Deist, F. D., & Winterton, J. (2005). What is competence? *Human Resource Development International, 8*(1), 27–46.
15. Pallant, J. (2010). *SPSS survival manual: A step by step guide to data analysis using SPSS*. Open University Press.

Part II
Exercise Science and Applied Performance

Acute Effects of Using Ricebag on Hip Range of Motion Among Backache Patients

Sylvia Augustine, Rahmat Adnan, Norasrudin Sulaiman, Shariman Ismadi Ismail and Ridzuan Azmi

Abstract Chronic back pain is a common injury in almost all age groups. Thermal therapy has been found to give beneficial effects for chronic injury. Since ricebags were found to hold heat better, it could be a good therapy modality for treating chronic back pain injury. The objectives of this study were to (1) determine the differences in temperature between the ricebags and the skin interface and (2) determine the effects of ricebags on the trunk and hip range of motion. To achieve the objectives of this study, two statistical analysis procedures were used. The analysis was conducted with the repeated-measure ANOVA and also the independent t test. A total of 40 subjects from Tuanku Mizan Military Hospital were recruited using the purposing sampling technique. Their mean age was 34.65 (±9.622) years with a mean weight of 72.23 (±11.44) kg. The results showed a significant difference in temperature between the ricebag and the hydrocullator ($p < 0.05$). However, there was no significant difference in temperature of the skin interface between the ricebag and the hydrocullator ($p = 0.732$). On the other hand, although there was improvement in trunk and hip range of motions, there was no significant difference found in trunk and hip range of motions for both groups. Hence, both the ricebag and hydrocullator gave the same effect on skin interface and on trunk and hip range of motions in the patients. Thus, the ricebag can be used as an alternative in chronic back pain management.

Keywords Ricebag · Hydrocullator · Backache · Heat treatment · Range of motion

---

S. Augustine · R. Azmi
Rehabilitation Department, Tuanku Mizan Military Hospital, Kuala Lumpur, Malaysia

S. Augustine · R. Adnan (✉) · N. Sulaiman · S. I. Ismail
Sports Science Center of Studies, Universiti Teknologi MARA, Shah Alam, Malaysia
e-mail: rahmatadnan@salam.uitm.edu.my; alangr3@gmail.com

R. Adnan · N. Sulaiman · S. I. Ismail
Faculty of Sports Science and Recreation, Universiti Teknologi MARA, Shah Alam, Malaysia

R. Adnan et al. (eds.), *Proceedings of the International Colloquium on Sports Science, Exercise, Engineering and Technology 2014 (ICoSSEET 2014)*,
DOI: 10.1007/978-981-287-107-7_12,

1 Introduction

Participation in physical activity, exercise, and sports increases the likelihood of injury, especially soft tissue injury. Pain originating from muscle, ligament, or tendon is defined as soft tissue pain [1]. One common soft tissue injury is back pain. In general populations, low back pain (LBP) will cause significant disability and loss of time from work and is one of the most common musculoskeletal ailments [2]. LBP constitutes a large proportion of public health problems as shown by national- and community-based surveys. General populations of both industrial and non-industrialized societies can suffer from LBP [2].

In managing soft tissue injury, heat and cold treatments using gel packs are commonly used [3]. Application of hot packs can cause stimulation of the thermoreceptors in the skin overlying the sore muscle and is thought to attenuate the perception of muscle pain [4]. Both hot and cold treatments have been shown to reduce pain in muscles. Moreover, according to Lin [5], the range of motions will increase as a result of the superficial heat which will raise local temperature of heated tissue and soften the surrounding tissue.

Applications of heat reduce pain and increase the joint mobility by reducing tissue viscosity, and increasing connective tissue extensibility [5]. Blood flow and the extensibility of connective tissues will increase, while reflexive muscle excitability and pain will decrease as a result of application of heat [6]. Hanson and Day [6] also added "Heat application is known to increase tissue extensibility, which should increase joint ROM without administration of a stretching protocol." Studies related to hot and cold treatments for other purposes was also carried out in 2003 and 2004, where the hot gel packs and reduction of delayed onset muscle soreness and application of hot and cold water immersions were used to enhance recovery. This study states that pain will be reduced with the application of hot treatment [4]. This shows the same mechanisms of heat applied to the muscles for reducing pain. Trunk and hip movements are most affected in backache patient [7, 8].

People with LBP will tend to compensate their movement in order to reduce pain [7]. The range of motions for the trunk is limited due to the protective mechanism of the spinal chord [9].

A recent study has proven that the ricebag has similar characteristics as ice and hot packs that can help in management of soft tissue injury [3]. Ricebags can hold both temperature at a better capacity and use the rib fabric to increase the effectiveness of the ricebag during certain periods of time [3]. The rib fabric is selected as the therapy tools for soft tissue injury as it can keep heat long enough for at least 20 min consistently. Acute effects that will be measured are the focal affect of the temperature and the increase in the range of motions of the trunk and hip. This study was designed to investigate the acute effects of using ricebags among backache patients.

2 Methodology

2.1 Sample

A total of 40 male patients who are diagnosed by the doctor with chronic back pain were recruited with a mean age of 34.65 (±9.622) years and weight of 72.23 (±11.44) kg using a purposive sampling technique.

2.2 Research Design

This study used a quasi-experimental method due to the nature of the participants or samples, who were chronic soft tissue injury patients. There were two parameters that were measured, including the difference in temperature between the ricebag and skin interface, and the effect of the heat transfer to the range of motions of trunk and hip. The differences of temperature were analyzed using repeated-measure ANOVA, while the effect of the ricebag was determined using independent *t* test (Fig. 1).

2.3 Tools

Ricebags were used as the experimental product as the purpose of the study was to evaluate the effectiveness of ricebag, while heat packs from the hydrocullator® unit were used in the control group. The temperature was measured using a digital thermometer (Hanna Instrument P/N: HI935005 and thermocouples probe which is Hanna *K*-Type Waterproof Thermocouples). A measuring tape was used to measure the range of motions of the trunk, while a goniometer was used to measure range of motions of the hip.

2.4 Data Collection Procedure

Subjects gave informed consent, before the measurements of range of motions for trunk and hip were determined. After the ricebag was heated, it was placed at the affected area which was at the lower back. The tip of digital thermometer, or the sensor, was placed at the ricebag or heat pack and the skin interface covered by the ricebag or heat pack. Both temperatures of the ricebag or heat pack and skin interface were measured and recorded at 1-min intervals from minute 0 to minute 20. The range of motions of trunk and hip was measured again after the treatment (Fig. 2).

Fig. 1 Research design

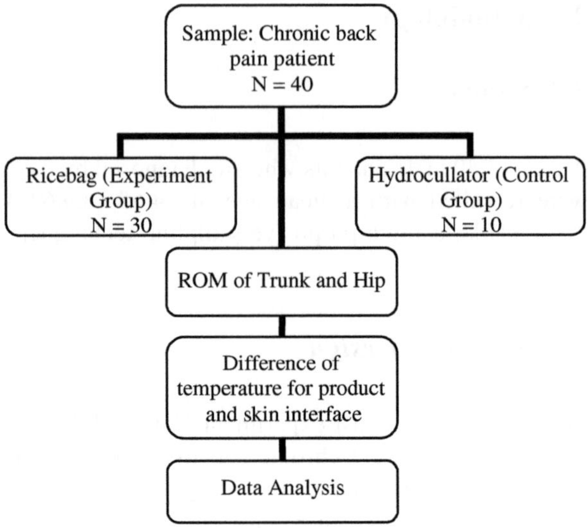

2.5 Data Analysis

The independent sample t test was used to determine the changes in the range of motions of the trunk and hip between different treatment groups. The Statistical Package for Social Sciences Software (SPSS) software (for Windows software 20.0) was used to analyze the parameters in this study. The two variables that were measured in this study were the differences in the temperature of the ricebag and the skin and the changes in the range of motions of the trunk and hip.

3 Result

3.1 Demographics Profile of the Subjects

See Table 1.

3.2 Demographic Data of Product and Skin

There was a significant difference in the temperature of the ricebag or hydrocullator and the skin interface within groups (Table 2). However, there was no significant effect of temperature on skin interface between groups.

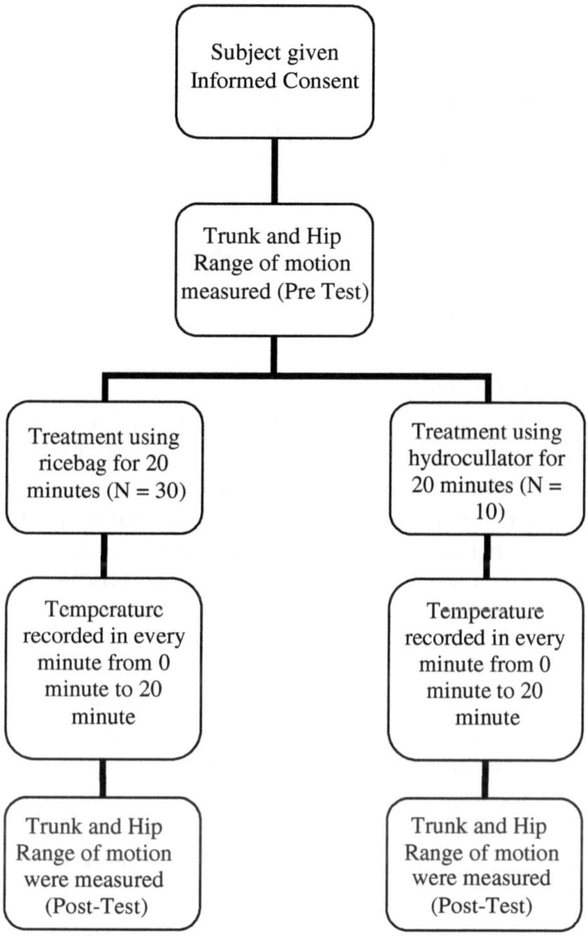

Fig. 2 Data collection procedure

Table 1 Mean of demographic profile of subjects

	N	Mean	Std. deviation
Age (years)	40	34.65	9.622
Weight (kg)	40	72.23	11.44

Table 2 Comparison of temperature between Groups

Measure	Df	Sig
Product	4.048	0.000
Skin	3.402	0.732

Product represents ricebag and heat pack

3.3 Temperature

The descriptive statistics on the means of temperatures of ricebag and hydroculator are presented in (Fig. 3), while the descriptive statistics on the means of skin interface temperatures are presented in Fig. 4.

3.4 Effect of Ricebags on Range of Motions

3.4.1 Trunk

(a) *Flexion*

The effects of the treatments (ricebag or hydrocullator) on the range of motions are presented in Table 3. There was a significant difference in the range of motions of trunk flexion between the pre- and post-treatments in both groups. However, it revealed no statistically significant differences in the range of motions of trunk flexion between the experiment group (54.8 ± 3.85°) and the control group (53.85 ± 4.85°). Overall, both groups showed improvement in the range of motions of trunk flexion.

(b) *Extension*

The effects of the treatments (ricebag or heat pack) on the range of motions are presented in Table 3. There was a significant difference in the range of motions of trunk extension between the pre- and post-treatments for both groups. However, it did not reveal any significant differences in the range of motions of trunk flexion between the experiment group (46.32 ± 3.49°) and the control group (53.85 ± 4.80°). Overall, both groups showed improvement in the range of motions of trunk extension.

(c) *Rotation*

The effects of the treatments (ricebag or heat pack) on the range of motions are presented in Table 3. There was a significant difference in the range of motions of trunk rotation between pre- and post-treatments in both groups. However, it did not reveal any significant differences in the range of motions of trunk rotation between the experiment group (62.52(L), 64.57(R) ± 6.31(L), 5.42(R)°) and the control group (61.35(L), 61.90(R) ± 5.83(L), 5.74(R)°). Overall, both groups showed improvement in their range of motions of trunk rotation.

(d) *Lateral Flexion*

The effects of the treatments (ricebag or heat pack) on the range of motions are presented in Table 3. There was a significant difference in the range of motions on trunk lateral flexion between the pre- and post-treatments in both groups. However, it did not reveal any significant differences on the range of motions of trunk lateral flexion between the experiment group (47.52(L), 47.25(R) ± 6.31(L),

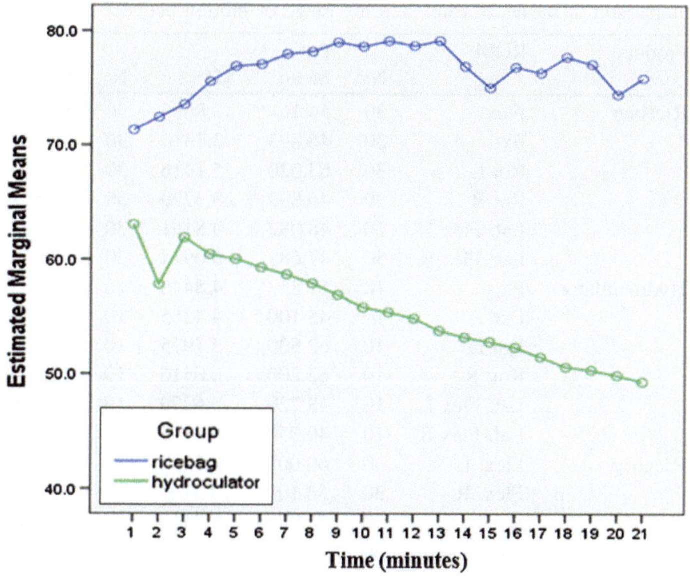

Fig. 3 Descriptive mean of ricebag and hydrocullator temperature

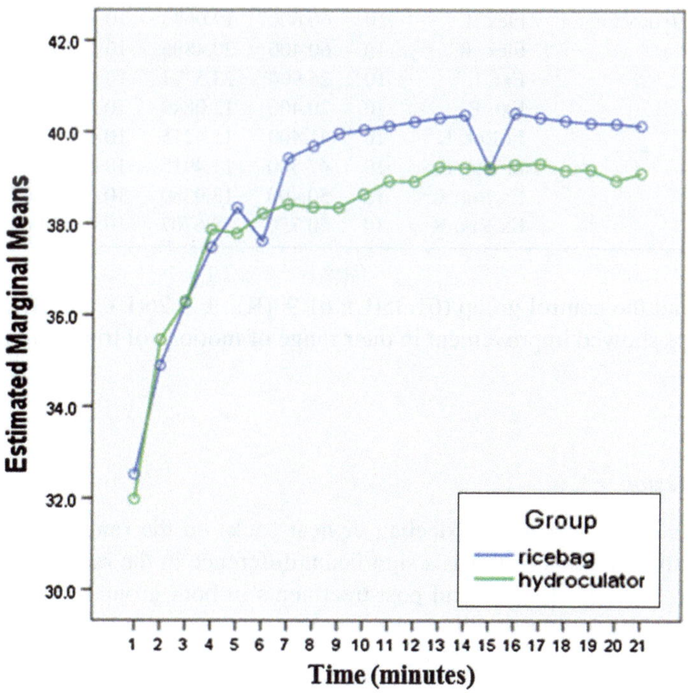

Fig. 4 Descriptive means of skin interface temperature

Table 3 Comparison of means of trunk and hip range of motion between two groups

Region	Product	ROM	Pre			Post		
			N	Mean	SD	N	Mean	SD
Trunk	Ricebag	Flex	30	54.16	3.8695	30	54.30	3.8542
		Ext.	30	46.893	3.1316	30	46.323	3.4916
		Rot L	30	63.020	5.1416	30	62.517	6.3088
		Rot R	30	46.893	5.3729	30	62.517	5.4179
		Lat. Flex L	30	48.082	4.8191	30	47.517	5.2596
		Lat. Flex R	30	47.683	5.0914	30	47.250	5.5966
	Hydrocullator	Flex.	10	51.85	4.8448	10	53.85	4.8480
		Ext.	10	45.100	4.4585	10	53.85	4.7912
		Rot L	10	62.500	5.7975	10	61.350	5.8312
		Rot. R	10	62.200	6.0516	10	61.900	5.7436
		Lat. Flex L	10	48.720	5.6279	10	47.450	5.2146
		Lat. Flex R	10	49.050	5.5199	10	47.400	5.5966
Hip	Ricebag	Flex. L	30	60.007	20.3944	30	64.867	19.7077
		Flex. R	30	58.100	19.3165	30	64.567	18.1729
		Ext L	30	20.983	9.3463	30	22.367	7.9545
		Ext. R	30	19.533	10.5919	30	19.533	9.4458
		In Rot. L	30	44.233	10.9974	30	44.817	10.4753
		In Rot. R	30	44.067	9.6951	30	47.650	8.8602
		Ex Rot. L	30	52.033	14.9008	30	53.950	13.2186
		Ex Rot. R	30	54.550	12.2646	30	56.400	9.8282
	Hydrocullator	Flex. L	10	60.600	19.0625	10	68.800	17.4789
		Flex. R	10	60.400	20.4896	10	59.600	25.1537
		Ext. L	10	26.900	23.5771	10	23.000	13.7840
		Ext. R	10	20.400	12.0849	10	22.200	13.2564
		In Rot. L	10	41.400	15.4215	10	45.800	12.5503
		In Rot. R	10	47.500	11.4915	10	51.100	8.7870
		Ex Rot. L	10	50.400	13.9380	10	54.800	9.4375
		Ex Rot. R	10	50.750	8.6707	10	54.000	6.2183

5.42(R)) and the control group (61.35(L), 61.90(R), ± 5.26(L), 5.21(R)). Overall, both groups showed improvement in their range of motions of trunk lateral flexion.

3.4.2 Hip

(a) *Flexion*

The effects of the treatments (ricebag or heat pack) on the range of motions are presented in Table 3. There was a significant difference in the range of motions of hip flexion between the pre- and post-treatments in both groups. However, it did not reveal any significant differences in the range of motions of hip flexion between the experiment group (64.87(L), 64.57(R) ± 19.71(L), 18.17(R)°) and the control group (68.80(L), 59.60(R), ± 17.48(L), 25.15(R)°). Overall, both groups showed improvements in the range of motions of hip flexion.

(b) *Extension*

The effects of the treatments (ricebag or heat pack) on the range of motions are presented in Table 3. There was a significant difference in the range of motions of hip extension between the pre- and post-treatments in both groups. However, it did not reveal any significant differences in the range of motions of hip extension between the experiment group (22.37(L), 19.53(R) ± 7.95(L), 9.45(R)°) and the control group (22.00(L), 23.20(R), ± 13.78(L), 13.26(R)°). Overall, both groups showed improvements in their range of motions of hip extension.

(c) *Internal Rotation*

The effects of treatments (ricebag or heat pack) on the range of motions are presented in Table 3. There was a significant difference in the range of motions of hip internal rotation between the pre- and post-treatments in both groups. However, it did not reveal any significant differences in the range of motions of hip internal rotation between the experiment group (44.82(L), 47.65(R) ± 10.48(L), 8.87(R)°) and the control group (45.80(L), 51.10(R), ±12.5(L), 8.79(R)°). Overall, both groups showed improvements in their range of motions of hip internal rotation.

(d) *External Rotation*

The effects of the treatments (ricebag or heat pack) on the range of motions are presented in Table 3. There was a significant difference in the range of motions of hip external rotation between the pre- and post-treatments in both groups. However, it did not reveal any significant differences in the range of motions of hip external rotation between the experiment group (M = 53.95(L), 56.40(R) SD = 13.21(L), 9.83(R) and the control group (M = 54.80(L), 54(R), SD = 9.44(L), 6.22(R)). Overall, both groups showed improvements in their range of motions of hip external rotation (Table 4).

4 Discussion

Heat treatment does help in the rehabilitation of chronic back pain. The present study found that the mean temperature was higher with the ricebag compared to the hydrocullator. This proved that the ricebag had similar characteristics to the regular cool pack and/or gel pack. Furthermore, it had the ability of retaining heat, which resulted in effective enhancement in the management of soft tissue injury during rehabilitation [3]. However, there was no significant difference in the skin temperature for both groups.

The hip movement is always affected in LBP patients [7]. In fact, the trunk also experiences an uncomfortable feeling such as stiffness due to the protective mechanisms to the injury [9]. This trunk stiffness affects the range of motions of the trunk. Thus, the heat treatment helps in reducing the stiffness, whereas raising the local temperature of the muscle can soften and enhance the range of motions of

Table 4 Comparison of trunk and hip range of motion

Range of Motion		Pre		Post	
		df	sig	df	sig
Trunk	Flexion	38	0.133	38	0.518
	Extension	38	0.168	38	0.157
	Rotation L	38	0.79	38	0.609
	Rotation R	38	0.579	38	0.760
	Lateral Flexion L	38	0.734	38	0.972
	Lateral Flexion R	38	0.476	38	0.938
Hip	Flexion L	38	0.936	38	0.578
	Flexion R	38	0.750	38	0.502
	Extension L	38	0.257	38	0.858
	Extension R	38	0.830	38	0.490
	In Rotation L	38	0.528	38	0.808
	In Rotation R	38	0.360	38	0.292
	Ex Rotation L	38	0.762	38	0.852
	Ex Rotation R	38	0.372	38	0.475

the joint. The presence of heat caused the joint to increase its mobility by reducing tissue viscosity, increasing connective tissue extensibility, and reducing pain [5]. The range of motions for both trunk and hip showed significant differences in improvement of the range of motions in the pre- and post-treatments, but there were no significant differences in change of range of motions in trunk and hip for both groups.

5 Conclusion

In conclusion, this study found that there were significant differences in the temperature of ricebags or heat packs and the skin interface within groups. However, this study found that there was no significant effect of temperature of skin between groups and no changes in the range of motions of trunk and hip for both groups.

Acknowledgments Researcher would like to express upmost gratitude to RAGS-RMI Grant for sponsoring this study. We would like to thank and to express my greatest gratitude to God, for giving us the strength and knowledge to accomplish this study. We would also like to thank all the staff and patients of Tuanku Mizan Military Hospital for their full commitment and support while the study was conducted.

References

1. Martin, D. J., & Palmer, S. (2005). Soft tissue pain and physical therapy. *Anaesthesia & Intensive Care Medicine, 6*(1), 23–25.
2. Pensri, P., et al. (2005). Physiotherapy management of low back pain in Thailand: A study of practice. *Physiotherapy Research International, 10*(4), 201–212.
3. Rahmat, A., et al. (2011). *The development of beany/grainy pack to treat soft tissue injury*. Penang.
4. Sumida, K. D., Greenberg, M. B., & Hill, J. M. (2003). Hot gel packs and reduction of delayed-onset muscle soreness 30 minutes after treatment. *Journal of Sport Rehabilitation, 12*(3), 221–228.
5. Lin, Y. H. (2003). Effects of thermal therapy in improving the passive range of knee motion: comparison of cold and superficial heat applications. *Clinical Rehabilitation, 17*(6), 618.
6. Hanson, M., & Day, J. (2012). Effects of Different Heating Modalities on Hip Flexion Passive Range of Motion. *International Journal of Athletic Therapy & Training, 17*(6), 27–30.
7. Van Dillen, L. R., et al. (2008). Hip rotation range of motion in people with and without low back pain who participate in rotation-related sports. *Physical Therapy in Sport, 9*(2), 72–81.
8. Dickey, J. P., et al. (2002). Relationship between pain and vertebral motion in chronic low-back pain subjects. *Clinical Biomechanics, 17*(5), 345–352.
9. van den Hoorn, W., et al. (2012). Mechanical coupling between transverse plane pelvis and thorax rotations during gait is higher in people with low back pain. *Journal of Biomechanics, 45*(2), 342–347.

Effects of Eccentric Training Using Theraband on Hamstring Flexibility in Elderly

Nur-Hasanah Ruslan, Wan Mohd Norsyam Wan Norman, Ayu Suzailiana Muhamad and Nursyaidatul Hafiza Madzlan

Abstract Reduce flexibility can put older adults at risk of pain and also can limit their movement. It is well established that resistance exercise improves flexibility. Thus, the aim of this study was to evaluate the effects of eccentric training using theraband on hamstring flexibility in elderly. This experimental study uses comparative analysis of pre- and post-training. A total of 29 female older adults (age: 78 ± 8 years; height: 1.57 ± 0.3 m) from the Nursing Care Centre Old folks Welfare Kuantan, Grannie Old folks Welfare and two branches from Grannie Old folks Centre were participated in this study. They were divided into experimental and control group. Experimental group underwent eccentric exercise program using yellow theraband two times per week for 6 weeks. Hamstring flexibility was measured using goniometer with 90/90 hamstring test. The experimental group shows the significant difference on left and right hamstring flexibility after the training regimen ($P < 0.05$). This study shows that eccentric training using theraband improves hamstring flexibility.

Keywords Eccentric training · Theraband · Older adults · Hamstring, flexibility · Range of motion

1 Introduction

Flexibility is the ability of a muscle to lengthen and allow one joint to move through a range of motion (ROM) [1]. Adequate flexibility is important to maintain balance, agility [2, 3], and musculoskeletal function [2]. However,

N.-H. Ruslan · W. M. N. Wan Norman · N. H. Madzlan
Faculty of Sport Science and Recreation, Universiti Teknologi MARA, Shah Alam, Pahang, Malaysia

A. S. Muhamad (✉)
Sports Science Unit, University Sains Malaysia, Kubang Kerian, Kelantan, Malaysia
e-mail: ayu_suzailiana@usm.my

R. Adnan et al. (eds.), *Proceedings of the International Colloquium on Sports Science, Exercise, Engineering and Technology 2014 (ICoSSEET 2014)*,
DOI: 10.1007/978-981-287-107-7_13, © Springer Science+Business Media Singapore 2014

age-induced loss of flexibility gives implication to daily life activities in elderly. Lacks of flexibility in all body joints will affect mobility and body functions. Previous studies have shown that people become more susceptible to injury when the joint flexibility is extremely low, high or when there is a significant imbalance between dominant and non-dominants sides of the body [4, 5].

It was reported that flexibility training is a vital component in reducing injury and also releasing pain [6]. American College of Sports Medicine (ACSM) and American Heart Association (AHA) also suggest that flexibility training is essential to maintain the ROM which is required for daily physical activities [7]. Most injuries occur in the eccentric phase of activity [8], where hamstring muscles mostly injured while working eccentrically during landing or decelerating. Hence, eccentric training could reduce injury rates, increase strength, prevent muscle damage, and improve flexibility [6, 9]. By definition, eccentric training is a contraction force that results in the lengthening of the muscle [5]. Thus, this study chose eccentric training as an intervention for elderly to increase their flexibility. Several types of eccentric training include eccentric push-up, eccentric chin-up, eccentric rollout, and eccentric slide board leg curl [10]. However, these eccentric training is not suitable and might be dangerous for elderly. Thus, eccentric training using theraband is the best choice. This is because using theraband, ones do not need to use their body weight to perform the exercise and only require a little movement.

Hamstring is the muscle at the posterior part of the thigh that is responsible for knee and hip joints movement. Another importance of hamstring muscle is that it can affect stride length, walking speed, and dynamics balance [11]. Low back pain, body posture problems, poor locomotion, risk of falling, and susceptibility to musculoskeletal injuries are related to lack of hamstring flexibility [11]. Thus, hamstring stiffness and short hamstring muscle–tendon unit is a potential risk to low back pain and jumper's knee (patellar tendinopathy). This shows that improving flexibility through exercise is very important to elderly.

Yet, despite numerous studies on the benefits of eccentric training, studies investigating the effects of eccentric training on elderly are scarce in the literature view. Most studies focus on athletes and adults population. Hence, the purpose of this study is to look into the effects of eccentric training on hamstring flexibility in elderly. The hypothesis of this study is that eccentric training will increase the passive range of motion of the knee joint and thus reduce injury rates and improve flexibility.

2 Methods

2.1 Research Design

This study uses a pre- and post-intervention study design. Figure 1 shows the flowchart of the study.

Fig. 1 Flowchart of the study

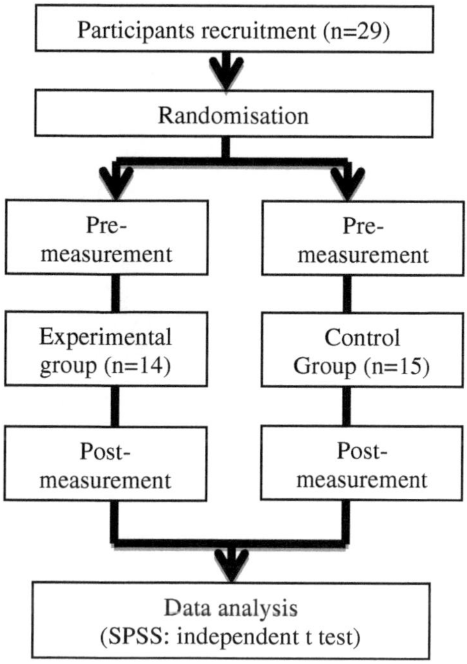

2.2 Participants

A total of 29 female older adults aged above 60 years old were included in the study. They were recruited from the Nursing Care Centre Old folks Welfare Kuantan, Grannie Old folk's welfare, and two branches from Grannie Old folks Centre. Another inclusion criterion is inability to achieve at least 20° of hamstring flexibility. On the other hand, the exclusion criterion includes any hip or knee replacement and history of pathology in the lower back, hip or knees.

2.3 Instruments

To perform eccentric training, leg press was done using the yellow theraband (TheraBand® manufactured by The Hygenic Corp. 1,245 Home Ave, Akron OH44310). Goniometer (Media Lifesport(M) Sdn Bhd, Malaysia) was used to measure ROM of the hamstring muscle [6, 7].

2.4 Training Protocol

Participants were randomly assigned to two groups: experimental group ($n = 14$) and control group ($n = 15$). In the experimental group, participants performed eccentric training by doing leg press using a yellow theraband. To implement leg

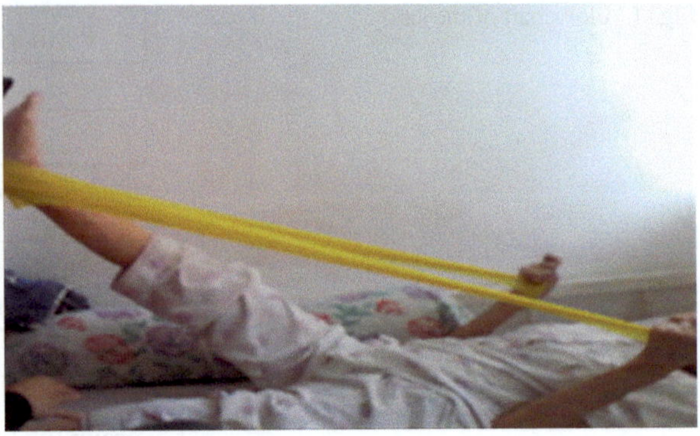

Fig. 2 Eccentric training

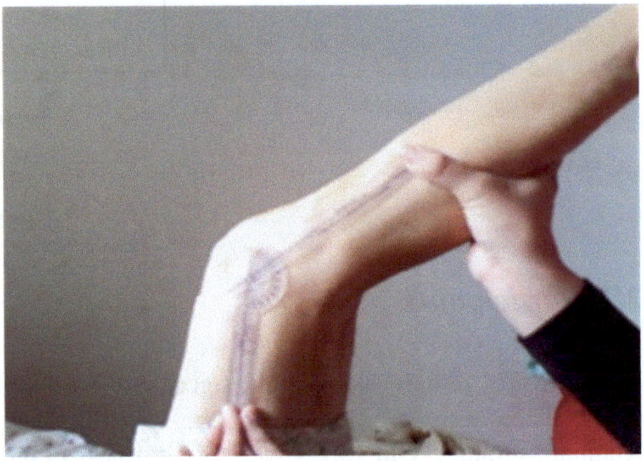

Fig. 3 Flexibility testing

press, participants were positioned supine with the hip and knee flexed to 90°. The yellow theraband was wrapped around the heel, and the ends of the theraband were held by hands (Fig. 2). This leg press was done on right and left hamstring two times per week for 6 weeks. The exercise program includes three repetitions with 15 s for each repetition [2]. On the other hand, control group not performed any exercises during the study period.

2.5 Data Collection Procedures

Goniometer was used to measure 90/90 hamstring flexibility before (pre-measurement) and after (post-measurements) the 6 weeks of the exercise program in both groups (Fig. 3).

2.6 Data Analysis

Data were analyzed using the Statistical Package for Social Sciences (version 20.0 for Windows, SPSS Inc., Chicago, IL, US). Descriptive analysis was used to describe the data. Paired t test was used to compare means between pretest and posttest. Independent t test was used to compare means between control and experimental group. All the data were expressed as mean ± standard deviation (SD). The accepted level of significance was $P < 0.05$.

3 Results

The mean physical characteristics of the participants are shown in Table 1.

Table 2 shows the results of independent t test, which was calculated to assess any significant differences existed in right and left hamstrings between groups. It was found that there was no significant difference between control and experimental groups pre- and posttest in the right and left hamstrings flexibility ($P > 0.05$). Tables 3 and 4 presented the flexibility comparisons within group for both right and left hamstrings, respectively. Pre- and posttest score for experimental group showed statistically significant difference in right and left hamstring flexibility. However, the control group showed no significant difference between pre- and post-measurement in both right and left hamstrings flexibility.

4 Discussion

After 6 weeks of eccentric training with theraband, experimental group showed significant improvement on hamstrings flexibility ($p < 0.001$; Tables 3 and 4), while control group did not show any improvement. Surprisingly, when comparing between groups, there was no significant difference found on hamstrings flexibility ($p > 0.05$; Tables 3 and 4) this nonsignificant finding between groups might be attributed to lower hamstrings flexibility (degrees) recorded in control group compared to experimental group at the start of this study (pretest) (Tables 3 and 4), although the difference is not significant. Thus, any improvement in experimental group is likely to be hindered by this lower pretest value in control group. It was speculated that this lower pretest value in control group is because of difference of mean age in both groups. As tabulated in Table 1, mean age in control group is lower than in experimental group. Although the difference is not much, it is likely to affect the study findings because it is well known that age-induced loss of flexibility. Hence, this is one of the limitations of this study that should be taking into consideration in the future.

Table 1 Physical characteristics of the participants

Group	Physical characteristics		
	Age (years)	Heights (m)	Weights (kg)
Experimental (n = 14)	81.14 ± 7.97	1.56 ± 0.34	52.14 ± 9.69
Control (n = 15)	76.53 ± 8.99	1.57 ± 0.44	54.80 ± 7.56

Table 2 Right and left hamstring flexibility comparisons between groups

Group	Hamstring Flexibility		
	Experimental (n = 14)	Control (n = 15)	Sig. (*P* value)
Pretest right hamstring flexibility (degree)	54.29 ± 11.07	49.33 ± 11.16	0.241
Posttest right hamstring flexibility (degree)	44.29 ± 11.24	49.33 ± 12.23	0.258
Pretest left hamstring flexibility (degree)	50.71 ± 13.58	48.67 ± 8.76	0.631
Posttest left hamstring flexibility (degree)	42.14 ± 12.04	48.33 ± 9.76	0.161

Table 13.3 Right hamstring flexibility comparisons within group

Group	Hamstring flexibility		
	Pretest	Posttest	Sig. (*P* value)
Experimental (n = 14)	54.29 ± 11.07	44.29 ± 11.24	0.001
Control (n = 15)	49.33 ± 11.16	49.33 ± 12.23	1.000

Table 13.4 Left hamstring flexibility comparisons within group

Group	Hamstring Flexibility		
	Pretest	Posttest	Sig. (*P* value)
Experimental (n = 14)	50.71 ± 13.58	42.14 ± 12.04	0.001
Control (n = 15)	48.67 ± 8.76	48.33 ± 9.76	0.792

Nonetheless, the positive finding (improvement in hamstrings flexibility in the experimental group) in this study is in agreement with previous findings, where eccentric training reduces injury rates and improves muscle flexibility [6, 12–14]. The improvement in hamstring flexibility may be due to eccentric contractions and change of morphology associated with the increase in sarcomeres which lead to large adaptation on hamstring muscle [15]. Another theory state that improvement in rom angle may be due to improvement of hamstrings muscle–tendon unit [6, 14] and increase in viscoelastic and function changes [16]. Furthermore, another hypothesis suggests that improvement of hamstring flexibility is due to sarcomerogenesis promoted by eccentric training [14]. Yet, the mechanism behind the sarcomerogenesis is still unclear.

It has been debated if the eccentric training is needed since static stretching also shows improvement in flexibility [17]. However, static stretching has little impact on injury risk or recurrence [17]. Thus, further investigation should be carried out on elderly to clarify the effects of eccentric training on flexibility compared with static stretching and other exercise interventions.

5 Conclusion

It can be concluded that the leg press eccentric training with yellow theraband improves hamstrings flexibility among elderly in 6-week training program.

Acknowledgment We would like to acknowledge all the participants involved in this study for their cooperation and kindness.

References

1. *ACSM's guideline for exercise testing and prescription*. Lippincott Williams & Wilkins. (9th ed.) Philadelphia.
2. Glickman, E. L., Collinsworth, T. A., & Leigh, M. (2006). Exercise testing and prescription in the enhancement of physiologic and overall well being in the older adult. *Forum on Public Policy: A Journal of the Oxford Round Table*.
3. Chodzko-Zajko, W. J., Proctor, D. N., Fiatarone Singh, M. A., Minson, C. T., Nigg, C. R., Salem, G. J., et al. (2009). Exercise and physical activity for older adults. *Medicine and Science in Sports and Exercise, 41*(7), 1510–1530.
4. Witvrouw, E., Danneels, L., Asselman, P., D'Have, T., & Dirk, C. (2003). Muscle Flexibility as a Risk Factor for Developing Muscle Injuries in Male Professional Soccer Players. *American Journal of Sports Medicine, 31*(1), 41–46.
5. Knapik, J., Jones, B. H., Bauman, C. L., & Harris, J. M. (1992). Strength, flexibility and athletic injuries. *Sports Medicine (Auckland, N. Z.), 14*(5), 277–288.
6. Nelson, R. T., & Bandy, W. D. (2004). Eccentric training and static stretching improve hamstring flexibility of high school males. *J Athl Train, 39*(3), 254–258.
7. Nelson, M. E., Rejeski, W. J., Blair, S. N., Duncan, P. W., Judge, J. O., King, A. C., et al. (2007). Physical activity and public health in older adults: Recommendation from the American College of Sports Medicine and the American Heart Association. *Medicine and Science in Sports and Exercise, 39*(8), 1435.
8. Eccles, J. C., Eccles, R., & Lindberg, A. (1957). Synaptic actions on moto neurons caused by episodes of Golgi tendon organ afferents. *Journal of Physiology, 138*, 227–252.
9. Aijaz, S. M., Hameed, U. A., & Quddus, N. (2011). A comparative study on eccentric training using thera-band and static stretching in improving triceps surae muscle flexibility. *Int J Sports Sci Eng, 5*(3), 155–162.
10. Gaglione, J. (2011). Eccentric training for athletes. Available at http://articles.elitefts.com/training-articles/eccentric-training-for-athletes/, 2011.

11. Jones, C. J., Rikli, R. E., Max, J., & Noffal, G. (1998). The reliability and validity of a chair sit-and-reach test as a measure of hamstring flexibility in older adults. *Research Quarterly for Exercise and Sport, 69*(4), 338–343.
12. Waseem, M., Shibili, N., Ram, C., & Ahmad, F. (2009). A comparative study: Static stretching versus eccentric training on popliteal angle in normal healthy Indian collegiate males. *International Journal Sports Science Engineering, 3*(3), 180–186.
13. Mahieu, N. N., McNair, P., & Cools, A. (2008). Effects of eccentric training on the pantar flexor muscle-tendon tissue properties. *Medicine and Science in Sports and Exercise, 40*, 117–123.
14. Potier, T. G., Alexander, C. M., & Seynnes, O. R. (2009). Effects of eccentric strength training on biceps femoris muscle architecture and knee joint range of motion. *European Journal of Applied Physiology, 105*, 939–944.
15. Daniela, N. F., Janaína, L. L., Michelle, F. S., Aikelton, F. S., Marco, T., Cristiane, G. P., Geraldo, F.M., & Natália, B. (2007). Analysis of the influence of static stretching and eccentric training on flexibility of hamstring muscles, XXV ISBS Symposium, Ouro Preto—Brazil, pp. 454–457, 2007.
16. Noobreg, A. C., Paula, K. C., & Carvalho, A. C. G. (2005). Interaction between resistance training and flexibility training in healthy young adults. *Journal Strength Conditioning Research, 19*(4), 842–846.
17. O'Sullivan, K., McAuliffe, S., & DeBurca, N. (2012). The effects of eccentric training on lower limb flexibility: a systematic review. *British Journal of Sports Medicine, 46*, 838–845.

Digitus Secundus and Digitus Medicinalis Ratio: Examination of Sporting Ability Predictor in Male Youth

Mohd Zulkhairi Mohd Azam, Wan Mohd Norsyam Wan Norman, Adam Linoby, Hanifa Sariman, Muhammad Sufyan Mohd Zaki, Azizul Afandi and Muhamad Noor Mohamed

Abstract The phalangeal ratio of the digitus secundus (index finger) and digitus medicinalis (ring finger), widely known as the digit ratio (2D:4D), primarily acts as a marker for the amount of hormone testosterone exposure during the prenatal period. Higher prenatal testosterone exposure leads to a low digit ratio, which has been found to be correlated with a good level of performance in sports. Evidence shows that testosterone attributes such as aggression, masculinity, and high physical fitness contribute to sports performance. The process of identifying sports talent among youth is very important as recognizing sporting ability early can be advantageous, providing a longer time frame to develop elite athletes in the future. We examined the digit ratio of male youth subjects aged 17.0 ± 0.69 years ($n = 60$) by putting them through a series of physical fitness measures to examine the relationship of the digit ratio with performance in the tests. We found negative correlations between aerobic capacity and muscular power and the 2D:4D ratio. There was a positive correlation between the 2D:4D ratio and agility. However, there was no significant correlation between 2D:4D ratio and abdominal core strength. The positive results present more opportunities to be explored in talent identification, especially in the younger population; establishing the value of using the 2D:4D ratio as part of the process of identifying sporting ability could be highly beneficial and useful to the sports development agenda.

Keywords 2D:4D · Digit ratio · Testosterone · Male youth · Sporting ability

This work was supported in part by the Fundamental Research Grant Scheme (FRGS), Ministry of Higher Education of Malaysia FRGS/1/2011/SKK/UiTM/03/30.

M. Z. Mohd Azam (✉) · W. M. N. Wan Norman · A. Linoby · H. Sariman ·
M. S. Mohd Zaki · A. Afandi
Faculty of Sports Science and Recreation, Universiti Teknologi MARA,
Bandar Pusat Jengka, Pahang, Malaysia
e-mail: zulkhairiazam@pahang.uitm.edu.my

M. N. Mohamed
Faculty of Sports Science and Recreation, Universiti Teknologi MARA,
Kuala Pilah, Negeri Sembilan, Malaysia

R. Adnan et al. (eds.), *Proceedings of the International Colloquium on Sports Science, Exercise, Engineering and Technology 2014 (ICoSSEET 2014)*,
DOI: 10.1007/978-981-287-107-7_14, © Springer Science+Business Media Singapore 2014

1 Introduction

The ratio of the length of the digitus secundus (second digit-2D) and the digitus medicinalis (fourth digit-4D) in humans has been a topic of considerable research interest in recent years. This is primarily because of its unique association with achievements in various fields, such as high success rates among financial traders, strong sports performance, the link between fertility and the high amount of sperm produced, career interests, and health aspects [3, 12, 22]. The main attributes of the digit ratio are sexually dimorphic with a lower digit ratio usually found among males and a higher digit ratio related to females [1, 12, 25].

Researchers in this field have proposed that digit length and the ratio of 2D and 4D are influenced by the concentration of androgens (hormone testosterone) during the early period of pregnancy [9, 12], pointing to the 2D:4D ratio as a gross amplitude with low 2D:4D ratios specifying greater testosterone susceptivity [5, 15, 22, 24]. According to study from [25], the sexual dimorphism in the digit ratio is stable at the age of two years, showing no further evidence of change at puberty level [11]. The 2D:4D ratio is expressed as the ratio of the lengths of the digitus secundus and digitus medicinalis, generally measured from the baseline of the finger crease to the end of fingertip. A study by [10] established that ratio values in men tend to be less than 1.00 mm, and in women, finger length either tends to be equal or a ratio of 1.00 mm higher than men.

Generally males tend to perform better in sports than females as males have superior physiques and physiological capabilities. Furthermore, testosterone is a vital defense system against any defect and impairment of the cardiovascular system in men. That statement is further supported by study [9] suggested that testosterone influences the development of the cardiovascular system and that a low 2D:4D ratio is an attribute of a high fetal testosterone concentration, resulting in a vascular system with good capability in adults [14]. This is further proved by Manning and Taylor [16], who found that prenatal testosterone concentration is affiliated with high sporting ability in men. There is further evidence pointing to various types of sports, such as rugby, football, and athletic disciplines [2, 9, 13, 21]. These findings support the reliability of the link between the 2D:4D ratio and performance in sports, resulting in part from testosterone susceptivity.

The digit ratio has recently been proven to be an early predictor of sporting ability and performance among adults. However, there is a lack of clarity and inconsistent data in relation to the relationship between the digit ratio and sporting ability among the youth population. Moreover, most recent studies regarding sporting potential among individuals have only focused on adolescent and adult target populations [2, 5, 11, 16]. Evidently, findings from [5] found that the 2D:4D ratio was more associated with sports performance in adults, not in youth. The studies that are closest to this in terms of subjects were undertaken by [13, 14], with one of the studies focusing on handwriting performance.

Other research has provided concrete evidence of genetic involvement in determining individual differences in sport performance attributes. Specifically, talented individuals are likely to receive genes from parents who are active in sports and have good physiques [4, 8]. However, good sporting ability is not solely due to genetic inheritance but also relates to other factors, such as effective training programs, nutrition intake, and physiological growth [6, 17].

The identification of talent is important in modern sports, and it is suggested that a systematic and effective talent identification process implementing scientific methods plays a vital role. These days, as can be seen in the international scene, competition has become more intense, and talent spotting has become common practice. An example of this is football, with professional football clubs prioritizing the identification of valuable young footballers with excellent prospects [4, 19, 20]. Finding the correct timing to nurture sports talent is vital: the younger the prospect the better. However, in team sports, such as football, [8] stated that young football players can still be nurtured at 15 years old. They consider there is still enough space and time for improvement as a result of the growth and maturational process, together with the expected improvement through the periodization and frequency of good training programs.

Detecting sporting potential by examining the digit ratio could help identify talented young individuals, which could be advantageous in developing elite athletes [17, 20, 24]. To measure the 2D:4D ratio, the right hand of the participants is used due to its attribute of being a far more accurate indicator in terms of sports performance and health aspects than the 2D:4D ratio of the left hand [11, 18]. In addition, the common characteristics of human masculinity are more evident in the right side of the body, the left side usually being synonymous with female characteristics [7, 11, 21].

2 Methodology

2.1 Participants

For investigation purposes, this study recruited male youth participants aged range from 16 to 18 years old ($n = 60$) were selected from the secondary schools in Pahang. Participants who reported deformities and injuries in their digitus secundus and digitus medicinalis were discarded from the analyses.

2.2 Measuring 2D:4D Ratio

Before the start of digit ratio measurement, the participants were asked to take out their accessories, jewelleries and cleaned their hand first. The participants stretched their fingers evenly and insert their hands with palm facing down on the scanner machine (Hewlett Packard Deskjet 1,050 Printer). The scanned diagrams of the hand were later processed into the MicroDicom software, in which the measuring technique was done by taking measures from the baseline of the finger crease to the end of fingertips; it provides much more accurate assessment with its computerized measurement [18]. Meanwhile, after that the 2D:4D ratios were calculated by dividing the measure of length from the digitus secundus and the digitus medicinalis. To avoid data inconsistency, the finger measurements and 2D:4D ratio calculations were completed two times by the same observer, and the average score was recorded. Ratio values that were lower than 1 mm indicated that 2D was longer than the 4D [23].

2.3 Physical Fitness Measures

There were four physical fitness measures conducted for this study. The measures were *t* test (agility), bleep test (cardiovascular endurance), standing broad jump (power), and 7-level sit-up (core strength). The subjects did the warm up session including the stretching session for 15 min before proceeding to the exact session of fitness testing and were supervised consistently.

2.4 Statistical Analysis

Data were analyzed using the International Business Machines Statistical Package for Social Sciences (version 21.0 for Windows). The normality of the data distributions was analyzed first, and Spearman correlation was used to find the association between low 2D:4D ratio with the performances on the physical fitness measures as P value was set <0.05.

3 Results

All subjects successfully completed the training protocols. The mean age of the subjects was 17.0 ± 0.69 years old, and they managed to complete all of the required physical fitness tests without any complications or injuries (Table 1).

3.1 Body Mass Index

Body Mass Index (BMI) was analyzed in order to know the body fat status of the subjects, and the BMI was 22.0 ± 3.08 kg/m^2 which still belongs in the acceptable normal zone with only one subject had the highest BMI of 31.64 kg/m^2. Based on the information gained from the subjects, all of them were in a good health condition and do not have any illness prior to the start of the study.

Table 1 The mean (SD) values for the male youth subjects

$n = 60$	Min	Max	Mean ± SD
Age (years)	16	18	17.0 ± 0.69
Height (cm)	148	186	165.7 ± 8.12
Weight (kg)	40	87.5	60.5 ± 10.30
BMI (kgm^{-2})	15.50	31.64	22.0 ± 3.08
2D:4D (mm)	0.85	0.99	0.93 ± 0.03

$^{*}P < 0.05$

Table 2 The mean (SD) values for the male youth subjects

Physical fitness	2D:4D	*p*-value
Aerobic capacity (bleep test)	−0.54	0.001**
Muscular power (standing broad jump test)	−0.30	0.019*
Agility (*t* test)	0.37	0.004*
Abdominal core strength (7-level sit-up test)	−0.25	0.053*

*Correlation is significant at the 0.05 level (2-tailed)
**Correlation is significant at the 0.01 level (2-tailed)

3.2 2D:4D Ratio and Physical Fitness

After the completion of the physical fitness testing, the data were analyzed in order to find association with the digit ratio. 2D:4D ratio was determined as 0.93 ± 0.03 mm among the subjects and all of them possessed low 2D:4D ratio with all calculated values were lower than 1.00 mm. The highest 2D:4D ratio was found at 0.99 mm and the lowest 2D:4D ratio was at 0.85 mm. The physical fitness testing conducted covered several important components of a good athlete including the agility, aerobic capacity, muscular power, and abdominal core strength. From the result, there were negative correlations found between 2D:4D ratio and aerobic capacity and muscular power. Moreover, positive correlation was observed between 2D:4D ratio and agility. However, there was no significant correlation found in 2D:4D ratio and abdominal core strength as can be seen in Table 2.

4 Discussion

This study managed to reveal the links between 2D:4D ratio and the results from the physical fitness testing in male youth subjects. The correlations were found in all measures of physical fitness except in abdominal core strength test with those tests were intended to include the aspect of aerobic and anaerobic efficiency. The negative correlation found in the bleep test was in accordance to the findings from [11, 16] in the capability of aerobic efficiency of low 2D:4D group to do endurance running. The better performances in endurance running were seen to be benefited from the influence of hormone testosterone concentration to the development of efficient cardiovascular system [9, 16]. Despite the effect of excessive manually intake of testosterone is known to expose an athlete to higher risk of heart diseases especially among the bodybuilders, while in contrast, the natural process of having high prenatal testosterone in a person seems to aid the growth of efficient vascular system and contributes to performances in endurance running in both men and women [13, 22].

In the testing of standing broad jump, the results were moderately correlated. This is however still in line with past studies that proved the idea of muscular power is significantly related to low 2D:4D ratio. In contrast to the present finding, the result of strength test was recorded by Manning and Hill [13], stated 2D:4D ratio was a weak indicator for strength and power, which we were not able to discover any relationship in the abdominal core strength test with 2D:4D. However, the strength test conducted by Manning and Hill [13] and [5] used the handgrip strength test, totally different compared to our abdominal core strength test, and at the same time, we could not find any similar past studies conducted testing on the abdominal part along with concerns to 2D:4D. On the other part, not to forget, this study also reported positive correlation observed in agility time, as it also in support to the previous findings [2, 11].

Therefore, to see the mixed relationship of 2D:4D ratio in both aerobic and anaerobic fitness components was intriguing. Referring to previous findings, 2D:4D ratio was evidently correlated to high performances in team sports of football and rugby [1, 2]. Those team sports included various skills and components of physical fitness in order for the players to perform well. Possessing good mixture of aerobic and anaerobic functionality and efficiency, then only the players is able to perform in playing football or rugby at the highest level. First team players or players of the league winning team were most to have greater physical ability and also low 2D:4D ratio compared to the reserve players or any players playing for teams that fighting to avoid league relegation [2, 19, 25]. Regard to this statement, the mixed findings may suggest that the factor of prenatal testosterone; 2D:4D is very much to be exploring on as prenatal testosterone which not the only influence athleticism and sports performances. The further exploration must not ignoring the factors of ethnicity, sex, races, genetic differences and training which may play an important role.

Concerning the current talent identification process which mainly focus on the physical fitness ability and the anthropometric measurement, there had been some arguments as some process did not implement the scientific methods and there were limited knowledge of talent identification on every sports available [20]. There were lots of young athletes that showed promising signs of high sporting potential. However, only a minority of them will attain sporting excellence due to the flaws in the ability of the talent identification process to detect those precious talents. That is why the importance put on the talent identification program is massive as nowadays the international competitions such as the Olympic Games have become more competitive, intense and mostly lead by younger athletes. Further to our knowledge, we found that researches on combining 2D:4D and talent identification were very limited. Integrating both 2D:4D with the current talent identification program will bring benefits to the sporting development in term of achievements, as it has been widely proved previously by researchers that high prenatal testosterone (low 2D:4D ratio) provided useful and significant predictor of a person's ability to perform well in sports or much more broader in other aspects of life such as success in managing finance and trades, high efficiency of visual–spatial and also even in achieving good health status [3, 9, 12, 22].

5 Conclusion

In summary, the relationship of 2D:4D with physical fitness among the male youth was monitored and evaluated in term of sporting ability or athleticism. The results from this study indicated that 2D:4D do have significant correlation which vastly contributed to the performances on those physical fitness components. These however cannot exclude the relationships of exercise frequency and other related factors such as nutrition intake, visual–spatial ability, and psychological aspect. Without the proper aid of a good training program, the sporting ability among the person itself would not develop well and may later on may influence the sports performances. Therefore, implementing low 2D:4D to be an indicator in identifying prospects with good sporting potential by adding it to the current talent identification system would have to be recommended for use in the near future in order to help sports development and athlete's improvement, especially among the younger population.

References

1. Bailey, A. A., & Hurd, P. L. (2005). Finger length ratio (2D:4D) correlates with physical aggression in men but not in women. *Biological Psychology, 68*, 215–222.
2. Bennett, M., Manning, J. T., Cook, C. J., & Kilduff, L. P. (2010). Digit ratio (2D:4D) and performance in elite rugby players. *Journal of Sports Sciences, 28*(13), 1415–1421.
3. Coates, J. M., Gurnell, M., & Rustichini, A. (2009). Second-to-fourth digit ratio predicts success among high-frequency financial traders. *Proceedings of the National Academy of Sciences of the United States of America, 106*(2), 623–628.
4. Falk, B., Lidor, R., Lander, Y., & Lang, B. (2004). Talent identification and early development of elite water-polo players: A 2-year follow-up study. *Journal of Sports Sciences, 22*(4), 347–355.
5. Fink, B., Thanzami, V., Seydel, H., & Manning, J. T. (2006). Digit ratio and hand-grip strength in german and Mizos men: CROSS-cultural evidence for an organizing effect of prenatal testosterone on strength. *American Journal of Human Biology : The Official Journal of the Human Biology Council, 18*(6), 776–782.
6. Georgopoulos, N. A., Roupas, N. D., Theodoropoulou, A., Tsekouras, A., Vagenakis, A. G., & Markou, K. B. (2010). The influence of intensive physical training on growth and pubertal development in athletes. *Annals of the New York Academy of Sciences, 1205*(1), 39–44.
7. Hill, R., Simpson, B., Manning, J. T., & Kilduff, L. (2012). Right–left digit ratio (2D:4D) and maximal oxygen uptake. *Journal of Sports Sciences, 30*(2), 129–134.

8. Hoare, D. G., & Warr, C. R. (2000). Talent identification and women's soccer: An Australian experience. *Journal of Sports Sciences, 18*(9), 751–758.
9. Honekopp, J., Manning, J. T., & Muller, C. (2006). Digit ratio (2D:4D) and physical fitness in males and females: Evidence for effects of prenatal androgens on sexually selected traits. *Hormones and Behavior, 49*(4), 545–549.
10. Lutchmaya, S., Baron-Cohen, S., Raggatt, P., Knickmeyer, R., & Manning, J. T. (2004). 2nd to 4th digit ratios, fetal testosterone and estradiol. *Early Human Development, 77*, 23–28.
11. Manning, J. T., Morris, L., & Caswell, N. (2007). Endurance running and digit ratio (2D:4D): Implications for fetal testosterone effects on running speed and vascular health. *American Journal of Human Biology : The Official Journal of the Human Biology Council, 19*(3), 416–421.
12. Manning, J. T. (2002). *Digit ratio: A pointer to fertility, behavior, and health.* New Brunswick, NJ: Rutgers University Press.
13. Manning, J. T., & Hill, M. R. (2009). Digit ratio (2D:4D) and sprinting speed in boys. *American Journal of Human Biology : The Official Journal of the Human Biology Council, 21*(2), 210–213.
14. Manning, J. T., Trivers, R. L., Thornhill, R., & Singh, D. (2000). The 2nd:4th digit ratio and asymmetry of hand performance in Jamaican children. *Laterality, 5*(2), 121–132.
15. Manning, J. T., Bundred, P. E., & Taylor, R. (2003). The ratio of 2nd and 4th digit length: A prenatal correlate of ability in sport. In T. Reilly & M. Marfell-Jones (Eds.), *Kinanthropometry* (Vol. VIII, pp. 165–174). London: Routledge.
16. Manning, J. T., & Taylor, R. P. (2001). Second to fourth digit ratio and male ability in sport: Implications for sexual selection in humans. *Evolution and Human Behavior : Official Journal of the Human Behavior and Evolution Society, 22*(1), 61–69.
17. Meyers, M. C., Stewart, C. C., Laurent, C. M., Leunes, A. D., & Bourgeois, A. E. (2008). Coping skills of Olympic developmental soccer athletes. *International Journal of Sports Medicine, 29*(12), 987–993.
18. Nicholls, M. E., Orr, C. A., Yates, M. J., & Loftus, A. M. (2008). A new means of measuring index/ring finger (2D:4D) ratio and its association with gender and hand preference. *Laterality, 13*(1), 71–91.
19. Pearson, D. T., Naughton, G. A., & Torode, M. (2006). Predictability of physiological testing and the role of maturation in talent identification for adolescent team sports. *Journal of Science and Medicine in Sport/Sports Medicine Australia, 9*(4), 277–287.
20. Pienaar, A. E., Spamer, M. J., & Steyn, H. S, Jr. (1998). Identifying and developing rugby talent among 10-year-old boys: A practical model. *Journal of Sports Sciences, 16*(8), 691–699.
21. Pokrywka, L., Rachon, D., Sucheka-Rachon, K., & Bitel, L. (2005). The second to fourth digit ratio in elite and non-elite female athletes. *American Journal of Human Biology, 17*, 796–800.
22. Rosana, G. M. C. (2000). Androgens and coronary heart diseases. A sex-specific effect of sex hormones? *European Heart Journal, 21*, 868–871.
23. Sophie, E. W., Annika, F., & Juergen, H. (2007). Associations between the second to fourth digit ratio and career interests. *Personality and Individual Differences, 4*(3), 485–493.
24. Stenstrom, E., Saad, G., Nepomuceno, M. V., & Mendenhall, Z. (2011). Testosterone and domain-specific risk: Digit ratios (2D:4D and rel2) as predictors of recreational, financial, and social risk-taking behaviors. *Personality and Individual Differences, 51*, 412–416.
25. Tester, N., & Campbell, A. (2007). Sporting achievement: What is the contribution of digit ratio? *Journal of Personality, 75*(4), 663–677.

Differences in Game Statistics Between Winning and Losing Teams in Inter-University Elite Male Sepak Takraw Tournament: A Pilot Study

Norasrudin Sulaiman, Rahmat Adnan and Shariman Ismadi Ismail

Abstract The purpose of the study was to analyze and investigate the differences in male Sepak Takraw game statistics between winning and losing teams. Data were collected from the 2012 elite male inter-university Sepak Takraw tournament. Seven (7) games starting from the quarter final matches until the final match were analyzed. There were 11 performance indicators selected as the variables in this study (service in, service out, service ace, success first ball, unsuccessful first ball, successful feeding, unsuccessful feeding, success strike, unsuccessful strikes, successful blocking and unsuccessful blocking). A descriptive and non-parametric analysis (Wilcoxon signed-ranks test) were performed with the significance value set at $p \leq 0.05$. The analysis showed only one (1) performance indicator to have a significant difference between the winning team and loosing team, which was service out ($z = -2.47$, $p = 0.013$, $p < 0.05$). The other ten (10) performance indicators did not show any significant differences between winning and loosing teams (success first ball, $z = -1.23$, $p = 0.218$, $p > 0.05$; unsuccessful first ball, $z = -1.91, p = 0.057, p > 0.05$; success feeding, $z = -1.75, p = 0.79, p > 0.05$; service in, $z = -1.09$, $p = 0.274$, $p > 0.05$; service ace, $z = -1.54$, $p = 0.122$, $p > 0.05$; unsuccessful feeding, $z = -0.60$, $p = 0.549$, $p > 0.05$; success spike, $z = -0.736$, $p = 0.462$, $p > 0$ .05; unsuccessful spike, $z = -0.736$. $p = 0.462$, $p > 0.05$; successful blocking, $z = -1.44$, $p = 0.149$, $p > 0.05$ and; unsuccessful blocking, $z = -0.983$, $p = 0.325$, $p > 0.05$). The values presented in this study are a useful reference for coaches to determine teams' weaknesses and strengths in

N. Sulaiman (✉) · R. Adnan · S. I. Ismail
Faculty of Sports Science and Recreation, Universiti Teknologi MARA, Shah Alam, Malaysia
e-mail: noras878@salam.uitm.edu.my

N. Sulaiman · R. Adnan · S. I. Ismail
Sports Science Center of Studies, Shah Alam, Malaysia

R. Adnan et al. (eds.), *Proceedings of the International Colloquium on Sports Science, Exercise, Engineering and Technology 2014 (ICoSSEET 2014)*,
DOI: 10.1007/978-981-287-107-7_15,

designing a training program, especially for higher level competitions. Coaches need to focus on the significant performance indicator during training in order to achieve success in Sepak Takraw inter-university tournaments.

Keywords Winning team · Losing team · Sepak Takraw

1 Introduction

Sepak Takraw is a traditional sport and became popular from year to year and to date, there are 30 countries across five continents who actively play this game. Instead, Sepak Takraw is a medal sport in South East Asia Games and the Asian Games. This game is played by three (3) players, namely the Server (Tekong), Feeder and the Spiker. The court size is similar to a badminton court and is divided by a net in the middle of the court. The game starts with the feeder feeding the ball by hand to the Server. The Server will serve using one of his legs and each opponent's team are allowed three touches for the rally. The point scoring system is similar to ping pong, where when any team makes a mistake, the opponent's team will get a point, and the team which manages to reach 15 points in the normal game will win the set. If both team were to draw at the point of 15, then two additional points are needed to separate the winning and loosing team. This game is played based on best of three sets and the maximum playing set is five.

Being among the earliest countries that plays this game, the performance of the Malaysian Sepak Takraw male team was not convincing when they failed to win any major trophies in the past 2 years. Currently, Malaysia is ranked as the second best team in the world after Thailand; however, the difference in terms of points is becoming bigger from year to year. Instead, the Indonesian team is catching up points with Malaysia, and is currently the third in world ranking. Something needs to be done to increase the performance of the Malaysian team. One possible solution is by performing an intensive game analysis by a performance analyst.

Performance analysis and sport performance are two things that are related to each other, and there are a lot of benefits in enhancing sport performance either for the individual or team sport. The analysis in team sports' performance such as basketball, rugby and football are actually a fundamental tool for coaches to collect valid and reliable information on both team performances, either their own team or opponents. The reasons why coaches and researchers depend on such of information are because its usable to identify the most valuable players and the importance of specific roles, assess the impact of rule changes, investigate the home advantage, evaluate the participation of the starting and reserve players in the game, and show how each player contributes to team performance [1].

Before performance analysts do a study on the team's performance, they will have to refer to the match statistics, either in training or in a match [2]. According to Lago-Peñas et al. [3], match analysis can be defined as the recording and examination of behavior that occurs and exists during competition. The study of sport performance and behavior through match analysis is important in evaluating and monitoring any team or individual performance [4], and to identify variables that can discriminate between the successful and unsuccessful team (Ibanez et al. 2008), where it will have a positive impact by involving modifications to the organization, design, teaching and training of team sports [5]. Currently, before performance, analysts can do a study on the team's performance, they will have to refer to the match statistics, either in training or in a match [2].

Lago-Peñas et al. [3] noted that match analysis can be defined as the recording and examination of behavior that occurs and exists during competition. The study of sport performance and behavior through match analysis is important in evaluating and monitoring any team or individual performance [4], and to identify variables that can discriminate between the successful and unsuccessful team (Ibanez et al. 2008), where it will have a positive impact by involving modifications to the organization, design, teaching and training of team sports [6]. Based on the above discussion, the current interest was to perform a notational analysis to investigate the differences in performance indicators among winning and loosing team.

2 Method

2.1 Selecting Performance Indicators

The data for this study was based on the data recorded at the Inter-University Elite Sepak Takraw Tournament. Seven (7) games starting from the quarterfinal match until the final match were analyzed using notational analysis. The Sport Code Elite software was used to analyze the data. There were 11 performance indicators that were selected as variables in this study and the selection was based on the discussion with four (4) state coaches and two (2) university coaches. The list of performance indicators are presented in Table 1.

Teams were divided into two, the winning team and the loosing team. All teams that won the games starting from the quarter final until the final were considered as the winning team, and team who lost were considered as the loosing team. The overall data collected from this study was analyzed and interpreted using the Statistical Package for Social Science version 17.0 (SPSS ver. 17). Data were analyzed using descriptive and inferential statistics. Non-parametric statistic and Wilcoxon signed-ranks test were used. All the statistical analysis was performed with the significance level set at $p \leq 0.05$.

Table 1 List of performance indicators

No.	Variables
1	Service in
2	Service out
3	Service ace
4	Success first ball
5	Unsuccessful first ball
6	Successful feeding
7	Unsuccessfully feeding
8	Success strike
9	Unsuccessfully strikes
10	Successful blocking
11	Unsuccessfully blocking

2.2 *Statistical Analysis*

As mentioned earlier, teams were divided into two, the winning team and the loosing team. All teams that won the games starting from quarter final until the final were considered as the winning team and teams which lost were considered as the loosing team. The overall data collected was analyzed and interpreted using the Statistical Package for Social Science version 17.0 (SPSS ver. 17). Data were analyzed using descriptive and inferential statistics. All statistical analysis was performed with the significance level set at $p \leq 0.05$.

3 Results

Results of descriptive analysis between the winning and losing teams are presented in Table 2.

Based on the results presented in Table 2, scoring in terms of good skills was nominated by the winning team scoring more than the loosing team. In terms of service in, the winning team performed 146 (mean ± SD, 20.9 ± 4.14) good services, 28 (4.00 ± 1.0) services were out, and 5 (0.71 ± 0.49) service aces compared to the loosing team, which only had 130 (18.6 ± 2.8) services in, 59 (8.4 ± 2.8) services out, and 2 (0.29 ± 0.49) service aces. In terms of percentages, the winning team performed well with 81.6 % of good serves, 15.6 % services out and 2.8 % service aces out of 179 total serves. The loosing team performed with 68.1 % of good serves, 30.9 % services out and 1 % service aces out of 191 services made.

The winning team performed with 138 (72.3 %) (138 ± 2.3) success first balls and 53 (27.7 %)(7.6 ± 2.3) of unsuccessful first balls, compared to the loosing team. The loosing team performed with 140 (77.8 %) (19.1 ± 4.1) success first balls and 40 (22.2 %) (5.7 ± 2.1) unsuccessful first balls. In terms of feeding, the

Table 2 Descriptive overall result for winning and loosing team

Variable	Winning			Loosing		
	Score	Mean	SD	Score (%)	Mean	SD
Service in	146 (81.6)	20.9	4.14	130 (68.1)	18.6	2.8
Service out	28 (15.6)	4.00	1.0	59 (30.9)	8.4	2.8
Service ace	5 (2.8)	0.71	0.49	2 (1.0)	0.29	0.49
Success first ball	138 (72.3)	18.3	2.3	140 (77.8)	19.1	4.1
Unsuccessful first ball	53 (27.7)	7.6	2.3	40 (22.2)	5.7	2.1
Successful feeding	132 (92.3)	18.9	3.9	117 (82.4)	16.7	1.9
Unsuccessfully feeding	11 (7.7)	1.6	1.1	25 (17.6)	3.6	4.1
Success spike	88 (66.2)	12.6	4.1	86 (79.6)	12.3	2.6
Unsuccessfully spike	45 (33.8)	6.4	2.9	22 (20.4)	5.1	1.7
Successful blocking	26 (38.8)	4.1	1.8	18 (33.3)	3.0	1.6
Unsuccessfully blocking	41 (61.2)	5.9	2.0	36 (66.7)	5.1	2.1

Table 3 Differences between winning and loosing teams based on variables

	SER IN	SER OUT	ACE	SUC-1ST	UNSUC 1ST	SUC FED	UNSC FED
Wilcoxon W	44.0	33.5	42.00	43.00	38.00	39.00	48.0
Z	−1.09	−2.47	−1.54	−1.23	−1.91	−1.75	−0.60
Asymp. sig. (2-tailed)	0.274	0.013	0.122	0.218	0.057	0.079	0.549
Exact sig. [2 * (1-tailed Sig.)]	0.32[a]	0.01[a]	0.209[a]	0.26[a]	0.07[a]	0.097[a]	0.620[a]

[a] Not corrected for ties

winning team performed 132 (92.3 %) (18.9 ± 3.9) successful feedings and 11 (7.7 %) (1.6 ± 1.1) unsuccessful feedings compared to the loosing team. The loosing team performed with 117 (82.4 %) (16.7 ± 1.9) successful feedings and 25 (17.6 %) (3.6 ± 4.1) unsuccessful feedings.

The winning team performed with 88 (66.2 %) (12.6 ± 4.1) success spikes and 45 (33.8 %) (6.4 ± 2.9) of unsuccessful spikes when compared to the loosing team. The loosing team performed with 86 (79.6 %) (12.3 ± 2.6) successful spikes and 22 (20.4 %) (5.1 ± 1.7) unsuccessful spikes. In terms of blocking, the winning team performed 26 (38.8 %) (4.1 ± 1.8) successful blockings and 41 (61.2 %) (5.9 ± 2.0) of unsuccessful blockings when compared to the loosing team. The loosing team performed with 18 (33.3 %) (3.0 ± 2.6) successful blockings and 36 (66.7 %) (5.1 ± 2.1) unsuccessful blockings.

Based on Table 3, however, one (1) performance indicator had shown a significant difference between the winning team and the loosing team, which is service out ($z = -2.47$, $p = 0.013$, $p < 0.05$). The other ten (10) performance indicators did not show any significant difference between the winning and the loosing team (success first ball, $z = -1.23, p = 0.218, p > 0.05$; unsuccessful first ball, $z = -1.91$, $p = 0.057$, $p > 0.05$; success feeding, $z = -1.75$, $p = 0.79$, $p > 0.05$; service in, $z = -1.09$, $p = 0.274$, $p > 0.05$; service ace, $z = -1.54$,

$p = 0.122$, $p > 0.05$; unsuccessful feeding, $z = -0.60$, $p = 0.549$, $p > 0.05$; success spike, $z = -0.736$, $p = 0.462$, $p > .05$; unsuccessful spike, $z = -0.736$. $p = 0.462$, $p > 0.05$; successful blocking, $z = -1.44$, $p = 0.149$, $p > 0.05$ and; unsuccessful blocking, $z = -0.983$, $p = 0.325$, $p > 0.05$).

4 Conclusion

The values presented in this study will be useful as a reference for coaches to determine their teams' weaknesses and strengths, and in designing a training program, especially for higher level competition. Coach only need to focus on four significant performance indicators during training in order to be successful in Sepak Takraw inter-university tournaments.

Acknowledgments We would like to express our gratitude to the Faculty of Sports Science and Recreation Universiti Teknologi MARA (UiTM) Shah Alam, Selangor, Malaysia for the support throughout the whole process of this study.

References

1. Ibáñez, S. J., García, J., Feu, S., Lorenzo, A., & Sampaio, J. (2009). Effects of consecutive basketball games on the game-related statistics that discriminate winner and losing teams. *Journal of Sports Science and Medicine, 8*, 458–462.
2. Miguel, A. G., Lorenzo, A., Ortega, E., Sampaio, J., & Ibáñez, J. S. (2009). Game related statistics discriminating between starters and nonstarters players in Women's National Basketball Association League (WNBA). *Journal of Sports Science and Medicine, 8*, 278–283.
3. Lago-Peñas, C., Lago-Ballesteros, J., Dellal, A., & Gómez, M. (2010). Game-related statistics that discriminated winning, drawing and losing teams from the Spanish soccer league. *Journal of Sports Science and Medicine, 9*, 288–293.
4. Vaz, L., Van Rooyen, M., Sampaio, J. (2010). Rugby game-related statistics that discriminate between winning and losing teams in IRB and super twelve close games. *Journal of Sports Science and Medicine*, 9, 51–55.
5. Lorenzo, A., Gómez, M.A., Ortega, E., Ibáñez, S.J., Sampaio, J. (2009). Game related statistics which discriminate between winning and losing under-16 male basketball games. *Journal of sports science and medicine*, 9(4), 664–668.
6. Ortega, E., Villarejo, D., & Palao, J. M. (2009). Differences in game statistics between winning and losing rugby teams in the six nations tournament. *Journal of Sports Science and Medicine, 88*(4), 523–527.

Digit Ratio [2D:4D] as Predictor of Body Composition Among School Children

Siti Nor Intan Nor Ali, Sarina Md. Yusof and Suhana Aiman

Abstract Digit ratio is known to be an indicator of prenatal testosterone and had been hypothesized that the length of the 2nd and 4th digits [2D:4D] is correlate with prenatal hormone and sexually dimorphic among children and stable with growth. This study aims to investigate the association of 2D:4D with body composition among school children. Digit ratio [2D:4D] had been shown previously to stimulate anabolic process in skeletal muscle and appears to be principal hormone negatively associated with changes in overall adiposity. It is not known whether these associations also extend to male and female school children body composition. In this study, 367 male [$n = 199$] and female [$n = 168$] school children aged 8.21 ± 0.826 years from 18 schools throughout Selangor state. An independent samples t test was run to determine whether there were differences between digit ratio and body composition. Pearson's correlation test indicates there were small negative correlations between 2D:4D and body mass index [$r = -0.228, p < 0.05$] and also skinfold percentile [$r = -0.142$, $p < 0.05$]. A very small negative association between 2D:4D and the body composition among school children were scarce to show further evidence of the possible connection between high testosterone levels with body composition thus resulting in inadequate link between prenatal testosterone which was indicated through high 2D:4D with body mass index and skinfold percentile.

Keywords Digit ratio (2D:4D) · Body composition · School children

S. N. I. N. Ali (✉) · S. Md. Yusof · S. Aiman
Faculty of Sports Science and Recreation, Universiti Teknologi MARA,
40450 Shah Alam, Malaysia
e-mail: sni_ayong@yahoo.com

R. Adnan et al. (eds.), *Proceedings of the International Colloquium on Sports Science, Exercise, Engineering and Technology 2014 (ICoSSEET 2014)*,
DOI: 10.1007/978-981-287-107-7_16,

1 Introduction

Based on past knowledge of major health problems and their risk factors among children, several field constructed fitness test series have been developed to evaluate fitness in this population. Due to this issues, major reviews covering the matter involved in obtaining accurate and reliable measurements have been published and this include the application of relative length of 2nd (index finger) and 4th (ring finger) as a predictive intervention to be used in analysis, prediction, and in initial lifestyle interventions which may suspended beginning of disease or assist its early detection [1].

Digit ratio have stimulated considerable amount of research in relation with various traits of putatively linked to sex hormone [2]. Also known as the ratio between the second digit, which is the index finger and the fourth digit, the ring finger had shown mixed results with several researcher reported significant correlations between digit ratio and such diverse traits as fertility, sexual attitudes and orientation, status, cognitive abilities, health, and athletic prowess [3–7].

Association between forms of digit formation and prenatal testosterone and estrogen may lie in the action of Homeobox or Hox genes. The Hox gene in vertebrates, including humans, is vital for the variation of both the urinogenital system (including the testes and ovaries) and the digit [8]. The sexual dimorphism is determined as initial as the 14th week of fetal lifespan, and remains unaffected at puberty [9]. In children, digit ratio has been testified to be linked with procedures of fetal growth, congenital adrenal hyperplasia, developmental psychopathology, autism, and Asperger's syndrome [10]. Therefore, if digit ratio is a correlate of prenatal testosterone and estrogen, then it must follow that much of the alteration in finger ratio is determined in utero. This means that children should demonstrate forms of digit ratio which are principally the same as those initiate in adults.

Testosterone level increases during early stage of puberty to rapid increase in mid-late puberty in boys, whereas high level of estrogen causes inhibition of muscle growth as a result of skeletal maturation [11]. Improvements on muscular strength among children are negatively associated with changes in overall adiposity [12]. Clinical and invasive muscle strength measurement of different muscle group is time consuming and a quick and simple measurement might deliver a good indication of overall muscle strength. Therefore, in this study, the implementation of digit ratio is predicted could be used as measuring tools for muscle strength as it had been found to be interrelated with prenatal testosterone which is found to be significant with strength among adult male [13].

Throughout the world, obesity is now being considered as a major medical problem besides under nutrition and diseases. Obesity is instigated by prolonged energy disproportion resultant in the storing of additional energy as adipose tissue [14]. Obesity generally is associated with several risk factors for later heart disease and other chronic disease [15]. As for obese children and adolescent, cardiovascular health problem (e.g., hypertension, hypercholesterolemia), endocrine system malfunction (e.g., hyperinsulinamia, insulin resistant), and mental health problem (e.g., depression, low self-esteem) are some of the common medical problem [16]. Therefore, digit ratio which is closely related to prenatal testosterone hormone is predicted in this study to have direct relationship toward body composition physical level among school children.

Over the last periods, a countless arrangement of attention has been dedicated to the capability of children and adolescents in health and fitness body of knowledge. As a consequence, numerous field-based investigation series have been developed to assess fitness in this population. Therefore, the purpose of this study was to examine the relationship of digit ratio which is found to be negatively correlating with prenatal testosterone levels toward body composition among school children in Selangor in order for it to act as a predictor to assess health-related fitness and physical activity level.

2 Methodology

2.1 Participants

A cross-sectional research design similar to the research done before by Trivers et al. [17] was used to determine the effectiveness of using digit ratio as a new method in predicting body composition among public school children in Selangor. 367 students (199 males and 168 females) from public school children in Selangor (aged 7–10) that have not reached pubertal was chosen with consideration of urban and rural school location.

The type of sampling technique will be used in this study is stratified randomized sampling which is based on the characteristic exists in a certain segment of a population, that was specifically public school children in Selangor (aged 7–9) that have not reached puberty.

24 students for each 18 primary schools of nine different districts in Selangor were chosen from total of 645 primary schools in Selangor to represent sample population of 562,095 students with consideration of urban and rural school location classification determined by Ministry of Education Malaysia.

All subjects were classified as 'school children' who was operationally defined as children that had registered into school (public school) and had started learning session. Exclusion was made due to certain unattainable criteria such as acute cardiovascular impairment, puberty, vestibular dysfunction or balance disorder, history in either lower extremity, and history of neurological disorder affecting upper or lower limbs.

2.2 Instrumentation

Body Composition To determine body composition, children height and weight was measured using stadiometer and weight scale. Two measurements of height and weight were performed and mean at each variable is taken. The BMI for each subject was calculated using BMI formula, and categorization of BMI in children was used in this study. Skinfold procedure includes triceps and calf skinfold measured using caliper [18]. Percentage of body fat was calculated through equations by Slaughter et al. [19].

Digit Ratio For indirect measurement technique, each subject hands were alternately placed with their palm facing down (supination) on the middle of the glass plate of photocopier until each hand had been photocopied twice. Careful attention was engaged to ensure that particular area of major crease could be seen on the palms. Digit length were later measured from the photocopies using Vernier callipers measuring to 0.01 mm with each digit was measured from tip to basal crease, and measurement were made twice with the second measurement made blind to the first. As for direct measurement technique, each subject hands were measured using Vernier callipers measuring to 0.01 mm with each digit was measured from tip to basal crease, and measurement were made twice with the second measurement made blind to the first. Both methods were used in this study as both methods had shown no significant differences [20].

2.3 Procedures

Present study was approved by UiTM Research Ethical committee. Participants were briefly explained on the purpose of present study and all procedures involved. Parental permission was obtained for the testing and the methodology was agreed by administration of Ministry of Education to undergo testing protocol as well as permission from Selangor Department of Education. All subjects had been informed to avoid food consumption 3 h before activity to provide valid skinfold measurement reading.

All subject need to submit parental consent letter that had been send earlier before testing day in order to inform parents regarding data collection involving their children as well as to explain to parents or guardian regarding the research benefits and data collection procedure. Data collection procedure involved several testing station which was at the first station, subject personal information such as age, height, and weight will be recorded.

Table 1 Independent sample *t* test table of digit ratio (2D:4D) and body composition

Variables	Male		Female		*t*-value	*p*
	Mean	SD	Mean	SD		
Digit ratio	0.955 (mm)	0.047	0.961 (mm)	0.046	−1.198	0.232
BMI	16.64	3.821	16.22	3.12	1.167	0.252
Skinfold	44.06 (%)	7.09	44.48 (%)	4.32	−0.115	0.109

2.4 Statistical Analyses

Descriptive analysis will be used to analyze age, gender, height, weight, score for handgrip strength test, bleep test score, and body composition measurement. Pearson's correlation will implement to analyze the relationship between body composition and motor skills proficiency of participant. Meanwhile, *t* test will be used to compare the measurement of those variables between both genders. All statistical analyses were performed with Statistical Package for Social Science (SPSS 16.0) program with an alpha level set at 0.05.

3 Result

Table 1 shows the independent *t* test scores for male and female student digit ratio, body mass index, and skinfold percentage. The mean for female digit ratio was 0.961 (±0.045) while the mean for male digit ratio was 0.955 (±0.046). Female digit ratio measurement seems to be slightly higher compared to male digit ratio measurement. An independent samples *t* test was conducted to compare the between male and female digit ratio. There was a no significant difference in digit ratio measurement for males ($M = 0.955$, SD = 0.047) and females ($M = 0.961$, SD = 0.045; $t\ (365) = -1.198$, $p = 0.232$). The magnitude of the differences in the means (mean differences = −0.005) was very small (eta squared = 0.003).

Similar outcomes was found for males and females body mass index as the independent sample t test indicates there was no significant differences in body mass index for males ($M = 16.64$, SD = 3.821) and females ($M = 16.22$, $D = 3.125$; t (365) = 1.148, $p = 0.252$). Mean differences (M = 0.423) between males and females body mass index is very small with eta squared = 0.003. Skinfold percentage also shown no significant differences between male ($M = 44.06$, SD 7.09) and female students ($M = 44.48$, 4.32; $t\ (365) = -0.115$, $p = 0.109$).

The relationship between digit ratio (2D:4D) was investigated using Pearson's product-moment correlation coefficient. Preliminary analyses were performed to ensure no violation of the assumption. It was found that there was a small negative correlation between one of the three variables, $r = -0.228$, $N = 367$, $p < 0.01$, with higher score of digit ratio (2D:4D) are associated with lower body mass index.

Table 2 Correlation analysis of digit ratio on body mass index and skinfold percentage

		BMI	Skinfold percentage
Pearson correlation	Digit ratio	−0.228	−0.142

A correlation matrix calculated between digit ratio and skinfold percentage is reported in Table 2. By using correlation analysis, it was discovered that there was a very small negative relationship between digit ratio and skinfold percentage with $r = -0.142$ and $p < 0.05$, which indicated that there are no significant relationship between digit ratio and skinfold percentage. Slight negative association between 2D:4D and the body composition among school children provide minor link on the effectiveness of 2D:4D and body mass index also skinfold percentage. The relatively small negative linear correlation found in this study, however, imply that 2D:4D may not be a strong alternative for measurement of body composition but might be a weaker predictor for body composition.

4 Discussion

This study highlighted the relationship of digit ratio and body composition among school children.

The development on the association between patterns of digit formation and prenatal testosterone which closely link with 2D:4D may lie in the action of Homeobox or Hox genes. In children, digit ratio had been reported to be associated with negative association between 2D:4D and such diverse traits as athletic prowess as well as social and physical dominance. An attempt to apply the similar impact toward school children in this study found that 2D:4D between male and female school children had not shown any differences. The results indicated that 2D:4D may be a weak predictor to determine male and female school children skinfold percentage with male school children indicated a low digit ratio with low level of skinfold percentage. Relationship between digit ratio and body mass index also skinfold percentage results indicated similar outcome to the research done by Hönekopp and Watson [21], thus female did not show higher 2D:4D than male with both male and female indicate a negative relationship toward physical fitness which closely related to health-related fitness.

From the results on correlation between digit ratio and body composition which was in this study were measured through body mass index and skinfold percentage. Pearson's correlation test indicates there were negative correlations between 2D:4D and body mass index and also skinfold percentile which was different from study done by Hönekopp et al. [22]. The study listing body mass index as part of the potential confounds found that it has no significance differences between male and female digit ratio. Regarding from the results, the differences between male and female school children skinfold percentage as well as handgrip strength support the possibility that an association between low 2D:4D and body composition component and muscular strength component at least in part from the action of prenatal testosterone on such condition.

From the study, it can be conclude that there are negative relationship among digit ratio toward body composition which involved body mass index and skinfold percentage of school children. The application of digit ratio in this study has revealed that 2D:4D is associated with male and female body composition or an alternative predictor of prenatal testosterone among school children. Although possible, these seem unlikely because there is only a small negative relationship of 2D:4D as a possible predictor of body composition differs by gender.

With the addition of this study, there is now considerable empirical evidence that links 2D:4D and body composition, even though the relationship in this study appears to be very weak. More data are required to clarify the situation. Overall, this study data provide some new support toward the advantages and disadvantages of implementing digit ratio as part of the alternative in predicting body composition.

A few limitation involve in this study was the consideration of ethnicity and on cross-sectional research design used. Since this was the first attempt to apply digit ratio on health-related fitness component, several matters to be considered involved types of testing procedure used (testing batteries) as well as the suitability to be implemented among school children should be considered. Future studies should consider the investigation of 2D:4D to be executed on children sports performance as it had been shown to be related in several sports among adult male and female [3, 4, 7].

Acknowledgments We would like to thank and express our gratitude to Research Management Institute (RMI) UiTM, Faculty of Sports Science and Recreation and Universiti Teknologi MARA (UiTM) for the never ending feedback, information and support through the whole process of finishing the study.

References

1. Manning, J. T., & Bundred, P. E. (2000). The ratio of 2nd to 4th digit length: A new predictor of disease predispostition?". *Medical Hypotheses, 54*(5), 855–857.
2. Putz, D. A., Gaulin, S. J. C., Sporter, R. J., & McBurney, D. H. (2004). Sex hormones and finger length: What does 2D:4D indicate? *Evolution and Human Behavior, 37*, 100–111.
3. Manning, J. T., Churchill, A. J., & Peters, M. (2007). The effects of sex, ethnicity, and sexual orientation on self-measured digit ratio (2D:4D). *Archives of Sexual Behavior, 36*, 223–233.
4. Coates, J. M., Gurnell, M., & Rustichini, A. (2009). Second-to-fourth digit ratio predicts success among high-frequency financial traders. *Proceedings of the National Academy of Sciences, 106*, 623–628.
5. Manning, J. T. (2002). *Digit Ratio: A pointer to fertility, behavior, and health.* New Brinswick, NJ: Rutgers University Press.
6. Manning, J. T., Morris, L., & Caswell, N. (2007). Endurance running and digit ratio (2D:4D): Implication for fetal testosterone effects on running speed and vascular health. *American Journal of Human Biology, 19*, 416–421.

7. Williams, T. J., Pepitone, M. E., Christensen, S. E., Cooke, B. M., Huberman, A. D., & Breedlove, N. J. (2000). Finger length patterns and human sexual orientation. *Nature, 404*, 455–456.
8. Herault, Y., Fradeau, N., & Zakany, J. (1997). Ulnaless (UI), a regulatory mutation inducing both loss-of-function and gain-of-function of posterior Hox genes. *Development, 124*, 3493–3500.
9. Fink, B., Neave, N., & Manning, J. T. (2003). Second to fourth digit ratio, body mass index, waist-to-hip ratio, and waist-to-chest ratio: their relationships in heterosexual men and women. *Annals of Human Biology, 30*(6), 728–738.
10. Manning, J. T., Stewart, A., Bundred, P. E., & Trivers, R. L. (2004). Sex and ethnic differences in 2nd to 4th digit ratio of children. *Early Human Development, 80*(2), 161–168.
11. Manning, J. T., Trivers, R. L., Thornhill, R., & Singh, D. (2000). The 2nd:4th digit ratio and asymmetry of hand performance in Jamaican children. *Laterality, 5*(2), 121–132.
12. Ruiz, J. R., Castro-Pinero, J., Ortega, F. B., Sjostrom, M., Suni, J., & Castillo, M. J. (2009). Predictive validity of health-related fitness in youth: a systematic review. *British Journal of Sports Medicine, 43*, 909–923.
13. Fink, B., Thanzami, V., Seydel, H., & Manning, J. T. (2006). Digit ratio and hand-grip strength in German and Mizos men: Cross-cultural evidence for an organizing effect of prenatal testosterone on strength. *American Journal or Human Biology, 18*, 776–782.
14. Goran, M. I., & Treuth, M. S. (2001). Energy expenditure, physical activity, and obesity in children. *Pediatric Clinics of North America, 48*, 931–953.
15. Berenson, G. S., Srinivasan, S. R., Bao, W., Newman, W. P., Tracy, R. E., & Wattigney, W. A. (1998). Association between multiple cardiovascular risk factors and atherosclerosis in children and young adults. The Bogalusa Heart Study. *New England Journal of Medicine, 338*(23), 1650–1656.
16. Christine, L. W., Laura, L. H., Stephen, R. D., Thomas, N. R., Julia, S., Stephen, P., et al. (2002). A statement for health professionals from the committee on atherosclerosis, hypertension, and obesity in the young (AHOY) of the council on Cardiovascular Disease in the Young, American Heart Association. *Journal of the American Heart Association, 106*, 143–160.
17. Trivers, R., Manning, J. T., & Jacobson, A. (2006). A longitudinal study of digit ratio (2D:4D) and other finger ratios in Jamaican children. *Hormones and Behavior, 49*, 150–156.
18. Poplawska, H., Dmitruk, A., & Wilczewski, A. (2006). Changes in body adiposity in girls and boys from the rural areas of East Poland over a time span of 20 years. *Annals of Human Biology, 33*(1), 78–88.
19. Slaughter, M. H., Lohman, T. G., Boileau, R. A., Horswill, C. A., Stillman, R. J., van Loan, M. D., et al. (1998). Skinfold equations for estimation of body fatness in children and youths. *Human Biology, 60*, 709–723.
20. Manning, J. T., Barley, L., Walton, J., Lewis-Jones, D. I., Trivers, R. L., & Singh, D. (2000). The 2nd:4th digit ratio, sexual dimorphism, population differences and reproductive success: evidence for sexually antagonistic genes? *Evolution and Human Behavior, 21*, 163–183.
21. Hönekopp, J., & Watson, S. (2010). Meta-analysis of digit ratio 2D:4D shows greater sex difference in the right hand. *American Journal of Human Biology, 22*, 619–630.
22. Hönekopp, J., Manning, J. T., & Müller, C. (2006). Digit ratio (2D:4D) and physical fitness in males and females: Evidence for effects of prenatal androgens on sexually selected traits. *Hormones and Behavior, 49*(4), 545–549.

Effects of Ricebag on Skin Interface and Pain in Chronic Back Pain Patients

Abdul Hadi Ruslan, Rahmat Adnan, Norasrudin Sulaiman, Shariman Ismadi Ismail and Ridzuan Azmi

Abstract Ricebags have been found to be good alternative therapy in the management of chronic back pain injury. Hence, the objectives of this study were: (1) to determine the difference in temperatures between the ricebag and skin interface, and (2) to determine the effect of heat compression using ricebags to pain sensation in chronic back pain patients. This study used the repeated measure design to achieve its objectives. A total of 40 male (30 experiments, 10 controls) subjects were recruited into the study where the mean age was 34.3 years old (SD = 8.11) and the mean weight was 72.63 kg (SD = 11.2). There was a significant difference in temperature between the ricebags and the hydrocullator ($p < 0.05$), but there were no significant differences in temperature on the skin interface between the ricebag and hydrocullator ($p = 0.716$). On the effects of pain sensation, there was a significant difference in visual analogue scale (VAS) within groups for both ricebags and hydrocullator ($p < 0.05$), but there was no significant difference in VAS between ricebags and the hydrocullator ($p = 0.303$). In conclusion, ricebags gave similar effects as the hydrocullator on skin interface temperature and a modest improvement in pain perception. Thus, ricebags can be considered as one of the treatment tools in back pain management.

Keywords Ricebag · Hydrocullator · Back pain injury · Skin interface temperature · Visual analogue scale

A. H. Ruslan · R. Azmi
Rehabilitation Department, Tuanku Mizan Military Hospital, Kuala Lumpur, Malaysia

A. H. Ruslan · R. Adnan (✉) · N. Sulaiman · S. I. Ismail
Sports Science Center of Studies, Universiti Teknologi MARA, Shah Alam, Malaysia
e-mail: rahmatadnan@salam.uitm.edu.my

R. Adnan · N. Sulaiman · S. I. Ismail
Faculty of Sports Science and Recreation, Universiti Teknologi MARA, Shah Alam, Malaysia

R. Adnan et al. (eds.), *Proceedings of the International Colloquium on Sports Science, Exercise, Engineering and Technology 2014 (ICoSSEET 2014)*,
DOI: 10.1007/978-981-287-107-7_17,

1 Introduction

The cause of lower back pain is mostly unknown, but it is one of the most frequent problems reported [1]. Heat is the most common treatment for chronic lower back pain. One of the benefits of heat therapy is that it will reduce the pain, accelerate the healing of tissue damage by increasing the supply of blood, oxygen and nutrients into the lesion site. Additionally, this mechanism helps remove from the injured area the products of cell metabolism that increase nociception, and hence reduce the pain [2].

Different types of injury need different types of treatment and management. Hardy and Woodall [3] suggest that heat treatment was suitable for chronic soft tissue injury. Although cold therapy is good for reducing swelling, pain and promote muscle spasm, but heat will help to increase blood flow and metabolic rate during chronic phase. By increasing blood flow and metabolic rate, the healing process of the injury will increase. Despite the benefits, heat treatment is not suitable for acute soft tissue injury.

There are various methods for heat therapy. It can be classified into two major categories which are superficial heat therapy and deep heat therapy, including active exercise that can increase intramuscular temperature [4]. Paraffin wax, warm whirlpool, infrared radiation and moist heat packs are considered as superficial heat therapy, while ultrasound and short-wave diathermy are considered as deep heat therapy [4].

The method to apply the heat varies from traditional to modern methods, including soft heated pack filled with grains [1]. Rahmat et al. [5] had discovered that the best grain for the heat pack was between rice, barley and mung bean. The discovery of ricebags with rib fabric that has similar properties to heat pack [5] can help in chronic soft tissue injury management. The use of heat pack is inexpensive and has no adverse effects to patients [2]; and hence it can be a very effective treatment. Rice is also very widely available.

The objectives of this research are twofold:

1. To determine the difference in temperature between ricebags and the skin interface.
2. To determine the effects of heat compression using ricebags to pain sensation in chronic back pain patients.

2 Research Background

There is limited evidence on the effects of application of heat [2]. The present study is designed to provide evidence on the effectiveness of the traditional method of applying heat, i.e., use of ricebags on lower back pain patients compared to the current method using heat packs.

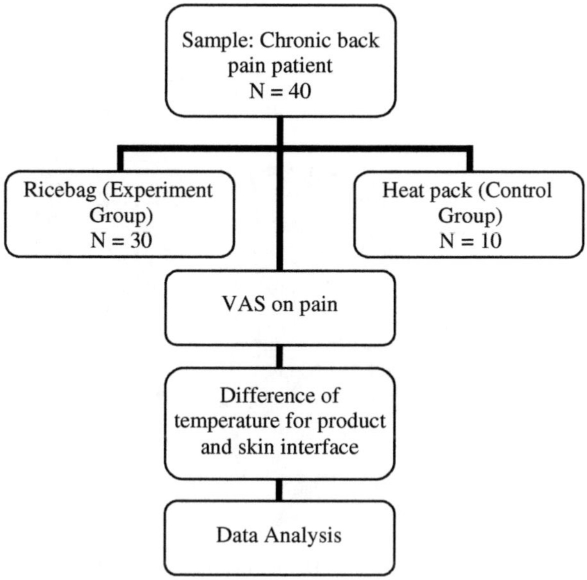

Fig. 1 Research framework

As pain is the first and most obvious symptom that tells our body that it hurts, the present study was aimed at finding the effectiveness of the ricebag as a heat pack in terms of pain management. Hence, this study will focus on determining heat transfer by observing the changes in temperature of the skin and ricebags, while applying ricebags in the management of chronic back pain.

3 Methodology

3.1 Research Design

This study used a quasi experimental method as randomization was not possible in choosing the participants or samples, who are chronic soft tissue injury patients. There were two parameters measured, including the difference in temperature between the ricebag and skin interface, and the effect of the heat transfer to the pain sensation on chronic soft tissue injury patients. The differences in temperature were analyzed using repeated measure, while the effect of heat transfer of the ricebags was determined using independent *t* test (Fig. 1).

3.2 Sample

Purposive sampling technique was used for this study in which all the 40 male participants were recruited from chronic back pain patients requiring heat therapy at Tuanku Mizan Military Hospital, Wangsa Maju (30 experiment groups,

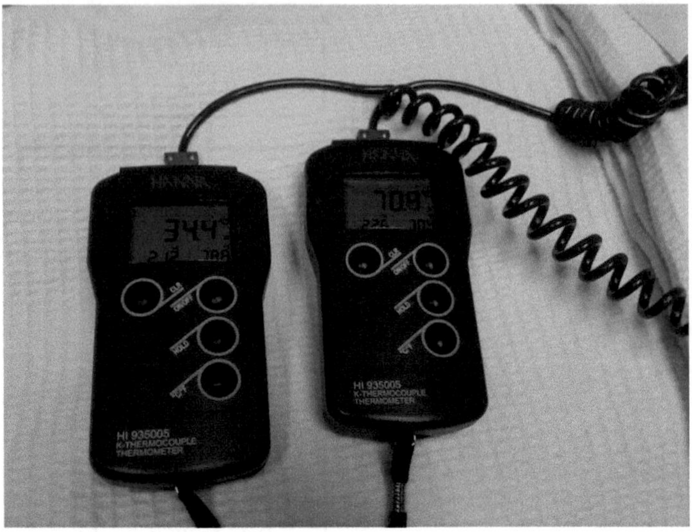

Fig. 2 Digital thermometer Hanna Instrument and Thermocouples Probe

10 control groups). All participants were identified as chronic back pain patients by a certified physiotherapist in the Physiotherapy Unit, Rehabilitation Department, Tuanku Mizan Military Hospital. Participants' exclusion criterion was set up prior to the test. These included participants who do not consume any pain killer medicine 24 h before the testing.

3.3 Tools

Ricebag was used as the experimental product as the purpose of the study was to evaluate the effectiveness of ricebag, while heat packs from the hydrocullator® unit was used in the control group. To measure temperature, a digital thermometer (Hanna Instrument P/N: HI935005 and thermocouples probe which is Hanna K-Type Waterproof Thermocouples) was used. Visual Analogue Scale (VAS) was used to measure the pain level of participants. VAS is a scale from 0 to 10, with ascending level of pain, in which 0 is no pain while 10 is unbearable pain [6] (Fig. 2).

3.4 Procedure

The testing was carried out over 20 min for each group. This test was performed in 20 min because maximal elevation of temperature in skin and subcutaneous tissues occurred with 20 min of superficial heating, whereas muscles require 30 min to

Table 1 Descriptive data of samples

	N	Mean	SD
Age	40	34.30	8.11
Weight	40	72.63	11.20

achieve their peak increase [3]. Ricebag or heat pack was placed at the site of injury which is at the lower back. The tip of digital thermometer, which is its sensor, was placed at the ricebag or heat pack and at the skin interface and was covered by the ricebag or heat pack. Both the temperature of ricebag or heat pack and skin interface were measured and recorded every minute from minute 0 to minute 20.

A closed environment with controlled room temperature and humidity was required to get the best results for this test, but such a room was not available. Hence, the same room was used throughout the testing, with a mean room temperature of 21.3 °C (SD = 1.49).

For recording pain level of participants, VAS for pain was measured before the ricebag or heat pack was placed at the site of injury (pre) and after 20 min of the treatment (post). VAS is very common and suitable to be used to evaluate acute pain effects [7].

4 Results

4.1 Demographic Data of Sample

The 40 participants with chronic lower back pain had a mean age of 35.7 years old (SD = 8.11) and a mean weight of 72.7 kg (SD = 11.2). Participants are divided into the experimental group ($N = 30$), which received treatment using ricebag and the control group ($N = 10$), which received treatment using heat pack from the hydrocullator (Table 1).

4.2 Demographic Data of Ricebag

For difference in temperature, the graph below shows that there is a significant different of temperature of ricebag or heat pack and skin interface within groups. But Table 2 shows that there is no significant effect of temperature of skin between groups.

The effect of the treatment (ricebag or heat pack) on pain sensation as evaluated by VAS of pain, showed significant differences in VAS of pain between the pre- and post-test for both groups (Table 4). But between groups, it showed no

Table 2 Comparison of temperature between groups

Measure	df	*p*
Product[a]	1.720	0.000
Skin interface	3.330	0.716

[a] Product represents ricebag and heat pack

Table 3 Descriptive statistic on VAS

		N	Mean	Standard deviation
VAS pre	Ricebag	30	4.23	1.331
	Heat pack	10	4.30	0.823
VAS post	Ricebag	30	2.60	1.248
	Heat pack	10	3.00	0.667

Table 4 Comparision of VAS between groups

Source	Df	F	*p*
VAS	1	84.367	0.000
VAS (between groups)	1	1.089	0.303

significant differences in VAS on pain between the experiment group (Mean = 2.6, SD = 1.25) and the control group (Mean = 3, SD = 0.67). Overall, both groups showed improvements in their VAS on pain (Table 3).

5 Discussion

The mean temperatures of the ricebags were high compared to the mean temperatures of the heat pack (Fig. 3). However, there was no significant difference in skin temperature between groups ($p < 0.05$) (Fig. 4). It shows that despite the ricebag's high-mean temperature, the distribution of heat was almost similar to the heat pack with a lower mean temperature. It means that the ricebag had a similar effect on heating superficial layer of skin compared with a heat pack by hydrocullator®. Thus, the hydrocullator® which has been widely used worldwide, can be a bench mark for the ricebag.

Both groups (ricebag and heat pack) had a positive effect of VAS on pain. However, the ricebag showed a better improvement in VAS (Mean = 2.6, SD = 1.25). These results indicate that heat therapy gives benefits in pain management of chronic lower back pain and other musculoskeletal problems in general. It is caused by the increase in nociception that is caused by a decrease in cell metabolism that reduces pain [2]. Apart from that, heat therapy will also increase blood flow, increase the extensibility of connective tissues and decrease reflexive muscle excitability (Figs. 5 and 6).

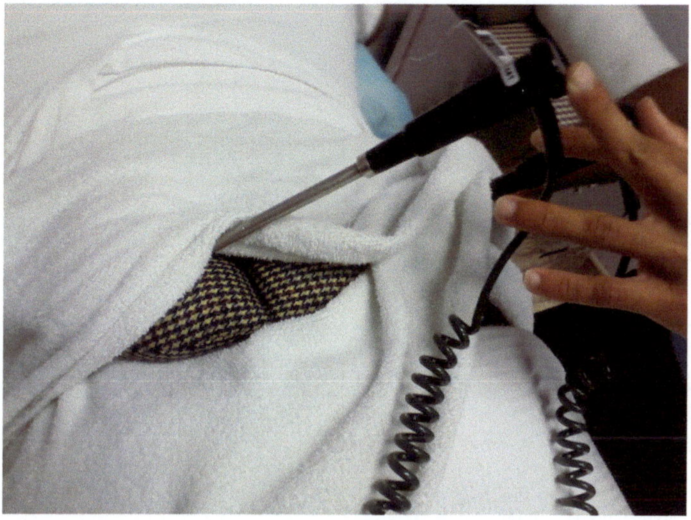

Fig. 3 Method of placing thermocouples to skin and ricebag

Fig. 4 Procedure

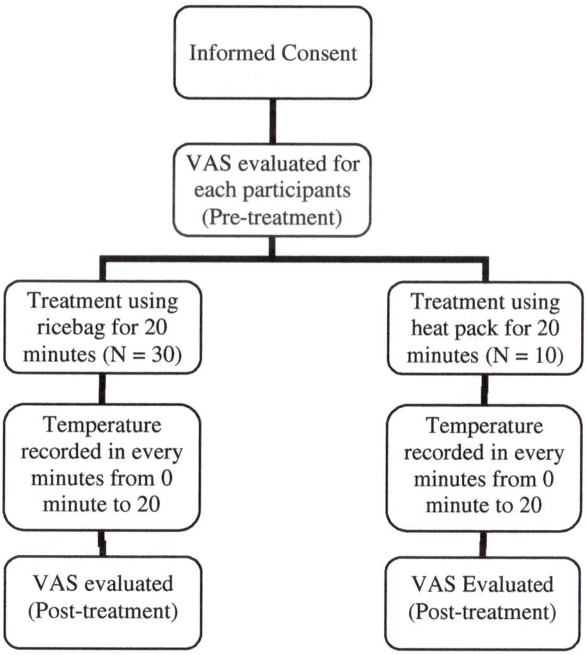

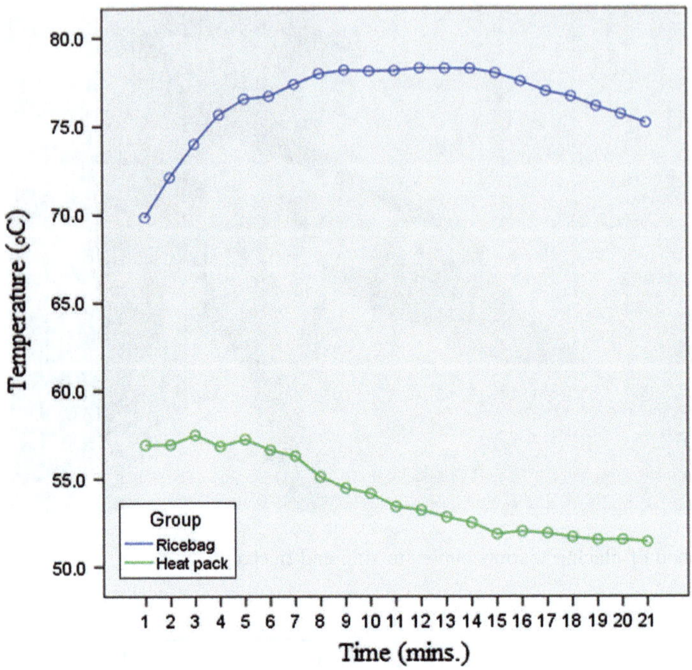

Fig. 5 Descriptive means of temperature for ricebag and hydrocullator

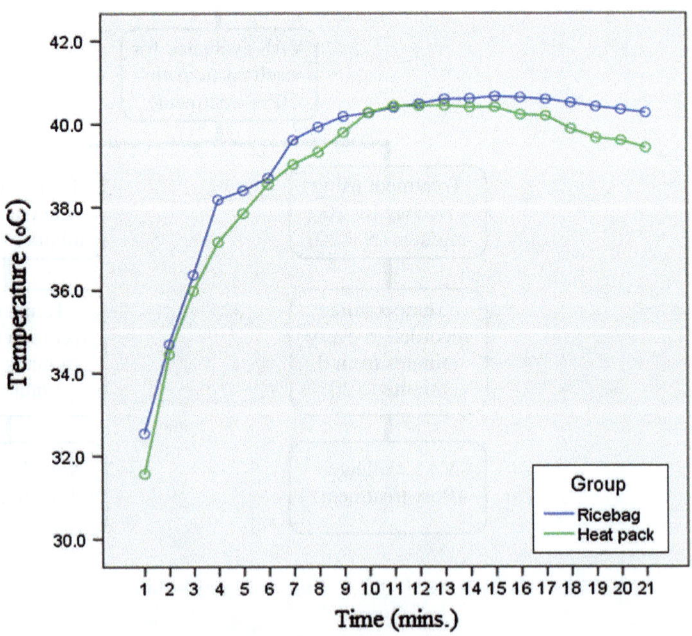

Fig. 6 Descriptive means of temperature for skin interface

6 Conclusion

In conclusion, the results showed that ricebags had similar properties to the heat pack from hydrocullator®, and that it will give similar temperatures on the skin interface as the heat pack. Ricebags also provide benefits in pain management as the results showed that VAS on pain is reduced in pre- and post treatments. Hence, ricebags are very suitable for heat therapy.

Acknowledgments Researcher would like to express upmost gratitude to RMI—RIF Grant for sponsoring this study. Thanks to Rehabilitation Department, Tuanku Mizan Military Hospital for giving us the opportunity to recruit participants. Last but not least thanks to everyone involved in producing this paper directly or nondirectly.

References

1. French, S. D., Cameron, M., Walker, B. F., Reggars, J. W., & Esterman, A. J. (2006). Superficial heat or cold for low back pain. *The Cochrane Library, 1*, 1–10.
2. Ramirez-Velez, R., et al. (2011). Effects of two methods of heat therapy on the acute vascular response and hemodynamics in healthy subjects. *Elsevier España, 54*(9), 513.
3. Hardy, M., & Woodall, W. (1998). Therapeutic effects of heat, cold, and stretch on connective tissue. *Journal of Hand Therapy, 11*, 148–156.
4. Hanson, M., & Day, J. (2012). Effects of different heating modalities on hip flexion passive range of motion. *International Journal of Athletic Therapy and Training, 17*(6), 27–30.
5. Rahmat, A., Zainon, M. N., Noriham, A., Marsinah, T. S., Norasrudin, S., & Mastura, M. (2011). The development of beany/grainy pack to treat soft tissue injury. In *IEEE Colloquium on Humanities, Science and Engineering Research* (pp. 398–401).
6. Couper, M. P., Tourangeau, R., Conrad, F. G., & Singer, E. (2006). Evaluating the effectiveness of visual analog scales. *Social Science Computer Review, 24*(2), 227–245.
7. Breivik, H., et al. (2008). Assessment of pain. *British Journal of Anaesthesia, 101*, 17–24.

The Differences Between Students with Intellectual Disabilities and Normal Students on the Physical Fitness Level

Nagoor Meera Abdullah, Norlizah Abdul Hamid, Wahidah Tumijan, Vincent Parnabas, Mohamad Rahizam Abdul Rahim, Sarimah Ismail and Rozita Abdul Latif

Abstract The aim of this study was to investigate the differences in the physical fitness level among student with intellectual disabilities (ID) and normal students. A total of 45 students that consist of ID students ($n = 25$) and 20 normal students ($n = 20$) from an integrated school were recruited voluntarily. The instruments used in the study are the handgrip test for muscular strength, sit-up and push-up test for muscular endurance, sit and reach test for flexibility, and AAHPERD shuttle run test for agility. The mean height and weight for the ID and normal student were 1.54 ± 6.01 m and 47.7 ± 8.15 kg, respectively. The mean value for sit and reach for student with ID was 9.2 ± 2.6 cm; handgrip (right) test was 21.9 ± 5.8 kg, handgrip (left) was 20.7 ± 4.8 kg, the sit-up test was 24.6 ± 8.6, the push-up was 20.7 ± 8.1, and for the shuttle run test was 9.2 ± 2.1 s. As anticipated, the normal students have demonstrated better physical fitness when compared to ID students. The present study findings reveal that the physical fitness level is almost similar to normal student; however, more attention and care must be given to the ID student with respect to health and normal growth development. Thus, more studies among ID individuals are required to better understanding on their physical performance and with such study; it will help the health care professional and special education teachers to formulate, design, and implement a well-structured physical fitness program for students with ID.

Keywords Physical fitness level · Students with intellectual disabilities · Normal students · Differences · Flexibility

---

N. M. Abdullah (✉) · N. Abdul Hamid · W. Tumijan · V. Parnabas
M. R. Abdul Rahim · S. Ismail · R. Abdul Latif
Faculty of Sport Science and Recreation, Universiti Teknologi MARA,
Shah Alam, Selangor, Malaysia
e-mail: nameera_ab@yahoo.com.my

R. Adnan et al. (eds.), *Proceedings of the International Colloquium on Sports Science, Exercise, Engineering and Technology 2014 (ICoSSEET 2014)*,
DOI: 10.1007/978-981-287-107-7_18,

1 Introduction

The term ID covers the same population of individuals whose conditions were diagnosed previously with mental retardation in definition, classification, and systems of support (1). Examples of syndromes and conditions associated with ID include, but are not limited to, attention deficit/hyperactivity disorder (ADHD), autistic syndrome disorder, Down syndrome, fragile X syndrome, and Klinefelter syndrome. Historically, there were limited studies that actually measured the PA patterns of youth with ID. Consequently, PA patterns of youth with ID were linked to or were based on their physical fitness status. It has been consistently reported that youth with ID have lower levels of cardiovascular fitness and muscular strength as well as high levels of adiposity [i.e., body mass index (BMI)] in comparison with their peers without ID [1, 2]. Therefore, it was assumed that youth with ID had lower PA patterns than youth without ID [3].

In general, studies have reported that individuals with intellectual disabilities (ID) demonstrate poor fitness levels on standard fitness tests. This has been reported on measures of cardiovascular endurance [4], body composition [5], and muscular strength and endurance [6, 7]. Young adults with ID (20–30 years of age) typically exhibit cardiorespiratory fitness levels that are 8–12 % lower than expected values [8] and show peak heart rates to be lower by about 15 beats/min compared with their peers without ID [9].

About 20 % of the adults with ID have been classified as obese with a strong inverse relationship between IQ and adiposity [10]. Comparison of the strength of elbow and knee extension and flexion between young adults with ID and sedentary, non-disabled individuals showed 35–40 % lower strength levels for individuals with ID [6]. No difference between individuals with ID and able-bodied individuals is reported with regard to flexibility [11], but there is a paucity of published information concerning this fitness component [12]. Physical fitness components are assessed in a variety of ways. Although most tests have good basic validity in the general population, this has not necessarily been shown in individuals with ID. It is, however, imperative to use tests that are validated for use in populations with ID [13].

Furthermore, tests developed to assess physical fitness measures in adults with ID do not necessarily apply to children with ID [14]. Also, too often, participants with various degrees of ID and/or various diagnoses are included in the same samples, and data often are collapsed across gender [9]. This is problematic when making inferences with regard to their physical fitness profile. For instance, including individuals with Down syndrome (DS) should not be done because their physical fitness level is inferior to that of peers with ID but without DS [8]. Some authors [14] have also noted that the degree of ID might influence initial fitness levels and trainability of participants.

Besides these methodological concerns about the research published so far, most research investigating physical fitness in individuals with ID until today is limited to participants that are relatively inactive [9]. Few investigations report findings in trained individuals with ID. For instance, [15] and [16] reported high VO_2 peak measures in individuals with ID, comparable with measures in well-trained able-bodied individuals, but very low sample sizes (19) in the [15] investigation, seven males, and two females in the Frey et al. investigation) limit the generalization of the findings. Furthermore, too often, one is left with the belief that the nature of ID renders it impossible for a person to engage in the level of training or mental preparation required for a high level of competition. As a consequence, one considers the lower physical fitness levels as "related to the disability," although "lifestyle" cannot be excluded as responsible for lower physical activity and fitness levels [17]. It is therefore unclear to what extent the physical fitness levels reported so far reflect the full potential of individuals with ID. To the authors' knowledge, the present investigation is the first of its kind in Malaysia to evaluate the physical fitness of the students with ID and compare them with the normal students.

2 Methodology

2.1 Sample

A total of 25 students with ID (15 males; 10 females) and 20 normal students (11 males; 9 females) participated in the study. They are studying in an integrated school where the school did accommodate both normal students and allocated a class for students with ID. The student's age is 14 years old. The normal students had no known disorder. All participants were informed of the study procedure, purposes, and all gave their inform consent.

2.2 Instrumentation

Standing height was recorded to the nearest half centimeter with the subject barefoot and with the back against a vertical wall. Body weight was measured to the nearest 0.5 kg with shoes, sweaters (SECA model 841).

Grip strength is an important prerequisite for good performance of the upper limb. In the study, handgrip strength was measured using a standard adjustable handgrip strength test (Takei model TKK5401). Maximum handgrip forces for dominant hand were recorded in kilograms as the highest of two trials. Before testing the participants individually, the researcher gave a brief orientation to the entire group. The dynamometer was adjusted to the size of the hand of participant. The arm, the hand, and the body position were standardized according to the suggestion of the American Society of Hand Therapists. Subject were sitting with shoulder adducted and neutrally rotated, elbow flexed at 90° resting on the table surface and the forearm in neutral and wrist in 0–30° extension. The test was performed by squeezing calibrated hand dynamometer as forcefully as possible with the dominant hand. Static strength was assessed.

Sit reach test use to measure the flexibility of the hamstrings, buttocks, and lower back [9]. The participants were instructed to reach as far as possible from a sitting position (Acuflex model 1).

The push-up test is use to measure upper body muscular endurance. The participant lies face down on the floor with the body straight, arm bent, and hands flat on the floor beneath the shoulders. The participants push upward to a straight-arm position. The participants lower the body until the chest touch the floor. Repeat the exercise as many times as possible, without rest. The body must stay rigid (not sag or pike upward) throughout the test. A sponge that is 2 inches high may be placed on the floor for the participants to touch with the chest.

Ten-millimeter AAHPERD shuttle run test is to measure agility. Mark two lines 10 m apart using marking tape or cones. The two blocks are placed on the line opposite the line they are going to start at. On the signal "ready", the subjects place their front foot behind the starting line. On the signal, "go!" the subject sprints to the opposite line, picks up a block of wood, runs back, and places it on or beyond the starting line. Then turning without a rest, they run back to retrieve the second block and carry it back across the finish line. Two trials are performed.

Sit-up test use to measure muscular endurance of the abdomen. The subjects need to lie down on the exercise mat, with both their leg bend at 90°. Upon receive the signal "Go," the subjects need to perform sit-up with their chest touching their leg as many as they can within 1 min. The score will be recorded.

2.3 Test Administration and Data Collection

In this study, the sequence of the test follows as such (a) handgrip strength test, (b) sit and reach test, (c) sit-up test, (d) push-up test, and (e) 10 m AAHPERD agility test been organized. The entire test been conducted at 10.00 a.m. to 4.00 p.m. at the school. Before conducting the test, the test administrators give the briefing for the students how the test was conducted. Besides that the weight and height the students been taken and the test administrators conduct the warm up and stretching for 15–20 min to prepare the body from injury. Then, the students perform the test and follow the procedure and the best score been recorded in the scores sheet.

2.4 Analysis of Data

The results are presented as means and the standard deviations. The SPSS package was used for the statistical analysis.

Table 1 Demographic data of the students with ID

	Mean	Standard deviation
Height	154.02	6.333
Weight	46.502	8.1667
Age	14.00	0.000

Table 2 Demographic data of the normal students

	Mean	SD
Height	154.15	6.335
Weight	50.520	6.1939
Age	14.00	0.001

3 Results

Table 1 shows the demographic data for the students with ID. The mean ± SD for gender were 0.58 ± 0.499. The mean height of the students was 154.02 ± 6.33 cm. Furthermore, the mean weight for the students was 46.50 ± 8.1667 kg. The mean age for these students is 14.00 ± 0.00.

Table 2 shows the demographic data for the normal students. The mean ± SD for gender were 0.55 ± 0.510. The mean height was 154.15 ± 6.34 cm. Furthermore, the mean weight for the students was 50.52 ± 6.19 kg. The mean age for these students is 14.00 ± 0.000.

The Table 3 shows the overall result of the achievement of the fitness test between students with ID and normal students. The mean for the sit and reach test for male normal students were 11.36 ± 1.75 and mean for the sit and reach test for the male students with ID were 10.0 ± 2.15. It shows that the male students with ID have poor flexibility than the male normal student and this is same with their female counterpart. The mean for the sit and reach test for the female normal students are 11.22 ± 1.89 compared to the female students with ID (10.0 ± 2.15).

The handgrip strength tests show the same results as the mean for the male students with ID for the right hand (24.42 ± 5.39) and the left hand (22.35 ± 4.83) are much lower compared to the male normal students for the right hand (28.54 ± 4.01) and for the left hand (25.92 ± 3.95). This shows the male students with ID have lower strength ability compared to their male counterparts. The results for female students with ID show the lower mean on the right hand and left hand, respectively (18.28 ± 4.32; 18.23 ± 3.64), compared to the normal female students on the right hand and the left hand (22.79 ± 4.87; 18.92 ± 4.17) exhibit that female student with ID also possess lower strength compared to the female normal students.

The sit-up test shows the male students with ID exhibit lower muscular endurance (29.13 ± 7.19) compared to the male normal students (34.09 ± 4.43), but they achieved equally the same results on push-up test (26.33 ± 3.84) compared to the normal students (26.91 ± 3.91). The female students with ID also

Table 3 The level of fitness among the students with id and the normal students according to the selected components and gender

Fitness test	Students with ID ($N = 25$)		Normal students ($N = 20$)	
	Male	Female	Male	Female
Sit and reach test (cm)	10.0 ± 2.15	8.0 ± 2.88	11.36 ± 1.75	11.22 ± 1.89
Handgrip strength test (Right) (Kg)	24.42 ± 5.39	18.28 ± 4.32	28.54 ± 4.01	22.79 ± 4.87
Handgrip strength test (Left) (Kg)	22.35 ± 4.83	18.23 ± 3.64	25.92 ± 3.95	18.92 ± 4.17
Sit-up test (num/min)	29.13 ± 7.19	17.80 ± 5.51	34.09 ± 4.43	29.11 ± 3.37
Push-ups (num/min)	26.33 ± 3.84	12.20 ± 4.47	26.91 ± 3.91	21.11 ± 3.98
AAHPERD shuttle run test (s)	7.75 ± 0.81	11.39 ± 1.34	6.41 ± 0.51	8.41 ± 0.66

show lower abdominal muscular endurance (17.80 ± 5.51) compared to the female normal students and also show lower deltoid, biceps, and triceps muscle endurance (12.20 ± 4.47) compared to the normal students (21.11 ± 3.98).

The AAHPERD shuttle run test shows male students with ID have lower ability on agility (7.75 ± 0.81) compared to the male normal students (6.41 ± 0.51). The female students with ID also show also lower ability agility (11.39 ± 1.34) compared to the female normal students (8.41 ± 0.66). Since this test is based on time, the male and the female students with ID took longer time to finish the test, and yet they are slower compared to the male and the female normal students.

4 Discussion

In the present study, the fitness level of students with ID is described and compared to standards of able-bodied students. All students took the opportunity to have their physical condition assessed on a voluntary basis.

The result of muscular strength and muscular endurance indicates that students with ID were poor in their muscular strength and endurance which is in agreement with [6] and [3]. The present study on flexibility also shows that student with ID demonstrates poor flexibility compared with normal student. This has been proven by previous research on adult by [18]. Results for agility test in the study also show that students with ID are also low in their agility, which agrees with [1]. Previous research has indicated ID people will demonstrate lower measures of muscular strength, muscular endurance, and flexibility [19]. According to [14], people with ID will exhibit poor fitness performance on standard fitness tests.

In the overall, it can be said that the health-related physical fitness of people with ID is low. This is agreed by [1–3]. The actual physical fitness status of athletes with ID is better than that of individuals without disabilities [20, 21], including individuals with lower activity rates, for flexibility, and upper body muscle endurance. For the other fitness components, even athletes with ID do not score better (i.e., abdominal muscle endurance and running speed) or even worse (i.e., speed of limb movement, explosive strength, handgrip strength) than individuals without disability. Percentage body fat for both male and female athletes is 3–4 % higher than the standards for percentage body fat in young active able-bodied adults (10 % body fat for men, 23 % body fat for women) [22], although ethnic differences need to be taken into consideration in the interpretation of this finding. Compared with physically active able-bodied individuals, all engaged in regular sports activities, but not necessarily competing at a high-performance level [23–25]; the cardiorespiratory endurance of elite athletes with ID is poor. If that is the case, one has to agree that it is true based on the present study proved that students with ID have lower fitness level compared to the normal students?

Even the male students with ID when compared, score better for flexibility and running speed, but have lower strength and muscle endurance levels. Female students with ID also have similar levels of flexibility, upper body muscle endurance, and running speed as female university students, but have lower explosive and handgrip strength levels and lower abdominal muscle endurance. These strength differences are in the range of 4–14 % for male students and 11–27 % for female students, whereas up to 40 % strength differences had been reported for non-trained individuals with ID. The overwhelming majority of published research on physical fitness of persons with ID indicates that there are significant differences that would place individuals with ID at risk for poor health over their lifetime. The present data refute that and indicate that with persistent and long-term training comparable profiles to able-bodied can be found. With the exception of strength, good levels of fitness seem to be possible, and likely the training effect influenced the data.

5 Conclusion

In summary, it can be said that high-performance athletes with intellectual disabilities reach fitness levels that are, in general, equal to or lower than those of their able-bodied sportive peers. The present data, in general, support the fact that individuals with ID indeed have limited physical capacities, especially with regard to strength measures, but much more scientific data are needed on high-performance athletes with ID. Further research should on the one hand include the investigation of training regimens, as to verify whether their trainings are optimal or can be optimized. On the other hand, further investigation should also focus on the nature of limited physical capacities in athletes with ID. The underlying mechanisms are not known at this time. In the meantime, the outcome of this investigation will certainly lead to the description of new physical fitness norms for individuals with ID. Next to that, the authors also hope that the present data contribute to the discussion of trainability and high-performance sports competition of individuals with ID.

Acknowledgments The authors wish to acknowledge the Faculty of Sport Science and Recreation, Universiti Teknologi MARA to provide funding for the presentation and the publication.

References

1. Pitetti, K. H., & Yarmer, D. A. (2002). Lower body strength of children and adolescents with and without mild mental retardation: A comparison. *Adapted Physical Activity Quarterly, 19*, 68–81.
2. Pitetti, K. H., Yarmer, D. A., & Fernhall, B. (2001). Cardiovascular fitness and body composition in children and adolescents with and without mental retardation. *Adapted Physical Activity Quarterly, 18*, 127–141.
3. Pitetti, K., Beets, M. W., & Combs, C. (2009). Physical activity levels of children with intellectual disabilities during school. *Medicine and Science in Sports and Exercise, 41*(8), 1580–1586.
4. Fernhall, B., & Pitetti, K. H. (2001). Limitations to physical work capacity in individuals with mental retardation. *Clinical Exercise Physiology, 3*, 176–185.
5. Rimmer, J. H., Braddock, D., & Fujiura, C. (1993). Prevalence of obesity in adults with mental retardation: implications for health promotion and disease prevention. *Mental Retardation, 31*, 105–110.
6. Horvat, M., Pitetti, K. H., & Croce, R. (1997). Isokinetic torque, average power, and flexion/extension ratios in nondisabled adults and adults with mental retardation. *Journal of Orthopaedic and Sports Physical Therapy, 25*, 395–399.
7. Mac Donncha, C., Watson, A. W. S., McSweeney, T., & O'Donovan, D. J. (1999). Reliability of Eurofit physical fitness items for adolescent males with and without mental retardation. *Adapted Physical Activity Quarterly, 16*, 86–95.
8. Fernhall, B., Pitetti, K. H., Rimmer, J. H., McCubbin, J. A., Rintala, P., Millar, A. L., et al. (1996). Cardiorespiratory. Capacity of individuals with mental retardation including down syndrome. *Medicine and Science in Sports and Exercise, 28*, 366–371.
9. Fernhall, B., & Pitetti, K. H. (2001). Limitations to physical work capacity in individuals with mental retardation. *Clinical Exercise Physiology, 3*, 176–185.
10. Hove, O. (2004). Weight survey on adult persons with mental retardation living in the community. *Research in Developmental Disabilities, 25*, 9–17.
11. Frey, G. C., McCubbin, J. A., Hannigan-Downs, S., Kasser, S. L., & Skaggs, S. O. (1999). Physical fitness of trained runners with and without mild mental retardation. *Adapted Physical Activity Quarterly, 16*, 126–137.
12. Chanias, A. K., Reid, G., & Hoover, M. L. (1998). Exercise effects on health-related physical fitness of individuals with an intellectual disability: A metaanalysis. *Archives of Physical Medicine and Rehabilitation, 15*, 119–140.
13. Pitetti, K. H., & Fernhall, B. (1997). Aerobic capacity as related to leg strength in youths with mental retardation. *Pediatric Exercise Science, 9*, 223–236.
14. Fernhall, B., Pitetti, K. H., Vukovich, M. D., Stubbs, N., Hensen, T., Winnick, J. P., et al. (1998). Validation of cardiovascular fitness field tests in children with mental retardation. *American Journal on Mental Retardation, 102*, 602–612.
15. Rintala, P., McCubbin, J. A., Downs, S. B., & Fox, S. D. (1997). Cross validation of the 1-mile walking test for men with mental retardation. *Medicine and Science in Sports and Exercise, 29*, 133–137.
16. Frey, G. C., McCubbin, J. A., Hannigan-Downs, S., Kasser, S. L., & Skaggs, S. O. (1999). Physical fitness of trained runners with and without mild mental retardation. *Adapted Physical Activity Quarterly, 16*, 126–137.
17. Draheim, C. C., Williams, D. P., & McCubbin, J. A. (2002). Physical activity, dietary intake, and the insulin resistance syndrome in nondiabetic adults with mental retardation. *American Journal on Mental Retardation, 107*, 361–375.
18. Montgomery, D. L., Reid, G., & Koziris, L. P. (1992). Reliability and validity of three fitness tests for adults with mental handicaps. *Canadian Journal of Sport Sciences, 17*, 309–315.

19. Fernhall, B., Pitetti, K. H., Vukovich, M. D., Stubbs, N., Hensen, T., Winnick, J. P., et al. (1998). Validation of cardiovascular fitness field tests in children with mental retardation. *American Journal on Mental Retardation, 102*(6), 602–612.
20. Lefevre, J., Beunen, G. P., Borms, J., & Vrijens, J. (1998). Sex differences in physical fitness in Flemish youth. In J. Parizkova & A. P. Hill (Eds.), *Physical fitness and nutrition during growth* (Vol. 43, pp. 54–67). Berlin: Karger Publishers. (Med Sport Sci).
21. Lefevre, J., Philippaerts, R. M., Delvaux, K., Thomis, M., Vanreusel, B., Vanden Eynde, B., et al. (2000). Daily physical activity and physical fitness from adolescence to adulthood: a longitudinal study. *American Journal of Human Biology, 12*, 487–497.
22. Lohman, T. G., Houtkooper, L. B., & Going, S. B. (1997). Body composition assessment: body fat standards and methods in the field of exercise and sports medicine. *ACSM Health Fitness Journal, 1*, 30–35.
23. Ramsbottom, R., Nevill, M. E., Nevill, A. M., & Hazeldine, R. (1997). Accumulated oxygen deficit and shuttle run performance in physically active men and women. *Journal of Sports Sciences, 15*, 207–214.
24. Stickland, M. K., Petersen, S. R., & Bouffard, M. (2003). Prediction of maximal aerobic power from the 20-m multi-stage shuttle run test. *Canadian Journal of Applied Physiology, 28*, 272–282.
25. Matsuzaka, A., Takahashi, Y., Yamazoe, M., Kumakura, N., Ikeda, A., Wilk, B., et al. (2004). Validity of the multistage 20-M shuttle-run test for Japanese children, adolescents, and adults. *Pediatric Exercise Science, 16*, 113–125.

23. Rintala, P., McCubbin, J. A., Downs, S. B., & Fox, S. D. (1997). Cross validation of the 1-mile walking test for men with mental retardation. *Medicine and Science in Sports and Exercise, 29*, [illegible].
24. [illegible]
25. [illegible]

Single- Versus Three-Set Resistance Training on Strength and Power Among Untrained Men

Zulkifli Abdul Kadir, Ali Md Nadzalan, Sarina Md Yusof, Suhana Aiman and Mohamad Nizam Mohamed Shapie

Abstract The purpose of this study was to compare the effects of single- versus three (3)-set resistance training on muscular strength and muscular power among untrained men. Thirty-six untrained men were recruited in this study. Participants were randomized into 3 groups; single set ($n = 12$, age $= 20.92 \pm 0.79$), 3 set ($n = 12$, age $= 21 \pm 0.74$), and control group ($n = 12$, age $= 21 \pm 0.67$). Muscular strength was measured by bench press and squat performance. Static strength was measured by handgrip strength test and muscular power by the vertical jump test. The intervention groups were engaged in 6 weeks of training three times per week. Common exercises were performed to the point of achieving muscular failure for every set. Significant improvement was recorded in muscular strength and muscular power performances, ($p < 0.05$) for both the single- and three-set groups. However, no significant differences were found for all the test performances between the invention groups ($p > 0.05$). In conclusion, single-set and three-set resistance trainings showed similar training effects on muscular strength and muscular power among untrained men.

Keywords Resistance training · Single set · Multiple set · Strength · Power · Untrained men

---

Z. Abdul Kadir (✉) · A. M. Nadzalan · S. Md Yusof · S. Aiman · M. N. Mohamed Shapie
Faculty of Sports Science and Recreation, Universiti Teknologi MARA, Shah Alam, Malaysia
e-mail: pmzulkadir@yahoo.com

R. Adnan et al. (eds.), *Proceedings of the International Colloquium on Sports Science, Exercise, Engineering and Technology 2014 (ICoSSEET 2014)*,
DOI: 10.1007/978-981-287-107-7_19,

1 Introduction

Resistance training is one of the best alternatives to achieve great improvement in muscular strength [1]. However, designing a resistance training program is a complex process that requires the recognition and manipulation of several program design variables [2].

One of the most important considerations within resistance training is the volume of exercise. Volume relates to the total amount of weight lifted in a training session derived from the number of sets, load lifted, and the number of repetitions performed.

For decades, debate has persisted regarding the volume or number of sets a person must perform to elicit maximal strength gains. Coaches were more adhered to the traditional training regimens which were based on the findings of Berger [3]. This early research found that a volume of three sets provided the best strength gains in individuals.

However, several studies had challenges the long held standards of strength training. In particular, many exercise scientists have found that single-set training volumes may be as effective as multiple-set protocols [4–7].

Propagators of one set strongly claimed that one-set resistance training would induce similar training effects for strength and power development as multiple-set resistance training. It is also time-saving and may help to prevent the onset of overtraining. However, multiple sets propagators insist on the superiority of multiple-set resistance training due to the increment of training volumes. Thus, the question of the number of sets that should be performed in a training session is still inconclusive.

In determining the answer for this question, the study needs to be conducted in a way that no methodological biasness could occur. Certain previous studies conducted were without controlling the intensity and repetitions performed, [8–10], thus raising the questions whether the results were affected by the number of sets or other variables.

A reduction in lifting volume from three sets per exercise to a single set would reduce the time needed for weight training by nearly one-third. However, it is still inconclusive whether performing single set is as effective as performing three sets.

In response to the current problem, the present study was designed to compare the effects of single- and multiple-set resistance training on muscular strength and muscular power among untrained men.

This study can help to provide greater understanding for the educationist, researchers, sports scientist, and strength and conditioning professionals on the effects of the different volume or number of sets performed in resistance training. This study is also important to the untrained populations in helping them to design effective training programs in resistance training. Having good muscular strength and muscular power is undoubtedly important to all concerned as these can benefit on their functionality and sporting performance.

2 Methodology

2.1 Participants

Thirty-six healthy and untrained men were selected and recruited for this study. Convenience sampling technique was used due to some specified criteria in selecting the sample. Participants selected were males aged between 18 and 25 years based on their year of birth. Participants had no experience in resistance training and were free from injury, endocrine and/or medical problems, and not consuming any performance-enhancing supplementation.

Participants were screened prior to testing using PAR Q and ECG test. Each participant had read and signed an informed consent for testing and training approved by the UiTM Research Ethics Committee. Participants were assigned into single-set, three (3)-set, and the control groups based upon randomized control trial of their bodyweight.

2.2 Static Strength Test

A calibrated Takei grip dynamometer with a reported reliability coefficient of 0.90 was used to measure handgrip strength. The participant performed the test in the standing position with head in the midposition (facing straight ahead). The grip size was adjusted so that the middle finger's (third digit's) midportion (second phalanx) is approximately at a right angle. The instructor then recorded the grip setting. The same setting was used for further tests on the same person. The participant's forearm was placed at 90° of the upper arm; the upper arm hanged in a vertical position. The participants were given three trials alternately with each hand, with at least 30 s or up to 1 min between trials for the same hand. The instructor reset the dynamometer's pointer to zero after each trial. The sum of the best of the left and right grip was divided by body weight for the relative strength score.

2.3 Upper Body Strength Test

Bench press performance was used to measure upper body strength. The equipments used were a Cybex bench station, a calibrated UESAKA (IWF approved) bar, calibrated UESAKA, and Olympic weight plates and calibrated locks. The testing procedures involved the protocols as mentioned by Baechle and Earle [11].

After adjusting the desired amount of weight on the bar, the participants assume a supine position on the bench and two spotters place the bar in his/her hands and across the chest. With the hands approximately shoulder-width apart, the participant extends the arm, pressing the bar to a “locked out” (elbow straight) position. The two spotters remove the bar on completion of the trial. The multiple RM achievement of not more than 8 repetitions to failure is then calculated to attain the 1RM score (in kg). It was divided by body weight for the relative strength score.

2.4 Lower Body Strength Test

Back squat exercise was used to measure lower body dynamic strength. All equipments were calibrated prior to the test. Calibrated UESAKA and Olympic weight plates and calibrated locks were used.

After adjusting the desired amount of weight on the bar, it was then placed on the shoulders behind the neck with the hands gripping the bar in pronated forearm position. The participant begins the lift from full knee extension and then bends at the knee until the top of the thigh is parallel with the ground. Once the thigh was parallel with the ground, the investigator gave a verbal signal and the subject lifted the weight to the starting position [12]. Similar scoring procedure as used in the bench press test was adopted.

2.5 Muscular Power

The vertical jump was measured by using a Vertec. Three vertical jump trials were performed with 1-min rest period between trials. The vertical jump was performed by having the participant standing upright with both arms fully extended overhead. Participants were to jump and touch the highest possible vane. Vertical jump score was measured by the difference of the standing height and the jumping height.

2.6 Resistance Training Program

Participants were assigned to three (3) groups; single-set ($n = 12$), three-set ($n = 12$), and the control group ($n = 12$). Single-set group trained one set for each exercise, while three-set group trained three sets for each exercise. Control group did not perform any training.

Both the experimental groups trained at 80 % of 1RM until momentary muscular failure is achieved at the eight repetitions. A rest interval of 2 min was fixed between sets, and exercises as 2 min were sufficient for muscular recovery [11].

All participants recorded their training details in a log book provided by the researcher. The training log helped to track the participants’ training compliance. Training was conducted three days per week, with a minimum of 48 h between

Table 1 Characteristics of participants based on group

Group	*n*	Age (years)	Height (m)	Weight (kg)
Single	12	20.92 ± 0.79	1.67 ± 0.04	65.28 ± 3.02
Multiple	12	21 ± 0.74	1.68 ± 0.04	65.27 ± 3
Control	12	21 ± 0.67	1.67 ± 0.04	64.71 ± 2.92

(mean ± SD; n = number of participants)

sessions, for 6 weeks subscribing to ACSM's recommendations to allow for full muscle recovery before the next training session is conducted. Three times per week was chosen because based on current ACSM position stand [13], adults should train each major muscle group two or three days each week using a variety of exercises and equipment.

2.7 *Statistical Analysis*

Descriptive statistic was used to describe the demographic data of the participants. Kolmogorov-Smirnov, descriptive statistic, and boxplot were used to check for the normality of data in each group. Mixed between within analysis of variance (ANOVA) were used to determine the significant interactions and main effect between the three resistance training protocols (single set, multiple set, control) on muscular strength and muscular power. Post hoc test with Tukey adjustment test was conducted to determine which intervention group showed more changes on variables tested.

3 Results

3.1 *Participant Characteristics*

Table 1 presented the physical characteristics of the samples selected including the age, height, and weight of each group.

3.2 *Muscular Strength and Muscular Power*

3.2.1 Upper Body Muscular Strength

Table 2 showed changes in bench press performances between groups across observation. Results showed single-set group demonstrated a 4.31 % improvement

Table 2 Changes in bench press performances between groups across observation

Variable	Time period	Single-set group $M \pm SD$	Multiple-set group $M \pm SD$	Control group $M \pm SD$
Bench press	Pre-test	0.95 ± 0.05	0.96 ± 0.05	0.93 ± 0.05
	Post-test	0.99 ± 0.05	1.03 ± 0.05	0.94 ± 0.05

of bench press performance, while multiple-set group demonstrated a 7.75 % improvement. Control group demonstrated a 0.21 % improvement in performance.

A significant main effect for time was obtained, Wilk's Lambda = 0.054, $F(1, 33) = 575.97$, $p < 0.001$, partial eta squared = 0.946 (large effect size), indicated that there was a significant increment in participants' strength across observations. A significant interaction was found between number of sets and time, Wilk's Lambda = 0.087, $F(2, 33) = 173.62$, $p < 0.001$, partial eta squared = 0.913 (large effect size). This shows that a significant improvement in participants' upper body strength scores across observations was influenced by the intervention. A significant main effect between groups, $F(2, 33) = 4.743, p < 0.05$, partial eta squared = 0.223 (small effect size), was also observed.

Post hoc test showed that both the single- and three-set groups showed significant difference compared to control group, $p < 0.05$. However, no significant difference was found between both of these groups, $p > 0.05$. Hence, the hypothesis is failed to be rejected.

3.2.2 Lower Body Muscular Strength

Table 3 showed changes in squat performances between groups across observation. Results showed single-set group demonstrated a 6.01 % improvement of squat performance, while multiple-set group demonstrated a 10.44 % improvement. Control group demonstrated a 0.33 % reduction in performance.

A significant main effect for time was obtained; Wilk's Lambda = 0.04, $F(1, 33) = 802.025$, $p < 0.001$, partial eta squared = 0.960 (large effect size), shows that there was a significant increment in participant's strength across observations. A significant interaction was found between number of sets and time, Wilk's Lambda = 0.057, $F(2, 33) = 273.23$, $p < 0.001$, partial eta squared = 0.943 (large effect size). This shows that a significant improvement in participant's lower body strength scores across observations was influenced by intervention. Besides, there was a significant main effect between groups, $F(2, 33) = 11.347$, $p < 0.001$, partial eta squared = 0.407 (moderate effect size).

Post hoc test showed that both the single- and multiple-set groups showed significant difference compared to control group, $p < 0.05$. However, no significant difference was found between both the groups, $p > 0.05$. Hence, hypothesis is failed to be rejected.

Table 3 Changes in squat performances between groups across observation

Variable	Time period	Single-set group $M \pm$ SD	Multiple-set group $M \pm$ SD	Control group $M \pm$ SD
Squat	Pre-test	1.25 ± 0.04	1.25 ± 0.05	1.23 ± 0.04
	Post-test	1.32 ± 0.05	1.38 ± 0.05	1.22 ± 0.05

Table 4 Changes in handgrip strength between groups across observation

Variable	Time period	Single-set group $M \pm$ SD	Multiple-set group $M \pm$ SD	Control group $M \pm$ SD
Handgrip strength	Pre-test	0.67 ± 0.05	0.67 ± 0.04	0.65 ± 0.04
	Post-test	0.74 ± 0.05	0.74 ± 0.04	0.65 ± 0.05

3.2.3 Static Strength

Table 4 shows changes in handgrip strength between groups across observation. Results showed single-set group demonstrated a 10.15 % improvement of handgrip strength, while multiple set group demonstrated a 10.57 % improvement. Control group demonstrated a 0.61 % reduction in performance.

A significant main effect for time was obtained; Wilk's Lambda = 0.065, $F(1, 33) = 475.22$, $p < 0.001$, partial eta squared = 0.935 (large effect size), shows that there was a significant increment in participant's strength across observations. A significant interaction was found between number of sets and time, Wilk's Lambda = 0.107, $F(2, 33) = 137.06$, $p < 0.001$, partial eta squared = 0.893 (large effect size). This shows that a significant improvement in participant's static strength scores across observations was influenced by the interventions. There was significant main effect between groups, $F(2, 33) = 6.318$, $p < 0.01$, partial eta squared = 0.277 (small effect size).

Post hoc test showed that both single- and multiple-set groups were significantly higher than control group, $p < 0.05$. However, no significant differences were found between both of the groups, $p > 0.05$. Hence, hypothesis is failed to be rejected.

3.2.4 Muscular Power

Table 5 showed changes in vertical jump between groups across observation. Results showed single-set group demonstrated a 12.63 % improvement of vertical jump performance, while multiple-set group demonstrated a 11.19 % improvement. Control group demonstrated a 1.2 % improvement in performance.

A significant main effect for time was obtained; Wilk's Lambda = 0.143, $F(1, 33) = 198.19, p < 0.001$, partial eta squared = 0.857 (large effect size), shows that there was a significant increment in participant's muscular power across

Table 5 Changes in vertical jump between groups across observation

Variable	Time period	Single-set group $M \pm SD$	Multiple-set group $M \pm SD$	Control group $M \pm SD$
Vertical jump	Pre-test	41.58 ± 2.15	42.42 ± 2.47	41.75 ± 2.30
	Post-test	46.83 ± 1.80	47.50 ± 1.45	42.25 ± 1.42

observations. A significant interaction was found between number of sets and time, Wilk's Lambda $= 0.310, F(2,33) = 36.81, p < 0.001$, partial eta squared $= 0.690$ (moderate effect size). This shows that a significant improvement in participant's muscular power scores across observations was influenced by the interventions. Besides, there was a significant main effect between groups, $F(2, 33) = 8.596$, $p < 0.01$, partial eta squared $= 0.343$ (small effect size).

Post hoc test showed that both the single- and multiple-set groups were significantly higher than control group, $p < 0.05$. However, no significant differences were found between the two groups, $p > 0.05$. Hence, hypothesis is failed to be rejected.

4 Discussion

Finding of the present study clearly indicated that resistance training was effective in improving upper and lower body strength, static strength, and muscular power and there were no significant differences in the measured parameters recorded from single- versus three-set trainings.

The study sought to compare the effects of single- versus multiple-set resistance trainings on muscular strength, muscular power, and physiological responses among untrained men. Participants underwent 6-week resistance training treatment three times per week for a total of 18 sessions.

The effects of resistance training on muscular performance recorded that the upper body strength, lower body strength, static strength, and muscular power were all significantly improved. This proved that both number of sets performed were effective in improving muscular strength and muscular power.

Resistance training had been shown to produce compensatory growth of skeletal muscle especially among initially untrained healthy subjects, which is potentially helpful in increasing the strength of the skeletal muscles. This hypertrophy is a result from the increment in the rate of protein synthesis over the rate of protein degradation which in turn produces a deposition of myofibrillar proteins within existing muscle fibers [14, 15]. This thus helps to explain the possible mechanism underlying the improvements of muscular performance.

The improvement can also be related to the neuromuscular adaptations. Apart from the quality and quantity of the involved muscles, neuromuscular performance also depends on the nervous system's ability to appropriately activate the muscles [16]. Resistance training had been proven to induce positive adaptations within the nervous system [17]. These positive adaptations include the increment in the firing rate of each motor unit, changes in the pattern of motor unit activation, and the recruitment of more motor units [18].

In attaining optimal increment of muscular performance, resistance training program had to be designed based on the specific objectives. The ACSM's recommendation that multiple-set resistance training will produce greater effects to the performance had been criticized in some of the reviews that claimed the recommendations had been made without strongly proven evidence and supported by earlier studies [1].

This present study found that the higher number of sets had not significantly induced more improvement of upper and lower body muscular strength, static strength, and muscular power but noted a very slight superiority of the three-set group in upper and lower body strength test achievements. Findings of this study is in line with Arthur Jones' earliest [19] to his final writing [20] who argued that the muscular strength and hypertrophy can be optimally improved by performing one set of resistance training carried to the point of momentary muscular failure and that further sets are therefore unnecessary. This was further supported by Smith and Bruce-Low [13].

Several other studies indicated and supported parallel findings that there were no additional benefits to perform multiple sets when training for strength [4–7, 21].

The similar gains in strength increment could be related to the same muscular size increment of participants. Ostrowski et al. [22] investigated the 10-week effects of 1, 2, or 4 sets of resistance training. Results showed that there were significant increases in body mass, rectus femoris circumference, rectus femoris hypertrophy, and tricep brachia thickness for the 1-, 2-, and 4-set groups, respectively, but there were no significant differences between the groups. This present study proved that in 6 weeks, the increment of number of sets failed to induce more muscle size increment compared to just performing one set.

The findings of the present study contradicts the guidelines of the ACSM [23] and Baechle and Earle [11] in which multiple sets have been recommended to produce more significant effects compared to single set. Further supporting the effectiveness of multiple set was a study by Kelly et al. [24] which concluded that performing 3 sets of isokinetic knee extensions was more effective than performing a single set for increasing peak torque.

Kemmler et al. [25] examined the effects of single- and multiple-set resistance trainings on untrained women. The participants were tested for strength gains in their shoulders, chests, arms, abdominals, backs, and legs. Results showed the multiple-set group increased their strength by 3.5–5.5 %. However, the single-set group showed 1.1–2.0 % decrements in strength. This finding suggested multiple-set resistance training is better for increasing strength for older women. The contradicting findings of the present study with the study by Kelly et al. [24] and Kemmler et al. [25] can be attributed to the methodology used in these studies. The participants in the present study were subjected to a total body workout, whereas the participants in the study by Kelly et al. [24] were only trained in isokinetic knee extensions. The participants used in the present study were untrained men compared with the study by Kemmler et al. [25] which recruited untrained women. The differences of exercises used and different populations might be the contributing factors to these contradicting results. Other than that, other contributing factor could be the intensity used in the study, rest between set, order of exercises, and many more.

Unlike the muscular strength effects, the study on the comparison of single- versus multiple-set resistance trainings on the static strength and muscular power had not been well documented. No significant differences of static strength and muscular power had been found between both groups. One study that could give explanation behind the finding of static strength was the study by Vincent et al. [26] which found that a single-set group resulted in similar improvement of peak isometric torque as multiple-set group. This showed that the isometric strength had not significantly been affected by the increment in the number of sets.

The rationale behind the insignificant difference of increment of muscular power between single set and multiple set could be related to the similar improvement of muscular strength between both groups. Strength and speed are the components that will make up the muscular power [11]. Without any speed training and the same muscular strength improvement, it was rationalized that the muscular power increment will also become insignificant.

In conclusion, findings from the present recommends that a single-set resistance training program suffice to promote strength and power gain among untrained. The minimum number of set may be more suited to the busy schedule of untrained population as it saves time for the same result.

References

1. Fisher, J., Steele, J., Bruce-Low, S., & Smith, D. (2011). Evidence-based resistance training recommendations. *Medicine and Sport, 15*(3), 147–162.
2. Fleck, S. J., & Kraemer, W. J. (1997). Designing resistance programs, 2nd ed. *Human Kinetics, Champaign, 3*(11), 83–115.
3. Berger, R. A. (1962). Optimum repetitions for the development of strength. *Research Quarterly for Exercise and Sport, 33*, 334–338.
4. Haas, C. J., Garzarella, L., De Hoyos, D., & Pollock, M. L. (2000). Single versus multiple sets in long term recreational weightlifters. *Medicine and Science in Sports and Exercise, 32*, 235–242.
5. Fincher, G. E. (2000). The effect of high intensity resistance training on peak upper and lower body power among collegiate football players. *Medicine and Science of Sports and Exercise, 32*, 657.
6. Fincher, G. E. (2003). The effect of high intensity resistance training on body composition among collegiate football players. *Medicine and Science in Sports and Exercise, 35*(Suppl. 5), 1793.
7. Baker, J. S., & Cooper, S. M. (2004). Strength and body composition: single versus triple set resistance training programs. *Medicine in Science of Sports and Exercise, 36*, 394.
8. Kraemer, W. J., Ratamess, N., Fry, A. C., Triplett-McBride, T., Koziris, L. P., Bauer, J. A., et al. (2000). Influence of resistance training volume and periodization on physiological and

performance adaptations in collegiate women tennis players. *American Journal of Sports Medicine, 28*, 626–633.
9. Marx, J. O., Ratamess, N. A., Nindl, B. C., Gotshalk, L. A., Volek, J. S., Dohi, K., et al. (2001). Low-volume circuit versus high-volume periodized resistance training in women. *Medicine and Science in Sports and Exercise, 33*, 635–643.
10. Sanborn, K., Boros, R., Hruby, J., Schilling, B., O'Bryant, H. S., Johnson, R. L., et al. (2000). Short-term performance effects of weight training with multiple sets not to failure vs. a single set to failure in women. *Journal of Strength and Conditioning Research, 14*, 328–331.
11. Baechle, T., & Earle, R. (2008). *Essentials of strength training and conditioning* (3rd ed.). Champaign: Human Kinetics.
12. LeSuer, D. A., McCormick, J. H., Mayhew, J. L., Wasserstein, R. L., & Arnold, M. D. (1997). Equations for estimating 1-RM performance in the bench press, squat, and deadlift. *Journal of Strength and The accuracy of prediction Conditioning Research, 11*(4), 211–213.
13. Smith, D., & Bruce-Low, S. (2004). Strength training and the work of Arthur Jones. *Journal of Exercise Physiology, 7*, 52–68.
14. Phillips, S. M., Tipton, K. D., Aarsland, A., Wolf, S. E., & Wolfe, R. R. (1997). Mixed muscle protein synthesis and breakdown following resistance exercise in humans. *American Journal of Physiology, 273*, E99–E107.
15. Tipton, K. D., & Wolfe, R. R. (2001). Exercise, protein metabolism, and muscle growth. *International Journal of Sport Nutrition and Exercise Metabolism, 11*, 109–132.
16. Häkkinen, K., Kallinen, M., Izquierdo, M., Jokelainen, K., Lassila, H., Mälkiä, E., et al. (1998). Changes in agonistantagonist EMG, muscle CSA, and force during strength training in middle-aged and older people. *Journal of Applied Physiology, 84*, 1341–1349.
17. Sale, D. G. (1988). Neural adaptation to resistance training. *Med Sci Sports, 20*, 135–145.
18. Häkkinen, K. (1989). Neuromuscular and hormonal adaptations during strength and power training: a review. *Journal of Sports Medicine and Physical Fitness, 29*, 9–26.
19. Jones, A. (1970). *Nautilus bulletin #1*. DeLand: Nautilus Sports/Medical Industries.
20. Jones, A. (2003). My first half-century in the iron game part 54. In The Arthur Jones collection, (pp. 740–741). Ontario: Bodyworx (originally published in Ironman magazine, 1996).
21. Wolfe, B. L., Valerio, T. A., Strohecker, K., & Szmedra, L. (2001). Effect of single versus multiple-set resistance training on muscular strength. *Medicine and Science in Sports and Exercise, 33*, 76.
22. Ostrowski, K. J., Wilson, G. J., Weatherby, R., Murphy, P. W., & Little, A. D. (1997). The effect of weight training volume on hormonal output and muscular size and function. *The Journal of Strength and Conditioning Research, 11*, 148–154.
23. American College of Sports Medicine. (2011). Position stand: Quantity and quality of exercise for developing and maintaining cardiorespiratory, musculoskeletal, and neuromotor fitness in apparently healthy adults: guidance for prescribing exercise. *Medicine and Science in Sports and Exercise, 43*(7), 1334–1559.
24. Kelly, S. B., Brown, L. E., Coburn, J. W., Zinder, S. M., Gardner, L. M., & Nguyen, D. (2007). The effect of single versus multiple sets on strength. *Journal of Strength and Conditioning Research, 21*(4), 1003–1006.
25. Kemmler, W. K., Lauber, D., Engelke, K., & Weineck, J. (2004). Effects of single- vs. multiple-set resistance training on maximum strength and body composition in trained postmenopausal women. *Journal of Strength and Conditioning Research, 18*, 689–694.
26. Vincent, K., De Hoyos, D., Garzarella, L., Hass, C., Nordman, M., & Pollock, M. (1998). Relationship between indices of knee extension strength before and after training. *Medicine and Science in Sports and Exercise, 30*, 163.

Motives of Malay, Chinese and Indian Football Players

Vincent Parnabas, Sarimah Ismail, Mohamad Nizam Mohamed Shapie and Julinamary Parnabas

Abstract Previous research in Malaysia found that the six important motives, given by student-athletes for sport participation, were achievement, body shape, physical fitness, teamwork, learning new skills and challenges. However, research comparing different ethnic on taking part in sports, especially football is limited. The aim of this study is to find the motives of taking part in sport and level of anxiety among Malay, Chinese and Indians. A hundred and thirty student-athletes from different ethnic background were selected, to respond Purpose of Sport Questionnaire. The sample was chosen from the Malaysian Sports Council, during Malaysian Inter-Schools Sports Competition. The result showed that the most highest motive for a Chinese ethnic taking part in football for physically active lifestyle, Malays ethnic for mastery or cooperation and Indians for social status or getting ahead. Sports psychologists, coach or sport officers can use this research to develop appropriate football environment for Malays, Chinese and Indians, to maintain their interest in football.

Keywords Motives · Malay · Chinese · Indian · Football

1 Introduction

Motives can be defined as the intentions that cause a person to move, act or behave in a certain way [1]. Motives for participating in sports are many [2]. Past studies showed that the major motives student-athletes have for their participation are

V. Parnabas (✉) · S. Ismail · M. N. Mohamed Shapie
Faculty of Sport Science and Recreation, Universiti Teknologi MARA (UiTM), 40450 Shah Alam, Selangor, Malaysia
e-mail: vincent@salam.uitm.edu.my

J. Parnabas
Institut Pendidikan Guru, Kampus Darulaman, Jitra, Kedah, Malaysia

R. Adnan et al. (eds.), *Proceedings of the International Colloquium on Sports Science, Exercise, Engineering and Technology 2014 (ICoSSEET 2014)*,
DOI: 10.1007/978-981-287-107-7_20, © Springer Science+Business Media Singapore 2014

enjoying, for arousal, thrills and excitement, achievement, fitness, energy release, skill development, friendship, fun and to become physically fit, healthy lifestyle [3–6].

Social learning theory explains that people are motivated to take part in sports because they want to copy the skills of their sports heroes [7]. Cognitive evaluation theory states that rewards can be divided into two types: controlling and informational. Controlling rewards, such as praise or trophies, are designed to influence (e.g., control) an individual's behaviour, while informational rewards convey information about an individual's competence at a task. If a player believes her coach's praise is designed to control her behaviour, the praise will lower her intrinsic motivation for the task. If, on the other hand, the player believes that the praise is designed to provide information about her successful performance, her intrinsic motivation for the task would increase [2].

There are many reasons why Malaysian athletes begin to participate in sports. For example, in a research study of 200 student-athletes from University Putra Malaysia, it was found that the six important motives, out of 30 motives, given by students for sport participation were achievement, body shape, physical fitness, teamwork, learning new skills and for challenges [8]. The same study also found that friendship and want to become popular are the lowest motive for sport participation among student-athletes at University Putra Malaysia.

A variety of cultural comparative studies have been conducted in the field of sport psychology. Culture may be defined as the changing patterns of learning behaviour and the products of learning behaviour (including attitudes, values, knowledge and material objects) that are shared by and transmitted among members of society, it is an ongoing social heritage [9]. Athletes from different cultural background show different motives for taking part in sport. Some progress has been made in addressing this problem in developed countries, especially comparing among blacks and whites, but a similar research is yet to be conducted in Malaysia. Since lack of research in this area, there is still uncertain on motives of different cultural groups in Malaysia.

A number of researches have been conducted on motives for sports participation in developed countries [2], but there is a lack of such research in developing countries, including Malaysia especially on different ethnics. Furthermore, research comparing different ethnic athletes directly is very limited. Comparing motives of different races on taking part in sports, especially on football players, can bring a lot of new knowledge for cultural studies. Furthermore, most studies on motives in sport done generally on sports without focusing on particular sports like football. The present research will evaluate the motives that encourage different background ethnics of football players to participate in sports. Three major ethics in Malaysia took part in this study; they are Malays, Chinese and Indians. Motives for taking part in sports include the following motives: mastery or cooperation, physically active lifestyle, good citizen, competitiveness, high-status career, enhance self-esteem and social status or getting ahead.

There is a need to plan and attract people to participate in football, but this will depend on their sport participation motives. Understanding the motives of those who are already involved as football players can help coaches, teachers and individuals to improve coaching, maintain motivation, prevent burn out and lower dropout rates in football. Lindner and Kerr [10] argued that the motivation to participate in a sport was both complex and multifaceted. Different motives could determine participation decisions at various times and in desperate circumstances.

Knowing motives of different ethnics for participating in sports will help sports psychologists to develop an appropriate football sport programme for those ethnics. In addition, identifying the range of motives given by different ethnic participants will help sports psychologists provide adequate and variety of football programmes to maintain interest among those ethnics. This is because understandings of why particular ethnic choose to take part in sports while others do not could be of great practical value. In particular, government officials and sports psychologists in Malaysia need to conduct sports programmes to encourage different ethnics in Malaysia to take part in football could use the research findings.

Malaysia consists of many ethnic groups with different cultural background, namely Malays, Chinese and Indians, to maintain separate cultural identities [11]. In [12] there are two ways to classify races biologically: (a) a scheme based on observable physical characteristics (e.g., skin colour, hair texture, eye shape, lip thickness) called phenotypes and (b) a scheme based on unobservable genetic features (e.g., blood type) called genotypes.

In Malaysia, Malay is the largest ethnic, comprised 50.4 %, followed by Chinese 23.7 %, indigenous 11 %, Indians 7.1 % and others 7.8 % (East Malaysia, especially Iban and Kadazan) [13]. Sometimes, it is hard to believe that they remain in the same country, but their way of living differs dramatically from the other.

2 Objectives

The aim of this study was to find the motives of taking part in football among Malay, Chinese and Indians. This research focuses of football players, since football is considered as the most preferred sport among Malaysians. One of the most influential variables that involved of taking part in sports is culture and belief. Thus, there is also a need for research to be carried out on the motives of different ethnic group of Malaysians for taking part in sport, especially in football.

3 Methodology

The sample consisted of 130 football players, with Malay ethnic athletes ($N = 54$), Chinese ethnic athletes ($N = 45$) and Indian ethnic athletes ($N = 31$). Based on the gender, there are 69 males and 61 females. The mean age for overall respondents was 22.08 years old. The age of male varied from 18 to 27 years, where the mean age was 21.31 years old. The age of females ranged from the minimum of 18 to the maximum of 26 years old. The mean age for female respondents was 21.29 years old.

The sample was chosen from the Malaysian Sports Council, athletes who took part in Malaysian Inter-Schools Sports Competition. The questionnaires were distributed to student-athletes during the sport event. Participants identify their racial group as 'Malay', 'Chinese' or 'Indians' in the demographic questionnaire.

A 46-item questionnaire called Purpose of Sport Questionnaire designed by Duda's [14] was used. Students will indicate their responses to Duda's [14] questionnaire on a Likert-typed scale ranging from strongly agree (5) and strongly disagree (1). Seven factors were derived from the questionnaire. They are mastery or cooperation, physically active lifestyle, good citizen, competitiveness, high-status career, enhance self-esteem, and social status or getting ahead.

4 Results

4.1 Cronbach Reliability Coefficients

In this study, Cronbach's alpha coefficients were found ranging from 0.64 to 0.82. In a similar study conducted by Duda, it was found that the Cronbach's alpha coefficients were relatively high, ranging from 0.75 to 0.83. Even though some of the reliability was found to be lower than that in Duda's study, [14] in the present study, they were still considered acceptable for analysis. In this study, coefficients of 0.70 and above were considered reliable, while those around 0.60 required some caution when the results were interpreted (Table 1).

4.2 Motives of Malay Ethnic Football Players Equations

The results showed that the motives of Malay ethnic respondents participated in sports were highest in mastery or cooperation (mean = 4.44), followed by physically active lifestyle (mean = 3.96), competitiveness (mean = 3.77), enhance self-esteem (mean = 3.06), good citizen (mean = 2.89) and high-status career (mean = 2.78). The Malay respondents indicated social status as the lowest motive (Table 2).

4.3 Motives of Chinese Ethnic Football Players

The results showed that the motives of Chinese ethnic respondents participated in sports were highest in physically active lifestyle (mean = 4.87), followed by mastery or cooperation (mean = 4.47), competitiveness (mean = 3.51),

Table 1 Cronbach reliability coefficients

Sport participation motives	Cronbach's alpha (n = 130)
Mastery or cooperation	0.8211
Physical active lifestyle	0.7810
Good citizen	0.7771
Competitiveness	0.7537
High-status career	0.7311
Enhance self-esteem	0.7121
Social status or getting ahead	0.6428

Table 2 Motives of Malay ethnic football players participate in sport (n = 54)

Motives	Mean	SD
Mastery or cooperation	4.44	0.49
Physically active lifestyle	3.96	0.55
Competitiveness	3.77	0.59
Enhance self-esteem	3.06	0.57
Good citizen	2.89	0.65
High-status career	2.78	0.68
Social status or getting ahead	2.61	0.69

high-status career (mean = 3.31), social status or getting ahead (mean = 2.57) and enhance self-esteem (mean = 2.26). The Chinese respondents indicated good citizen as the lowest motive (Table 3).

4.4 Motives of Indian Ethnic Football Players

The results showed that the motive of Indian ethnic respondents participated in sports was highest in social status or getting ahead (mean = 4.41), followed by physically active lifestyle (mean = 4.27), mastery or cooperation (mean = 4.07), enhance self-esteem (mean = 3.92), good citizen (mean = 3.71) and high-status career (mean = 3.13). The respondents indicated competitiveness as the lowest motive (Table 4).

4.5 Comparison of Ethnics

Table 5 shows the F scores for the motives of taking part in football sport among the athletes of different ethnicity:

1. Mastery or cooperation, F (2, 130) = 4.001, $p = 0.05$.
2. Physically active, F (2, 130) = 4.017, $p = 0.05$.

Table 3 Motives of Chinese ethnic football players participate in sport ($n = 45$)

Motives	Mean	SD
Physically active lifestyle	4.87	0.47
Mastery or cooperation	4.47	0.44
Competitiveness	3.51	0.51
High-status career	3.31	0.52
Social status or getting ahead	2.57	0.57
Enhance self-esteem	2.26	0.54
Good citizen	2.13	0.58

Table 4 Motives of Indian ethnic football players participate in sport ($n = 31$)

Motives	Mean	SD
Social status or getting ahead	4.41	0.35
Physically active lifestyle	4.27	0.34
Mastery of cooperation	4.07	0.36
Enhance self-esteem	3.92	0.41
Good citizen	3.71	0.39
High-status career	3.13	0.44
Competitiveness	2.07	0.54

Table 5 Comparison of motives of ethnic ($n = 130$)

Motives	Ethnic	Mean	F value
Mastery or cooperation	Malay	4.44	4.001*
	Chinese	4.47	
	Indian	4.07	
Physically active	Malay	3.96	4.017*
	Chinese	4.87	
	Indian	4.27	
Competitiveness	Malay	3.77	3.179*
	Chinese	3.51	
	Indian	2.07	
High-status career	Malay	2.78	3.181*
	Chinese	3.31	
	Indian	3.13	
Social status or getting ahead	Malay	2.61	4.003*
	Chinese	2.57	
	Indian	4.41	
Enhance self-esteem	Malay	3.06	3.799*
	Chinese	2.26	
	Indian	3.92	
Good citizen	Malay	2.89	3.332*
	Chinese	2.13	
	Indian	3.71	

$^*p < 0.05$

3. Competitiveness, F (2, 130) = 3.179, $p = 0.05$.
4. High-status career, F (2, 130) = 3.181, $p = 0.05$.
5. Social status or getting ahead, F (2, 130) = 4.003, $p = 0.05$.
6. Enhance self-esteem, F (2, 130) = 3.799, $p = 0.05$.
7. Good citizen, F (2, 130) = 3.332, $p = 0.05$.

5 Discussions

5.1 Motives of Malay Ethnic Football Players

The results showed that the motive of Malay ethnic respondents participated in football was highest in mastery or cooperation. According to the competence motivation theory, people engage in sports as a means of mastering their surroundings [15]. This is achieved through the learning of sports skills [16]. The successful mastery of sport skills results in the experiencing of pleasant emotions and increased self-confidence. Furthermore, successful attempts for Malays at mastery promote self-efficacy and feeling of personal competence, which in turn fosters high competence motivation. Cooperation in football teams involves Malays to fulfil their desire to raise the performance of all as a team experience; cares for others, empathizes, congratulates others; and looks forward to the next contest as an opportunity to partake in a competition.

According to the competence motivation theory, people engage in sports as a means of mastering their surroundings [15]. This is achieved through the learning of sports skills [16, 17]. The successful mastery of sport skills results in the experiencing of pleasant emotions and increased self-confidence [17]. Furthermore, these pleasant feelings will serve Malay athletes as an incentive to continue to take part in sports. In other words, successful attempts at mastery promote self-efficacy and feeling of personal competence, which in turn fosters high competence motivation for Malay athletes. As competence motivation increases, the Malay athletes are encouraged to make further mastery attempts [15].

5.2 Motives of Chinese Ethnic Football Players

The results showed that the motive of Chinese ethnic respondents participated in football was highest in Physically Active Lifestyle. Physically active lifestyle is the motive to take part in sports for physical reasons, notably health, weight or keeping fit [2, 18]. Taking part in physical activity is important to reduce the risk of coronary heart disease, cancers, obesity and many other health-related problems [19, 20]. Taking part in football can increase the level of physical activity, as an ideal way to reduce health problems. Taking part in football as a physically active lifestyle enhances the fitness level; improves the immune system, mental health, maintaining healthy bones, reducing health problems; and increases the effective functioning of the body organ.

Study done by [21] showed that body shape satisfaction was significantly higher in Chinese ethnic compared to Korean. In Malaysia, Chinese look much fitter in their body shape compared to other races.

5.3 *Motives of Indian Ethnic Football Players*

The results showed that the motive of Indian ethnic respondents participated in football was highest in social status or getting ahead. A social motive is an interest in people and the ways a person prefers to relate to other people [2]. Study of [22, 23] found that the interaction among their colleagues encourages people to participate in sports. Indian, as a minority ethnic in Malaysia has a strong desire to be with others and to function as members of a team. For Indians, the social aspects of football are an important reason for their participation. Indians found that their motivation to take part in football was derived from something relating to a social motive, like to enjoy camaraderie and to mix socially with others who enjoy the same activity as a team. Social reasons for taking part in football range from meeting new people to deal with loneliness and social isolation. Research of [24] showed that social support was the most important motives of African-American and American Indian women choosing to be active in sport.

6 Conclusion

This research showed that the motive of taking part in sport among different ethnic groups is various. A Malay ethnic group takes part in sport for mastery or cooperation, Chinese for physically active lifestyle, while Indians for social status or getting ahead.

Sports psychologists, coach or sport officers can use this research to develop appropriate football environment for Malays, Chinese and Indians, to maintain their interest in football. In addition, identifying the range of motives given by Malays, Chinese and Indians will help sports psychologists provide adequate and variety of football programmes regarding their motives in football.

References

1. Brehm, B.A. (2004). Your role in the exercise adherence story. Successful fitness motivation strategies. Champaign, IL: Human Kinetics.
2. Wann, D. L. (1997). *Sport psychology*. New Jersey: Prentice-Hall.

3. Bennett, R., Mousley, W., Kitchin, P., & Ali-Choudhury, R. (2007). Motivations for participating in charity-affiliated sporting events. *Journal of Customer Behaviour, 6*(2), 155–178.
4. Devine, R., & Lepisto, L. (2005). Analysis of the healthy lifestyle consumer. *Journal of Consumer Marketing, 22*(5), 275–283.
5. White, S. A., & Duda, L. D. (1994). The relationship of gender, level of sport involvement, and participation motivation to task and ego orientation. *International Journal of Sport Psychology, 25*(1), 4–18.
6. Gill, D. L., Gross, J. B., & Huddleston, S. (1983). Participation motivation in youth sports. *International Journal of Sport Psychology, 14*, 1–14.
7. Anderson, D., South, D. (1993). Racial differences in collegiate recruiting, retention and graduation rates. In D. Brooks, R. Althouse (Eds.), Athletics:The African-American athlete's experience (pp. 79–100). Morgantown: Fitness Information Technology.
8. Chuen, S.P. (1994). Motif penglibatan atlit University Putra Malaysia dalam kegiatan sukan. Unpublished degree's thesis. University Putra Malaysia, Serdang, Malaysia.
9. Soyinka, W. (1991). Africa's cultural producers. *Society, 28*, 32–40.
10. Lindner, K., & Kerr, J. (2001). Predictability of sport participation motivation from metamotivational dominances and orientations. *Personality and Individual Differences, 30*(6), 759–773.
11. Kahn, J. S. (1998). *Southeast Asian identities: Culture and the politics of representation in Indonesia, Malaysian, Singapore and Thailand*. Singapore: Institute of Southeast Asian Studies.
12. Leonard, W. M. (1998). *A sociological perspective of sport*. London: Allyn and Bacon.
13. Malaysia Demographics Profile (2012). http://www.indexmundi.com/malaysia/demographics_profile.html.
14. Duda, J. L. (1989). Relationship between task and ego orientation and the perceived purpose of sport among high school athletes. *Journal of Sport and Exercise Psychology, 11*(3), 318–335.
15. Cox, R. H. (2011). *Sport psychology: Concepts and application*. Dubuque: Brown & Benchmark.
16. Anshell, M. H. (2011). *Sport psychology from theory to practice*. Scottsdale: Gorsuch Scarisbrick.
17. Harter, S. (1981). A model of intrinsic mastery motivation in children: Individual differences and developmental change. In A. Collins (Ed.), Minnesota symposium on child psychology (vol. 14, pp. 215–255). Hillsdale: Erlbaum.
18. Kilpatrick, M., Hebert, E., & Bartholomew, J. (2005). College students' motivation for physical activity: Differentiating men's and women's motives for sport participation and exercise. *Journal of American college health, 54*, 87–94.
19. Adair, L. S. (2008). Child and adolescent obesity: Epidemiology and developmental perspectives. *Physiology & Behavior, 94*, 8–16.
20. Allender, S., Cowburn, G., & Foster, C. (2006). Understanding participation in sport and physical activity among children and adults: a review of qualitative studies. *Health Education Research, 21*, 826–835.
21. Ro, Y., & Hyun, W. (2012). Comparative study on body shape satisfaction and body weight control between Korean and Chinese female high school students. *Nutrition research and practice, 6*(4), 334–339.
22. Wang, L., Myers, D., & Yanes, M. (2010). Creating student-centered learning experience through the assistance of high-end technology in physical education. *Journal of Instructional Psychology, 37*(4), 352–356.
23. Sabiston, C. M., & Crocker, P. R. (2008). Exploring self-perceptions and social influences as correlates of adolescent leisure-time physical activity. *Journal of Sport & Exercise Psychology, 30*(1), 22–33.
24. Henderson, K., & Ainsworth, B. (2001). Researching leisure and physical activity with women of color: Issues and emerging questions. *Leisure Sciences, 23*(1), 21–34.

The Effects of Eight-Week Integrated Training Program on Malaysian Junior Tennis Players' Performance

Mohamad Rahizam Abdul Rahim, Balbir Singh, Vincent Parnabas, Rezian-na Muhammed Kassim and Nagoor Meera Abdullah

Abstract The purpose of this study was to determine the effects of an eight-week integrated training program on functional performance among Malaysian junior tennis players. Twenty-four ($n = 24$) nationally ranked junior tennis players between the age of twelve to sixteen years old (18 males and 6 females, mean age 14.29 ± 1.65 year) were selected using purposive sampling method and were equally divided into experimental and control groups. The experimental group performed four sessions of intervention every week for eight weeks, while the control group performed normal training sessions and match play for the same frequency and duration. All participants were tested on four different performance areas, namely the physical, psychological, technical, and tactical profiles. Normality test showed that all data were normally distributed (Shapiro–Wilk Test). The results of sample paired t test analysis subscales with every performance attribute showed significant improvement in all performance attributes with level of significant differences of $\alpha < 0.05$ after the intervention. These methods correlate with the objectives of this study that focused on the effects of an intervention program, and the results proved that the participants showed an improvement in their performance based on the four major areas of physical, psychological, technical, and tactical skills. The experimental group showed higher improvement results that dictate the optimistic effect of intervention training program developed in this study, while the control group also showed significant difference in some of the variables. The results of this study show that an integrated training program can be an effective training module to improve player's performance.

Keywords Performance · Profile · Training · Tennis · Coach · Player

M. R. Abdul Rahim (✉) · V. Parnabas · R. Muhammed Kassim · N. M. Abdullah
Faculty of Sports Science and Recreation, Universiti Teknologi MARA,
Shah Alam, Malaysia
e-mail: mrahizam@salam.uitm.edu.my; rahizamrahim@gmail.com

B. Singh
Sports Center, University of Malaya, Kuala Lumpur, Malaysia

R. Adnan et al. (eds.), *Proceedings of the International Colloquium on Sports Science, Exercise, Engineering and Technology 2014 (ICoSSEET 2014)*,
DOI: 10.1007/978-981-287-107-7_21,

1 Introduction

It is an aspiration of any coach to train and to produce a successful athlete. Effective training is about providing not only quantity but also quality programs. Athletes must also be able to demonstrate excellent levels of technical and tactical abilities as well as to persist longer than the opponents physically and mentally. A simple subjective view of how well the athletes perform is no longer applicable as a foundation of training and competing. The need to record, analyze, and evaluate information on key performance areas such as physical, psychological, technical, and tactical has become fundamental to elite performance. The information on several characteristics of sports performance is the foundation for providing feedback on how the athlete is performing. This feedback will lead to the development of up-to-date coaching interventions or methods centered on evidence-based training programs. If athletes are to attain world-class performances, information from continuous assessments of training and competition must be made available in order to assist in the evaluation of athletes' performance and progress. Therefore, it is important for coaches to possess a solid framework of sport science support system that is designed to help foster the skills of athletes and improve their performance.

In order to play tennis at competitive level, certain standards need to be achieved in all major components of performance such as physical, psychological, technical, and tactical areas of skills or abilities. A high level of performance can only be achieved through a meticulously planned and well-controlled training system based on scientific knowledge and proven methods of training fundamentals. As such, many sports scientist have been researching methods and studying new approaches or tools to help train tennis players in the laboratory and on the court. This is to facilitate tennis players to reach their full potential during competitions as well as in their playing career. Many high-performance coaches agree that there are four main components that are required to develop the complete athlete: psychological, physical, tactical, and technical [1]. Each of these components has its own development path and methods. Developing all of the components in an integrated way is known as a holistic approach and all are interconnected. According to Bangsbo [2], performance of an athlete at elite level depends on the athlete's technical, tactical, physiological, and psychological/social characteristics. Improvement in performance, particularly at the elite level, largely depends on applying scientific knowledge to increase the quality of training programs. Therefore, it is always recommended that tennis players train in a specific manner to improve tennis-specific performance. It has been proven that the development of specific areas such as techniques and tactics as well as mental and physical skills will contribute to the success and improvement of athletes.

Training tennis players require effective integration of tactical, technical, physical, and psychological components in order to ensure successful performance. Technical development will improve players in various strokes such as forehand, backhand, serve, and volleys; tactical development will assist in the use of different strategies or approaches, while mental preparation will provide stability of the players in critical situations. Physical prowess will help players to execute skills at optimum level. The central involvement of the athlete in the process is the forte that may increase their motivation and encourage adherence to any intervention strategies devised. Additionally, the profile can be utilized as a monitoring device to assess the effectiveness of any intervention and highlight areas of good and poor progress. The comprehensive monitoring of athletes' performance is necessary particularly during phases of training overload in order to allow a coach to make a decision regarding the effects and consequent planning of training. The principle of individualization suggests that every athlete will react differently to any training stimulus [3].

The purpose of the study is to examine the effect of an integrated training program on the performance of national elite junior tennis players aged between 12 and 16 years old and to provide evidence-based recommendations that can be adopted by players, parents, coaches, and national associations in the training programs with the objective of upgrading the performance level and achievements of Malaysian junior tennis players. The findings of this study will provide important information, knowledge, and guidelines for coaches, parents, and players on a systematic approach to become a better player and to reach the maximum potential of the player. Tennis players looking forward to reach their full potential deserve to be trained using the most up-to-date or systematic approaches that have been scientifically proven to enhance their performance. It is the responsibility of the coach to impart the latest information and findings into a successful evidence-based training program to enhance players' performance and to achieve their fullest potential. The aim of this study is to seek a realistic and feasible solution in order to improve and to enhance the performance of Malaysian junior tennis players as well as to upgrade the standard of their achievement at international level.

The objectives of this study are:

1. To explore the performance characteristics of Malaysian junior players.
2. To examine the effects of an intervention program on key performance areas of Malaysian junior tennis players.

2 Methodology

2.1 Participants

A total of twenty-four junior tennis players ($n = 24$) were selected for this study using the purposive sampling method where all the participants were the top tennis players in Malaysia between the ages of 12 and 16 years old (18 males and 6 females, mean age 14.29 ± 1.65 years) and were ranked in the National Junior Ranking System. The ranking system was based on the accumulated points collected by the players in various national-level tennis tournaments. The points collected determine the ranking achievement of each of the players who participated in the national tennis tournaments. All testing and training procedures were fully explained, and written parental consent was obtained for each participant, all of whom agreed to participate in the study.

2.2 Procedures of Study

All subjects were tested on four different performance variables, namely the physical, psychological, technical, and tactical profiles. These four performance components were tested utilizing established and proven reliable instruments as well as reliable procedures. Each of the performance profiling was assessed using a standard scoring methodology. These methods correlate with the objective of this study that focuses on effects of an intervention program on the four selected performance components. The participants were inducted into an intervention program that was employed for eight weeks. An integrated training program employed as the intervention was adopted and adapted from existing training program used by many professional tennis players and recognized by panel of qualified coaches. The participants were tested on the four criterion performance variables before and after the completion of the intervention training program. Example of the weekly training program is shown in Fig. 1.

During a two-day testing phase of pre- and posttest sessions, four experimental sessions were conducted as described below:

(a) A series of fitness test batteries based on United States Tennis Association Fitness Protocol [4]. The test batteries should have the following:

- Aerobic endurance—using bleep test, a 20-m shuttle run test to maximal exhaustion.
- Flexibility—using sit-and-reach test.
- Speed—20-m sprint using timing lights.
- Grip strength—using dynamometer.
- Upper-body strength—using 1-min sit-up test.
- Leg power—double leg counter movement vertical jumps
- Body composition—using body mass index (BMI).
- Agility—using Illinois run test.
- Balance—using 1-min stork stand test.

(b) International Tennis Number (ITN) on-court assessment, a tennis skill assessment method developed by International Tennis Federation [5].
(c) Mental performance inventory [6]. A 42 items self-report inventory with seven subscales, designed to measure factors that reflect mental toughness in an athlete.
(d) Tactical performance profile. Components of tactical skills based on coach's observation of player's on-court performance.

Fig. 1 Weekly integrated training plan

Monday, Wednesday, Friday
3.00PM – 7.30PM
Warm-up Technical warm-up ***TECHNICAL & TACTICAL DRILLS*** Killer forehand Ballistic backhand Baseline game Serve & Return Approach shots & volleys Passing the net player Consistency & Control ***PSYCHOLOGICAL DRILLS*** Modified match play. Emphasise pace of play and visualization ***PHYSICAL DRILLS*** Sprinting (15 minutes) Short explosive (15 minutes) Ladder drills (15 minutes) Upperbody strength using Medicine ball (15 minutes) Cool down and stretching

3 Significance of Study

The development of a champion should start from the early age. This is because in general, it takes eight to twelve years of training to build a talented player to become a world-class player. Profiling of players' performance appears to be particularly useful in aiding the design of specific mental, physical, and technical training programs. The central involvement of the athletes in the process is a key of strength that may boost motivation and promote adherence to any intervention strategies devised [7]. Additionally, the performance profiling can be used as a monitoring device to assess the effectiveness of any interventions.

The findings of this study will provide important information, knowledge, and guidelines for coaches, parents, and players on the systematic approach to reach optimum potential and to achieve the performance target. Tennis players looking forward to reach their full potential deserve to be trained using the most up-to-date or systematic approaches that have been scientifically proven to enhance players' performance. It is the responsibility of the coach to impart latest information and findings into a successful evidence-based training program to enhance players' performance.

4 Results

4.1 Statistical Analyses

The results of this study were analyzed using the Statistical Package Social Sciences (SPSS) 18.0. Results showed that all data collected in pretest and posttest were normally distributed (Kolmogorov–Smirnov test), and therefore, a two-tailed paired t test was used to detect differences for each test between the pretest and posttest periods. Effect sizes were calculated and interpreted according to Cohen's standards [8]. A level of $p < 0.05$ was considered to be statistically significant. A reliability analysis for tactical profile data was conducted, and it has shown that Cronbach's alpha for experimental group was 0.98 and was more than 0.70 as suggested by Nunnally [9]. Therefore, they were adequately reliable and appropriate to be used for further analysis.

4.2 Data Analysis on Physical Profile

Table 1 shows the t test results on each component of the physical profile. The obtained t test values were statistically significant at $\alpha = 0.05$. Hence, it also clearly shows that most of attributes in the physical profile by participants have produced an improvement with level of significant differences of ($\alpha < 0.05$) on cardiovascular endurance ($0.000 < 0.05$), flexibility ($0.00 < 0.05$), speed ($0.003 < 0.05$), handgrip strength ($0.02 < 0.05$), leg explosive power ($0.001 < 0.05$), muscular endurance ($0.002 < 0.05$), body composition ($0.01 < 0.05$), agility ($0.00 < 0.05$), and balance ($0.001 < 0.05$).

4.3 Data Analysis on Psychological Performance

Table 2 shows the t test results on each component of the psychological performance. The obtained t test values were statistically significant at $\alpha = 0.05$. Hence, it also clearly shows that most of attributes in the psychological profile by the participants have produced improvement for all attributes that were tested among participants with a significant difference of $\alpha = 0.05$. The results of the p values for each attributes on this profile were self-confidence ($0.00 < 0.05$), negative energy ($0.002 < 0.05$), attention control ($0.00 < 0.05$), visual and imagery ($0.00 < 0.05$), motivational level ($0.00 < 0.05$), positive energy ($0.00 < 0.05$), and attitude control with p value ($0.01 < 0.05$). The results have confirmed the existence of an improvement for all attributes that had been tested among the participants in the experimental group with a significant difference of $\alpha = 0.05$.

Table 1 Significance of means between pre- and post test on physical performance

Variables	Pre-test	Post-test	Diff	t test	*p* value*
Cardiovascular endurance	43.73	46.13	2.39	−14.01	0.00
Flexibility	33.37	35.89	2.53	−6.52	0.00
Speed	3.60	3.52	−0.09	3.73	0.003
Strength (handgrip)	64.50	65.15	0.65	−2.88	0.02
Leg power	27.00	29.08	2.08	−4.31	0.001
Muscular endurance	26.42	27.58	1.17	−3.92	0.002
BMI	20.07	20.08	0.01	−3.03	0.01
Agility	17.59	17.21	0.38	5.20	0.00
Balance	17.83	19.33	1.50	−4.78	0.001

$^{*}p < 0.05$

Table 2 Significance of means between pre- and post test on psychological performance

Variables	Pre-test	Post-test	Diff	t test	*p* value*
Self-confidence	18.33	21.08	2.75	−9.87	0.00
Negative energy	17.00	17.92	0.92	−4.01	0.00
Attention control	17.17	19.50	2.33	−9.11	0.00
Visual and imagery control	18.33	20.83	2.50	−9.57	0.00
Motivational level	20.67	23.17	2.50	−10.86	0.00
Positive energy	17.42	19.67	2.25	−8.07	0.00
Attitude control	18.83	19.75	0.92	−3.53	0.01

$^{*}p < 0.05$

4.4 Data Analysis on Technical Performance

Table 3 displays the summaries of result of significant differences for each attribute on technical profile performance for the experimental group. The *p* values for each attribute were forehand ground stroke (crosscourt) ($0.00 < 0.05$), backhand ground stroke (crosscourt) ($0.00 < 0.05$), forehand ground stroke (down the line) ($0.00 < 0.05$), backhand ground stroke (down the line) ($0.002 < 0.05$), forehand ground stroke (crosscourt) ($0.00 < 0.05$), backhand ground stroke (crosscourt) ($0.002 < 0.05$), first serve ($0.00 < 0.05$), and forehand volley ($0.00 < 0.05$), and backhand volley ($0.00 < 0.05$). Hence, the results confirmed a significant improvement for every tested attributes because the p values were less than the significance level ($\alpha = 0.05$).

4.5 Data Analysis on Tactical Profile

Table 4 shows the results of significance of means difference analysis of each attribute on tactical profile. The obtained *p* values for t test results to be significant at $\alpha = 0.05$ and the *p* values for each attributes in tactical performance were the

Table 3 Significance of means between pre- and post test on technical performance

Variables	Pretest	Posttest	Diff	t test	*p* value*
FH GS (depth)	26.00	31.33	5.33	−14.18	0.00
BH GS (depth)	24.92	30.50	5.58	−13.40	0.00
FH GS (DLine)	13.33	15.75	2.42	−8.40	0.00
BH GS (DLine)	14.00	14.83	0.83	−4.02	0.002
FH GS (crosscourt)	14.75	16.00	1.25	−6.97	0.00
BHGS (crosscourt)	14.83	15.75	0.92	−4.01	0.002
First serve	73.75	78.83	5.08	−9.13	0.00
FH volley	20.67	23.83	3.17	−6.09	0.00
BH volley	21.17	23.92	2.75	−6.17	0.00

*$p < 0.05$

Table 4 Significance of means between pre- and post test on tactical performance

Variables	Pretest	Posttest	Diff	t test	p value*
Attacking shots	5.67	6.58	0.92	−6.17	0.00
Defensive shots	6.42	7.25	0.83	−5.00	0.00
Court positioning	5.92	6.75	0.83	−5.00	0.00
Anticipation	5.83	6.67	0.83	−7.42	0.00
Game plan	6.17	6.67	0.50	−3.32	0.01
Depth	6.25	7.42	1.17	−10.39	0.00
Decision making	6.08	6.92	0.83	−7.42	0.00

*$p < 0.05$

following: attacking shots (0.00 < 0.05), defensive shots (0.00 < 0.05), court positioning (0.00 < 0.05), anticipation (0.00 < 0.05), game plan (0.01 < 0.05), depth (0.00 < 0.05), and decision making (0.00 < 0.05). Therefore, the *p* values showed less than 0.05, and it can be concluded that there was a significant improvement in all the components on tactical performance.

5 Discussion

The findings proved that participants showed an improvement in their performance based on four major areas of physical, psychological, technical, and tactical skills as a key in engaging a better quality of training program. The findings indicated that participants from the experimental group have shown significant improvement in all performance components based on four major areas of physical, psychological, technical, and tactical skills. Tennis is an open-loop sport that requires constant decision making, response organization, spatial awareness as well as wide range of physical, psychological, technical, and tactical abilities [10]. It also validates the positive effect of the training intervention program as sample paired t test analysis subscales with every performance profile results showed a significant difference after the intervention. The participants showed a higher improvement results that dictate the optimistic effect of intervention training program developed in this study. The findings on physical profile were supported by previous study [11] who found that sprint, power abilities, and handgrip strength can be good predictors in tennis performance. According to Cook [12], tennis is an impact game that needs overall fitness, which comprise of several components. Psychological skills are of considerable advantage to tennis players. Some psychological characteristics can be developed independently, while some of the skills can be coached [13]. Technical skill also plays important role in successful performance of tennis players as pointed out by Crespo and Higueras [14] that ability to produce good ground stroke is a distinguishing feature of the modern tennis. The findings on tactical skill are also supported by Scully and O'Donoghue [15] who demonstrated the interaction between knowledge and decision making in which their study showed that winners in tennis matches made similar changes in tactics. This demonstrates that elite tennis players are able to combine their knowledge of the game and situation that arise during competition to make decision on tactical moves. In conclusion, the practicality of information gathered from this study can be applied when designing training programs for elite tennis players. The need analysis on player's performance will help coaches to identify the strengths and weaknesses of players and to design the proper training program for performance enhancement of the players accordingly. Throughout the above in-depth discussion of the research findings, the researcher has wind up certain important elements as well as attributes of a successful tennis player that should be adopted in tennis training intervention programs as a benchmark of many previous studies. It can be concluded that all the four performance components, namely physical, psychological, technical, and tactical skills, are important attributes of successful tennis players that should not be ignored in any coaching methodology. This information can be used as guideline to develop the training schedules in order to help the athletes to improve on relatively weak areas of performance. In conclusion, the practicality of information gathered from this study can be applied when designing training programs for elite tennis players. Effective planning and training programs will help in designing an effective, productive, reliable, and safe program design to help in optimizing the performance.

6 Practical Applications

An eight-week training program for competitive junior tennis players was adopted and adapted in this study and used as an intervention. The supervised training sessions are conducted four times a week for 3.5 h per session. The training program covers all four areas of performance profile: physical, psychological, technical, and tactical skills. All training and testing procedures are performed on the tennis court. Further research is recommended to determine whether this training program can reduce the incidence of injuries among junior tennis players.

Acknowledgment The authors would like to thank Selangor Tennis Association, the coaches and players for the support and cooperation in completing this study.

References

1. Elderton, W. (2004). Situation training: Key to training in a game-based approach. ITF. Coaching and Sports Science Review; *44*(4).
2. Bangsbo, J., Mohr, M., Poulsen, A., Perez-Gomez, J., Krustrup, P. (2006). Training and testing the elite athlete. J Exerc Sci Fit, *4*(1).
3. Bompa, T. (1999). *Periodization: Theory and methodology of training* (4th ed.). Champaign: Human Kinetics.
4. Roetert, E.P., Ellenbecker, T.S. (1998). Testing tennis fitness. In: Usta (Eds), Complete conditioning for tennis (pp. 7–31). Champaign: Human Kinetics.
5. International Tennis Federation (ITF) (2004). The ITF international tennis number manual.
6. Loehr, J. E. (1986). *Mental toughness training for sport: achieving athletic excellence.* Lexington: Stephen Greene Press.
7. Crust, L. (2002). Performance profiling: A coaching tool for pinpointing strengths and weaknesses, designing training strategies and building better communication with athletes. *Peak Performance, 183*, 7–9.
8. Cohen, J. (1988). *Statistical power analysis for the behaviorial sciences* (2nd ed.). Hilsdale: Erlbaum.
9. Nunnally, J. C. (1978). *Psychometric theory* (2nd ed.). New York: McGraw-Hill.
10. MacCurdy, D. (2006). Talent identification around the world and recommendations for the Chinese Tennis Association. ITF Coaching.
11. Girard, O., & Millet, G. P. (2009). Physical determinants of tennis performance in competitive teenage players. *The Journal of Strength & Conditioning Research, 23*(6), 1867–1872.
12. Cook, L.R. (2001). Speed and agility training for volleyball.
13. Williams, J. M., & Krane, V. (2001). Psychological characteristics of peak performance. In J. M. Williams (Ed.), *Applied sport psychology: personal growth to peak performance* (4th ed., pp. 137–147). Mountain View: Mayfield.
14. Crespo, M., & Higueras, J. (2001). Forehands. In: P. Roetert, J. Groppel (Ed.), World-class tennis technique (pp. 147–171). Champaign: Human Kinetics.
15. Scully, D. & O'Donoghue, P.D. (1999). The effect of score line on tennis strategy in Grand Slam men's singles. Journal of Sports Sciences.

Athlete Overtraining Monitoring System

Mohamad Asyraf Faris Abdol Aziz, Ahmad Faizal Salleh, Sukhairi Sudin, Fezri Aziz, Ali Yeon Mohamad Shakaff, Mohammad Shahril Salim and Norasmadi Abdul Rahim

Abstract This paper introduces the design and development of a system that can early detect overtraining problem during training activities. These problems can affect athletes' physiological and psychological conditions as well as reducing their performance. Maximum heart rate (MHR) is the limitation heart rate of athlete and can be indicator for overtraining exercise. Heart rate is usually used to detect and prevent overtraining by coaches and athlete. In this study, we use electrocardiograph (ECG) for amplifying and filtering the signal from the body. National instrument (NI) DAQ is used to acquire the real-time signal from the sensors circuit and pass the data to LabVIEW for real-time monitoring and analysis. The software displays heart rate as well as detecting the abnormality if present. Furthermore, it also features a simple yet comprehensive user interface where the athlete data, date and time for the data collection are saved in the specified txt file for future reference.

Keywords Overtraining · Heart rate · Electrocardiograph · National instrument DAQ · LabVIEW

1 Introduction

The objective of training competitive athlete is to provide effective training in terms of improving performance. While athlete undergoing hard training, they are exposed to a situation that can affect their performance such as overtraining.

M. A. F. Abdol Aziz (✉) · A. F. Salleh · F. Aziz · M. S. Salim · N. Abdul Rahim
School of Mechatronics (Biomedical Electronics Engineering), University Malaysia Perlis (UniMAP), Kampus Ulu Pauh, 02600 Arau, Perlis, Malaysia
e-mail: asyraf_faris87@yahoo.com

S. Sudin · A. Y. Mohamad Shakaff
Center of Excellence Advance Sensor and Technology, University Malaysia Perlis (UniMAP), Kampus Ulu Pauh, 02600 Arau, Perlis, Malaysia

R. Adnan et al. (eds.), *Proceedings of the International Colloquium on Sports Science, Exercise, Engineering and Technology 2014 (ICoSSEET 2014)*,
DOI: 10.1007/978-981-287-107-7_22,

Overtraining is a physical, behavioral, and emotional condition that occurs when the volume and intensity of an individual's exercise exceeds their recovery capacity [1–3]. These conditions can influence the athlete's physiological and psychological state and thus causes underperformance, muscle weakness, chronic fatigue, and less motivated. In some cases, the term overtraining may not be appropriate, as other stress factor (e.g., psychological, lifestyle, malnutrition, and infection) may be responsible for underperformance [4]. Overtraining can be monitored by using many indicators. In this study, we are using electrocardiograph (ECG) to trace heart rate and maximum heart rate (MHR) are use to indicate overtraining exercise. The MHR is the highest heart rate an individual can achieve without severe problems through exercise stress.

Over the past years, coaches and sport scientist have been monitoring athlete heart rate in their training. This method was used since they are relatively simple, cheap, and more importantly able to prevent and detect overtraining [5]. An increased demand for oxygen by working muscles is the primary factor that influences heart rate during exercise.

As explained in Fig. 1, ECG is used, respectively, for amplifying and filtering the signal from the body. NI DAQ is used to acquire the real-time signal from the ECG circuit and digitalize the analog signal. LabVIEW is used for displaying real-time raw ECG waveform, computing and showing the normal heart rate and as well as detecting the abnormality if present.

2 Methodology

2.1 Electrocardiograph

The ECG is an interpretation of the heart's electrical activity over time recorded by skin electrodes externally. It is recorded noninvasively by a diagnosis tool known as electrocardiographic device which is used to measure heart rate, heart rhythm, and analysis of several abnormalities of heart signal. The electrocardiographic data acquisition system can be used to diagnose heart disease as well as detect the occurrence of sudden cardiac arrest in a person by detecting the heart signal abnormalities [6].

2.2 Preamplifier

Preamplifier or first-stage amplifier, which was the head stage of the circuit, is needed to provide high input impedance and high common-mode rejection ratio. The INA321 is a low power, general purpose instrumentation amplifier offering excellent accuracy. Its versatile 3-op amp design and small size make it ideal for a

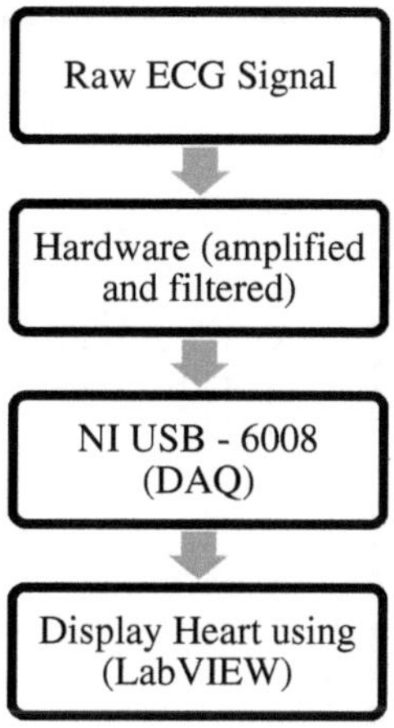

Fig. 1 Block diagram of overall system

wide range of applications. Current-feedback input circuitry provides wide bandwidth even at high gain (200 kHz at $G = 100$). A single external resistor sets any gain from 1 to 10,000. The INA321 is laser trimmed for very low offset voltage (50 μV), drift (0.5 μV/°C), and high common-mode rejection (120 dB at $G \geq 100$). It operates with power supplies as low as ±2.25 V, and quiescent current is only 700 μA—ideal for battery-operated systems. Internal input protection can withstand up to ±40 V without damage. The gain can be calculated using Eq. 1.

$$G = \frac{50\ \text{k}\Omega}{R_G} + 1 \tag{1}$$

2.3 Filter

The second stage is filtering. The frequency of ECG ranges from 0.05 to 150 Hz. A combination of passive high-pass and passive low-pass filter is constructed to only allow the signals with required frequency range [6] (Fig. 2).

2.4 NI-USB-6008 DAQ System

NI-USB-6008 is a 12 bit, 10KS/s low-cost multifunction DAQ. This device has 8 analog input and 2 analog outputs. The NI-DAQmx driver and NI LabVIEW software were installed on the computer to allow device interface. The basic data acquisition functionality is provided in NI-USB-6008 for various applications.

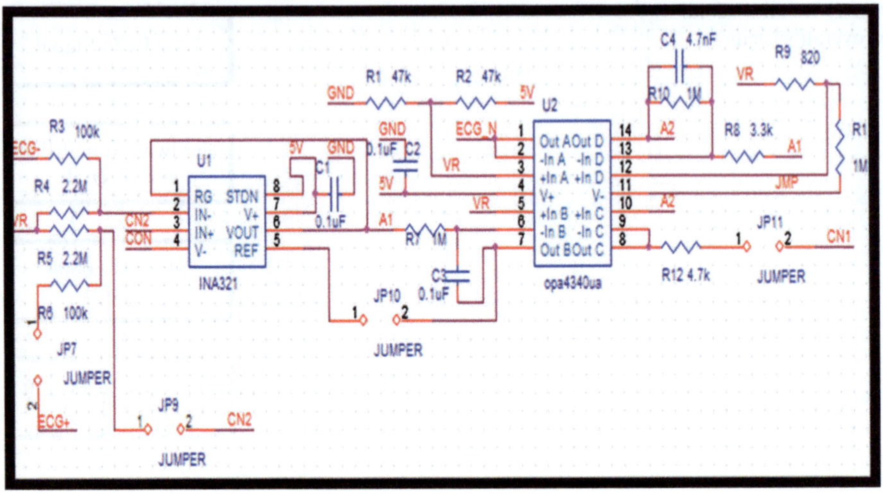

Fig. 2 Circuit design using or card

For this project, the DAQ system is utilized for acquiring the real-time signal from the ECG circuit, digitalize the analog signal, and present and save the data in the chosen file.

2.5 LabVIEW

The LabVIEW is used to control and monitor the whole system. The software is utilized as it features a user-friendly graphical programming and provides variety of functions. The DAQ assistant function is used to acquire ECG and Flex sensor analog signal from the circuit. This function will only be available after the DAQmx driver is installed into the computer. Once the DAQ icon pulled out on the block diagram, a window command will appear to fill the requirements and setting. After done with the requirements, the DAQ Assistant will emerge on the block diagram. The ECG output from the circuit will be available from DAQ Assistant where the data can be stored or displayed in a graph [7].

2.6 Signal Processing

Preprocessing is done to eliminate the unnecessary signals from the hardware such as noise from the ECG sensor and circuit. The ECG signal is multiplied with a gain of 2.5 to amplify the signal and for the purpose of better displayed. The gain in this program is adjustable; enabling the user to change it according to the size of signal

Fig. 3 The ECG circuit

amplitude. The ECG signal can also contain power line interference, Electromyography noise, patient-electrode motion, and baseline wandering. The power line interference is among the noises that can affect the ECG signal analysis. This type of noises typically exists at around 50–60 Hz and produced by the power supply [6]. Hence, in order to obtain a quality signal, digital filters are implemented in the software. The band-stop filter is used to remove frequency in between 50 and 60 Hz. Then, a bandpass filter with cutoff frequency of 0.05–150 Hz is applied to ensure the signals are within the valid frequency range.

The prototype of ECG amplifier circuit was design as shown in Fig. 3. By using ECG hardware, the raw ECG signals in range of 0.05–150 Hz.

Figure 4 shows the front panel of the program. The hardware sampling rate and threshold must be set by the user to get the reading. In this project, the sampling rate was set to be 1,000 Hz. The threshold value can be adjusted according to the amplitude of the ECG signal. The software also displays the real-time graphs of filtered sensors data as well as indicators of the heart rate. Three-level indicators are used to make sure the athlete stop training before they reach the limit of heart rate: level 1 for 50 % of MHR, level 2 for 70 % of MHR, and level 3 for 85 % of MHR.

Simulations were done to check the reliability of the system. ECG simulator was used to provide ECG signals for the circuit. Figure 5 shows the graphs, and heart rate calculations perform in the LabVIEW software upon receiving different frequencies of signals.

It was found from the results that the accuracy of heart rate calculations were 99 %.

Next, the system was also tested on real subjects (athletes). In this experiment, the subjects were asked to perform star jump exercise until exhaustion. The ECG electrodes were attached to their body as shown in Fig. 6. All the signals produced during the exercise were processed by the system in real time, and the outcome particularly the heart rate and overtraining level were analyzed. The result is explained in Sect. 3.

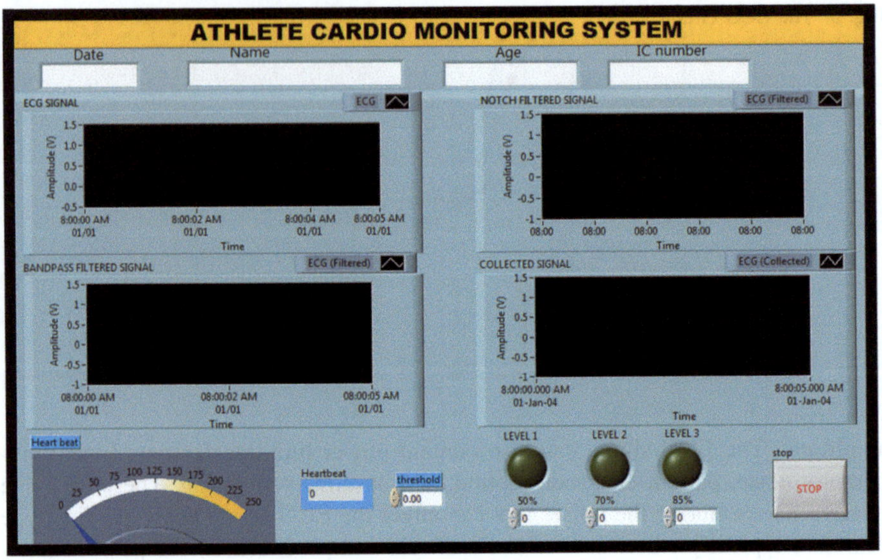

Fig. 4 Front view of GUI system using LabVIEW

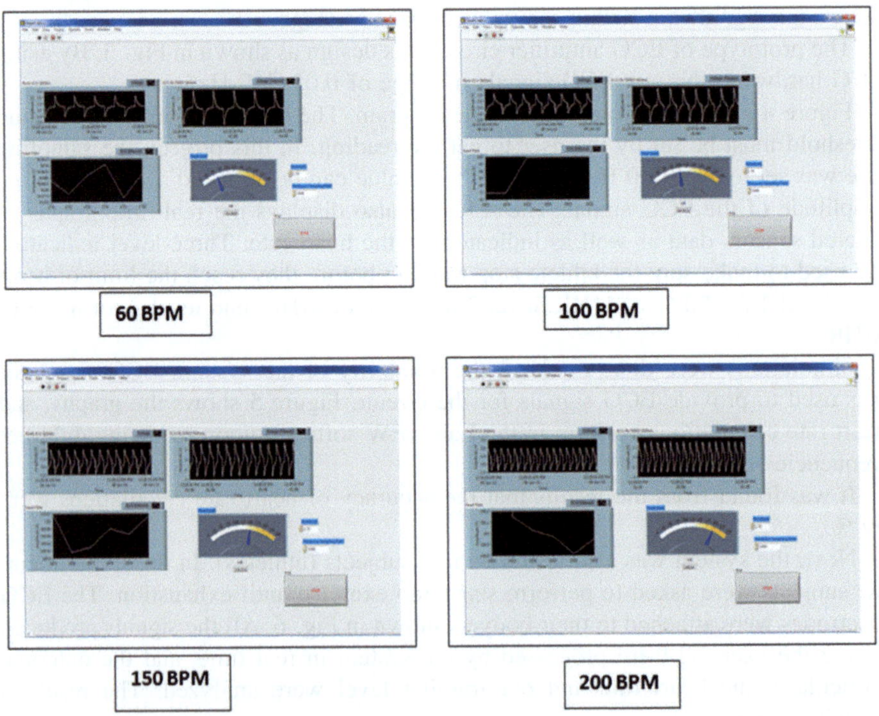

Fig. 5 The heart rate value (in bpm) using ECG simulator

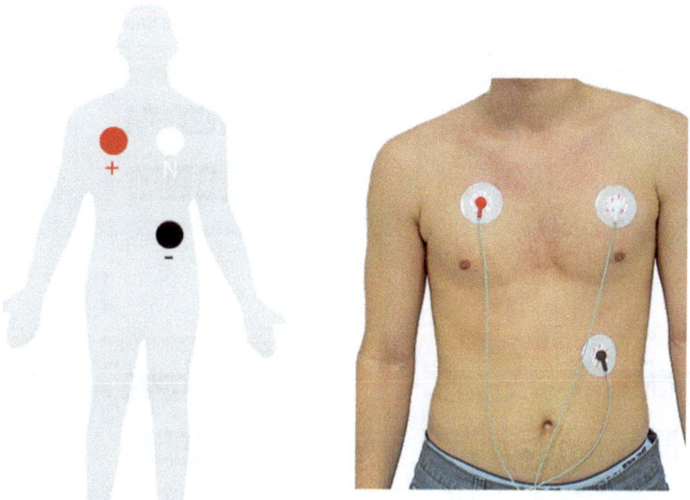

Fig. 6 ECG electrode placement

2.7 Maximum Heart Rate and Overtraining Calculation

The normal heart beat of a human usually lies within 60–90 beat/min. MHR during exercise can be calculated using [8].

$$\mathrm{MHR}_{\mathrm{male}} = 220 - \mathrm{age} \tag{2}$$

$$\mathrm{MHR}_{\mathrm{female}} = 206 - (0.88 \times \mathrm{age}) \tag{3}$$

If the heart rate during the training exceeds 85 % of MHR, it is classified as overtraining.

3 Result

Figure 7 shows the raw ECG signals obtained during the star jump activity, real-time heart rate, and overtraining level. At the beginning of the exercise, the subject is still energetic and the heart rate is low. So, the three indicators did not light up. When the heart rate of the subject had exceeded 50 % (at around 3 min), the level 1 indicator lighted up. After 8 min, the second indicator lighted up showing that the heart rate had exceeded 70 %. Upon heart rate reaching 85 % (at around 12 min), the third indicator brightened to alarm that the subject has been exhausted, and thus, further exercise can be considered as overtraining.

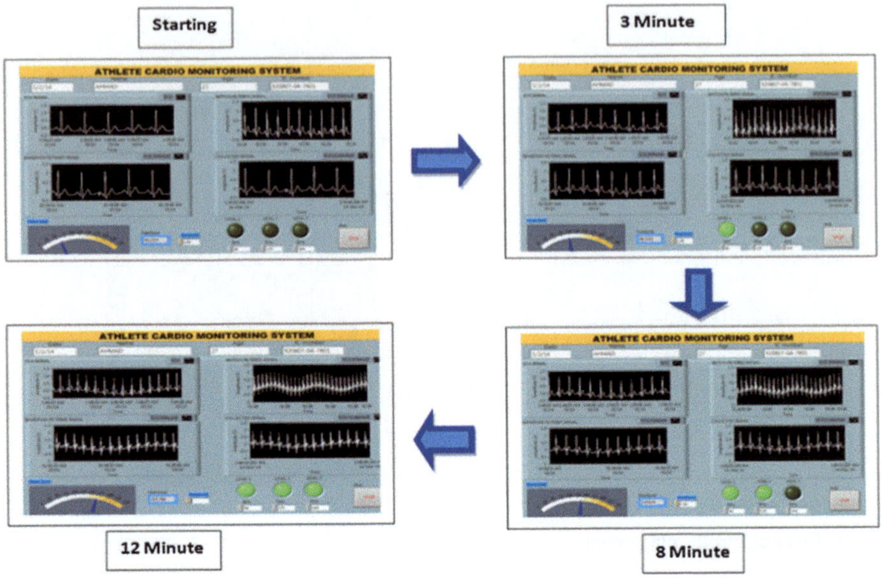

Fig. 7 Subject level performance

4 Discussion and Conclusion

The ECG signal acquisition were designed and implemented in this project. From the result, it can be clearly said that ECG signal can be acquired and processed by the implemented hardware and software.

The quality of the ECG signal collected can be reduced due to large size of electrode deviates to noise artifacts. Hence, the signal is more susceptible to errors, making it difficult for further analysis.

The main challenge in analog circuit is to amplify the weak ECG signal without noise. The frequency range chosen for ECG signal is 0.05–150 Hz. Besides the ECG waveform, the signal picked up by the electrodes also contains several kinds of noise. A low-frequency (<0.05 Hz) noise produced by respiration that result in a baseline drift of the ECG signal, high frequency (>150 Hz), and 50–60 Hz noise from power line interference. The magnitudes of these noises are comparable to that of the ECG waveform.

In conclusion, we have successfully designed and developed a system using the ECG sensor, which able to detect the overtraining phase during an exercise. We also managed to implement the system using low-cost components and devices as well as designed user-friendly software for signal processing.

References

1. Budgett, R. (1998). Fatigue and underperformance in athletes: the overtraining syndrome. *British Journal of Sports Medicine,* 32, 107–110.
2. MacKinnon, L. T. (2000). Special feature for the olympics: effects of exercise on the immune system (2000): overtraining effects on immunity and performance in athletes. *Immunology and Cell Biology, 78*(5), 502–509.
3. Raglin, J. S., Wilson, G. S. (2000). Overtraining in athletes. In Hanin, Y.L (Ed.), *Emotions in sport*, (pp. 191–207). Champaign (IL): Human Kinetics.
4. Budgett, R. (1990). Overtraining syndrome. *British Journal of Sports Medicine, 24*, 231–236.
5. Acthen, J., & Jeukendrup, A. E. (2003). Heart rate monitoring: Applications and limitations. *Medicine and Science in Sports and Exercise, 33*(7), 517–538.
6. Palanivel, S., Sukanesh, R., Vijayprasanth, S. (2012). Design and development of mobile based smart tele-health care system for remote patient, 7, 148–158.
7. Data Acquisition and Signal Conditioning Course Manual; National instrument Corporation, 1995 2003.4.
8. GULATI, M., et al. (2010). Heart rate response to exercise stress testing in asymptomatic women. *Exercise Physiology.*

Part III
Sports Physiology and Behaviour

Blood Profile Alterations in Overweight Females After Aerobic Interventions

Maisarah Shari, Suhana Aiman and Sarina Md Yusof

Abstract Worldwide there is a sharp increment in obesity including Malaysia. It was estimated that 62 and 48 % of female and male adults, respectively, were overweight and obese in the year 2010. This leads to a rapid increment in medical costs outage and mortality due to disease globally. Aerobic exercises are crucial in improving overall physiological health of oneself, and reducing risks of chronic diseases. The current study was designed to determine the effects of land-based and water-based aerobic dance exercises on blood glucose and blood lipid parameters among overweight young female adults. A total of 75 overweight females participated in this study. They were randomly assigned to interventions and control groups (with 25 for land-based aerobic dances, 25 for water-based aerobic dances, and 25 controls). Intervention groups were engaged in aerobic exercise workouts (land-based aerobic dances and water-based aerobic dances) for 9 weeks, three times per week and 1 h per session. Blood glucose (BG), triglycerides (TG), cholesterol (CHO), high density lipoprotein (HDL), and low density lipoprotein (LDL) were measured at baseline, week 5, and postexercise. Significant reductions were found in both land-based and water-based aerobic dance exercise groups ($p < 0.05$) on BG (4.18 ± 0.39, 4.01 ± 1.28), TG (1.31 ± 0.78, 1.22 ± 0.38), and HDL (1.55 ± 0.30, 1.67 ± 0.37) across three observation periods. However, only TG and HDL showed significant differences between the two intervention groups ($p < 0.05$). Water-based aerobic dance group elicited

M. Shari (✉) · S. Aiman · S. Md Yusof
Faculty of Sports Science and Recreation, Universiti Teknologi MARA,
Shah Alam, Selangor, Malaysia
e-mail: maisarah_shairi@yahoo.com

S. Aiman
e-mail: suhana083@salam.uitm.edu.my

S. Md Yusof
e-mail: sarin864@salam.uitm.edu.my

R. Adnan et al. (eds.), *Proceedings of the International Colloquium on Sports Science, Exercise, Engineering and Technology 2014 (ICoSSEET 2014)*,
DOI: 10.1007/978-981-287-107-7_23,

greater improvement compared to land-based aerobic dance ($p < 0.05$) on TG (24.22, 18.63 %) and HDL (32.54, 20.16 %). In conclusion, the results showed that aerobic exercises altered blood profiles positively.

Keywords Aerobic exercise · Blood parameters · Overweight females

1 Introduction

Obesity is defined as a state of excess body fat or body energy stores in excess of physiological needs [1]. Based on current statistics, obesity is a growing problem not only in the USA, but also in developing countries including Malaysia. It was estimated that 62 % of females in Malaysia are overweight and obese and that 38 % of them are young adults. This population faces a greater risk in many types of metabolic and musculoskeletal diseases as they age [2].

A significant amount of scientific information has been published regarding acute and chronic health effects gained by women from various forms of aerobic exercises, and it has been verified that regular participation in aerobic exercises among woman is important in preventing chronic health, mental, and mobility ailments [3–7]. The vital key to lessen all risks of fatal diseases in obese people is via removing excess body mass through utilization of subcutaneous and visceral fat molecules as source of energy [8].

It has been clearly noted that a long duration of aerobic workouts with moderate to high intensity that utilizes a majority of muscle groups, modifies plasma lipids and lipoprotein profiles especially in women through improvement in the levels of high density lipoprotein (HDL), low density lipoprotein (LDL), triglycerides (TG), and total cholesterol (CHO) [9]. Moreover, glucose homeostasis is achieved through continuous insulin secretion by pancreatic β cells and increased GLUT4 transporter activities after consistent aerobic training. Consistent improvement in lipid parameters and serum glucose due to regular aerobic activities significantly decreases body fat and body weight, hence reducing the incidence of mortality-related diseases [10–12].

Previous studies had shown that both aerobic and aqua dances are categorized as enjoyable and beneficial aerobic activities in reducing body weight, increasing cardiovascular fitness, and also in the toning of body structure [13]. Even though, both land-based and water-based aerobic dances are proven to be valuable exercises to promote overall health in oneself, however, only limited studies had been conducted to compare the effects between land-based and water-based aerobic dances on blood modifications in overweight young female adults.

Hence, this study was designed to compare the effects of land-based and water-based aerobic dance exercises in modifying blood lipids and blood glucose (BG) levels, and to determine which type of aerobic exercises was most effective in improving the health of young adult overweight females.

2 Methodology

2.1 Participants

A total of 75 volunteer females aged 20–25, sedentary, and overweight (body mass index 25–30) participated in this study. All volunteers with chronic disease (hypertension, low blood pressure, diabetes, any type of heart disease, cancer, and kidney failure), musculoskeletal disease or injury, abnormal menstruation, categorized as obese and morbidity obese (body mass index 30), and taking any type of drug or traditional medication were excluded. Prior to the intervention, all volunteers completed their medical history and physical activity readiness (PAR-Q) questionnaire, and underwent electrocardiogram (ECG) screening. They were randomly assigned into land-based aerobic ($n = 25$), water-based aerobic ($n = 25$), or control groups ($n = 25$). Participants followed the structured training program assigned to them and they were reminded to maintain their daily diet patterns throughout the intervention. Written consent letters was obtained from participants prior to their participation. The study was approved by the Ethics Committee of Universiti Teknologi MARA (600-RMI: 5/1/6/01).

2.2 Outcome Measures

Participants' BG and lipids were measured at baseline, week 5, and at postexercise.

2.2.1 Body Composition

Participants were weighed on a digital weighing scale (SECA780, Germany) in the standing position with hands placed at the side, barefooted and wearing a non-sleeved light clothe with short tight pants. Their weights were measured to the nearest 0.1 kg. Participants' height was measured to the nearest 0.1 cm using a body meter (SECA206, Germany). The values of weights and heights were used in determining the participants' body mass index. The body mass index was calculated as follows:

$$\text{Body mass index} = \text{weight (kg)} / \text{height} \left(\text{m}^2\right)$$

2.2.2 Serum Glucose and Lipids

Participants' venous blood (6 mL) samples were drawn after 10 h fasting via an antecubital vein with participants seated in an armchair. The venipuncture procedure was performed by qualified medical personnel. Participants' blood samples were collected into two vacutainer tubes with K_2EDTA (lipid profiles), potassium oxalate, and sodium fluoride (fasting BG). Fasting serum lipids and serum glucose of participants were assayed enzymatically by using the VISTA kit analyzer (SIEMENS, USA). Serum glucose was measured using hexokinase (HK) method, while TG was calculated based on an enzymatic procedure. The Abell-Kendall method was used to measure HDL (SIEMENS, USA). The total CHO was measured after precipitation with horseradish peroxide (HPO), and the LDL was calculated using Friedwald's formula.

2.3 Aerobic Dance Intervention Programmes

Training groups were conducted over 9 weeks of aerobic dance exercises, with three sessions per week and 60 min per session. Each session was led by a group of qualified fitness instructors. Aerobic dance training (land-based aerobic dance class and water-based aerobic dance class) was performed on three different days in a week with at least 1 day of rest between sessions. The water-based aerobic dance group performed the exercise routines in an indoor swimming pool with the water level fixed at chest or near chest level with an average water temperature of 24 °C. The land-based aerobic dance group performed the exercise routines in an air-conditioned gymnasium at an average of 25 °C. The exercise programs consisted of stretching and warm-up exercises (10 min), aerobic conditioning (40 min), and cool-down or muscular relaxation exercises (10 min). The warm-up session consisted of static stretching followed by slow-paced dynamic dance movements which involved simultaneous arm and leg motion. The water-based aerobic dance group performed the static stretching before entering the pool followed by slow-paced dynamic dance routines in the water. For the aerobic conditioning, combinations of walking, jogging, and jumping with various dance movements in different rhythms and movement direction were performed from moderate to fast tempo or with low to high repetitions. The training intensity was gradually increased from 50 % in the first week to 75 % in the final week (Fig. 1). The heart rate maximum was used as an indicator of the prescribed training intensity. Participants' heart rate was monitored continuously during training sessions through carotid palpation to ensure training intensity was maintained as prescribed.

2.4 Dietary Control

Dietary intake was monitored throughout the 9 weeks intervention period using 3 day food records. Participants were asked to record everything they consumed for two consecutive weekdays and 1 weekend. The time, the type, and the quantity of food taken using standard household measures were recorded. A computerized nutrient database software (MyFitnessPal) was used to calculate the daily food intake.

Fig. 1 Periodization scheme of aerobic dance training groups

1st – 2nd week	3rd – 5th week	6th – 9th week
10 min warm-up (HR > 120)	10 min warm-up (HR > 120)	10 min warm-up (HR > 120)
40 min conditioning 50-55% Max HR	40 min conditioning 55-65% Max HR	40 min conditioning 65-75% Max HR
10 min cool-down HR < 120	10 min cool-down HR < 120	10 min cool-down HR < 120

2.5 Statistical Analysis

SPSS statistical program (version 17.0) was used to determine means and standard deviations (SD). Kolmogorov Smirnov test ($p < 0.05$) was used to check normality of dependent measures; meanwhile, the Levene test ($p < 0.05$) was used to compare baseline differences between groups of dependent variables included in the study.

A 3 × 3 repeated measures ANOVA was used to (a) compared the three aspects, (b) differentiate significant changes within the times of treatment (at baseline, week 5, and after 9 weeks of intervention) for each group (land-based aerobic dance group, water-based aerobic dance group, and control group), (c) differentiate significant changes in main effects between groups, and (d) differentiate significant changes in the interaction effects among groups, across observations.

The changes in mean scores were used to calculate the percentage changes between baseline and post intervention. The effects brought about by the aerobic exercise training will be indicated by the time-by-group interaction. The criterion for significance of all analyses was set at an alpha level of $p < 0.05$.

3 Results

3.1 Participants Characteristics

No statistically significant differences were observed between the groups (land-based aerobic dance, water-based aerobic dance, control) in terms of age, weight, height, body mass index, serum glucose, and serum lipids at baseline (Table 1). Thus, a normal distribution was evident for the scores of each variable.

Table 1 Characteristics of participants at baseline

Variable	Land-based	Water-based	Control
Age (year)	23.08 ± 1.35	23.08 ± 1.35	22.78 ± 1.51
Weight (kg)	66.92 ± 3.75	69.01 ± 4.00	68.62 ± 2.44
Height (cm)	157.6 ± 3.93	158.6 ± 3.31	156.9 ± 4.14
BMI (kg/m^2)	27.04 ± 1.20	27.02 ± 1.37	27.70 ± 1.39
BG (mmol/L)	5.09 ± 0.33	5.15 ± 0.34	5.11 ± 0.35
TG (mmol/L)	1.61 ± 0.32	1.61 ± 0.17	1.65 ± 0.28
CHO (mmol/L)	5.44 ± 0.49	5.15 ± 0.46	5.26 ± 0.45
HDL (mmol/L)	1.29 ± 0.43	1.26 ± 0.26	1.27 ± 0.31
LDL (mmol/L)	3.79 ± 0.63	3.70 ± 0.57	3.74 ± 0.42

Values are presented as mean ± SD

3.2 Serum Glucose

Statistically significant changes ($p < 0.05$) were noted in the reduction of serum glucose in aerobic training groups across the 9 weeks of exercise duration. Serum glucose decreased by 18.87 % from 5.09 (±0.33) to 4.18 (±0.39) in the land-based aerobic dance group and by 22.14 % from 5.15 (±0.34) to 4.01 (±1.28) in the water-based aerobic dance group. The control group showed no significant differences ($p < 0.05$) in serum glucose across observations. In terms of percentages, the results revealed that the water-based aerobic dance group had 3.27 % greater reduction in serum glucose compared to the land-based aerobic dance group; however, there were no significant differences in the main effect for either group ($p < 0.05$). Summary of the results are presented in Table 2.

3.3 Serum Lipids

The effects of the two aerobic dance intervention programmes and the control group on blood lipid parameters at baseline, week 5, and postexercise are presented in Table 3. The effects were observed on TG, CHO, HDL, and LDL as all scores on the blood lipid parameters measured showed positive changes following the 9 weeks of aerobic dance intervention program. Significant differences in the time effect were observed only on TG and HDL following 5 weeks of land-based and water-based aerobic dance exercises, and this continued to decline after 9 weeks of intervention ($p < 0.05$). In contrast, no significant time effect was observed in all blood lipid variables in the control group ($p < 0.05$). Significant main effects of the three groups (land-based aerobic dance, water-based aerobic dance, and control group) on TG and HDL were noted ($p < 0.05$). The water-based aerobic dance demonstrated greater reduction in TG and increment in HDL compared to land-based aerobic dance and control groups.

Table 2 Effects of aerobic exercises on modifications of serum glucose in young adult overweight females

Groups Variables	Land-based	Water-based	Control
BG (mmol/L) Baseline	5.09 ± 0.33	5.15 ± 0.34	5.11 ± 0.35
Week-5	4.73 ± 1.10	4.90 ± 0.32	5.01 ± 1.36
Post	4.18 ± 0.39	4.01 ± 1.28	5.15 ± 1.24
Change (%)	−18.87*	−22.14*	+0.78

Values are presented as mean ± SD
*Significant changes ($p < 0.05$) across three observations

Table 3 Alterations of blood lipids measures between groups after exercise interventions

Groups Variables	Land-based	Water-based	Control
TG (mmol/L)			
Baseline	1.61 ± 0.32	1.61 ± 0.17	1.65 ± 0.28
Week-5	1.49 ± 0.32	1.32 ± 0.32	1.63 ± 0.74
Post	1.31 ± 0.78	1.22 ± 0.38	1.70 ± 1.56
Change (%)	−18.63*	−24.22*	+3.03
CHO (mmol/L)			
Baseline	5.44 ± 0.49	5.15 ± 0.46	5.26 ± 0.45
Week-5	4.69 ± 0.86	4.11 ± 0.87	5.10 ± 0.60
Post	4.33 ± 0.85	3.89 ± 0.73	5.65 ± 0.82
Change (%)	−20.40	−24.47	+7.41
HDL (mmol/L)			
Baseline	1.29 ± 0.43	1.26 ± 0.26	1.27 ± 0.31
Week-5	1.31 ± 0.32	1.58 ± 0.27	1.28 ± 0.34
Post	1.55 ± 0.30	1.67 ± 0.37	1.23 ± 0.30
Change (%)	+20.16*	+32.54*	−3.15
LDL (mmol/L)			
Baseline	3.79 ± 0.63	3.70 ± 0.57	3.74 ± 0.42
Week-5	3.49 ± 0.74	3.43 ± 0.83	3.79 ± 0.58
Post	3.32 ± 0.77	3.29 ± 0.66	3.79 ± 0.64
Change (%)	−12.40	−11.08	+1.34

Values are presented as mean ± SD
*Significant changes ($p < 0.05$) across three observations

4 Discussion

Overall, the findings obtained in this study demonstrated that 9 weeks of aerobic and aqua dance exercise modified blood parameters. The present findings were consistent with past studies [3, 11, 14]. Fasting BG was reduced significantly in both aerobic training groups and was slightly increased in the control group. These results are in agreement with previous studies which found a significant decline in fasting BG levels after eight to 12 weeks of aerobic exercise training [7, 12, 15]. The mechanism associated with the reduction of fasting BG is due to a decrease in insulin resistance due to an increase in the number of insulin receptors in the muscles which increase insulin binding to monocyte sites during exercise and postexercise [16, 17]. Further reduction in fasting BG is due to the increment of postexercise adipose tissue oxidation resulting from increased glucose utilization to aid in the combustion of fat molecules [12, 15]. In addition, positive changes in BG were due to an increment in the body basal metabolic rate resulting from aerobic interventions [18–20].

Even though significant reductions in BG were observed in both training groups, but no significant difference was noted between these two groups. However, it is interesting to note that water-based aerobic dance demonstrated a better trend in BG reduction compared to land-based aerobic dance. In this present study, it was possible for the water-based aerobic dance exercises to indirectly seize a higher intensity than land-based aerobic dance exercises due to water resistance, even though the actual intensity implemented were the same in both groups (50–75 % HRmax). The cooler temperature in the aquatic exercises may have further enhanced number of insulin receptors in the muscles [20, 21].

The present study also indicated that 9 weeks of aerobic exercises reduced TG, CHO, and LDL, and increased HDL in both aerobic training groups. These present findings are consistent with past reports [7, 11, 22–24] which also showed alterations in these lipid parameters after 8–24 weeks of regular aerobic exercise programmes in overweight, obese, and healthy populations.

A significant decrease in plasma TG resulted from an enhanced TG catabolic capacity due to an increase in lipoprotein lipase enzyme activity in skeletal muscles and adipose tissues [12, 25, 26]. The significant increment in HDL demonstrated in this present study was also in agreement with previous findings [26, 27]. These changes are in response to the decrease in triglyceride levels due to some of the remaining CHO, phospholipid, and apoprotein from the catabolic activity of lipoprotein lipase on chylomicrons and very LDLs are transferred to HDL particles, thereby increasing the plasma HDL mass [25, 26]. Other possible reasons that lead to an increase in HDL are a decrement in hepatic lipase as well as CHO ester transfer protein, and an increase in lecithin CHO acyltransferase following chronic exercise training [25–28].

Although the changes in total CHO and LDL were not significant, it is interesting to note that there were decrements in blood lipid parameters observed in both training groups. The possible biological mechanism highlighted was due to changes in the CHO content of various lipoproteins. The increase in HDL CHO content was compensated by a decrease in the CHO content of other lipoproteins, thus leading to a no significant change in total CHO. Meanwhile, the potential cause for small changes in LDL after aerobic training might be related to insufficient exercise frequency [9–12]. Besides, significant reductions in LDL among obese sedentary individuals seem to occur with greater exercise frequency (five times a week) [26].

The present study found that water-based aerobic dance exercises showed greater improvement in TG and plasma HDLs compared to land-based aerobic dance exercises. Besides the water environment decreases compression stress on weight-bearing joints, bones, and muscles in obese people who carry excess weight. The water resistance from aquatic exercise indirectly increases the exercise intensity as the density of water is approximately 800 times that of air, thus leading to a higher drag force and energy cost during limb movements [29, 30]. Previous studies [25, 26] had also reported that the intensity of the exercise bout is the determining factor for the degree of changes in plasma lipids and lipoproteins. Higher exercise intensity, not exceeding 85 % of maximum heart rate, resulted in greater expansion in blood lipid parameters due to an increase in energy expenditure by fat utilization [11, 22].

5 Conclusion

It can be concluded that 9 weeks of aerobic exercises positively modified serum glucose and serum lipids among young adult overweight females. In addition, performing aerobic exercises in the water resulted in better effects compared to land-based aerobic exercises. Therefore, it is recommend that water-based aerobic exercises are beneficial as a means to enhance weight reduction and modify lipid profiles in overweight or obese populations interested in losing weight and improving overall health. Furthermore, water not only provides resistance, but also offers a buoyancy effect as well. This buoyancy effect is beneficial for reducing strain on joint extremities, and hence water-based exercises are strongly recommended for obese people.

References

1. Park, M. I., & Camilleri, M. (2005). Gastric motor and sensory functions in obesity. *Obesity Research, 13*(3), 491–500.
2. Letchuman, G. R., Wan Nazaimoo, W. M., Wan Mohamad, W. B., Chandran, L. R., Tee, G. H., Jamaiyah, H., et al. (2010). Prevalence of diabetes in the malaysian national health morbidity survey iii. *Medical Journal of Malaysia, 65*(3), 180–186.
3. Katsura, Y., Yoshikawa, T., Ueda, S., Usui, T., Sotobayashi, D., Nakao, H., et al. (2010). Effects of aquatic exercise training using water-resistance equipment in elderly. *European Journal of Applied Physiology, 108*(5), 957–964.
4. Çakmakçi, O. (2011). The effects of 8 plates exercise on body composition in obese women. *Collegium Antropologicum, 35*(4), 1045–1050.
5. Ranavolo, A., Donini, L. M., Mari, S., Serrao, M., Silvetti, A., Iavicoli, S., Cava, E., Asprino, R., Pinto, A., & Draicchio, F. (2013). Lower-limb joint coordination pattern in obese subjects. *Biomed Research International*, 2013.
6. Thomas, E. N., & Blotman, F. (2010). Aerobic exercise in fibromyalgia: a practical review. *Rheumatology International, 20*, 1143–1150.

7. Akdur, H., Sozen, A. B., Yigit, Z., Balota, N., & Guven, O. (2007). The effect of walking and step aerobic exercise on physical fitness parameters in obese women. *Istanbul Tip Fakiiltesi Dergisi Cilt, 30*(3), 64–69.
8. Slentz, C. A., Duscha, B. D., Johnson, J. L., Ketchum, K., Aiken, L. B., Samsa, G. P., et al. (2004). Effects of the amount of exercise on body weight, body composition, and measures of central obesity stride: a randomized controlled study. *Archives of Internal Medicine, 164*(1), 31–39.
9. Lawrence, A. (2010). *ACSM's guidelines for exercise testing and prescription* (8th ed., pp. 253–256). Philadelphia: Lippincott Willian & Wilkins.
10. Bonds, D. E. (2012). Targeting high-density lipoprotein and triglycerides in statin-treated patients with diabetes. *Current Cardiovascular Risk Reports, 6*(1), 102–107.
11. Vatansev, H., & Çakmakçi, E. (2010). The effects of 8-week aerobic exercises on the blood lipid and body composition of the overweight and obese females. *Ovidius University Annals, Series Physical Education and Sport/Science, Movement and Health, 2*, 814–820.
12. Rahmani-nia, F., Rahnama, N., Hojjati, Z., & Soltani, B. (2008). Acute effects of aerobic and resistance exercises on serum leptin and risk factors for coronary heart disease in obese females. *Sport Science for Health, 2*(3), 118–124.
13. Gappmaier, E., Lake, W., Nelson, A. G., & Fisher, A. G. (2006). Aerobic exercise in water versus walking on land: Effects on indices of fat reduction and weight loss of obese women. *The Journal of Sports Medicine and Physical Fitness, 46*(4), 564–569.
14. King, D. E., Carek, P., Mainous, A. G, I. I. I., & Pearson, W. S. (2003). Inflammatory markers and exercise: Differences related to exercise type. *Medicine and Science in Sports and Exercise, 35*(4), 575–581.
15. Barbosa, T. M., Marinho, D. A., Reis, V. M., Silva, A. J., & Bragada, J. A. (2006). Physiological assessment of head-out aquatic exercises in healthy subjects: A qualitative review. *Journal of Sports Medicine and Physical Fitness, 465*(4), 564–569.
16. Nayak, S., Maiya, A., & Hande, M. (2005). Influence of aerobic treadmill exercise on blood glucose homeostasis in noninsulin dependent diabetes mellitus patients. *Indian Journal of Clinical Biochemistry, 20*(1), 47–51.
17. McArdle, W. D., Katch, F. I., & Katch, V. L. (2010). *Essentials of exercise physiology* (7th ed., pp. 458–470). Philadelphia: Lippincott Willian & Wilkins.
18. Shishkova, A., Petrova, P., Tonev, A., Softov, O., Kalchev, E., & Staykova, Y. (2008). Multiple approaches in initial obesity treatment. *Trakia Journal of Sciences, 6*(4), 22–28.
19. Cornelissen, V. A., & Fagard, R. H. (2005). Effects of endurance training on blood pressure, blood pressure–regulating mechanisms, and cardiovascular risk factors. *Hypertension, 46*, 667–675. (American Heart Association Inc).
20. Greene, N. P., Lambert, B. S., Greene, E. S., Carbuhn, A. F., Green, J. S., & Crouse, S. F. (2009). Comparative efficacy of water and land treadmill training for overweight or obese adults. *Medicine and Science in Sports and Exercise, 41*(9), 1808–1815.
21. Raffaelli, C., Lanza, M., Zanolla, L., & Zamparo, P. (2010). Exercise intensity of head-out water-based activities (water fitness). *European Journal of Applied Physiology, 109*(5), 829–838.
22. Shaw, K. A., Gennat, H. C., O'Rourke, P., & Mar, C. D. (2006). Exerise for overweight or obesity. *Journal of Obesity, 10*(9), 530–535. (Review).
23. Nuttamonwarakul, A., Amatyakul, S., & Suksom, D. (2012). Twelve weeks of aqua-aerobic exercise improve physiological adaptations and glycemic control in elderly patients with type 2 diabetes. *Journal of Exercise Physiology Online, 15*(2), 64–70.
24. Veronique, O., Weisnagel, S. J., Marois, J., Bergeron, J., Julien, P., Gougeon, R., et al. (2005). Dietary cod protein reduces plasma c-reactive protein in insulin-resistant men and women. *The Journal of Nutrition, 138*(12), 2386–2391.
25. Thompson, P. D., Crouse, S. F., Goodpaster, B., Kelley, D., Moyna, N., & Pescatello, L. (2001). The acute versus the chronic response to exercise. *Medicine and Science in Sports and Exercise, 33*(6), 438–445.

26. Tladi, D. M., Phanton, L. N., & Moffatt, R. J. (2006). Acute effects of exercise on plasma lipids and lipoprotein of obese women. *American College of Sports Medicine, 14*(4), 187–196.
27. Durstine, J. L., Grandjean, P. W., Davis, P. G., Ferguson, M. A., Alderson, N. L., & DuBose, K. D. (2001). Blood lipid and lipoprotein adaptations to exercise. *Sports Medicine, 31*(15), 1033–1062.
28. Ferguson, M. A., Alderson, N. L., Trost, S. G., Essig, D. A., Burke, J. R., & Durstine, J. L. (1998). Effects of four different single exercise sessions on lipids, lipoproteins, and lipoprotein lipase. *Journal of Applied Physiology, 85*(3), 1169–1174.
29. Takeshima, N., Rogers, M. E., Islam, M. M., Yamauchi, T., Watanabe, E., & Okada, A. (2004). Effect of concurrent aerobic and resistance circuit exercise training on fitness in older adults. *European Journal of Applied Physiology, 93*(1), 173–182.
30. Kravitz, L., & Mayo, J. (2006). The physiological effects of aquatic exercise. *International Journal of Medical, 33*(2), 305–311.

The Obesity Awareness and Perception Among Obese People

Mazlifah Omar, Mazapuspavina Md Yasin, Hashekin Mokhtar, Johan Jitos, Nur Afifah Ab Halikun, Nur Farhana Ahmad Fisol and Siti Aishah Mat Yaacob

Abstract Behavioral issues play an important factor in the management of obesity in order to maintain adherence to intervention. Objective of this study was to determine the awareness and perception on obesity among obese people. Thirty obese people ($M = 21$, $F = 9$) were identified during health screening program. Parallel back-to-back translation, validated, and self-administered questionnaire was given out to subjects, which is available in Bahasa Malaysia and English. Twenty-three (77 %) respondents were aware of being obese and had their body mass index (BMI) correctly calculated. Twenty-four (80 %) respondents were aware that diet and exercise were the main factors for obesity. Nineteen (63 %) subjects perceived obese people as lazy and 14 (46 %) perceived as untidy. These information highlighted that obese people has considerable insight into their obesity problems, which can act as a platform for further affective evaluation such as family dynamics, work environment, and psychological issues.

Keywords Awareness · Perception · Obesity

M. Omar (✉)
Discipline of Rehabilitation Medicine, Faculty of Medicine,
Universiti Teknologi MARA, Shah Alam, Malaysia
e-mail: mazlifahomar@salam.uitm.edu.my

M. Md Yasin
Primary Care Medicine, Faculty of Medicine, Universiti Teknologi MARA,
Shah Alam, Malaysia

H. Mokhtar · J. Jitos · N. A. Ab Halikun · N. F. Ahmad Fisol · S. A. Mat Yaacob
Faculty of Medicine, Universiti Teknologi MARA, Sungai Buloh Campus,
Shah Alam, Malaysia

R. Adnan et al. (eds.), *Proceedings of the International Colloquium on Sports Science, Exercise, Engineering and Technology 2014 (ICoSSEET 2014)*,
DOI: 10.1007/978-981-287-107-7_24, © Springer Science+Business Media Singapore 2014

1 Introduction

Obesity can be defined as abnormal or excessive fat and body weight accumulation that may impair health. Scientifically, World Health Organization (WHO) defined obesity as body mass index (BMI) of more or equal to 30 kg/m^2. However, for Asia-Pacific region, the definition has been redefined as BMI of more or equal than 27.5 kg/m^2 [1].

Obesity is a major health problem in Malaysia, and it has been a global phenomenon with serious implications such as diabetes, heart disease, and even cancers [2]. Based on the WHO survey in 2010, Malaysia has been ranked as the 6th highest adult obesity rate in Asia [3]. A total of 60 % of Malaysians aged 18 and above has a BMI over 25 kg/m^2, which are overweight.

The main mechanism in the development of obesity is energy imbalance. It can be due to metabolic disorder or simply by poor eating or lifestyle habits. Behavioral treatment in psychological intervention primarily significant during dietary restrain from poor eating habits and motivating physical activity. Adherence to the modified lifestyle is a challenging component during the treatment as well as during the maintenance of weight after weight loss. Throughout the management of obesity, behavioral, or cognitive therapy plays an important component in the program for lifestyle modification. Therefore, multidisciplinary approach apart from biological factors, which also include psychological, social, and environmental factors, has the best practice outcomes.

In relation to above factors, the health care providers understanding on how obese people perceived their status of health is deemed important for behavioral and motivational intervention approach. Hence, this study is to determine obese subjects' awareness and perception on their obesity status and factors.

2 Methodology

2.1 Participants

Obese subjects of BMI of 27.5 kg/m^2 and above, who were identified during the health screening program and had given their consent, were included. Other inclusion criteria were adult age 18 years and above and able to read and understand the self-administered questionnaire.

2.2 Outcome Measures

A parallel back-to-back translation and validated questionnaire regarding awareness on obesity status, understanding of BMI and perception toward obesity problem, which available in Bahasa Malaysia and English, was used in the study. Questionnaire was self-administered by the study subjects.

3 Result

3.1 Demographic Data

Thirty obese person participated in the study, 21 (70 %) males and 9 (30 %) females with mean age of 47.7 ± 6.3 years. Mean BMI were 34.5 kg/m^2 ± 5.1. All participants had completed secondary level of education and more.

3.2 Obesity Awareness

1. Self-body weight: Twenty-three (77 %) of them were aware of their current weight. Four (13 %) of the respondent were not aware of their current weight, and 3 (10 %) of them were unsure. Those who unsure of their weight thought that their weight were in normal range.
2. Body mass index: Twenty-four (80 %) of the respondents informed that they know the BMI formula, while only 3 (7 %) of them stated that they do not know the BMI formula. The other 3 (7 %) told that they were unsure. However, 23 (77 %) respondents provide the correct BMI formula calculation, and only 18 calculated their own BMI correctly.
3. Factors causing their obesity: Twenty-four (80 %) respondents answered that diet and lack of exercise are the main factors which causes their obesity, and 12 (40 %) of them answered marriage is one of the factors (Fig. 1).
4. Methods used to lose weight: Twenty-three (73 %) respondents had attempted to lose weight and only 9 (31 %) succeeded to lose weight, however unable to sustain the weight maintenance. Twenty-three (77.2 %) respondents had tried diet management and exercise as their methods to lose weight, 11 (36.4 %) had tried complimentary and alternative medicine, and only 1 (4.5 %) tried commercial weight loss program as well as in surgery (Fig. 2).

3.3 Self-perception on Obesity

Twenty (66.7 %) respondents disagree with the statement overweight was the symbol of richness and prosperity. Twenty (66.7 %) respondents perceived that they were not able to live a normal life due to their obesity. Nineteen (63.3 %) respondents perceived obese people were lazier, and 14 (46.7 %) perceived that obese people were untidy (Table 1).

Nineteen (63.3 %) respondents perceived workplace as part of the stress factors and causes prolonged sitting leads to obesity, 16 (53.3 %) respondents perceived socioeconomic condition may affect obesity. The respondents also perceived that family and friends expect them to be normal instead of overweight (18.60 %).

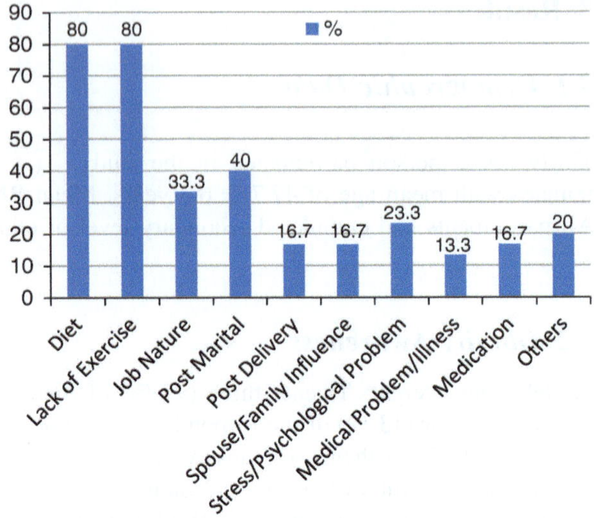

Fig. 1 Factors causing obesity

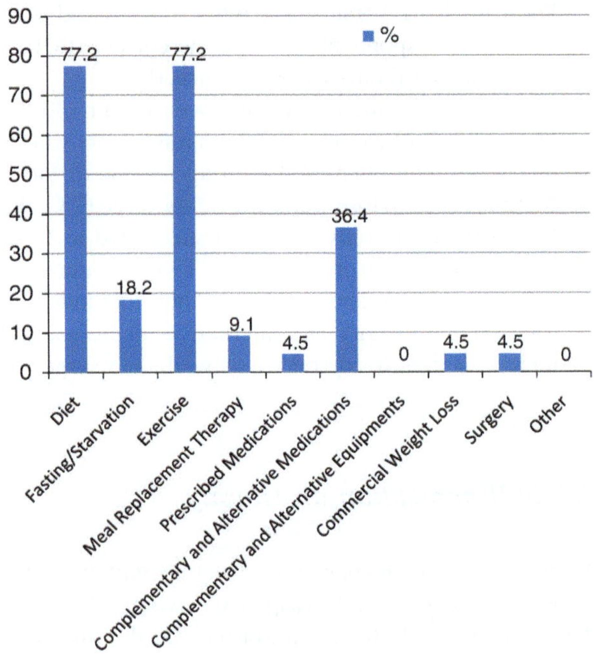

Fig. 2 Methods attempted in reducing weight

Table 1 Perception toward obesity

Statements	Agree (n, %)	Not sure (n, %)	Disagree (n, %)
Overweight is symbol of richness and prosperity	3 (10)	7 (23.3)	20 (66.7)
Stress at work and long periods of sitting leads to being obese	19 (63.3)	6 (20)	5 (16.7)
Obese people should not expect to live normal lives	20 (66.7)	5 (16.7)	5 (16.7)
Socioeconomic conditions affect body weight	16 (53.3)	8 (26.7)	6 (20)
Overweight people tend to be lazier than normal people	19 (63.3)	3 (10)	8 (26.7)
Normally obese people are untidy	14 (46.7)	3 (10)	13 (43.3)
Family and friends expect you to be normal rather than being overweight	18 (60)	3 (10)	9 (30)

4 Conclusion

This study showed that people who are obese were aware of their obesity status, and the importance of diet control and increase in physical activity can optimize weight reduction.

Studies have shown that the best method to lose weight is by increase physical activity and controlled diet. Aadland and Robertson [4] found that physical activity causes 55.8 and 5.6 % of weight change for men and women, respectively. However, many have not been aware of their obesity (60 %) due to lack of appropriate education on outcome measure of the characteristics of obesity. Besides, education also found that socioeconomic status also influence awareness in obesity [5].

Many studies have shown the effect of obesity on health, such as with cardiovascular diseases, diabetes mellitus, hyperlipidemia, and metabolic syndrome. This study also shows that health care providers play a main role in motivating or initiating people to lose weight. Therefore, health professionals have to have certainly acceptable knowledge on educating people regarding obesity and at least diet and exercise prescription.

Negative experiences during the weight reduction process are important in determining a successful weight lose intervention. Childhood experience in poor eating habit has been a negative influence to weight reduction. Further study regarding family background and even genetic changes in this group of obese people would contribute an additional knowledge in management of obesity. Kruger et al. [6] found that only one-third (30.96 %) were successful at losing weight and keeping their weight off due to influenced of negative experience.

Media also played an important role in providing information on method available in the market for weight loss which usually non-evidence based for effectiveness but expensive. Control of this widely spread weight reduction techniques especially causing inevitable side effects should be implemented.

Motivating people in changing to healthy lifestyle is a huge challenge. Adherence to the healthy lifestyle changes is also the main issue. Wing and Phelan

[7] stated that a large part of weight regain may be attributable to an inability to maintain healthy eating and exercise over time. They also emphasized on the importance of maintaining behavioral changes in the long-term maintenance of weight loss.

5 Suggestion

Human behavior study toward the management of obesity should be conducted widely in order to overcome this problem. A detail and specific-outcome measure on assessment of psychological cause and what triggers the motivation should be available [10, 11].

After the text edit has been completed, the paper is ready for the template. Duplicate the template file by using the Save As command, and use the naming convention prescribed by your conference for the name of your paper. In this newly created file, highlight all of the contents and import your prepared text file. You are now ready to style your paper; use the scroll down window on the left of the MS Word Formatting toolbar.

References

1. WHO Expert Consultation. (2004). Appropriate body-mass index for Asian populations and its implications for policy and intervention strategies. *Lancet, 363*, 157–163.
2. McGinnis, J. M., & Foege, W. H. (1993). Actual causes of deaths in the United States. *Journal of the American Medical Association, 270*(18), 2207–2222.
3. Lee, C. M. Y. (2007). The burden of overweight and obesity in the Asia-Pacific region. *Obesity Reviews, 8*, 191–196.
4. Aadland, E., & Robertson, L. (2012). Physical activity is associated with weight loss and increased cardiorespiratory fitness in severely obese men and women undergoing lifestyle treatment. *Journal of Obesity, 2012*.
5. Hammond, R. A., & Levine, R. (2010). Review: The economic impact of obesity in the United States. *Diabetic, Metabolic Syndrome and Obesity: Target and Therapy, 3*, 285–295.
6. Kruger, J., Blanck, H. M., & Gillespie, C. (2006). Dietary and physical activity behaviors among adults successful at weight loss maintenance. *International Journal of Behavioral Nutrition and Physical Activity, 3*(17) doi:10.1186/1479-5868-3-17.
7. Wing, R. R., & Phelan, S. (2005). Long-term weight loss maintenance. *The American journal of clinical nutrition, 82*(1), 222S–225S.
8. Nelissen, R. M. A., De Vet, E., & Zeelenberg, M. (2011). Anticipated emotions and effort allocation in weight goal striving. *British Journal of Health Psychology, 16*, 201–212.
9. West, D. S., Gorin, A. A., Subak, L. L., Foster, G., Bragg, C., Hecht, J., et al. (2011). A motivation-focused weight loss maintenance program is an effective alternative to a skill-based approach. *International Journal of Obesity, 35*, 259–269.

Relationship Among Repeated Ability Tests with Aerobic Power and Blood Lactate in Soccer

Annisaa Basar, Sarina Md Yusof, Muhammad Sufyan Mohamad Zaki, Suhana Aiman and Zulkifli Abdul Kadir

Abstract Soccer involves sprinting, jumping, running, and tackling as anaerobic components. It is significant to players to be able to perform high-intensity intermittent exercise and sprint repeatedly with short recovery either with or without ball. Therefore, the purpose of the present study was to determine the relationships between the repeated sprints ability (RSA) or repeated dribbling ability (RDA) with aerobic power and blood lactate of Malaysian University soccer players. A total of 52 soccer players participated (age 21.73 ± 1.82 years, height 171.31 ± 5.27 cm, weight 63.78 ± 7.30 kg, BMI 21.69 ± 2.29 kg m^{-1}) and performed all four tests: VO_{2max}, RSA, RDA, and peak blood lactate. The VO_{2max} was measured by 20-m multistage shuttle run (MST). The RSA and RDA performance indices were total time (TT), fastest time (FT), mean time (MT), and performance decrement (PD). The RSA and RDA tests involved sprinting or dribbling 7 × 34.2 m with 25 s rest intervals. Significant difference was found in peak blood lactate from baseline to posttest for RSA and RDA ($p < 0.05$). RSA showed low negative correlation of TT ($r = -0.28$, $p < 0.05$) and RDA, a negative moderate correlation ($r = -0.303$, $p < 0.05$) with VO_{2max}. Significant negative correlation was found between MT and VO_{2max} ($r = -0.280$, $p < 0.05$). Negative significant correlation between FT and VO_{2max} ($r = -0.311$, $p < 0.05$)

This work was supported in part by the Fundamental Research Grant Scheme (FRGS), Ministry of Higher Education of Malaysia FRGS/1/2011/SKK/UITM/03/32.

A. Basar (✉) · S. Md Yusof · M. S. Mohamad Zaki · S. Aiman · Z. Abdul Kadir
Faculty of Sports Science and Recreation, Universiti Teknologi MARA,
Shah Alam, Malaysia
e-mail: annisaa_basar@yahoo.com

A. Basar
Faculty of Sports Science and Recreation, Universiti Teknologi MARA,
Jengka, Pahang, Malaysia

R. Adnan et al. (eds.), *Proceedings of the International Colloquium on Sports Science, Exercise, Engineering and Technology 2014 (ICoSSEET 2014)*,
DOI: 10.1007/978-981-287-107-7_25, © Springer Science+Business Media Singapore 2014

was found in RDA protocol. The VO_{2max} was correlated in RSA (TT, MT) and RDA (TT, FT) protocols. The 25 s rest intervals between sprinting and dribbling can extent the resynthesis of phosphocreatine (PCr) and maintain performance.

Keywords VO_{2max} · Peak blood lactate · Repeated sprint ability · Repeated dribbling ability

1 Introduction

Soccer team is judged on their ability and capability to win a match. Thus, during match each of the players has to perform an effective skills and team cohesion in order to score goals. This complex characteristic of players plays an important role to pursue goals and the team has to strengthen their tactical skills in creating successful attacks. Soccer is an intermittent sport which requires high aerobic [1] and anaerobic capacity [2]. Soccer comprised of 19.5 % standing, 41.8 % walking, 16.7 % jogging and 16.8 % running, 1.4 % sprinting, and 3.7 % other activities [3]. Although sprinting is less than 10 % of total distance covered in a game [3, 4], it is one of the most important skills which will determine the success of a game [5]. Studies have shown the anaerobic component in soccer involves 35–40 sprints, tackles, and jumps [3] in one game.

The match play will be interspersed with shortest rest interval performed in high-intensity activities. Duration of 90 min of soccer game is important for the recovery from high-intensity activities [1, 6] and this due to high-intensity activities may increase blood lactate and affect blood flow to tissue. The accumulation of the blood lactate will lead to fatigue and decreases player's performance. The physiological responses to intermittent activities depend on the player's ability to recover from activities [7]. Therefore, the relevance of aerobic power highly predominant in soccer as well as players' performance has been confirmed to delay the concentration of blood lactate [8].

Repeated sprint ability is required to be performed in multidirectional either with ball or without ball [9]. Games as in team sports require high-level technical skills of dribbling. Dribbling a ball is a high-intensity activity which represents 16–30 % of competition time [10]. Dribbling or running with ball requires player to control the possession of the ball as well as maintain high-running speed [11]. To control the possession of ball while running requires good skill and advantages over opponents.

The ability to produce fastest sprint in shortest time separated by recovery known as repeated sprint ability (RSA), [12]. The repeated dribbling ability (RDA) was developed to measure dribbling skill performance in soccer [2, 13]. Training studies showed that RSA training increase aerobic power performance [14]. Most RSA protocols utilized either short- or long-sprint intervals. Only few studies reported on repeated dribbling activity when dribbling is noted as discriminator between novices and elite players. Therefore, the purpose of the present study was to determine the relationships between the repeated ability (RSA and RDA) that involved repeated high-intensity action in running with aerobic power of Malaysian University soccer players and to compare peak blood lactate between RSA and RDA.

2 Methods

2.1 Participants

Fifty-two well-trained soccer players (age 21.73 ± 1.82 years, height 171.31 ± 5.27 cm, weight 63.78 ± 7.30 kg, BMI 21.69 ± 2.29 kg m^{-1}) participated in the study were enrolled university students. All were members of top three teams (UiTM, UM, and UKM) from the first division of Malaysian Higher Education Institution soccer league, 2013. These teams were consistently ranked among the top 3 teams in the league for the last 2 years. The players trained 5 days every week (~90 min per session) with their university team for at least 4 months and have been active in sports within 1 year. Competitions were held on the weekend. At the time of tests execution, all teams were in the middle of the soccer season (May–June 2013). Players who were free from any injuries were included in this study. Exclusion criteria consist of history of any cardiovascular, metabolic disorders, musculoskeletal, and neurological problem. Prior to that, screening tests consists of PAR-Q and ECG test were conducted two days before the actual tests. All the procedures were approved by the Research and Ethic Committee of Universiti Teknologi MARA, Malaysia. The testing procedures were explained to participants and a written informed consent was obtained before testing started. All participants were fully familiar and clearly understood with the procedures used in this testing. They were informed that they could withdraw from the testing at any time without penalty.

2.2 Procedures

Tests in all teams were conducted between 1630 and 1900 as their routine training time and to avoid any circadian variability. During RSA and RDA testing, the temperature and wind velocity were recorded to ensure both factors were controlled. In the evenings, during the tests temperature was about 33 °C and wind velocity 1.2 mps. The reliability of RDA was validated in a pilot study, TT: $r = 0.744$ and FT: $r = 0.857$. All tests were conducted in two sessions after at least 48 h after last performed match or vigorous exercise, strength or polymeric training to prevent unnecessary fatigue and to ensure participants fully rest. First session was aerobic power test (20 m MST). Second session comprised of RSA and RDA tests. All tests were conducted outside on grass training field, using regular soccer shoes to replicate playing condition. Before testing a standardized warm up consists of jogging and stretching for 20 min were conducted.

2.3 20 m Multistage Shuttle Run Test

The 20 m multistage shuttle run (MST) test was conducted according to established procedures. When the participants were ready, the CD was started and the participant commenced the test. Participants had to follow the procedures and performed their maximal effort. Participant ran follow the running tempo. If the participants failed to reach the end of the shuttle before the beep they should be allowed two or three further shuttles to attempt to regain the required pace before being withdrawn. At the beginning the running, tempo of speed was 8.5 km/h and it increased 0.5 km/h in every minute. The time interval between beep would decrease and tempo increase. Researcher recorded the level and number of shuttles completed at that level either when the subject withdraws voluntarily from the test failed to be within three meters of the end lines on two consecutive tones [15]. Later, the level and the shuttles would be converted to VO_{2max} based on provided table [16].

2.4 Repeated Sprint and Repeated Dribbling Ability Tests

In order to stimulate the actual activities in the match play, the RSA and RDA was developed appropriately. The RSA and RDA tests were conducted after 48 h from the first session. Blood sample was taken at rest as baseline prior to warm up. Before RSA test, each participant was allowed to perform two sprints and for RDA test subject allowed to performed two repetitive dribbling skills along the test course for familiarization purposes. Although the absence of any learning effects from this type of multiple sprint protocol [17], all participants completed a familiarization trial of both protocols.

Testing was started with RSA followed by RDA test. Same participant would perform the next testing after 1 h rest interval. The protocol comprised of seven maximal 34.2 m sprints [1, 5, 18]. Participant performed in track along a distance of 34.2 m involved in sprinting/running and change direction between obstacles left or right (10 m sprints, 14.2 m agility) as showed on Fig. 1. Similar protocols were used to perform RDA test except participant instead of running would dribble a ball. Seven balls with the same pressure and same brand were used for each sprint. A photoelectric timing gate (Brower-Speed Trap) linked to a digital device recorded each sprints with an accuracy of 0.001 s was used to measure participant's performance. Following each sprint, there was a period of active recovery (25 s to cover a distance of 40 m), which consists of jogging to return to the initial point. Recovery was timed using stop watch in order to ensure participant's return to

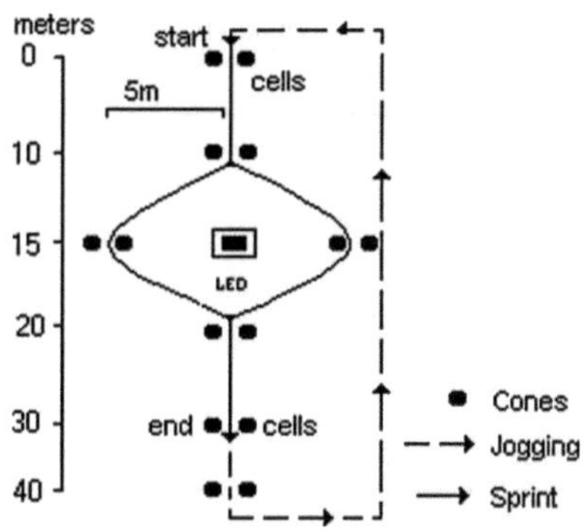

Fig. 1 Bangso test protocol [4], adapted by Abrantes et al. [5]

initial point of course between 23rd and 24th second. Verbal feedback was given by assistants at 5, 10, 15, 20, and 25 s of recovery [5, 18]. In the present study, we determined the total time (TT), fastest time (FT), mean time (MT), and performance decrement (PD) from RSA and RDA tests. The sum of all sprints times over 34.2 m was assumed as TT [14]. The FT was calculated as the best 34.2 m sprint time of each player and MT was average of seven sprints times [19]. The consequences of repeated sprints contributed to fatigue and limited the ability of muscle to perform. This indicator was the PD calculated by dividing the sum of sprints times for seven sprints by the best possible TT and then multiplying by 100 [19].

2.5 Peak Blood Lactate

Blood sample was collected from fingertip at the end of RSA and RDA tests (zero time recovery), and at 1, 3, 5, 7, 9, and 12 min of recovery [19]. The participant's finger was washed with water and dried, then it was disinfected by alcohol-contained cotton, using single use of disposable lancet devise subject's finger was pricked. Disposable lancet device offered three adjustable depths (1.3, 1.8, and 2.3 mm). After participant's finger was pricked laboratory assistant would swab alcohol-contained cotton again at the finger to ensure the hygiene of blood sample. The blood sample was collected into lactate strip and analyzed using an automated lactate analyzer (Accutrend). After 1 min, the reading was recorded. The highest concentration of blood lactate was considered as peak blood lactate.

2.6 Statistical Analyses

All the results are reported as mean ± standard deviation (SD). A two-tailed Pearson's product-moment correlation was used to determine the strength of relationship between VO_{2max} and RSA, RDA indices and peak blood lactate. Independent *t* test was conducted to compare the peak blood lactate between the two protocols. The significance level was set at $p < 0.05$.

Table 1 Anthropometric characteristics and vertical jump and the 20 m MST ($N = 52$)

Characteristics	Mean ± SD
Age (years)	21.7323 ± 1.82304
Height (cm)	171.3119 ± 5.26937
Weight (kg)	63.7754 ± 7.30318
BMI (kg m^{-2})	21.6940 ± 2.28635
VO_{2max} (ml min^{-1} kg^{-1})	47.91 ± 6.378

Table 2 Performance indices of the RSA and RDA (Mean ± SD) ($N = 52$)

Indices	RSA	RDA
Total time (s)	41.41 ± 3.88	61.01 ± 6.53
Fastest time (s)	5.63 ± 0.53	7.99 ± 0.96
Mean time (s)	5.92 ± 0.55	8.64 ± 0.90
Performance decrement (%)	4.82 ± 2.11	8.91 ± 4.07
Peak blood lactate (mmol L^{-1})	12.62 ± 1.92	11.21 ± 2.69

Table 3 Relationship between VO_{2max} and performance indices of the RSA and RDA

Protocol	Performance indices	Maximal oxygen (*r*)
RSA	Total time (s)	−0.280*
	Fastest time (s)	−0.242
	Mean time (s)	−0.280*
	Performance decrement (%)	−0.128
RDA	Total time (s)	−0.303*
	Fastest time (s)	−0.311*
	Mean time (s)	−0.268
	Performance decrement (%)	0.174

*Correlation is significant at the 0.05 level (2-tailed)

3 Results

The anthropometric characteristics of the study subjects and result of the 20 m MST are presented in Table 1. Performance indices of the RSA and RDA are presented in Table 2. The strength of relationship would follow the guidelines in reference [19]: low ($r = 0.10–0.29$), moderate ($r = 0.30–0.49$), and high ($r = 0.50–1.0$).

The correlations between VO_{2max} and performance indices of the RSA and RDA are presented in Table 3. A significant negative correlation was found between TT and VO_{2max} in both protocols. The RSA showed a low negative correlation of TT ($r = -0.28$, $p < 0.05$) and RDA indicated negative moderate correlation ($r = -0.303$, $p < 0.05$) with VO_{2max}. Significant low negative correlation was found between MT and VO_{2max} ($r = -0.280$, $p < 0.05$) in RSA. No correlations were found between FT or PD and VO_{2max} in RSA protocol. However, there was

Table 4 Relationship between peak blood lactate and performance indices of the RSA and RDA

Protocol	Performance indices	Peak lactate (*r*)
RSA	Total time (s)	−0.036
	Fastest time (s)	−0.080
	Mean time (s)	−0.037
	Performance decrement (%)	0.100
RDA	Total time (s)	0.152
	Fastest time (s)	0.082
	Mean time (s)	0.152
	Performance decrement (%)	0.068

significant correlation between FT and VO_{2max} ($r = -0.331$, $p < 0.05$) in RDA protocol. No correlations were found between MT or PD and VO_{2max} in RDA protocol.

Significant difference was found in peak blood lactate in both protocols. The correlations between performance indices of RSA or RDA and peak blood lactate are presented in Table 4. No significant correlations were found between RSA and RDA performance indices and peak blood lactate.

4 Discussion

The present study examined the relationships between the repeated ability (RSA and RDA), aerobic power, and blood lactate in a group of university soccer players. The main finding from the present study was significant moderate negative correlations were found between VO_{2max} (ml kg min^{-1}) and RDA indices as TT and FT. Significant low negative correlations were found between TT and MT in RSA. The results support the theory of aerobic power, greater VO_{2max} tends to delay accumulation of hydrogen ions (H^+) in working muscle [2]. The rest intervals between sprinting and dribbling [5] can extent the resynthesis of phosphocreatine (PCr) [21] and maintain performance. It was suggested only the intensity to perform RSA and RDA was slightly different. The appropriate training for RSA related increase VO_{2max} [7, 19] with different recovery time and protocols.

No significant correlations were found between RSA and RDA performance indices and peak blood lactate. However significant difference was found in peak blood lactate in both protocols. Increasing the blood lactate would affect player's performance due to an increase in hydrogen ions (H^+) [19]. Previous studies reported that high-intensity interval training with longer rest intervals improves the PCr resynthesis [22] compared to shorter rest interval.

Intervals between sprints has been attributed to an increase H^+ and degradation the sources of adenosine triphosphate (ATP) [23]. The metabolic changes contribute to fatigue and the appropriate rest interval speed up the resynthesis of PCr [21]. The recovery during high-intensity intermittent dependent on two factors: (1) adequate of blood flow for transport hydrogen ions (H^+) to remove, (2) adequate oxygen to supply muscle fibers during oxidation process [8].

5 Conclusion

The VO_{2max} was correlated in both RSA and RDA protocols. This suggests that to improve RSA and RDA, it is important to develop appropriate sprinting and dribbling skills training. Future research with variation of repeated dribbling should be developed to identify the best training and replicate soccer sport. The ability to recover between high-intensity activities in soccer is the parameter to determine athlete's performance. The quality of performance judged by the improvements in distance covered number of sprints and number of ball possessions in match play [5].

Acknowledgments The authors wish to thank Tan Choew Yum for her contributions and to the dedicated participants who participated in this study.

References

1. Bangsbo, J., Mohr, M., & Krustrup, P. (2006). Physical and metabolic demands of training and match play in the elite football players. *Medicine of Science Sports and Exercise, 24*, 665–674.
2. Sporis, G., Ruzic, L., & Leko, G. (2008). The anaerobic endurance of elite soccer players improved after a high-intensity training intervention in the 8 week conditioning program. *Journal of Strength and Conditioning Research, 22*(2), 559–566.
3. Mohr, M., Krustrup, P., & Bangsbo, J. (2003). Match performance of high standard soccer players with special reference to develop fatigue. *Journal of Sport Science, 21*, 519–528.
4. Bangsbo, J., Norregaard, L., & Thorssoe, F. (1991). Activity profile of competition soccer. *Canadian Journal of Sports Sciences, 16*, 110–116.
5. Abrantes, C., Macas, V., & Sampaio, J. (2004). Variation in football players' sprint test performance across different ages and levels of competition. *Journal of Sport Medicine, 3*, 44–49.
6. Stolen, T., Chamari, K., Castagna, C., & Wisloff, U. (2005). Physiological on soccer: an update. *Sports Medicine, 35*, 501–536.
7. Meckel, Y., Machnai, O., & Eliakim, A. (2009). Relationship among repeated sprint tests, aerobic fitness, and anaerobic fitness in elite adolescent soccer players. *Journal of Strength and Conditioning Research, 23*(1), 163–169.
8. Krustrup, P., Mohr, M., Armstrup, T., Johansen, J., Steensberg, A., Pedersen, P., et al. (2003). The Yo-Yo intermittent recovery test: physiological response, reliability and validity. *Journal of Science and Medicine in Sport, 35*, 697–705.
9. Hamit, C., Ibrahim, C., & Mustafa, S. (2012). Comparison of recovering times and aerobic capacity according to playing positions of elite football players. *Journal of Physical Education and Sport Science, 6*.

10. Lothian, F., & Farrally, M. (1994). A time motion analysis of women's hockey. *Journal of Human Movement Study, 26*, 255–265.
11. Lemmink, K. A. P. M., Elferink-Gemser, M. T., & Visscher, C. (2004). Evaluation of the reliability of two field hockey specific sprint and dribbles tests in young field hockey players. *Journal of Sport Medicine, 38*, 138–142.
12. Bishop, D., Girard, O., & Mendez-Villanueva, A. (2011). Repeated-sprint ability—part ii recommendations for training. *Sports Medicine, 41*(9), 741–756.
13. Helgerud, J., Engen, L. C., Wisloff, U., & Hoff, J. (2001). Aerobic endurance training improves soccer performance. *Medicine of Science Sports and Exercise, 33*, 1925–1931.
14. Chaouachi, A., Manzi, V., Wong, D. P., Chaalali, A., Laurencelle, L., Chamari, K., et al. (2010). Intermittent endurance and repeated sprint ability in soccer players. *Journal of Strength and Conditioning Research, 24*(10), 2663–2669.
15. Paliczka, V. J., Nichols, A. K., & Boreham, C. A. G. (1987). A multi-stage shuttle run as a predictor of running performance and maximal oxygen uptake in adults. *British Journal of Sports Medicine, 21*(4), 163–165.
16. Leger, L. A., & Lambert, J. (1982). A maximal multistage 20 m shuttle run test to predict VO_{2max}. *European Journal of Applied Physiology, 49*, 1–5.
17. Glaister, M., Howats, G., Lockey, R. A., Abraham, C., Goodwin, J., & Mcinnes, G. (2007). Familiarization and reliability of multiple sprint running performance indices. *Journal of Strength and Conditioning Research.*
18. Aziz, A. R., & Chuan, T. (2004). Correlation between tests of running repeated sprint ability and anaerobic capacity by Wingate cycling in multi-sprint sports athletes. *International Journal of Applied Sports Sciences, 16*, 14–22.
19. Da Silva, J. F., Guglielmo, L. G. A., & Bishop, D. (2010). Relationship between different measures of aerobic fitness and repeated sprint ability in elite soccer players. *Journal of Strength and Conditioning Research, 24*(8), 2115–2121.
20. Cohen, J. W. (1988). *Statistical power analysis for the behavioral sciences* (2nd ed.). Hillsdale: Lawrence Erlbaum Associates.
21. Brooks, G., Fahey, T., White, T., & Baldwin, K. (1999). *Exercise physiology: Human bioenergetics and its applications* (3rd ed.). New York: McGraw Hill.
22. Bishop, D., Edge, J., & Thomas, C. (2008). Effects of high intensity training on muscle lactate transporters and postexercise recovery of muscle lactate and hydrogen ions in women. *Journal of Physiology-Regulatory, Integrative and Comparative Physiology, 295*, 1991–1998.
23. Gaitons, G., Williams, C., Boobis, L., et al. (1993). Human Muscle metabolism during intermittent maximal exercise. *Journal of Applied Physiology, 75*(2), 712–719.

Single Versus Two Sets of Resistance Training on Muscular Endurance, Strength and Fat Percentages Among Recreationally Trained Men

Mohd Aizzat Adnan, Zulkifli Abdul Kadir, Sarina Md Yusof, Mardiana Mazaulan and Mohd 'Aizat Abdul Razzaq Mohamed

Abstract Several design programme variables are very important in resistance training. Until now, the comparison of the effectiveness of one versus two sets or multiple sets of resistance training on trained population was still inconclusive. This study was conducted to compare the effects of one versus two sets of resistance training on upper and lower body muscular strength, upper and lower body muscular endurance and body fat percentages among recreationally trained men. A total of 32 trained men were recruited in this study. They had been randomized into two groups (one set, $n = 16$ or two set, $n = 16$) based on their body weight. The muscular strength was measured by bench press and squat using multiple-RM assessment. Muscular endurance had been measured by push-up and 70 % 1 RM on leg extension, while body fat percentages were measured using 8-site skinfold measurement. Both groups engaged in 6 weeks of training. Significant improvement was noted throughout 6 weeks of training for both groups in all variables ($p < 0.05$). No significant differences between single set and two sets were found for all the variables ($p > 0.05$) except for leg extension muscular endurance test between both groups ($p < 0.05$). In conclusion, single set had been found to produce similar effects as two sets on muscular strength, upper body muscular endurance and fat percentages among recreationally trained men.

Keywords Single set · Multiple sets · Muscular strength · Muscular endurance · Fat percentages

M. A. Adnan (✉)
Faculty of Sport Science and Recreation, Universiti Teknologi MARA, Shah Alam, Malaysia
e-mail: aizzatadnan88@gmail.com

M. A. Adnan · Z. Abdul Kadir · S. Md Yusof · M. Mazaulan · M. 'A. A. R. Mohamed
FSR Strength Training and Conditioning Laboratory, Universiti Teknologi MARA, Shah Alam, Malaysia

R. Adnan et al. (eds.), *Proceedings of the International Colloquium on Sports Science, Exercise, Engineering and Technology 2014 (ICoSSEET 2014)*,
DOI: 10.1007/978-981-287-107-7_26,

1 Introduction

Weight training or resistance training has now been considered as an important aspect of a larger training routine with the objective of increasing muscle strength, muscular endurance and lowering body fat [1]. In addition to increasing the strength and endurance of muscles, reducing body fat and thus reducing the risk of obesity, strength training can bring a wide variety of positive advantages that can also improve both mental and physical health [2]. Endurance training on the other hand is beneficial as it can reduce blood pressure when the body is at rest, increase the utilization of blood glucose, increase bone mineral density and improve blood lipid profile [3]. One of the best alternatives to achieve great improvement in muscular strength was by doing resistance training [4]. Muscular strength can be defined as the ability to exert the maximum force to overcome the most resistance in one effort by a muscle or a muscle group [5]. Muscular endurance is the ability of a muscle group to execute repeated contractions over a period of time until fatigue [6]. The study being undertaken had focused on the effect of different volume on the strength and muscular endurance and body fat percentages of recreationally trained individuals. Volume, defined as the total amount of weight lifted in a single session [7]. Sets, repetitions and loads are the sub-variables that are included in the volume variable. Sets defined as a number of repetitions in a group are sequentially performed before the person stops for a rest [8]. There is a slight difference between single set, two sets and multiple sets of exercise. Single set means that the person performs only one set of exercise (not including the warm-up) after which he will continue to do another set of a different exercise. If the person repeats the same exercise before proceeding to a different exercise, this is called two sets. Many researches disagreed and are still debating on the most effective method in doing resistance training [9]. Some scientific research suggested that multiple sets are more superior to a single set [10]. However, there are also researchers who state that single set of exercise performed until muscular failure produces equal results as multiple sets [11]. However, none of the previous researches compare the effect of single set versus two sets of exercise on muscular strength and muscular endurance.

2 Methodology

2.1 Participants

A total number of 32 participants were recruited, which recorded an effect size of 0.48 [12]. The study had created 2 groups and assigned 16 participants to each group according to their body weight. Taking into consideration the drop-out rate of 20 %, the study had planned for 3 additional participants to be added to each group.

2.2 Instruments

Accuracy is of an utmost importance in this study. All instrumentations used were of the highest quality, properly calibrated and certified by the relevant certification bodies. The weight of all bars, the weight plates and collars including the weight of all the participants were measured using the Tanita BSC 200 G digital scale. For the skinfold measurement, the Fitness ASSIST Harpenden Calipers were used with the correlation coefficient of 0.83–0.89. The muscular strength of the participants was measured by using the multiple-RM test as suggested by American College of Sports Medicine [6]. The test used 45° incline bench, Olympic bar, Olympic plate and power rack. For lower body muscular endurance, the proposed study utilized the leg extension at 70 % load of 1 RM and the test used the leg extension machine.

2.3 Data Collection Procedure

2.3.1 Anthropometric Measurement

Before any physical testing was undertaken, the participants' anthropometric measurements were prepared taken. The measurement of the weight (measured in kilogram and represented by "kg"), height (measured in centimetre and represented by "cm") and skinfold of the subjects was made. The weight and height of the participants were measured to the nearest 100 g and 0.1 cm by using the Tanita BSC 200 G digital scale. The study utilized eight-site system for male subjects to determine the body fat percentages by using Harpenden Caliper. The 8 sites are as follows: biceps, triceps, subscapular, iliac crest, supraspinal, abdomen, thigh and calf according to the International Society for the Advancement of Kinanthropometry (ISAK) guidelines. The calculation of body fat percentages was using equations from [14].

$$\text{Male}: \%\text{BF} = -1.268 + 0.142^{*}(\text{sumofeightskinfolds}) + 0.270^{*}(\text{age})$$

2.3.2 Multiple-RM (Strength Testing)

In order to determine the dynamic strength on upper and lower body, the study used the multiple repetition test not exceeding eight repetitions. The upper body dynamic strength was tested and determined by using 45° incline press while the lower body dynamic strength was established with the squat test. Multiple repetition test method was chosen over the single-RM method for safety and practical reasons and was documented to be highly correlated to the single-RM test. To convert the multiple-RM score, the formula was used [15].

$$[(0.03 \times \text{ weight lifted } \times \text{ repetition}) + \text{ weight lifted}]$$

All test scores were then converted into the relative strength scores [16].

2.3.3 Multiple-RM Testing Protocols

Participants were instructed to warm-up with a light resistance that easily allows 5–10 repetitions. Participants were provided 1-min rest period. Participants were required to lift a load that he estimates can be performed for 8 RM or less based upon the best previous load that was lifted in training. If the participants were able to lift more than 8 RM, the load was increased accordingly; 4–9 kg or 5–10 % for upper body exercise or 14–18 kg or 10–20 % for lower body exercise. An important factor being considered and clearly defined in the study was the point where a repetition is considered to be complete and the next repetition begins. The point of failure is also important as this will determine the end of a repetition. Accordingly, failure is defined as the time or at the point when the bar ceased to move; also, if the participant paused for more than 1 s as soon as the leg or arm was in the extended position, or if the subject was unable to complete a repetition's full range of motion as required.

2.3.4 Muscular Endurance

The test for muscular endurance for both the upper and lower body was conducted differently. For lower body muscular endurance, the study utilized the leg extension at 70 % load of 1 RM. The number of repetitions to volitional fatigue was counted and then is reported [17]. A much simpler test was conducted for the upper body muscular endurance, which is push-up test, and this was determined by the maximal number of push-ups that completed in a minute [6].

2.3.5 Training Programme

The training programmes conducted in the study subscribed to the guidelines as provided by ACSM and NSCA. The study was designed such that each participant will undergo the resistance training programme at a frequency of three times per week. Both groups trained at 80 % intensity level (or at eight repetitions to failure) or until momentary muscular failure reached the prescribed repetition. This is in accordance with the current recommendation which stated that 80 % of intensity is sufficient to improve muscular strength [6]. The participants were given a rest interval of 2 min between set (for two-set group) and additionally another 2 min between each exercise [18]. The resistance training programme consisted of the warm-up phase, conditioning phase (main part) and the cooling down phase. The exercises prescribed were as follows: chest press, lateral pull down, leg extension, leg flexion, biceps curls, triceps extension, lunges and torso flexion. Other common groups of exercises for the targeted muscle group were introduced to provide variation to the training programme after every 2 weeks. A minimum rest of 2 days (48 h) between each session was imposed to provide sufficient rest and to allow for super compensation to take place. The intensity of training was progressively increased after three consecutive workouts.

2.4 Statistical Analysis

Descriptive statistics was used to determine the descriptive height, age and weight of participants. Mixed between–within Anova was used to compare the effects of pre- and post-test and the comparisons of score between the two groups. The statistical significant was set at alpha level $p < 0.05$.

3 Result

Table 1 showed the comparison of bench press effects between groups. There are no significant differences in the bench press performance for the two groups, $p > 0.05$. Thus, the null hypothesis failed to be rejected. The main effect of comparing the two groups was not significant, $F(1, 30) = 0.709$, $p > 0.05$.

Table 2 showed the comparison of squat effects between groups. There are no significant differences in the squat performance for the two groups, $p > 0.05$. Thus, the null hypothesis failed to be rejected. The main effect of comparing the two groups was not significant, $F(1, 30) = 0.059$, $p > 0.05$.

Table 3 showed the comparison of push-up effects between groups. There are no significant differences in the push-up performance for the two groups, $p > 0.05$. Thus, the null hypothesis failed to be rejected. The main effect of comparing the two groups was not significant, $F(1, 30) = 0.670$, $p > 0.05$.

Table 4 showed the results of the test between subjects. There is a significant difference in the leg extension performance for the two groups, $p < 0.05$. It shows that group two (sets) has better performance as compared to group one (set). So, the null hypothesis was rejected. The main effect of comparing the two groups was significant, $F(1, 30) = 6.04$, $p < 0.05$.

Table 5 showed the comparison of fat percentages effects between groups. There are no significant differences in the fat percentages performance for the two groups, $p > 0.05$. Thus, the null hypothesis failed to be rejected. The main effect of comparing the two groups was not significant, $F(1, 30) = 0.160$, $p > 0.05$.

Table 1 Comparison of bench press effects between groups

Variable	Effect	df1	df2	*F* value	*P* value
Bench press	Group	1	30	0.709	0.406

Table 2 Comparison of squat effects between groups

Variable	Effect	df1	df2	*F* value	*P* value
Squat	Group	1	30	0.059	0.810

Table 3 Comparison of push-up effects between groups

Variable	Effect	df1	df2	*F* value	*P* value
Push-up	Group	1	30	0.67	0.42

Table 4 Comparison of leg extension effects between groups

Variable	Effect	df1	df2	*F* value	*P* value
Leg ext	Group	1	30	6.04	0.02

Table 5 Comparison of fat percentage effects between groups

Variable	Effect	df1	df2	*F* value	*P* value
Fat percent	Group	1	30	0.160	0.69

4 Discussion

Present study provide strong data that shown there is no significant difference for recreationally trained men in doing single set of resistance training or two sets of resistance training in improving upper and lower body muscular strength, upper body muscular endurance and body fat percentages. In contrast to upper body muscular endurance, which is the lower side, two sets of resistance training exerted a slightly superior effect to single set of training.

4.1 Difference Between One Set Versus Two Sets of Resistance Training on Muscular Strength

Results showed that there were no significant differences of performing one or two sets of resistance training on upper body strength and lower body strength. While both groups were successful in increasing the upper and lower body strength after 6-week intervention, there were no significant differences between the two groups in the bench press and squat test. Why the finding is such, because the most important criteria for improvement in strength is to have total recruitment of the motor unit and the muscles fibres and this can only be achieved when the muscle is brought to the point of momentary muscular failure in the training, so that the full stimulation will take place. If training multiple sets without training to momentary muscular failure, it will not institute a significant quality improvement in strength because fail to recruit the nerve and muscle that suppose to be recruiting. Doing single set to two sets training has very little volume difference, and thus, doing the exercise until failure makes not much impact to the nerve or to the muscle. The increment in strength was because of the increment in motor unit activation. Motor unit coordination also showed to be improved following resistance training, and some qualitative changes in the muscle have been suggested as possible mechanism in which the participants increase their muscular strength in response to resistance training [19].

4.2 Difference Between One Set Versus Two Sets of Resistance Training on Muscular Endurance

The study found that there is no significant difference in upper body muscular endurance between the two groups. However, following 6 weeks of training, both one-set group and two-set group report a significant improvement in upper body muscular endurance. Improvement in muscular endurance could be because of decreasing in the activity of glycolytic enzymes, but increases intramuscular substrates stores, oxidative enzymes activities and capillary as well as mitochondrial density [20]. In contrast to the non-significant result of muscular strength and upper body endurance, results showed that there is a significant difference of performing one or two sets of resistance training on lower body muscular endurance. While both groups were successful in increasing the upper and lower body muscular endurance after 6-week intervention, two-set groups had significantly shown more increment in lower body endurance compared to one set. The possible reasons might be due to the total volume of exercises performed among two-set groups that cause their muscles able to sustain more repetitions of exercises performed. Lack of previous researches had been conducted to study the differences of one- versus two-set resistance training on muscular endurance. Overall, results showed that muscular endurance as measured by the push-up and leg extension can be increased throughout 6 weeks of training but the two set had shown better performance compared to just one set in lower body part.

4.3 Difference Between One Set Versus Two Sets of Resistance Training on Body Fat Percentages

The results showed that there were no significant differences on performing one or two sets of resistance training on fat percentages. While both groups were successful in decreasing fat percentages after 6-week intervention, there were no significant differences between the two groups in the 8-site skinfold test. These showed that exercising with two sets was unable to show significant different of fat percentages compared to performing just one set. Just like the muscular endurance, the study on one versus two sets of resistance training on body composition especially fat percentages was still lacking. The duration of the intervention itself might be the determining factor on the effects that emerged from the conducted exercise [6]. A total of 4–6 weeks of training may be too short that mostly results in fluctuation in the effects gained from the training. Another explanation of fat loss due to the desire of the participant to give restrictions in the diet is given in [21].

4.4 Recommendation

A possible cause to the implementation of single-set strength training is its impact on muscle endurance. Some practitioners may be hesitant to implement a single-set training programme because of fears that a reduction in training volume will reduce endurance. Further research is necessary to examine the changes in muscular endurance that occur with single-set strength training.

Future research should consider more body composition measurement such as muscle mass and bone density. Knowing the impacts of different training protocols to the body composition will provide more information to all populations, especially those that are interested in having the best shape. This will also contribute more to the body of knowledge.

Future research on the topic of single-set resistance training should also examine the effects of single-set training on various populations such as to old people, young people, genders, special populations and many more. Also, different lifting intensities with different repetitions should be studied to find the best load and repetitions to be included in single-set resistance training. While the bench press and leg press are common lifts to be studied because of its big muscle involved, an examination of the effectiveness of single-set training using other exercises could prove beneficial. It would be interesting to see whether small muscle groups, such as biceps and triceps, are affected similarly after single-set training. Most of previous study focuses on big muscle groups and very limited test for small muscle groups.

5 Conclusion

Data from present study provide strong evidence to show that there is no difference between single set of resistance training and two sets of resistance training in improving muscular strength, muscular endurance, body fat percentages and blood pressure. However, two sets of resistance training exerted a slightly superior effect to single set of training to lower body muscular endurance. Due to the reduction in training time and the reduced physiological training load provided by single-set training, trained populations that are considering whether to perform one- or two-set resistance training are suggested to perform just one set if muscular endurance is the primary target. If the lower body muscular endurance had become a serious target, two sets are recommended as it has been shown that two-set resistance training is significantly greater in improving lower body muscular endurance compared to just performing one set.

The results of this study could have great implications on the way strength and conditioning practitioners approach on designing resistance training programmes. This study shows that muscular strength, upper body muscular endurance, fat percentages and blood pressure benefit through the use of one-set lifting protocols that can be just as effective as two-set protocols when athletes are strictly monitored and adhered to training procedures. The implementation of one-set training protocols would decrease the amount of time spent on strength training allowing for busy-scheduled individuals to perform resistance training. Aside from simply freeing up more time for the student athlete, single-set training can also help to minimize the possibility of overtraining and fatigue.

Acknowledgments First of all, I would like to thank Allah the Al-Mighty who has given me the strength, courage and motivation to accomplish these pieces of work. Besides, I would like to thank and express my gratitude to Professor Madya Zulkifli Abd Kadir as my supervisor in his guidance and has given full support throughout in finishing this research paper no matter how busy he was. Without his guidance, patience and understanding, I would never have done these pieces of work.

I would also want to thank all lecturers of Faculty of Sport Science and Recreation, UiTM Shah Alam for their time in giving me the explanations and guidance in finishing this research. Last but not least, thanks to my lovely family especially my mother, Puan Zaruna bt Abd Rahman and my father, Adnan b Hj Abu Bakar who always gave me all their supports in any situations. Not forgotten to all postgraduate students of Sport Science Faculty, UiTM Shah Alam who have helped me in finishing this research. I would like to appreciate for all of the help towards the completion of this study.

References

1. Geliebter, A., Maher, M. M., Gerace, L., Gutin, B., Heymsfield, S. B., & Hashim, S. A. (1997). Effects of strength or aerobic training on body composition, resting metabolic rate, and peak oxygen consumption in obese dieting subjects. *Journal of Clinical Nutrition, 66*(3), 557–563.
2. Westcott, W. (1992). A new look at youth fitness. *Am Fitness Q, 11*, 16–19.
3. Harris, K., & Holly, R. (1987). Physiological response to circuit weight training in borderline hypertensive subjects. *Med Science Sports Exercise, 19*(3), 246–252.
4. Fisher, J., Steele, J., Bruce-Low, S., & Smith, D. (2011). Evidence-based resistance training recommendations. *Medicina Sportiva, 15*(3), 147–162.
5. McArdle, W. D., Katch, I., & Katch, L. (2000). *Essentials of exercise physiology* (2nd ed.). Philadelphia: Lippincott Williams & Wilkins.

6. American College of Sports Medicine. (2006). In P. D. Thompson, L. S. Pescatello & D. Riebe (Eds.), *ACSM's guidelines for exercise testing and prescription* (7th ed.). Philadelphia: Lippincott Williams & Wilkins.
7. Fleck, S. J., & Kraemer, W. J. (2003). *Designing resistance training programs* (3rd ed.). Champaign: Human Kinetics.
8. Byrd, R., Chandler, T. J., Conley, M. S., Fry, A. C., Haff, G. G., Koch, A., et al. (1999). Strength training: Single versus multiple sets. *Sports Medicine, 27*, 409–416.
9. Paulsen, G., Myklestad, D., & Raastad, T. (2003). The influence of volume of exercise on early adaptations to strength training. *The Journal of Strength and Conditioning Research, 17*, 113–118.
10. Rhea, M. R., Alvar, B. A., Burkett, L. N., & Ball, S. D. (2003). A meta-analysis to determine the dose response for strength development. *Medicine and Science in Sports and Exercise, 35*(3), 456–464.
11. Hass, C. J., Garzarella, L., De Hoyos, D., & Pollock, M. L. (2000). Single versus multiple sets in long-term recreational weightlifters. *Medicine and Science in Sports and Exercise, 32*, 235–242.
12. Ostrowski, K. J., Wilson, G. J., Weatherby, R., Murphy, P. W., & Little, A. D. (1997). The effect of weight training volume on hormonal output and muscular size and function. *The Journal of Strength and Conditioning Research, 11*, 148–154.
13. Kagawa, M., Kerr, D., & Binns, C. (2003). Ethnic differences in the BMI-%BF relationships between young Japanese and Australian-Caucasian men living in Australia using dual-energy x-ray absorptiometry. *Asia-Pacific Journal of Public Health, 15*, S27–S32.
14. Brzycki, M. (1993). Strength testing—predicting a one-rep max from reps-to-fatigue. *Journal of Physical Education, Recreation & Dance, 64*(1), 88–90.
15. National Strength and Conditioning Association. (2010). NSCA's essentials of personal training. In R. W. Earle & T. R. Baechle (Eds.) Champaign, IL: Human Kinetics.
16. Franks, B. D., & Howley, E. T. (1989). *Fitness leader's handbook.* Champaign, IL: Human Kinetics.
17. Baechle, T. R., Earle. R., & Wathen. D. (2008). Resistance training. In Essentials of strength training and conditioning (3rd ed.). Champaign: Human Kinetics.
18. Faigenbaum, A. D., Westcott, W. L., Loud, R. L. R., & Long, C. (1999). The effects of different resistance training protocols on muscular strength and endurance development in children. *Pediatrics, 104*.
19. Hirofumi, T., & Thomas, S. (1998). Impact of resistance training on muscular endurance. *Sports Medicine, 259*(3), 191–200.
20. Harmon, C. B., & Jack, H. W. (1974). The effect of maximal resistance training on strength and body composition on women athletes. *Medicine and Science in Sport, 6*(3), 174–177.

Relationship Between Handgrip Strength on Muscular Strength Among Racquet Sport Athletes

Mohd 'Aizat Abdul Razzaq Mohamed, Zulkifli Abdul Kadir, Sarina Md Yusof, Mardiana Mazaulan and Mohd Aizzat Adnan

Abstract Handgrip strength has been popularly used to predict overall body strength in the general and athletic populations. The purposes of this study were to compare handgrip strength between both hands (dominant and non-dominant hand) as well as selected arm position (flexion and extension). This study also explores the relationship between handgrip strength on muscular strength among racquet sport athletes. Ninety male racquet athletes aged 18–26 years old from Public Institutes of Higher Learning (IPTA) in Klang Valley area participated in this study (tennis = 30, badminton = 30, and squash = 30). Result of this study revealed that there was a significant difference between dominant and non-dominant hand in handgrip strength among all three racquet sport athletes ($p < 0.05$). The tennis athletes recorded the highest handgrip strength score for both the dominant and non-dominant hand. A significant difference was also reported between extension and flexion arm positions among racquet sport athletes ($p < 0.05$) except for badminton athletes ($p > 0.05$). The tennis and squash athletes showed more superior handgrip strength score in the extension arm position. However, the badminton athletes showed superior handgrip strength score in the flexion arm position. Moderate linear correlation was recorded between handgrip strength with upper body strength among all three racquet sport athletes, with ($r = 0.374 - 0.529$). Result of the study also showed the significant correlation was recorded between handgrip strength with lower body strength among three racquet sport athletes. The badminton athletes showed highest correlation among the three racquet sport athletes ($r = 0.543$; $r = 0.604$). Based upon the findings of this study, handgrip strength can be used as a predictor for upper body strength and lower body strength among racquet sport athletes.

Keywords Handgrip strength · Muscular strength

M. 'A. A. R. Mohamed (✉) · Z. Abdul Kadir · S. Md Yusof · M. Mazaulan · M. A. Adnan
Faculty of Sports Science and Recreation, Universiti Teknologi MARA,
Shah Alam, Malaysia
e-mail: aizatrazzaq_88@yahoo.com

R. Adnan et al. (eds.), *Proceedings of the International Colloquium on Sports Science, Exercise, Engineering and Technology 2014 (ICoSSEET 2014)*,
DOI: 10.1007/978-981-287-107-7_27,

1 Introduction

Hand was much more than a machine in the factory of human body. The hand was irreplaceable when it comes to perform any kind of movement, either it was gross or skilled. The prime function of hand was grip. The importance of a normally functioning hand needs no emphasis in living, practicing a hobby, or allowing independence in daily activities [1].

Handgrip strength was correlated with the strength of the upper extremity, general strength of the body, and some anthropometric measurements [2] and therefore was often adopted in clinical practice as an objective measure of upper extremity function [3]. Handgrip strength test is usually used a convenient measurement of muscle function in the laboratory and hospital setting because it is simple to use and the handgrip dynamometer is a compact tool and easy to care and handle [4].

Upper body and lower body strength can be measured by the maximum amount of weight that can be lifted in one repetition maximum (1RM), or a given number of repetitions using free weights or resistance machine.

According to Filingeri et al., [5] showed that the handgrip strength might be associated to the individual maximal strength performance and it might be considered as a valid predictor of upper body strength performance within a sport fitness context.

Sport can be categorized into racquet and non-racquet sport. In many types of sport, the characteristics of the hand are important. A good gripping strength is important to the racquet player for maintenance of rackety control for possibly five games in a match. Besides that, Pereira et al. [6] stated that in tennis, the handgrip strength comparisons between any game categories did not present significant difference, regardless of the hand assessed or the protocol for assessing handgrip strength. This study was conducted upon Public Institutes of Higher Learning (IPTA) athletes in the Klang Valley who are actively competing in racquet sports at Malaysian University Sport Council (MASUM) championship. The racquet sports consist of tennis, squash, and badminton.

2 Methodology

2.1 Participants

Participants for this study were selected from athletes that involved in MASUM. All of them are athletes from different IPTA in the Kang Valley that were UiTM, UKM, UM, and UPM. The criteria of the participants are racquet sport athletes, namely tennis, badminton, and squash. Ninety male athletes aged between 18 and 26 were recruited.

2.2 Instruments

1. Handgrip Dynamometer

In this study, Takei A5401 handgrip dynamometer digital was used to measure handgrip strength test. The reliability of digital hand dynamometer for right hand ($r = 0.976$) and left hand ($r = 0.986$) [7].

2. 45° Incline Bench Press

In this study, 45° Cybex incline bench press was used for upper body strength test with three spotter assistance.

3. Olympic Bar

An International Weight-Lifting Federation (IWF) Olympic bar was used for the 45° incline press test. The bar was 7.2 feet (2.2 m) long and weight 44 lbs. or (20 kg).

4. Olympic Plate (Uesaka)

IWF approved Uesaka bumper weight plates were used together with the Olympic bar. The weight plate ranges from 10 to 25 kg in weight.

5. Free Standing Rack

A free standing rack was used for lower body strength test with 3 spotter assistance.

2.3 Data Collection Procedure

Participants was briefed and explained on procedures involved for this study. Participants signed an informed consent letters and provided important personal details as an agreement between the researcher and the participants. The participants also underwent the resting electrocardiogram test (ECG test) as a screening where it checked for abnormalities regarding the electrical activity of participant's heart.

A briefing and demonstration was given to the participants a week before the test. During the briefing, the researcher will be explained about the all procedures involved in the test and demonstrated it. The briefing and demonstration was useful to make all the subject understand and know what they do in the real test.

For the first phase in a week, the handgrip strength test was measured using handgrip dynamometer, Takei A5401. Three trials were given for each dominant and non-dominant hand, and the highest result will be taken as a data. The position of elbow for the participants during the handgrip strength test was in flexion and extension positions. The second phase was implemented by indirect 1RM 45° incline bench press and indirect 1RM squat test. Before the 1RM performed, all participants were required to do some warm-up session. The 1RM test was measured by using the indirect repetition maximum method to determine their 1RM. The formula to calculate indirect repetition is as follows: [(0.03 × weight lifted × repetition) + weight lifted] [8]. The result of the 1RM prediction was then divided with the body.

2.4 Statistical Analysis

Independent *t* test was used to analyze the strength difference between dominant and non-dominant hand and also between arm flexion and extension positions among racquet sports athletes. Pearson correlation was used to analyze the relationship between handgrip strength and upper body strength, the relationship between handgrip strength and the lower body strength. The statistical significant was set at an alpha level $p < 0.05$.

3 Result

Table 1 showed for overall dominant and extension handgrip for tennis mean score was ($M = 45.640$, SD = 1.341), badminton mean score was ($M = 43.183$, SD = 2.355), and squash mean score was ($M = 44.130$, SD = 2.865). Tennis showed higher mean score for non-dominant handgrip ($M = 43.589$, SD = 1.476), badminton mean score was ($M = 40.877$, SD = 3.528), and squash mean score ($M = 41.030$, SD = 2.596). For flexion handgrip, the higher mean score was tennis ($M = 44.823$, SD = 1.473), followed by badminton mean score was ($M = 43.550$, SD = 1.403), and squash mean score ($M = 42.683$, SD = 2.205). Bench press for tennis mean score was ($M = 0.965$, SD = 0.053), badminton mean score ($M = 0.955$, SD = 0.042), and squash mean score ($M = 0.940$, SD = 0.042). For squat, the higher mean score was badminton ($M = 1.171$, SD = 0.069), followed by tennis mean score was ($M = 1.166$, SD = 0.788), and squash mean score ($M = 1.158$, SD = 0.065) (Table 2).

Based on the result, it showed a significant and strong correlation between handgrip strength and 1RM bench press among tennis athletes ($r = 0.529$, $p < 0.05$). For badminton athletes, it showed a significant and moderate correlation between handgrip strength and 1RM bench press ($r = 0.374$, $p < 0.05$). Squash showed a significant and moderate correlation between handgrip strength and 1RM bench press ($r = 0.404$, $p < 0.05$) (Table 3).

Based on the result, it showed a significant, moderate, and negative correlation between handgrip strength and 1RM squat among tennis athletes ($r = -0.483$, $p < 0.05$). For badminton athletes, it showed a significant, strong, and negative relationship between handgrip strength and 1RM squat ($r = -0.543$, $p < 0.05$). Squash showed a significant, moderate, and negative relationship between handgrip strength performance and 1RM squat ($r = -0.417$, $p < 0.05$).

Table 1 Descriptive statistics of participants' dependent variables

		M	SD
Dominant extension	Tennis	45.640	1.341
	Badminton	43.183	2.355
	Squash	44.130	2.865
Non-dominant	Tennis	43.589	1.476
	Badminton	40.877	3.528
	Squash	41.030	2.596
Flexion	Tennis	44.823	1.473
	Badminton	43.550	1.403
	Squash	42.683	2.205
Bench press	Tennis	0.965	0.053
	Badminton	0.955	0.042
	Squash	0.940	0.042
Squat	Tennis	1.166	0.788
	Badminton	1.171	0.069
	Squash	1.158	0.065

Table 2 Correlation between handgrip strength with upper body strength

	Indirect 1RM incline bench press		
	N	Pearson correlation	Sig.(2-tailed)
Tennis HG	30	0.529**	0.003
Badminton HG	30	0.374*	0.042
Squash HG	30	0.404*	0.027

* 0.05
** 0.01

Table 3 Correlation between handgrip strength with lower body strength

	Indirect 1RM squat		
	N	Pearson correlation	Sig.(2 tailed)
Tennis HG	30	−0.483**	0.007
Badminton HG	30	−0.543*	0.002
Squash HG	30	−0.417*	0.022

* 0.05
** 0.01

4 Discussion

4.1 Relationship Between Handgrip Strength with Upper Body Strength

This current study described that handgrip strength with upper body strength has a significant correlation, among racquet sport athletes, tennis ($r = 0.529$, $p < 0.05$), badminton ($r = 0.374$, $p < 0.05$), and squash ($r = 0.404$, $p < 0.05$). Tennis athletes score higher correlation compared to badminton and squash athletes. This result has similarities with previous study by Caterisano et al. [9] indicated that the existence of a significant correlation between grip strength and 1RM bench press may depend on the type of resistance training that have being tested.

4.2 Relationship Between Handgrip Strength with Lower Body Strength

Relationship between handgrip strength with lower body strength stated significant and negative correlation among racquet sport athletes. The result indicated that tennis ($r = -0.483$, $p < 0.05$), badminton ($r = -0.543$, $p < 0.05$), and squash ($r = -0.417$, $p < 0.05$). Newton and Kraemer [10] revealed the difference between strength of upper and lower body. Several researchers have suggested that certain resistance training exercises are used as mainly strength exercises, probably depending upon the kinetic profiles of the exercise.

4.3 Relationship Between Handgrip Strength with Lower Body Power

The present study, it showed a significant, strong, and negative correlation between handgrip strength with lower body power among racquet sport athletes, tennis ($r = -0.532$, $p < 0.05$), badminton ($r = -0.604$, $p < 0.05$), and squash ($r = -0.512, p < 0.05$). The current study demonstrates some similarities with the previous studies [11] and has reported that peak power is generally found at one-third MGS in the case of simple muscle power output effort. Measurements of muscle power exertion by explosive gripping may have higher reliability than those of isometric muscle power exertion.

4.4 Recommendation

The researcher has come out with a few suggestion and recommendation as follows: for future research, should increase number of sample size and choose the high standard competition such ASEAN University Games (AUG) athletes. Research should be done for different population of athletes, to get the easiest method to predict total body strength. Power test for upper extremities should be done to find whether there is a relationship between handgrip strength test and upper body power test. Researcher suggest to include more exercises that represent other parts of the body particularly exercise that isolates body parts.

5 Conclusion

In conclusion, the current study provides strong and sufficient evidence to prove that there is significant correlation between handgrip strength on upper body strength. Grip strength showed higher score in tennis athlete compared to badminton and squash. This possibly explains due to athletes' strength and different muscle recruitment for athlete. The results indicate that there is significant correlation between handgrip strength on lower body strength and higher score in badminton athletes followed by tennis and squash.

The result of present study showed that tennis athletes score higher mean in dominant hand for overall racquets sport athletes, followed by squash and badminton. For hand position, tennis athletes score higher mean in arm extension for overall racquet sport athletes, followed by squash and badminton. The present study showed significant evidence to prove that handgrip strength can be used to predict upper and lower body strength, whereby, extension hand position showed more significant as compared to flexion hand position.

Acknowledgments Highest appreciation and I cannot finish thanking my supervisor, Associate Professor Zulkifli Abd Kadir for his unflinching support, encouragement, and invaluable contributions to the study, he is too much. To the coaching staff of tennis, badminton, and squash of the UiTM, UKM, UM, and UPM, thank you for the support and cooperation especially in the data collection stage.

I am also greatly indebted to the many people who in some way contributed to the progress in this study. I thank to all my lecturers in Faculty of Sports Science and Recreation, Universiti Technologi MARA, for their support. Their constructive criticism and insight have been a tremendous help in making this study possible.

Finally, my deepest appreciation for the unconditional love and encouragement provided by my family during the hard and easy times. Thank you to all who had contributed their sweat and tears toward the completion of this study. Thank you very much.

References

1. Wynn Parry, C. B. (1981). *Rehabilitation of Hand* (4th ed.). London: Butterworth.
2. Rantanen, T., Guralnik, J. M., & Foley, D. (1999). Midlife hand grip strength as a predictor of old age disability. *The Journal Medicine Association, 281*, 558–560.
3. Balogun, J. A., Akomolafe, C. T., & Amusa, L. O. (1991). Grip strength: Effects of testing posture and elbow position. *Archives of Physical Medicine and Rehabilitation, 72*, 280–283.
4. Jakobsen, L. H., Rask, I. K., & Kondrup, J. (2010). Validation of handgrip strength and endurance as a measure of physical function and quality of life in healthy subjects and patients. *Journal of Nutrition, 26*(5), 542–550.
5. Filingeri, D., Thomas, E., Raccuglia, M., Paoli, A., Bianco, A., & Palma, A. (2012). Handgrip strength and 1RM bench press performance: a novel approach to evaluate upper body maximal strength. *Journal Sport Science Health, 8*, 1–70.
6. Pereira, H. M., Menacho, M. O., Takahashi, R. H., & Cardoso, J. R. (2011). Handgrip strength evaluation on tennis players using different recommendations. Locomotor apparatus in exercise and sports.

7. Moreira, M. R., & Machado, P. M. (2011). Assessment of handgrip strength—Validity and reliability of the Saehan dynamometer. *Journal of Physiotherapy Research, 18*(2), 1809–2950.
8. Brzycki, M. (1993). Strength testing—predicting a one-rep max from reps-to-fatigue. *Journal of Physical Education, Recreation & Dance, 64*(1), 88–90.
9. Caterisano, A., Yurich, S., Bonfiglio, C., Fowler, A., Greer, B., & Brown, C. W. (2001). The correlation between maximal grip strength and 1RM bench press: The effects of Sport specific resistance training and gender. *Medicine & Science in Sports & Exercise, 33*(5(S1)), S244.
10. Newton, R. U., & Kraemer, W. J. (1994). Developing explosive muscular Strength Implications for a mixed methods training strategy. *Strength Conditioning, 16*, 20–31.
11. Kaneko, M. (1988). *Science of power-up: Power and efficiency of a human body engine in Japanese*. Tokyo: Asakurasyoten.

Advancing Recreational Studies: An Analysis of Mental Toughness in Outdoor Adventure Program

Mohd Shariman Shafie and Hisyam Che Mat

Abstract The purpose of this study was to examine the effects of outdoor adventure program toward participants' mental toughness. This study utilized pre- and post-quasiexperimental with control group design. The 42 items of psychological performance inventory (PPI; Loehr 1986) was administered among 200 ($n = 100$, experimental group; $n = 100$, control group) diploma level intervarsity athletes and age in a range of 17–30 years old. The study was conducted during 7 days program of outdoor recreation camp 2013. The camp exposed students to variety of outdoor adventure activities and challenges in natural setting. The paired sample t test analysis showed that experimental group recorded statistically significant changes in their mental toughness score [t (1, 264) = −4.050, $p = 0.001$] compared to the control group [$t\,(1, 99) = -0.853, p = 0.396$]. As to further confirm these findings, independent t test result showed that significant differences for pre-test scores [t (1, 264) = 2.777, $p = 0.006$] between the experimental and control groups. On the post-test, significant differences also recorded [t (1, 264) = 4.063, $p = 0.001$] between both groups. The primary finding of this study supported that outdoor adventure camp is effective in improving participants' mental toughness. The study also provides an empirical data for future study in this field.

Keywords Mental toughness · Outdoor adventure · Athletes

1 Introduction

Issues on athletes' mental toughness in sporting performance are popular topic among general public [22]. People believe that mental toughness is crucial in order to achieve excellent sporting performance. Supporting the idea, Norris [23]

M. S. Shafie (✉) · H. Che Mat
Faculty of Sports Science and Recreation, Universiti Teknologi MARA, Shah Alam, Malaysia
e-mail: shariman_shafie@salam.uitm.edu.my; sharimanshafie@yahoo.com

R. Adnan et al. (eds.), *Proceedings of the International Colloquium on Sports Science, Exercise, Engineering and Technology 2014 (ICoSSEET 2014)*,
DOI: 10.1007/978-981-287-107-7_28,

highlighted that sport practitioners, coaches, athletes and sport fans all around the world to acknowledge the importance of mental toughness in sporting performance. However, despite the awareness on the issues, mental toughness remains a relatively growing area of sport psychology research.

1.1 Mental Toughness

Middleton et al. [22] claimed that "mental toughness is remaining inadequately defined and conceptualize" [p. 2]. Middleton et al. [23] defined mental toughness as a consistency or persistency to achieve the goal despite in the tough, pressure or difficult situation. Expanding the definition, Middleton et al. [22] also stated that mental toughness is "unshakeable perseverance and conviction toward some goals despite pressure or adversity" [p. 1]. Supporting the idea, Gucciardi et al. [9] described mental toughness as a superior mental quality of an athletes and it is one of the important psychological constructs for performance in athletics context.

In Malaysia, issues on athletes' mental toughness are not rare to be heard. As reported in our local newspaper (Harian Metro 2011), National Sport Council (NSC) and National Sport Institute (NSI) are working hard to improve Malaysian athletes' tough character for Olympics London 2012. Based on these scenarios, arguably mental toughness today is becoming a focus by the sports practitioners in sporting performance. Several previous researchers also stated that in achieving excellent sporting performance, athletes could not only rely on skills and strategies, but also must be complemented with excellent emotionally, spiritually and psychologically skills (Loehr 14; [9, 23, 22].

In the fields of sport psychology, mental toughness is one of the character buildings that have been focused by NSC and NSI particularly among youth athletes. Numerous programs have been organized by NSC and NSI with the purpose of improving youth athletes' mental toughness. Interestingly, several programs were conducted based on outdoor adventure program. For an example, NSC has organized several outdoor adventure camps aiming at developing youth athletes' mental toughness. However, there is lack of studies relating outdoor adventure program and athletes' mental toughness.

1.2 Outdoor Adventure

On the other hand, outdoor adventure programs are greatly participated by peoples all around the world including athletes (Simmons 18). Virden (21) defined outdoor recreation as a "recreation behavior, activities and experiences that occur or depend on natural setting" [p. 310].

Specific to the study, Priest and Gass [22] and Virden (2006) stated that the use of outdoor for recreation is called outdoor recreation pursuit or human powered outdoor adventure. They also defined outdoor adventure as self-propelled activities performed in outdoor settings, and involved the elements of: (1) danger; (2) risk taking; (3) challenge; (4) sensation seeking; (5) achievement; (6) competence; and (7) one's skills.

Over decades, outdoor recreation researchers found recreation programs such as outdoor adventure camps effect positively on participants' psychological aspects (e.g., [2, 5–9, 23–23]; Ewert 6; Allain 2006). Most of the studies focused on common psychological constructs such as: (1) self esteem; (2) self confidence; (3) team cohesion; (4) leadership; (5) self actualization; (4) team trust; (5) communication; and (6) self efficacy.

1.3 Intervention of Outdoor Adventure in Enhancing Athletes' Mental Toughness

The researcher found relationship between outdoor adventure and mental toughness definitions. In brief, outdoor adventure is commonly defined as the act of engaging in any adventure activities that performed in outdoor setting and involve element of challenge, risk, pressure, competence and one's skill.

On the other side, mental toughness commonly defined as the ability to stay calm despite in a tough, pressure or difficult situation. Several definitions also stressed that mental toughness exist in any kind of difficulties, hardiness or tough situation [9, 22]. By reviewing both definitions, the researcher believed that the outdoor adventure program able to produces physical and mental pressures as well the activities exposed participants to the rough and open weather condition. As a result, mental toughness can be trained and improves in outdoor adventure program.

Unfortunately, there is an absence of study that specifically relates to mental toughness and outdoor adventure program. Previous studies on mental toughness were conducted based on sport itself. In addition, it is also difficult to find a study that specifically examines the effects of outdoor adventure program on participants' mental toughness. After all, the effectiveness of outdoor adventure program on participants' mental toughness is remains to be questioned.

2 Method

2.1 Subject

The study recruited 200 first year students of Universiti Teknologi MARA (UiTM) from Pahang and Perlis campus (intake of November—March 2010/2011). Most of the students are athletes and aged in the range of 17–30 years old. The total of

200 students was divided into two groups: experimental groups and the control group. The experimental group was comprised of second semester students ($n = 100$), while the control group was among first semester students ($n = 100$).

2.2 Setting

The 7 days program in natural setting involved several outdoor adventure activities included land-based (e.g., jungle trekking, navigation, survival, camp craft, initiative games, and orienteering) and water-based (e.g., kayak expedition, water confidence and water rescue). The camp venues were at Tasik Chenderoh, Perak and Pulau Langkawi, Kedah. Similarly, both venues had a campsite area which was flat and grassy. The campsite had basic infrastructure such as halls, toilets and A-frame cabin. But there were no built-in facilities for outdoor activities and most of the activities were conducted and engaged with natural setting. The SPS133's final camp exposed students to variety of outdoor adventure activities and challenges. Participation in all activities was compulsory. The activity rotation schedule was developed as to give every participants opportunity to experience the activities.

2.3 Instrumentation

The study utilized psychology performance inventory (PPI: Loehr 1986) as study instrument. The questionnaire was divided into two sections (A–B). Section A has a set of questions on student's demographic. Section B has a psychology performance inventory (PPI: Loehr 1986) that was used for analyzing participants' mental toughness.

2.3.1 Data Analysis

The data collected from the questionnaires was analyzed using paired sample *t* test and independent *t* test (pre- and post-test). All data gathered was statistically analyzed as to answer the research questions in quantitative manner.

3 Findings

The following sections describe the overall effect of the camp based on paired Sample *t* test analysis and followed by independent *t* test analysis to examine the participants' mental toughness scores at pre- and post-test (Table 1).

Table 1 Overall scores: paired sample *t* test

Result		Paired sample *t* test				
		n	$\bar{x}$	SD	*t**	*p*
Experimental	Pre	100	147.738	16.369	−4.050	0.001
	Post	100	152.118	16.514		
Control	Pre	100	141.859	17.243	−0.853	0.396
	Post	100	143.569	16.755		

**t* test significant is at 0.05 (2-tailed)

Table 2 Pre- and post-test: independent *t* test

Result		Independent *t* test				
		n	$\bar{x}$	SD	*t**	*p*
Group						
Experimental	Pre	100	147.738	16.369	2.777	0.006
Control	Pre	100	141.859	17.243		
Experimental	Post	100	152.118	16.514	4.063	0.001
Control	Post	100	143.569	16.755		

**t* test significant is at 0.05 (2-tailed)

Overall, a paired sample *t* test analysis suggested the Outdoor Recreation (SPS133) Final Camp 2011 in this study had recorded statistically significant effects on the improvement of mental toughness of the experimental group.

Table 1 presents the result from the paired sample *t* test comparing the experimental group's mental toughness with that of control group. The experimental group recorded statistically significant changes in their mental toughness score (t (1, 264) = −4.050, $p = 0.001$). Before the camp, their mean score was 147.738, and after the camp, the mean score was significantly increased to 152.118.

The control group did not record any statistically significant changes in their mental toughness score (t (1, 99) = −0.853, $p = 0.396$). At pre-test, their mean score was 141.859, and at the post-test, their mean score was 143.569. Even though mean score was slightly increased, but the increase was not statistically significant.

As to further confirm these findings, independent *t* test analyses were conducted to examine the score of experimental and control group for pre-test and post-test (Table 2).

Table 2 showed that there were statistically significant differences for pre-test scores (t (1, 264) = 2.777, $p = 0.006$) between the experimental and control groups. In essence, the experimental group recorded higher mental toughness scores for pre-test than control group which suggested that experimental group already had greater mental toughness level than control group.

On the post-test, there were also statistically significant differences (t (1, 264) = 4.063, p = 0.001) that were recorded between the experimental and control groups. Specifically, the scores differences between both groups had widened. After the completion of the camp, experimental group' mental toughness score was increased higher than that of the control group.

4 Conclusion

4.1 Overall Findings

Overall, there were evidences that Outdoor Recreation (SPS133) Final Camp 2011 had resulted on the improvement of participants' mental toughness scores. After the completion of the camps, paired sample t test analyses indicated that the experimental group had recorded statistically significant improvement on their mental toughness scores (p = 0.001). In contrast, the control group did not record any statistically significant changes in their mental toughness scores (p = 0.396).

In details of the findings, analyses on the pre-test results showed that the experimental group exhibited higher mental toughness scores ($\bar{x}$ = 147.738) than the control group ($\bar{x}$ = 141.859).

By the end of the SPS133 camp, the t test analyses showed that the experimental group's score remained higher ($\bar{x}$ = 152.118) than the control group ($\bar{x}$ = 143.569). There was statistically significant improvement on the experimental group' mental toughness scores (p = 0.001).

Based on the findings, the researcher suggests that the SPS133 camp had significant influences on experimental group's mental toughness improvement. These outcomes were in line with Gatzemann et al. [9] and Pasarelli et al. [16] studies. They found the intervention of outdoor adventure program contributes toward positives psychological outcomes.

5 Discussion

In focus to the study, the researcher proposed several plausible reasons contributing significant improvement in mental toughness development, such as the influence of: (1) camp's activities and (2) camp's environment. To discuss these ideas in further detail, the researcher elaborated each factors separately.

5.1 Influence of Camp's Activities

Overall, this recent study's findings strongly supported the effectiveness of the camp in enhancing experimental group' mental toughness. Firstly, the study proposed that engagement in outdoor adventure activities during the camp had significantly influenced participants' mental toughness improvement. The camp provided participants real-life exposure to several outdoor adventure activities (e.g., survival, kayaking, jungle trekking, camp craft) in challenging natural settings.

Based on the researchers' observation, all of the respondents participated actively in each of camp's activities. Participation in camp activities seems to foster participants' positives development on several aspects of: (1) psychological (e.g., self confidence, self esteem, mental toughness); (2) physical (e.g., fitness, improve cardiovascular, strength); and (3) social (e.g., communication, unity, trust, cohesion). Previous studies also revealed the benefits of engaging in outdoor adventure activities on the improvement of psychological, physical and socio-psychological aspects ([2, 9, 23–23]; Allain 2006; Tracey et al. 2008).

Moreover, the intensive camp's schedule also acted as pressures to the participants. During the camp, each group participated in two to three activities per day. Each of the activity's duration was about two to three hours (except for survival, 12 h). This activities were conducted continuously while to ensure the participant have minimal rest time during the intensive camp. At night, they had limited time to sleep because they were assigned to night watch duty around the campsite and preparing a communal breakfast in rotation. As a result, participants tended to be sleepless, exhausted and fatigue. The researcher proposed that participant's adaptations to these extensive pressures might have contributed to the improvement of their mental toughness. Supporting the idea, Middleton et al. [22] found that mental toughness is a result of individual response to adversity including high level physical effort or pressure.

Through observation, each camp's activity might have different effects on participants' mental toughness. For example, activity such as survival (12 h), kayaking (approximately 20 km) and jungle trekking seemed to have extra contributions toward the improvement. These kinds of activities require higher physical and mental pressures if compared to other short-duration physical activities. In fact, Connole [4] and Gucciardi et al. [9] claimed different type of activities or sports demanded level different of mental toughness.

Based on these arguments, the researcher proposed that different type of adventure activities may have different effects on the improvement of mental toughness. Therefore, camp's coordinator should be creative in programing their camp's activities in order to maximize the improvement of mental toughness.

5.2 *Influence of Camp's Environment'*

Secondly, the researchers believe the camp's environment might exert some influences on the improvement of participants' mental toughness. The camp utilized minimum technological intervention thus encouraged the participants to experience traditional camping method, such as cooking in a group, making a fire and building groups' shelter. During the camp period, this traditional camping method was believed to create a challenging situation especially when the participants were mostly accustomed to technology-dependent lifestyles. According to Gucciardi et al. [23], challenging situation does influence the development of mental toughness. Ability to overcome these challenging situations seemed to contribute in participants' mental toughness enhancement.

Moreover, participants were exposed to direct contact with the open weather and the natural environment. During the camp, participants experienced challenging environment such as heavy rain and high temperatures (up to 38 °C). These situations heightened the challenges, and at the same time, increase the element of risk during the camp. These challenging situations might have influence on the improvement of participants' mental toughness. Researcher's idea was supported by Bull et al. [3]. The study proposed environmental influence as the most important factor in the development of mental toughness. Based on this claim, the researcher believed challenging environment had contributed positively on individual' mental toughness development.

However, it is also important to note that the improvement of mental toughness (with regard to environmental influences) is not limited to remote and natural setting. Some outdoor adventure programs were also organized indoors or in urban areas. For example, adventure activities such as ropes courses and wall climbing can be held in the indoor setting and yielded similar outcome in mental toughness improvement ([1]; Taylor et al. 1989).

These contrasting findings of indoor versus outdoor and nature versus urban setting had challenged the researcher's claim on the influence of open environment setting on mental toughness since closed environmental setting also produced the same result. Eventually, the researcher strongly pointed that challenging outdoor environment has greater potential to influences participants' mental toughness improvement than indoor setting. This claim was based on a study Bull et al. [3], which stated that challenging environment as the most influential factor to the development of mental toughness. The ways participants responded and coped with challenging environment was believed to greatly influence their mental toughness improvement.

Acknowledgments We would like to express our gratitude to the Faculty of Sports Science and Recreation Universiti Teknologi MARA (UiTM) Shah Alam, Selangor for support throughout the whole processes of conducting this study.

References

1. Allain, M. (1996). *The effect of adventure-based experiential training (ABET) on team cohesion*. Halifax: Dalhouise University, University Microfilms International.
2. Boyle, I. (2002). The impact of outdoor adventure-based training on team cohesion and psychological skills development in elite sporting teams. Unpublished Ed.D., University of Wollogong, Wollogong.
3. Bull, S., Shambrook, C., James, W., & Brooks, E. (2005). Towards an understanding of mental toughness in elite English cricketers. *Journal of Applied Sport Psychology, 17*, 209–227.
4. Connole, I. (2009). *Towards an understanding of mental toughness in high school boys' basketball*. Fullerton: California State University, ProQuest LLC.
5. Darst, P. W., & Armstrong, G. P. (1980). *Outdoor adventure activities for school and recreation programs*. Prospects Height, IL: Wave land.
6. Ewert, A. W. (1989). Outdoor adventure pursuits.
7. Gatzemann, T., Schweizer, K., & Hummel, A. (2008). Effectiveness of sports activities with an orientation on experiential education, adventure-based learning and outdoor-education. *Effectiveness of Sports: Kinesiology, 40*(2), 146–152.
8. Godfrey, G. (2009). Outdoor recreation, health, and wellness: Understanding and enhancing the relationship. *Resources for the Future*, 2–29.
9. Gucciardi, D. F., Gordon, S., & Dimmock, J. A. (2008). Towards an understanding of mental toughness in Australian football. *Journal of Applied Sport Psychology, 20*, 261–281.
10. Gucciardi, D. F., Gordon, S., & Dimmock, J. A. (2009). Advancing mental toughness research and theory using personal construct psychology. *International Review of Sport and Exercise Psychology, 2*, 54–72.
11. Loehr, J. E. (1986). Mental toughness training for sports: Achieving athletic excellence. S. Greene Press.
12. Middleton, S. C., Marsh, H. M., Martin, A. J., Richards, G. E., & Perry, C. (2004). Discovering mental toughness: A qualitative study of mental toughness in elite athletes. *Psychology Today, 22*, 60–72.
13. Middleton, S. C., Marsh, H. M., Martin, A. J., Richards, G. E., & Perry, C. (2004). Mental toughness: Is the mental toughness test tough enough? *International Journal of Sport & Exercise Science, 35*, 19–108.
14. Mohamad, N. A., Omar, F. M. S., & Abu, S. B. (2009). The affect of higher score of mental toughness in the early stage of the leagues towards winning Malaysian football players. *Research Journal of International Studies, 12*, 68–78.
15. Norris, E. K. (1999). *Epistemologies of champions: A discursive analysis champions' retrospective attributions; looking back and looking within*. Michigan: Michigan University Microfilms International.
16. Passarelli, A., Hall, E., & Anderson, M. (2010). A strengths-based approach to outdoor and adventure education: Possibilities for personal growth. *Journal of Experiential Education, 33*(2), 120–135.
17. Priest, S., & Gass, A. M. (2003). *Effective leadership in adventure programming* (2nd ed.). USA: Human Kinetics, University of New Hampshire.
18. Simmons, D. (2004). American outdoor recreation participation continues six-year ascent. (Press release: July 28, 2004). Boulder, CO: Outdoor Industry Association. www.outdoor industry.org/press.ola.php?news_id=708&sort_year=2004.
19. Sheard, M., & Golby, J. (2006). The efficacy of an outdoor adventure education curriculum on selected aspects of selected positive psychological development. *The Journal of Experiential Education, 29*(2), 187–209.
20. Taylor, F. L. (1989). *The influence of outdoor adventure recreation class on personality type, locus control, self esteem, and selected issues of identity development of college students*. San Francisco, Ann Arbor: University of San Francisco, University Microfilms International.

21. Virden, R. J. (2006). Outdoor and adventure recreation. Introduction to recreation and leisure, 307–332.
22. Wang, J. C. K., Liu, W. C., & Kahlid, A. (2006). Effects of a five-day outward bound course on female students in Singapore. *Australian Journal of Outdoor Education, 10*(2), 20–28.
23. Webb, D. J. (1999). Recreational outdoor adventure programs. In J. C. Miles & S. Priest (Eds.), *Adventure programming*. State College, PA: Venture.

The Relationship Between 20-m Multistage Fitness Test and Yo-Yo Intermittent Fitness Test in Measuring Cardiovascular Fitness Among Kuching Sarawak Rugby Player

Wahidah Tumijan, Abdul Shaqir Rahit, Nagoor Meera Abdullah, Rahmat Adnan and Vincent Parnabas

Abstract Most coaches will use 20-m multistage fitness test to determine cardiovascular level in rugby players. Yo-Yo intermittent fitness test also can measure the cardiovascular fitness level among athlete, but lack of study showed which test should be specified implemented to rugby player. The objective of this study is to determine whether the result of the cardiovascular fitness level of 20-m multistage fitness test has a relationship with the Yo-Yo intermittent fitness test among rugby players. A sample of 60 rugby players was purposively sampling from rugby player. The result of this test will be analyzed by Pearson's product moment coefficient correlation using statistical package for social science (SPSS) version 21. The descriptive result showed that mean (SD) 20-m multistage fitness test [33.31 (5.92)] was lower compared to Yo-Yo intermittent test [47.48 (1.06)]. There was positive strong relationship between 20-m multistage fitness test and Yo-Yo intermittent fitness test [$p < 0.001$, $r = 0.652$]. The study showed that there is relationship between 20-m multistage fitness test and Yo-Yo intermittent test.

Keywords 20m Multistages fitness test · Yo-Yo Intermittent Fitness test · Cardiovascular Fitness · VO_{2max} · Rugby Athletes

1 Introduction

Rugby is an intermittent high-intensity physical activity that requires a well-developed aerobic and anaerobic fitness based on [1]. They also said that although rugby performance is thought to be mainly dependent on player ability, high aerobic fitness is also important to improved performance. Maximal aerobic power (VO_{2max}) is

W. Tumijan (✉) · A. S. Rahit · N. M. Abdullah · R. Adnan · V. Parnabas
Faculty of Sport Science and Recreations, Universiti Teknologi MARA,
Shah Alam, Malaysia
e-mail: wahidah06@salam.uitm.edu.my; am_id8@hotmail.com

R. Adnan et al. (eds.), *Proceedings of the International Colloquium on Sports Science, Exercise, Engineering and Technology 2014 (ICoSSEET 2014)*,
DOI: 10.1007/978-981-287-107-7_29, © Springer Science+Business Media Singapore 2014

considered to improve the ability to recovery from the anaerobic efforts during game. Furthermore, aerobic conditioning had been suggested to be important in preparing players able to sustain an appropriate training load volume for rugby. According to [2], cardiovascular fitness is important to sport with long time duration of playing. In rugby, all energy sources such as ATP-PC, anaerobic, and aerobic will be used, but the most important energy source that is needed to develop is aerobic. This is because rugby is played for 80 min on the field consisting of 40 min each half. Every player needs to develop aerobic capacity to play and win the game.

There are several tests that can determine cardiovascular fitness that are usually used by coaches to determine their athlete's fitness level. Firstly is the 20-m multistage fitness test that had been developed by Léger and Lambert [3]. The test is developed to monitor the development of the athlete's maximum oxygen uptake (VO_{2max}). The second test that will be used is Yo-Yo intermittent fitness test. This test had been developed by the Danish soccer physiologist [4] for soccer sport. This test evaluates an individual's ability to repeatedly perform intense exercise. This study is mainly about testing the relationship between the Yo-Yo intermittent fitness test results and the 20-m multistage fitness test result. Both of the tests are field test that can be measured the cardiovascular fitness level among athlete. Even the purpose of the two tests is to determine the cardiovascular fitness, but the method and the result of the test is different. Lack of the study showed the specific test used for the rugby player. Are both of the tests can be used for the rugby player or not? This study will try to figure the relationship between the two tests using the equation to convert the result in VO_{2max} value.

2 Method

2.1 Participant

Sixty male rugby players aged between 14 and 17 years who actively taking part in the competitions had been purposively sampling. 33.3 % of respondent was played at division level, 30 % at district level and others played at school, state, and international level (36.7 %). They have to do both 20-m multistage fitness test and the Yo-Yo intermittent fitness test. Participants had been informed the purpose, benefits, and risks of the study, and all provide written inform consent. The inform consent had been completed by their parents.

2.2 Instrument

A total of 20-m multistage fitness test involves 20-m continuous running between lines to record beeps of each times. Subjects need to stand behind lines and begin running with instructed from CD or tape. The speed will increase each level. If the line is not reached in time for each beep, the subject must run to the line turn and try to catch up with the pace within 2 more ‘beeps’. Also, if the line is reached before the beep sounds, the subject must wait until the beep sounds. The test is stopped if the subject fails to reach the line (within 2 m) for two consecutive ends. The formula below had been used to calculate estimates VO_{2max} [5].

$$VO_{2max} = 31.025 + (3.238 \times \text{velocity}) - (3.248 \times \text{age}) + (0.1536 \times \text{age} \times \text{velocity})$$

Yo-Yo intermittent fitness test level 1 (YYIT1) had been used because it aimed for well-trained and elite athletes and starting at a faster speed. The test used cones to mark out three lines: 20 m and 2.5 m (endurance test) or 5 m (recovery test) apart. The subject begins running 20 m when instructed by the CD. The subject turns and returns to the starting point when signaled by the recorded beep. The active recovery period (5 and 10 s, respectively, for the endurance and recovery versions of the test) interjected between every 20-m (out and back) shuttle, where the subject must walk or jog around the other cone and return to the starting point. This is test procedures that all subjects need to follow to ensure the validity of the test. Hence, to disqualify the subject on the which level of test the subject will complete, a warning was given for the first time and if the subject did not pass the shuttle for the second time then the test is done. The formula below had been used to calculate estimates VO_{2max} [4].

$$VO_{2max}(\text{mL/min/kg}) = \text{IR1 distance(m)} \times 0.0084 + 36.4$$

In both fitness tests, subjects had been given a familiarization trial on a day before the test. Prior the test, warm-up exercise is needed. It includes the 5-min free running on ground at self-select pace and then 10-min stretching exercises. 72 h of recovery interval time had been implementing between the tests [6].

2.3 Statistical Analysis

The result of this test had been analyzed using the statistical package for social science (SPSS) version 21. Descriptive data analysis had been done to identify the mean and standard deviation of cardiovascular fitness. Inferential analysis Pearson’s product moment coefficient correlation had been used to determine the relationship between the 20-m multistage fitness test and the Yo-Yo intermittent fitness test.

3 Result

Table 1 showed the descriptive result of demographic data of the respondents.

Table 2 showed the mean (SD) was lower to 20-m multistage fitness test [(33.31(5.92)] compared to Yo-Yo intermittent fitness test [47.48 (1.06)].

Table 1 Descriptive result of demographic data

Socio-demographic characteristic	Frequency (%)
Age (year)	
Mean (SD)	15.12 (1.17)
• 14	25 (41.67)
• 15	15 (25)
• 16	8 (13.33)
• 17	12 (20)
Level played	
• School	10 (16.7)
• Division	20 (33.3)
• District	18 (30)
• State	7 (11.7)
• International	5 (8.3)
Position	
• Not specific	10 (16.7)
• Prop	11 (18.3)
• Hooker	6 (10)
• Second row	3 (5)
• Flanker	3 (5)
• No. 8	2 (3.3)
• Fly half	5 (8.3)
• Stand off	3 (5)
• Center	11 (18.3)
• Wing, blind, and full back	6 (10)

Table 2 Descriptive result of cardiovascular fitness level

	N	Min	Max	Mean	SD
20-m multistage fitness test	60	22.19	46.83	33.31	5.92
Yo-Yo intermittent fitness test	60	45.71	49.79	47.48	1.06

Table 3 Relationship between 20-m multistage fitness test and Yo-Yo intermittent fitness test

	Yo-Yo intermittent fitness test
20-m multistage fitness test	0.652**

$^{**}p < 0.001$

Is similar goes to minimum and maximum where 20-m multistage fitness test was 22.19 and 46.83 compared to Yo-Yo intermittent fitness test [min = 45.71, max = 49.79].

The relationship between these two tests had been analyzed using the Pearson's correlation. The result showed in Table 3 that there is positive strong relationship between 20-m multistage fitness test and Yo-Yo intermittent fitness test $p < 0.001$ ($p < 0.05$), $r = 0.652$ (Fig. 1).

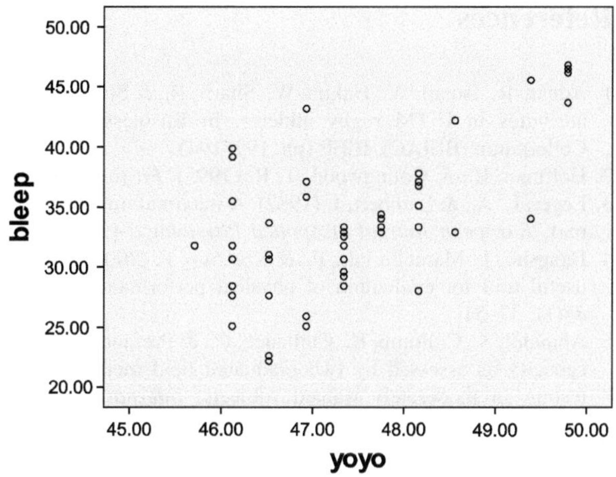

Fig. 1 Relationship between 20-m multistage fitness test and Yo-Yo intermittent fitness test

4 Discussion

Based on the finding, it showed that there was a significant relationship between the results of the 20-m multistage fitness test and the Yo-Yo intermittent fitness test ($r = 0.652$, $p < 0.05$, $n = 60$). According to the study of [7] which determine the reliability of Yo-Yo test and association with 20-m multistage fitness test, result showed strong correlation between the two tests' results ($r = 0.80$). According to the study by Abdul Rashid Aziz [8, 9] significant correlation was observed between the 20-m multistage fitness test and the Yo-Yo test ($r = 0.65$, $p < 0.05$). Based on the studies, it can be proof that there is relationship between 20-m multistage fitness test and the Yo-Yo intermittent fitness test. These tests would present similar result to identify cardiovascular fitness level. So, either using 20-m multistage fitness test or the Yo-Yo intermittent fitness test, it will present similar result.

This study conclude that both test shows positive correlation in measuring VO_{2max} among rugby player test were r = 0.652. It showed that coaches122 can use either 20-m multistage fitness test or Yo-Yo intermittent fitness test to determine athlete cardiovascular level because both tests will present the same result. Which test can be used depends on coach itself. Even both tests using difference method, equation, and norm, both tests are ideally suited for testing aerobic endurance capacity for the athlete.

Acknowledgments The authors wish to express their sincere gratitude and appreciation to Universiti Teknologi MARA (UiTM) research ethical committee for accepting and approving this study. Further gratitude to the participants for agreeing takes part in this study. No sources of funding were used to assist in the preparation of this article. We have no conflicts of interest that are directly relevant to the content of this study. Thanks to Faculty of Sports Science and Recreation for given permission and funding to present the paper.

References

1. Adnan, R., Ismail, W., Hakim, W., Sharir, R., & Sulaiman, N. (2013). Determinants of physical attributes in UiTM rugby athletes. In Business Engineering and Industrial Applications Colloquium (BEIAC), IEEE (pp. 192–194).
2. Hoffman, R., & Colingwood, T. R. (1995). *Fit for duty*. Human Kinetics Publishers Inc.
3. Léger, L. A., & Lambert, J. (1982). A maximal multistage 20-m shuttle run test to predict VO_2 max. *European Journal of Applied Physiology, 49*, 1–5.
4. Bangsbo, J., Marcello Iaia, F., & Krustrup, P. (2008). The Yo-Yo intermittent recovery test: A useful tool for evaluation of physical performance in intermittent sports. *Sports Medicine, 38*(1), 37–51.
5. Ahmaidi, S., Collomp, K., Caillauce, C., & Prefaut, C. (1992). Maximal and functional aerobic capacity as assessed by two graduated field methods in comparison to laboratory exercise testing in moderately trained subjects. *International Journal of Sports Medicine, 13*(3), 243–248.
6. Grant, S., Corbett, K., Amjad, A. M., Wilson, J., & Aitchison, T. (1995). A comparison of methods of predicting maximum oxygen uptake. *British Journal of Sports Medicine, 29*(3), 147–152. doi:10.1136/bjsm.29.3.147.
7. Thomas, A. D. (2006). The Yo-yo test: Reliability and association with a 20-m multistage fitness test run and VO_2 max. *International Journal of Sport Physiology Performance, 1*(2), 137–149.
8. Abdul Rashid Aziz, F. H. (2005). A pilot study comparing two field tests with the treadmill run test in soccer players. *Journal of Sports Science and Medicine*, 105–112.
9. Adnan, R., Sharifudin, S. A., Tumijan, W., & Sulaiman, N. (2011). The VO_{2max} differences based on The Position Roles amongst Malaysia Leagues Soccer Athletes. In Humanities, Science and Engineering (CHUSER), IEEE Colloquium on (pp. 290–293).

Criterion Validity of Selected Cardiovascular Field Based Test Among Healthy Male Adults

Norasrudin Sulaiman, Izudin Idrus, Muhammad Zulqarnain Muhamad Nasir, Rahmat Adnan, Shariman Ismadi Ismail and Mohd Hafdzam Osman

Abstract The purpose of this study was to determine the criterion validity of selected cardiovascular field-based tests in estimating VO_{2max} among healthy male adults. A sample of 20 male adults [(mean ± SD), age 21.4 years (±2.4), height 165.9 cm (±5.9), weight 66.6 kg (±6.6), and BMI 24.2 kg/m^2 (±2.1)] was willing to participate in this study. Subjects were required to perform 20-m multistage shuttle run (MST) and Cooper's 12-min run (12MR) as a test and treadmill-graded Bruce protocol as a criterion test in different weeks. VO_{2max} level and last heart rates were recorded during the test. The data were analyzed using the Statistical Package for Social Sciences (SPSS) version 20.0 with the significance level set at $p \leq 0.05$. The agreement for both tests was analyzed using the Bland–Altman method. This study found that the difference between the mean (±SD) VO_{2max} values of direct measurement ($VO_{2max} = 40.42 \pm 4.49$ ml/kg/min) and both direct measurement tests (MST = 37.52 ± 4.62 ml/kg/min; 12MR = 41.27 ± 4.2 ml/kg/min) was statistically significant. Both tests showed a strong relationship with the criterion test (MST, $r = 0.83$, $p = 0.01$, $p < 0.05$; 12MR, $r = 0.95$, $p = 0.01$, $p < 0.05$). However, the 12MR test showed a good limit of agreement (−3.59 and 1.88) with a small bias value (−0.86). In conclusion, both tests are valid and can be used to measure aerobic capacity, but the Cooper's 12-min run test is more suitable in measuring VO_{2max} among healthy male adults.

Keywords Component · VO_{2max} · MST · 12MR · Criterion validity · Field-based test

N. Sulaiman (✉) · I. Idrus · M. Z. Muhamad Nasir · R. Adnan · S. I. Ismail
FSR Performance Analysis Laboratory, Universiti Teknologi MARA, Shah Alam, Malaysia
e-mail: noras878@salam.uitm.edu.my

N. Sulaiman · M. Z. Muhamad Nasir · R. Adnan · S. I. Ismail
Sports Science Center of Studies, Universiti Teknologi MARA, Shah Alam, Malaysia

M. H. Osman
Sports Development Divisions, Ministry of Youth and Sport (KBS), Putrajaya, Malaysia

R. Adnan et al. (eds.), *Proceedings of the International Colloquium on Sports Science, Exercise, Engineering and Technology 2014 (ICoSSEET 2014)*,
DOI: 10.1007/978-981-287-107-7_30,

1 Introduction

Assessment of VO_{2max} allows clinicians to identify individuals who may have difficulty and limitations in performing various functional tasks, which require energy supply to working muscles. Participation in aerobic exercise can improve health and prevent excess disability in persons during their daily-life activities. Aerobic exercise is recommended for comprehensive health management and physical fitness as an important outcome to measure cardiovascular disease. Field-based assessments depend on the prediction techniques by the assessor and thus are prone to error when compared to the laboratory [1, 2]. Due to these difficulties, researchers have developed field-based tests such as the 20-m shuttle run test, 2.4-km run test, and the 6-km walk test, while laboratory tests such as the Bruce test, Astrand test, and Harvard step test are commonly used to predict the VO_{2max}. Some researchers have indicated that coaches and fitness trainers may face difficulty in the administration of laboratory tests due to high costs and limited time to gain access to laboratory facilities although laboratory tests provide good internal validity and reliability compared with field tests [3].

Researchers have developed a number of tests to quantify VO_{2max} outside the laboratory. These field tests are generally more variable than their laboratory counterparts, as they estimate VO_{2max} from the performance of a physical task, rather than measuring oxygen uptake, and are conducted in a less-controlled environment than the laboratory tests. The valid test to measure cardiovascular fitness is not necessarily a valid test for sedentary people because of differences in the nature of movement. In order to counter this problem, a valid test to measure cardiovascular fitness among healthy male adults is essential.

In an extensive review [4], it was found that specificity of the assessment is needed during conducting field-based tests. In their study, the researcher used the 20-m multistage fitness test (MFT) where the subjects need to perform shuttle running between two parallel lines set 20-m apart and as many as possible until exhaustion. This study was focused on the validity of repeatability of MFT in predicting levels among active young men. The results showed strong correlation between continuous incremental treadmill run test (laboratory test) and the 20-m MFT (field-based test). There are various types of field-based tests that can be administered by the researcher in order to evaluate the aerobic capacity as an option from conducting in a laboratory setting. However, the most appropriate field-based test among healthy male adults is still unclear. Hence, there was a need for research to get clear evidence. Thus, this study attempts to investigate the criterion validity of selected cardiovascular fitness tests in estimating aerobic capacity among healthy adults.

2 Method

2.1 Subject

A sample of 20 male adults [(mean ± SD), age 21.4 years (±2.4), height 165.9 cm (±5.9), weight 66.6 kg (±6.6), and BMI 24.2 kg/m^2 (±2.1)] was willing to participate in this study. The participants were undergraduates who were still pursuing their study at UiTM Shah Alam campus. They were selected purposively based on their participation in physical activity and exercise regularly at the UiTM campus.

2.2 Procedures

The selection of instruments for this study will help in conducting the tests efficiently in order to obtain good results. The most important aspects were that the instruments must be appropriate, reliable, and valid. There were three tests that had been chosen, which included field-based and laboratory tests. There were two field-based tests, which consist of the 20-m multistage shuttle run test and Cooper's 12-min run test. The Bruce test was the laboratory test. These tests were chosen based on the literature review, as most of these tests were useful for testing healthy male adults on their aerobic capacity performance.

2.2.1 20-m Multistage Shuttle Run Test

The test involves continuous running between two lines 20 m apart in time to recorded beeps. The participants stand behind one of the lines facing the second line as a start position. Participants start running when instructed by the cassette disk or tape. The initial speeds were quite slow. The participants continue running between the two lines, turning when signaled by the recorded beeps. If the line is reached before the beep sounds, the subject must wait until the beep sounds. If participants did not reach in time for each beep, the participants had to run for two more beeps, and if they cannot reach, the participants are called out. The test was stopped if the subject fails to reach the line within 2 m for two consecutive ends. The scoring was based on the estimated VO_{2max} formula by Leger and Gadoury 1989. The equation is as follows:

$$Y = 4.77 + 5.67X - 1.637A + 0.01AX$$

where $Y = VO_{2max}$ (ml/kg/min), $X =$ maximal shuttle run speed (km/hr), and $A =$ age (yr).

2.2.2 Cooper's 12-min Run Test

Participants were given the opportunity to stretch before the starting of the test. Prior to starting the test, it was clearly explained to participants either to walk or to run as fast as possible within 12 min. Start the stopwatch as the run or walk is initiated, and record subjects' time as they finish the tests in the individual data sheets. Testing was then completed, and the subject was allowed to cool down, followed by static stretching. Note down the distance covered by the subject based on the marker cone that has been set every 100 m on the track. The scoring is based on the estimated VO_{2max} using the formula [5]. The equation is as follows:

$$Y = (\text{Distance in meters} - 504.9)/44.73$$

where $Y = VO_{2max}$ (ml/kg/min).

2.2.3 Bruce Incremental Treadmill Run Test

Participants were given the opportunity to stretch before the starting of the test. Participants used the Borg scales of rated perceived exertion to know how the level can go. Participants were encouraged to go on as long as possible during the test to obtain the best possible measurement of VO_{2max}. Prior to each session of testing, the metabolic chart was calibrated and each participant was measured and weighed to ensure accuracy in the data entered into the computer for the necessary computations.

Participants were asked to report RPE during each stage. With increasing of intensity as RPE reached 10 scale, or until the subject signaled to end of the test. Testing was then completed, and the subject was allowed to cool down on the treadmill. The equation is as follows:

$$Y = 14.8 - (1.379 \times T) + (0.451 \times T) - (0.012 \times T)$$

where $Y = VO_{2max}$ (ml/kg/min), T = total time on the treadmill.

3 Findings

The summaries of descriptive data for the subjects used in this study are presented in Table 1. The total number of subjects in this study was 20 ($n = 20$). Statistical analysis showed that the mean (±standard deviation) age was 21.3 (±2.4) years. The mean (±standard deviation) height was 165.9 (±5.9) cm. The minimum height of subjects was 158 cm, and the maximum height of subjects was 178 cm. The mean (±standard deviation) weight was 66.6 (±6.6) kg. The minimum weight of subjects was 57 kg, and the maximum weight of subjects was 79 kg. The mean

Table 1 Descriptive analysis

	Min	Max	Mean	SD
Age (years)	19	28	21.3	2.4
Height (cm)	158	178	165.9	5.9
Weight (kg)	57	79	66.6	6.6
BMI (kg/m2)	21.5	29.4	24.2	2.1

Table 2 Correlation analysis between tests

Correlation analysis			
			Bruce protocol
Pearson	MST	Correlation coefficient	0.826**
		Sig. (two-tailed)	0.000
		N	20
	12MR	Correlation coefficient	0.951**
		Sig. (two-tailed)	0.000
		N	20

**Correlation is significant at the 0.01 level

(±standard deviation) BMI of subjects was 24.2 (±2.1) kg/m^2. The minimum BMI of subjects was 21.5 kg/m^2, and the maximum BMI of subjects was 29.4 kg/m^2.

Based on the analysis as stated in Table 2, it can be concluded that Cooper's 12-min run tests and 20-m multistage shuttle run test showed a significant correlation with the criterion test. However, Cooper's 12-min run tests showed a higher correlation ($r = 0.95$, $p = 0.01$, $p < 0.05$) compared with the 20-m multistage shuttle run test ($r = 0.83$, $p = 0.01$, $p < 0.05$).

This study found a high correlation in VO_{2max} value between the selected cardiovascular field-based tests and the criterion test. However, in order to strengthen the evidence of relationship between selected field-based tests and the laboratory tests, the Bland–Altman analysis was used to determine the limit of agreement between methods, which were selected as suitable tests for measuring aerobic capacity among healthy male adults. This analysis was used to determine how much the new method is likely to differ from the old, and if this causes problem in clinical interpretation, it can replace the old method with the new or use the two interchangeably [6]. The agreement is acceptable when the bias value is close to zero, and the difference between measurements using the two assay methods should lie within the limits of agreement approximately 95 % of the time [7].

Current study states that the 95 % limit of agreement should reflect the actually observed nature of the differences. The present study showed that there was a good limit of agreement for Cooper's 12-min run test as it was small (−3.59 and 1.88) with the bias value −0.86, which was close to zero. Meanwhile, the 20-m multistage shuttle run showed a wider range of limit of agreement where the bias values were larger than the Cooper's 12-min run test (−2.38 and 8.17) with the bias value of 2.89 (Figs. 1 and 2) (Table 3).

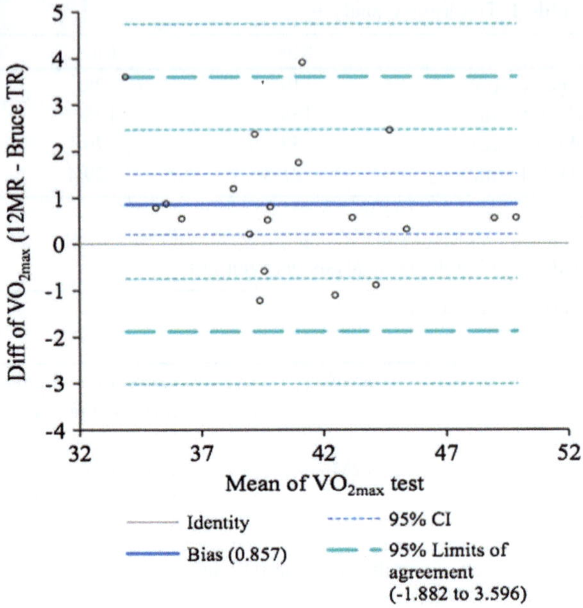

Fig. 1 Limit of agreement of VO_2max between Cooper's 12-min run test with Bruce treadmill run test

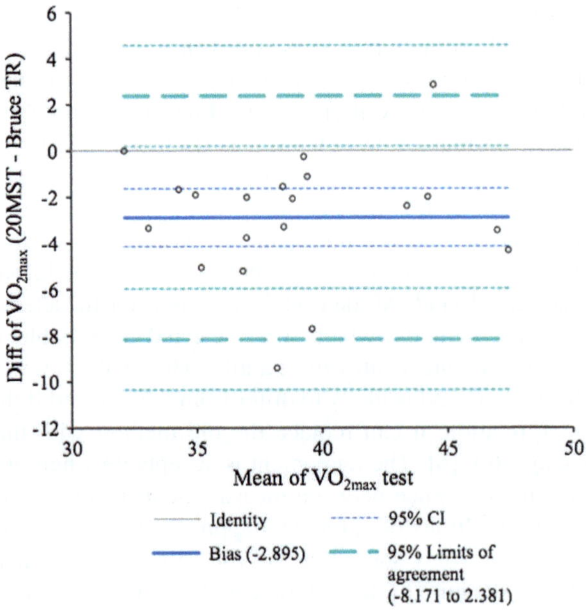

Fig. 2 Limit of agreement of VO_2max between 20-m multistage shuttle run test with Bruce treadmill run test

Table 3 Agreement analysis

Bland–Altman analysis						
Method	95 % CI		SD diff bias		95 % limits of agreement	
	Mean diff				Upper limits	Lower limits
MST-TR	1.64	4.16	2.69	2.89	−2.38	8.17
12MSR-TR	−1.51	−0.2	1.39	−0.86	−3.59	1.88

4 Discussion

The statistical analyses conducted consisted of descriptive and correlation analysis. The descriptive analysis reports on the age, height, weight, and BMI of the subjects. A Pearson's product-moment correlation coefficient was computed to analyze the relationship between selected aerobic capacity fields based on laboratory tests measuring maximal oxygen uptake of the subjects. The main result of this study was that there was a positive relationship between field-based tests and laboratory tests in measuring maximal oxygen uptake (VO_{2max}) among healthy male adults.

In this study, there was a strong, positive correlation between field-based tests and laboratory tests used in measuring aerobic capacity of healthy male adults. This result is supported by the finding [8] that states a strong relationship of field-based tests with laboratory tests for measuring VO_{2max} among healthy males. This study used Cooper's 12-min run test, the 20-m multistage shuttle run test, and the sub-maximal cycle ergometer test to be compared with laboratory treadmill running to measure aerobic capacity of healthy males. The Cooper's 12-min run test produced a strong relationship ($r = 0.92$, $p < 0.05$) when compared to the 20-m multistage shuttle run test and the sub-maximal cycle ergometer test ($r = 0.86$, $p < 0.05$). It can be concluded that the Cooper's 12-min run test can be used as a parameter of healthy individuals to assess their cardiorespiratory fitness.

The aerobic system is the most important component that needs to be controlled by sedentary people and also high-level athletes. The ability of aerobic systems to provide energy to the working muscles can be assessed through VO_{2max} scores. As the VO_{2max} scores increase, the energy supply to the working muscles also increases. This statement is supported by other studies [4] which showed a higher degree of agreement of repeatability ($-0.007 \pm +0.025$), for the 20-m multistage shuttle run test among young active men, and no significant bias was found in this study.

5 Conclusion

From the results of this research, it can be concluded that the selected aerobic capacity field-based tests in measuring maximal oxygen uptake (VO_{2max}) were valid for the study on healthy male adults. Therefore, the field-based tests can be used to determine the cardiorespiratory fitness as an option to the gold standard test, which is the Bruce incremental sub-maximal test. This statement was supported by a comparative study on field-based and laboratory tests for the evaluation of aerobic capacity among soccer players [9].

Acknowledgments We would like to express our gratitude to the Faculty of Sports Science and Recreation, Universiti Teknologi MARA (UiTM) Shah Alam, Selangor, Malaysia, for support throughout the whole process of conducting this study.

References

1. Castro-Piñero, J., Artero, E. G., España-Romero, V., Ortega, F. B., Sjöström, M., Suni, J. & Ruiz, J. R. (2010). Criterion-related validity of field-based fitness tests in youth: a systematic review. *British Journal of Sports Medicine, 44*(13), 934–943.
2. Clerk Maxwell, J. (1892). *A treatise on electricity and magnetism* (3rd ed., Vol. 2, pp. 68–73). Oxford: Clarendon.
3. Norasrudin, S., Anis, M., Rahmat, A., Rezian-na, K., & Mohd Norizham, S. (2012). Validity of selected aerobic capacity field-based test among rugby players. *IEEE Colloquium on Humanities, Science and Engineering Research*, 854–857.
4. Cooper, S. M., Baker, J. S., Tong, R. J., Roberts, E., & Hanford, M. (2005). The repeatability and criterion related validity of the 20-m multistage fitness test as a predictor of maximal oxygen uptake in active young men. *British Journal of Sports Medicine, 39*(4), 19.
5. Cooper, K. K. (1968). A means of assessing maximal oxygen uptake. *Journal of the American Medical Association, 203*, 201–204.
6. Bland, J. M., & Altman, D. G. (1986). Statistical method for assessing agreement between two methods of clinical measurements. *Lancet, 1*, 307–310.
7. Dewitte, K., Fierens, C., Stöckl, D., & Thienpont, L. M. (2002). Application of the Bland–Altman plot for the interpretation of method-comparison studies: a critical investigation of its practice. *Clinical Chemistry, 48*, 799–801.
8. Grant, S., Corbett, K., Amjadt, A. M., Wilson, J., & Aitchson, T. (1995). A comparison of methods of predicting maximum oxygen uptake. *British Journal of Sports Medicine, 29*(3), 147–152.
9. Metaxas, T. I., Koutlianos, N. A., Kouidi, E. J., & Deliagiannis, A. P. (2005). Comparative study of field and laboratory tests for the evaluation of aerobic capacity in soccer players. *Journal of Strength and Conditioning Research, 19*(1), 79–84.

Level of Cognitive and Somatic Anxiety on Performance of University Kebangsaan Malaysia Athletes

Vincent Parnabas, Nagoor Meera Abdullah, Mohamad Nizam Mohamed Shapie, Julinamary Parnabas and Yahaya Mahamood

Abstract The main purpose of this study was to examine the levels of anxiety of somatic and cognitive among student-athletes. The participants of this study were recruited from University Kebangsaan Malaysia (UKM). The instrument used for the study comprised of a 27-item Competitive State Anxiety Inventory-2 and Sport Performance Questionnaire (SPQ), which had been distributed during MASUM (sport between Universities) competition to the student-athletes. The participants of this study were recruited from University Kebangsaan Malaysia (UKM). The sample consisted of 106 athletes, with national athletes ($N = 33$), state athletes ($N = 35$), university athletes ($N = 21$), and district athletes ($N = 17$). The results showed that elite or national athletes exhibited lower levels of somatic and cognitive anxiety than non-elite athletes. The result also showed that there exits negative correlation between cognitive anxiety and sport performance, and somatic anxiety and sport performance. Sport psychologists, sport counselors, and coaches should use the present findings to recommend coping strategies to university- and district-level athletes that are appropriate for dealing with their athletes' cognitive and somatic anxiety.

Keywords Cognitive · Somatic · Student-athletes · Categories of athletes

V. Parnabas (✉) · N. M. Abdullah · M. N. Mohamed Shapie
Faculty of Sport Science and Recreation, Universiti Teknologi MARA,
Shah Alam 40450, Selangor, Malaysia
e-mail: vincent@salam.uitm.edu.my

J. Parnabas
Institut Pendidikan Guru, Kampus Darulaman, Jitra, Kedah, Malaysia

Y. Mahamood
Faculty of College of Art and Science, University Utara Malaysia (UUM),
Kuala Lumpur, Malaysia

R. Adnan et al. (eds.), *Proceedings of the International Colloquium on Sports Science, Exercise, Engineering and Technology 2014 (ICoSSEET 2014)*,
DOI: 10.1007/978-981-287-107-7_31,

1 Introduction

The transition from high school to college can be stressful for any student [1], but recent evidence suggests that athletes may experience even greater levels of stress or anxiety due to the dual demands of athletics and academics placed on them [2]. In today's MASUM (sport between Universities) competitive sports, a lot of pressure is placed on collegiate athletes to perform the excellence level.

Researchers have reported that over 50 of consultations among athletes at an Olympic festival were related to stress or anxiety problems [3]. Anxiety, as a negative emotion, affects perceptions in sport competitions, where a large majority of athletes consider anxiety to be debilitative toward performance, which may result in decreases in performance [4–6] found "sports psychologist have long believed that high levels of anxiety during competition are harmful, worsening performance, and even leading to dropout."

Many researches showed that winning in a competition depends on how an athlete can control their anxiety level [7]. Anxiety consists of two subcomponents: cognitive and somatic anxiety, which influence performance [8, 9]. The cognitive is the mental component, which characterized by negative expectations about success or self-evaluation, negative self-talk, worries about performance, images of failure, inability to concentrate, and disrupted attention [8, 9]. The somatic is the physiological element, which related to autonomic arousals, negative symptoms such as feelings of nervous, high blood pressure, dry throat, muscular tension, rapid heart rate, sweaty palms, and butterflies in your stomach [8–10].

The level of anxiety has the tendency to change during competition by becoming higher or lower [4, 11–14] because the cognitive and somatic component changes according to the time and situation [15].

Recent investigation found that male and female athletes suffering stresses resulted pressure to win, excessive anxiety, frustration conflict, irritation and fear, which significantly affected their mental or emotional health [16]. Heavy playing schedules, competition for team places, the media, and fans as well as the pressure to win trophies all play a part in players developing high stress and anxiety levels [17]. Research has indicated that most successful athletes using more positive coping strategies than less successful athletes [18, 19]. Many times athletes do not handle anxiety properly by not using coping strategies, which deteriorate their performance.

Most of the previous research, focused on elite athletes, while ignoring less successful athletes. This was confirmed by [20] that research on competitive anxiety mainly focused on elite athletes. The extant literature also shows that there is a limited research comparing on competitive anxiety among athletes of state, district, and school level.

The present research aims to determine the level of somatic and cognitive anxiety between different categories of athletes, which consists national-, state-, district-, and university-level athletes. It sought to correlate the relationship between competitive anxiety and athletes of different skills (national, state, district, and university levels) on performance.

2 Methods

The participants of this study were recruited from UKM. The instrument used for the study comprised of a 27-item Competitive State Anxiety Inventory-2 (CSAI-2), which had been distributed during MASUM (sport between Universities) competition to the student-athletes. CSAI-2 measures somatic anxiety (9 items), cognitive anxiety (9 items), and self-confidence (9 items). Besides that, SPQ which contain 15 items evaluated athletes' performance based on their confident, satisfaction, enjoyment, and concentration level.

The data were collected during competition. The sample consisted of 106 athletes, with national athletes ($N = 33$), state athletes ($N = 35$), university athletes ($N = 21$), and district athletes ($N = 17$), and the higher level of achievement in sport declared as their category in sport.

3 Results

3.1 Respondents' Profile

Frequency, percentage, mean, and standard deviation are presented in Table 1, which shows the overall results of the respondents' profile. The profile of the respondents described the ethnics, gender, age, level of performance, and type of sports taken part. There are 71 male and 35 female athletes participated in this study.

Based on the ethnic, the majority of athletes belong to Malays ($n = 61$) since they are the majority race in Malaysia. The second largest ethnic in Malaysia are Chinese. There are 25 athletes from Chinese ethnic. Indians are the lowest ethnic in Malaysia. There are 20 athletes from Indian ethnic.

The mean age for overall respondents was 22.18 years.

The age of male varied from 19 to 27 years, where the mean age was 22.05 years. The age of females ranged from the minimum of 19 to the maximum of 27 years. The mean age for female respondents was 22.17 years.

The variable "types of sport" is gathered through the studies. This variable is categorized into two big sports. The highest respondent involved in individual sports (61.32 %), followed by team sports (38.68 %). Based on the level of performance, the majority of the respondents obtain the highest performance in sport, state level (33.01 %), followed by national (31.13 %), university (18.81 %), and district level (16.03 %).

3.2 Cronbach Reliability Coefficients

In this study, Cronbach's alpha was found ranging from 0.72 to 0.81. Coefficients of 0.70 and above were considered reliable (Table 2).

Table 1 Profile of the respondents

Variables	Frequency	Percentage	Mean	SD
Ethnics				
Malay	61	57.55		
Chinese	25	23.58		
Indian	20	18.87		
Gender				
Male	71	66.98		
Female	35	33.02		
Age				
Male			22.51	2.23
Female			22.78	2.07
Overall			22.12	1.63
Types of sport				
Individual	65	61.32		
Team	41	38.68		
Level of performance				
State	35	33.01		
National	33	31.13		
University	21	18.81		
District	17	16.03		

Table 2 Cronbach reliability coefficients

Questionnaires	Cronbach's alpha ($n = 106$)
Competitive State Anxiety Inventory-2 (CSAI-2)	
Cognitive anxiety	0.8101
Somatic anxiety	0.7237
Self-confidence	0.7708
Sport Performance Questionnaire (SPQ)	
Confident	0.7309
Satisfaction	0.8153
Enjoyment	0.7710
Concentration	0.7252

3.3 Level of Cognitive Anxiety

Table 3 shows the mean scores for the cognitive anxiety among the athletes of different skills, $F\ (3, 106) = 15.227$, $p = 0.01$. Apparently, significant differences emerged for the athletes having different skills at competition. Overall, the mean score obtained for the national athletes was lower than those in other categories.

Post hoc Tukey test (Table 4) showed that the level of cognitive anxiety of university-level athletes was higher than those of state-level ($p = 0.05$) and national ($p = 0.05$)-level athletes, but no significance difference with district-level

Table 3 Level of cognitive anxiety according to the skills

Skills of athletes	Mean	*F*-value	*P*-value
National	17.4132	15.227**	0.000
State	17.7812		
District	21.5531		
University	22.5121		

**$p = 0.01$

Table 4 Post hoc Tukey test: level of cognitive anxiety according to the skills

Skill of athletes	National	State	District	University	*N*
National		X	*(2.331)	*(2.412)	33
State			*(1.457)	*(1.041)	35
District				X	17
University					21

*$p = 0.05$

athletes ($p > 0.05$). Furthermore, the level of cognitive anxiety of district-level athletes was higher than those of state-level ($p = 0.05$) and national-level athletes ($p = 0.05$), but no significance difference with university-level athletes ($p > 0.05$). In addition, the level of cognitive anxiety of state-level athletes was lower than those of district-level ($p = 0.05$) and university ($p = 0.05$)-level athletes, but no differences with national-level athletes ($p > 0.05$). Lastly, the level of cognitive anxiety of national-level athletes was lower than district-level ($p = 0.05$) and university-level athletes ($p = 0.05$), but no differences with state-level athletes ($p > 0.05$).

3.4 Level of Cognitive Anxiety and Sport Performance

The correlation coefficient of −0.018 was noted between the level of cognitive anxiety and sport performance in the evaluation of 106 UKM athletes, which is significant ($P < 0.01$). In other words, the negative relationship existing between these variables is statistically significant (Table 5). Negative correlation indicates that either variables increase or decrease contradictory.

3.5 Level of Somatic Anxiety

Table 6 shows the mean scores for the somatic anxiety among the athletes of different skills, F (3, 106) = 17.2157, $p < 0.01$. Apparently, significant differences emerged for the athletes having different skills at competition. Overall, the mean score obtained for the national athletes was lower than those in other categories.

Table 5 The relationship between the level of cognitive anxiety and sport performance

Subject	Sport performance
The level of cognitive anxiety	−0.018** (0.000)

$^{**}p < 0.01$

Table 6 Level of somatic anxiety according to the skills

Skills of athletes	Mean	*F*-value	*P*-value
National	16.4277	17.2157**	0.000
State	19.4711		
District	22.7241		
University	24.1157		

$^{**}p = 0.01$

Table 7 Post hoc Tukey test: level of somatic anxiety according to the skills

Skill of athletes	National	State	District	University	*N*
National		*(2.712)	*(2.355)	*(2.791)	33
State					35
District					17
University					21

$^{*}p = 0.05$

Post hoc Tukey test (Table 7) showed that the level of somatic anxiety of university were higher than those of district-level ($p = 0.05$), state-level ($p = 0.05$), and national ($p = 0.05$)-level athletes. Furthermore, the level of somatic anxiety of district-level athletes were higher than those of state-level ($p = 0.05$) and national-level ($p = 0.05$), but lower than those of university-level athletes ($p = 0.05$). In addition, the level of somatic anxiety of state were higher than those of national-level ($p = 0.05$), but lower than those of district-level ($p = 0.05$) and university ($p = 0.05$)-level athletes. Lastly, the level of somatic anxiety of national were lower than those of state-level ($p = 0.05$), district-level ($p = 0.05$), and university-level athletes ($p = 0.05$).

3.6 Level of Somatic Anxiety and Sport Performance

The correlation coefficient of −0.019 was noted between the level of somatic anxiety and sport performance in the evaluation of 106 UKM athletes, which is significant ($p = 0.01$). In other words, the negative relationship existing between these variables is statistically significant (Table 8). Negative correlation indicates that either variables increase or decrease contradictory.

Table 8 The relationship between the level of somatic anxiety and sport performance

Subject	Sport performance
The level of somatic anxiety	−0.019** (0.000)

**$p = 0.01$

4 Discussion

4.1 Level of Cognitive Anxiety

The result showed that athletes representing their university and district exhibited higher cognitive anxiety level than those in state, whereas national athletes showed the lowest level of cognitive anxiety. In Malaysia, no research involving the four categories of skills has been conducted so far, therefore this research has failed to compare these with the findings of previous research.

Several researches indicated that low level of athletes easily can become "victim" of sources of anxiety than elite athletes. According to Drive theory, less successful athletes or those who have not mastered their skill perfectly tend to increase their anxiety level with an audience present [4, 21]. The more audience appears in the sport venue, there is a tendency to increase cognitive anxiety on low-skill athletes. The higher the cognitive anxiety increases the probability of making more errors and mistakes are unavoidable. More aptly, research showed elite athletes use higher number of coping strategies than less successful athletes, which reduce their cognitive anxiety [9, 10, 18, 22, 23]. Therefore, national athletes show less cognitive anxiety than low-skill athletes.

Furthermore, a few researches have been indicated that high level of self-confidence has the tendency to reduce the level of cognitive anxiety. Self-confidence generates positive emotions, such as, optimism, less anxiety, relaxation, and happiness, in athletes [7, 24, 25]. Research of [26] showed that athletes in the categories of university and district skill exhibited lower level of self-confidence than categories of state and national athletes. Low level of self-confidence contributes the level of cognitive anxiety to increase.

4.2 Level of Somatic Anxiety

The result showed that athletes representing their university exhibited higher somatic anxiety level than those in state and district categories, whereas national athletes showed the lowest level of somatic anxiety. In Malaysia, no research involving the four categories of skills has been conducted so far; therefore, this research has failed to compare these with the findings of previous research.

The explanation for somatic anxiety is similar with cognitive anxiety. According to Drive theory, the presence of audience for low-skilled athletes, during the sport competition, not only increases their cognitive anxiety but also the somatic anxiety. Somatic anxiety refers to athletes' changes in their physiology, such as increased perspiration, difficulty in breathing, increased heart beat, changes in the brain wave, elevated blood pressure, increased urination, butterflies in the stomach, less saliva in the mouth, and muscle tension. The sympathetic nervous system is stimulated by fear perception in the cerebral cortex, prompting an immediate stress response. Athletes, who have learned anxiety management skills, often respond to a greater degree to an anxiety symptom but return to their resting rate sooner than those athletes, who are not trained in anxiety management.

A few research indicated that cultural background or ethnic may play an important role in increasing the level of somatic anxiety. For example, cross-cultural studies showed that Latins have higher somatic anxiety than any other ethnic groups [27]. In this study, there are three types of ethic taken part (Malay, Chinese, and Indian); most probably, one or two ethnics contribute in the increase of somatic anxiety. Anyway, further research on ethnic and somatic anxiety can determined this.

A few research indicated, types of food, especially salty food can contribute to somatic anxiety. A little research indicated that there is a positive relationship between salt and somatic anxiety. Some people are genetically susceptible to sodium and will develop high blood pressure when they ingest too much of it. As a known fact that Malaysian foods are much saltier compared with any other country. Besides that, saltier food will reduce the level of water in our body. Research of [28] showed low level of the water in our body has the tendency to increase the anxiety and tension. Furthermore, Malaysians are famous of drinking tea compared with any other country. Tea contains theobromine and theophylline, which are sympathomimetics. These substances increase metabolism, make one high alert, and result in the release of stress hormones, which elevate the heart rate and blood pressure [29].

4.3 Level of Anxiety and Sport Performance

The result revealed that there exists negative correlation between cognitive anxiety and sport performance. Similar result also detected between somatic anxiety and sport performance. It means the higher the level of somatic or cognitive anxiety experience by athletes, the lower sport performance.

The relationship between somatic and cognitive anxiety was explained best in Multidimensional Anxiety Theory. This theory explains both cognitive and somatic anxiety effect performances. The basic premise of multidimensional conceptualizations of anxiety is that the two components of anxiety are independent because they have different antecedents and consequences, particularly that they differently influence behavior [8]. If an athlete worried about competition (cognitive anxiety), his or her performance will be poor. The relationship between somatic anxiety, where an athlete experiences physiological changes, such as increases in the levels of muscle tension, nervousness, sweating, and heartbeat, and performance is, however, similar to the inverted-U theory [24]. When increases in somatic anxiety are recorded in an athlete, it can result in arousal at an optimal level that results in the best performance results. However, an increase in arousal beyond or below the optimal level of arousal will lead to a decrease in athletic performance.

5 Conclusion

The findings of the research determined that there are differences in the level of anxiety of cognitive and somatic anxiety, showed by different categories of Malaysian athletes. These differences were related to their level of skill. The results showed that elite or national athletes exhibited lower levels of somatic and cognitive anxiety than non-elite athletes. Low anxiety levels are very important in high sport performance. This study also showed that there exists negative correlation between cognitive anxiety and performance. Similarly, the same result exists between somatic anxiety and performance. Sport psychologists, sport counselors, and coaches should use the present findings to recommend coping strategies to university-level and district-level athletes that are appropriate for dealing with their athletes' cognitive and somatic anxiety.

Future research should identify the most prevalent sources of somatic and cognitive anxiety among different skill of athletes. Initial evidence suggests that among the sources of cognitive and somatic anxiety are fear of injury, presence of audience, past unpleasant experiences, fear of lose, negative evaluation, knowledge of the opposition team, uncertainty, playing at the opposition's place, high hope, and perceived sport events as very important. Seeking sources of somatic and cognitive anxiety should be a great value to reduce the level of anxiety. Furthermore, types of coping strategies can be used to reduce the level of somatic and cognitive anxiety among athletes, much depend on the sources of anxiety. For example, positive self-talks are effective to apply for athletes with fear of lose and past unpleasant experiences. In addition, imagery technique are suitable for the athletes who fear of injury, presence of audience and high hope.

References

1. Pritchard, M. E., Wilson, G. S., & Yamnitz, B. (2007). What predicts adjustment among college students? A longitudinal panel study. *Journal of American college Health*, *56*(1), 15–22.
2. Wilson, G., & Pritchard, M. (2005). Comparing sources of stress in college student athletes and non-athletes. *Athletic Insight*.
3. Murphy, S. M. (1988). The on-site provision of sport psychology services at the 1987, US Olympic Festival. *The Sport Psychologist*, *2*(4), 337–350.

4. Weinberg, R. S., & Gould, D. (2011). *Foundations of sport and exercise psychology* (p. 2011). Champaign: Human Kinetics.
5. Raglin, J. S., & Hanin, Y. L. (2000). Competitive anxiety. In L. H. Yuri (Ed.), *Emotions in sport* (pp. 93–111). Champaign: Human Kinetics.
6. Hann, Y. L. (2000). *Emotions in sports*. Champaign: Human Kinetics.
7. Humara, M. (2001). The relationship between anxiety and performance: A cognitive-behavioral perspective. *Athletic Insight 1*(2), 1–14.
8. Martens, R., Vealey, R. S., & Burton, D. (1990). *Competitive anxiety in sport*. Champaign: Human Kinetics.
9. Jarvis, M. (2002). *Sport psychology*. New York: Routledge.
10. Jones, G. (2000). Stress and anxiety. In S. J. Bull. Sport Psychology: A self-help guide. Ramsbury, Marlborough: Crowood.
11. Weinberg, R. (1989). Anxiety, arousal, and motor performance: Theory, research, and applications. In Dieter Hackfort and Charles, D. Spielberger, Anxiety in sports. New York: Hemisphere.
12. Cashmore, E. (2002). *Sport psychology*. London: Routledge.
13. O'Neil, H. F., & Abedi, J. (1992). Japanese children's trait and state worry and emotionality in a high-stakes testing environment. *Anxiety, Stress, and Coping, 5*, 253–267.
14. O'Neil, H. F., Baker, E. L., & Matsuura, S. (1992). Reliability and validity of Japanese trait and state worry and emotionality scales. *Anxiety, Stress, and Coping, 5*, 225–239.
15. Caruso, C. M., Dzewaltowski, D. A., Gill, D. L., & McElroy, M. A. (1990). Psychological and physiological changes in competitive state anxiety during noncompetitive and competitive success and failure. *Journal of Sport and Exercise Psychology, 12*, 6–20.
16. Humphrey, J. H., Yow, D. A., & Bowden, W. W. (2000). *Stress in college athletics: Causes, consequences, coping*. Binghamton: The Haworth Half-Court Press.
17. B. Heather. Psychology: motivation, anxiety, confidence and goal setting. (http://www.soccernh.org/LinkClick.aspx?fileticket=z-jKZpam25k%3D&tabid=4766), 2010.
18. Orlick, T., & Partington, J. (1988). Mental links to excellence. *The Sport Psychologist, 2*, 105–130.
19. Gould, D., Finch, L. M., & Jackson, S. (1993). Coping strategies used by national champion figure skaters. *Research Quarterly for Exercise and Sport, 64*, 453–468.
20. Krane, V. (1995). Anxiety and stress: reflection of the past and visions of the future. Paper presented as the Dorothy Harris Young Scholar—Practitioner lecturer at the meeting of the Association for the Advancement of Applied Sport Psychology. New Orleans, LA.
21. Anshel, M. H. (2003). *Sport psychology: from theory to practice*. New York: Benjamin Cummings.
22. Dale, G. A. (2000). Distractions and coping strategies of elite decathletes during their most memorable performances. *The Sport Psychologist, 14*, 17–41.
23. Park, J. (2000). Coping strategies by Korean national athletes. *The Sport Psychologist, 14*, 63–80.
24. Ampofo-Boateng, K. (2009). *Understanding sport psychology*. Selangor: UPENA.
25. Vealey, R. S., Hayashi, S. W., Garner-Holman, M., & Giacobb, P. (1988). Sources of sport-confidence: Conceptualization and instrument development. *Journal of Sport and Exercise Psychology, 20*, 54–80.
26. Parnabas, V., Mahamood, Y., Parnabas, J., & Ebenez, J. S. (2013). Level of self-confidence among University Sains Malaysia (USM) Athletes. *International Journal of Physical and Social Sciences, 3*(10), 142–150.
27. Canino, G. (2004). Are somatic symptoms and related distress more prevalent in Hispanic/Latino youth? Some methodological considerations. *Journal of Clinical Child and Adolescent Psychology, 33*(2), 272–275.
28. Ganioa, M. S., Armstronga, L. E., Casaa, D. J., McDermotta, B. P., Leea, E. C., Yamamotoa, L. M., et al. (2011). Mild dehydration impairs cognitive performance and mood of men. *British Journal of Nutrition, 106*(10), 1535–1543.
29. Greenberg, J. S. (1999). *Stress management*. New York: McGrew-Hill.

Cognitive Anxiety and Performance on Team and Individual Sports Athletes

Vincent Parnabas, Tumijan Wahidah, Nagoor Meera Abdullah, Mohamad Nizam Mohamed Shapie, Julinamary Parnabas and Yahaya Mahamood

Abstract Cognitive anxiety has the capability to threaten a person's well-being because it can increase worries and doubt. Athletes will achieve the best performance when their levels of cognitive anxiety were low. The rationale for this study was designed to examine the levels of cognitive anxiety before and during competition between team and individual athletes. In addition, this paper had examined the relationship between cognitive anxiety and sports performance, during competition. Cognitive anxiety was measured using the "Competitive State Anxiety Inventory-2" (CSAI-2) and the "Sports Performance Scale" (SPS). The participants for this research were 121 individuals and 103 team sports athletes. The study was conducted during sports events between schools. The players completed the CSAI-2 twice: before and during the sports competition. The result showed that the cognitive anxiety was higher on team compared to individual sports athletes, before and during competitions. Besides that, the results also revealed the tendency of cognitive anxiety increases during competition on team compared to individual sports athletes. Furthermore, the result showed a negative correlation between cognitive anxiety and sports performance during competition on individuals and team sports. Sport psychologists, sport counselors, and coaches should use the present findings to recommend coping strategies to team and individual athletes that are appropriate for dealing with their athletes' levels of cognitive anxiety.

V. Parnabas (✉) · T. Wahidah · N. M. Abdullah · M. N. Mohamed Shapie
Faculty of Sport Science and Recreation, Universiti Teknologi MARA, 40450 Shah Alam, Selangor, Malaysia
e-mail: vincent@salam.uitm.edu.my

J. Parnabas
Institut Pendidikan Guru, Kampus Darulaman, Jitra, Kedah, Malaysia

Y. Mahamood
Faculty of College of Art and Science, University Utara Malaysia (UUM), Sintok, Malaysia

R. Adnan et al. (eds.), *Proceedings of the International Colloquium on Sports Science, Exercise, Engineering and Technology 2014 (ICoSSEET 2014)*,
DOI: 10.1007/978-981-287-107-7_32,

Keywords Cognitive anxiety · Sports performance · Team sports · Individual sports

1 Introduction

Anxiety, as a negative emotional state, can affect athletes' performance by displaying cognitive symptoms [1, 2], and this leads majority of athletes to consider anxiety to be debilitative toward performance. The cognitive is the mental component, which is characterized by negative expectations about success or self-evaluation, negative self-talk, worries about performance, images of failure, inability to concentrate, and disrupted attention [3, 4]. In general, cognitive anxiety has the capability to threaten a person's well-being because it can increase a person's worries and doubts [5]. Athletes will achieve the best performance when their levels of cognitive were low [4, 6]. This was well illustrated in Catastrophe Model by Fazey and Hardy [7]. According to this model, once an athlete experiences high level of cognitive anxiety as in a sport competition, where an athlete was dominated with worries and doubts, there will be a quick or catastrophic decrease in performance. Therefore, this theory is different than other theories of anxiety, where this theory focuses on the importance of cognitive anxiety in determining the performance of athletes. However, since not much research had been done to investigate this theory, it received very poor research support [8].

Even though quite a number of researches had been done on the levels of anxiety, it focused on one or two sports or all the sports in general. For example, the research done by Kamsi [9] on the level of competitive anxiety only focused on badminton players. Hence, this present research focused on the effect of cognitive anxiety on team and individual sports. Besides that, Aufenanger [10] conducted a study using unequal samples of individuals and team athletes based on 88 team sports athletes and 40 individual sports athletes. Therefore, there is a need to determine the level of cognitive anxiety with bigger and quite equal samples.

Thatcher et al. [11] and Martin and Hall [12] claimed that the level of anxiety of individuals and team sports athletes is not clear, since the result of most of the research done is contradictory to each other. Therefore, more research is needed, to determine the levels of cognitive anxiety on individuals and team sports athletes especially on Malaysian athletes from the view of different culture compared to Western culture. The coach and sports psychologist can only help the athletes if they know the levels of cognitive anxiety on individuals and team sports athletes, since it affects their performance. The level of cognitive anxiety can be reduced using cognitive coping strategies. However, many coaches and athletes are using ineffective coping strategies since they fail to determine the levels of cognitive anxiety on athletes [2]. Therefore, it is important to know the levels of cognitive anxiety among athletes especially on individuals and team sports athletes.

In sports, the levels of anxiety before competition deteriorate the performance [13]. On the contrary, according to Raglin [14], the levels of anxiety during the competition deteriorate the performance compared to the level of anxiety before competition. According to Mahoney and Meyers [15], athletes exhibited different levels of anxiety before and during competitions. The levels of anxiety have the tendency to change during competitions by becoming higher or lower [1, 16–18] because the cognitive component changes according to the time and situation [19]. According to Hanton et al. [20] and Jones [21], the level of anxiety before and during competition is not clear because of the contradictory findings. Different athletes reported different levels of cognitive anxiety from high to low [2]. Therefore, the findings are contradictory to each other [22]. Hence, this research intends to determine the levels of cognitive anxiety before and during competitions among athletes. This research will also evaluate the correlation of cognitive anxiety and performance, during competitions on individuals and team sports. Besides that, in Malaysia, the level of cognitive anxiety of athletes before and during competition is yet to be identified through research.

The rationale for this study was to examine the levels of cognitive anxiety before and during competitions between team and individual athletes. In addition, this paper had examined the relationship between cognitive anxiety and sports performance, during competitions. The team sports included in this study were hockey, football, basketball, volleyball, tennis, rugby league, cricket, and handball. On the contrary, the individual sports are swimming, track and field, and badminton.

2 Methodology

Cognitive anxiety was measured using the "Competitive State Anxiety Inventory-2" (CSAI-2) and the "Sport Performance Scale" (SPS). This study specifically measured the levels of cognitive anxiety and sports performance on individuals and team athletes. The participants for this research were 121 individuals and 103 team sports athletes. The study was conducted during sports events between schools. The players completed the CSAI-2 twice: before and during the sports competitions.

3 Result

3.1 The Levels of Cognitive Anxiety on Individuals and Team Sports

t test showed significant differences on levels of cognitive anxiety among types of sports before competitions, t (224) = 16.212, $p < 0.01$ and during competitions, t (224) = 17.745, $p < 0.01$ (Table 1).

Table 1 Cognitive anxiety among types of sports before competitions

Types of sport	Cognitive anxiety (before competitions)			Cognitive anxiety (during competitions)		
	Mean	Value *t*	Value *p*	Mean	Value *t*	Value *p*
Individual	17.1983	16.212**	0.000	28.1612	17.745**	0.000
Team	24.8231			26.2731		

**$p < 0.01$

Table 2 Cognitive anxiety among types of sports

Subject	Sports performance (individual sports)	Sports performance (team sports)
Cognitive anxiety (during competition)	−0.013** (0.000)	−0.019** (0.000)

**$p < 0.01$

3.2 Correlation of Cognitive Anxiety and Performance

Pearson correlation was used to determine the relationship between cognitive anxiety and sports performance during competitions (Table 2). The results indicated a negative correlation between cognitive anxiety and sports performance during competitions on individuals ($r = -0.013$; $p < 0.01$) and team sports ($r = -0.019$; $p < 0.01$).

4 Discussion

4.1 The Levels of Cognitive Anxiety on Individuals and Team Sports

The main aims of the study were to describe and compare the levels of cognitive anxiety differences between individuals and team sports. The result showed that the cognitive anxiety was higher on team compared to individual sports athletes, before and during competitions. Therefore, different types of sports can show different levels of anxiety [10]. Furthermore, Cerin et al. [23] stated that teams and individual sports athletes had the tendency to show different levels of anxiety.

As we know, team sports like football and hockey are the most favorable sports among Malaysians compared to most of the individual sports like badminton, track and field, and swimming. Therefore, the audiences who presented at the football and hockey matches are far more than the individual sports. The stadium was full house during the team sports event. This might increase the levels of cognitive anxiety of the team athletes. This situation can be explained by using drive theory.

According to the drive theory, athletes' level of anxiety has a tendency to get higher as the audience increases [24].

Furthermore, the pressure to win in football and hockey matches from coaches, universities, and fans is higher in the individual sports. In Malaysia, the image of a university in sports mostly depends on football matches since it is considered as the most favorite sports of Malaysians.

In addition, individuals and team sports athletes might use different kinds of coping strategies which can affect their levels of cognitive anxiety. Coping strategies play an important to role on anxiety [10]. Therefore, in future, more researches need to be done to determine the levels of cognitive anxiety and the coping strategies being used on individuals and team sports.

The result of this study contrary to a few studies related cognitive anxiety is higher on individuals compared to team sport athletes. Results of Martens et al. [4] and Smith and Smoll [25] showed individual athletes experienced higher level of anxiety than team. However, this depends on the individual differences and situation of the competitions.

Besides that, the results also revealed the tendency of cognitive anxiety increased during competitions on individuals and team sport athletes. However, it was found higher on teams compared to individual sports athletes. The result consisted of the findings of Hatzigeorgiadis and Biddle [26] that the levels of cognitive anxiety were not stabile before and during competitions. The data were collected during the interval period of team and individual sports. Team sports like football and hockey had a tough and crucial period just before the interval compared to the individual sports; thus, this may cause the cognitive anxiety to increase drastically on team athletes.

The result of this research is concurrent with Swain and Jones [27], where the level of cognitive anxiety has the tendency to increase at the middle of the match. Athletes considered the football and hockey matches, like "life and death" matter. Therefore, this might contribute to the increment of cognitive anxiety during competitions.

4.2 Correlation of Cognitive Anxiety and Performance

The result showed a negative correlation between cognitive anxiety and sports performance during competitions on individuals and team sports. This means when the levels of cognitive anxiety get higher, the level of sports performance decreases. The result of this research consisted that of Hanton and Connaughton [28], where they found that the performances of elite athletes also deteriorate when the level of cognitive anxiety increases. However, the result of this research is contradictory to the research done by Hanton et al. [20] and Ntoumanis and Biddle [29], where they found that athletes' performances increase when the level of cognitive anxiety gets higher.

The result (Table 1) showed that the level of cognitive anxiety of team and individuals increased at the middle of the match. Furthermore, Table 2 showed that there is a strong negative correlation between cognitive and sports performance, during competitions. This is concurrent with the theory of Catastrophe Model by Fazey and Hardy [7] that there will be catastrophic decreases in performance when the athletes experienced worries and doubts.

It is important for the coaches or sports psychologist to determine the level of cognitive anxiety since the performance of athletes depends on the coping strategies used. According to Jones [21], those athletes who experienced cognitive anxiety should use cognitive coping strategies to enhance their performance. Examples of cognitive coping strategies are imagery, positive self-talk, self-confidence, and goal setting. If athletes used somatic coping strategies like biofeedback and autogenic relaxation, it will not be effective. Therefore, a few researches done in anxiety are to determine which type of coping strategy is needed by athletes based on the type of anxiety. According to multidimensional theory, there are two types of anxieties, namely cognitive and somatic [1, 8]. Furthermore, there are two types of coping strategies: cognitive and somatic. Cognitive coping strategy is more successful if used by athletes who experienced cognitive anxiety, and somatic coping strategy is more effective if used by athletes with somatic anxiety. For example, research of Maynard and Cotton [30] on hockey players showed that the effect of coping strategies depends on the types of anxiety experienced by athletes. Furthermore, many researchers agreed that it is important to use the right and effective technique of coping strategies to enhance athletes' performances [2, 31, 32].

5 Conclusion

Overall, the results showed a tendency for performance to decrease when cognitive anxiety increases. This showed that cognitive anxiety has impacted performance. This confirmed that high levels of cognitive anxiety were the main barrier that inhibits athletes from gaining higher achievements in sports. Furthermore, the result also showed that team athletes experienced higher level of cognitive anxiety than individual athletes. Sports psychologists, sports counselors, and coaches should use the present findings to recommend coping strategies to team and individual sport athletes that are appropriate for dealing with their athletes' level of cognitive anxiety.

The contributions of this study are to lay ways to further investigation on levels of cognitive anxiety and sport performances on a particular sport like football, rugby, badminton, baseball, basketball, and so on. Since coping strategies contribute to the level of cognitive anxiety and performance, future research should focus on the correlation between coping strategy and sports performance.

References

1. Weinberg, R. S., & Gould, D. (2011). *Foundations of sport and exercise psychology*. Champaign, IL: Human Kinetics.
2. Anshel, M. H. (2012). *Sport psychology: from theory to practice*. New York: Benjamin Cummings.
3. Jarvis, M. (2002). *Sport psychology*. New York: Routledge.
4. Martens, R., Vealey, R. S., & Burton, D. (1990). *Competitive anxiety in sport*. Champaign, Illinois: Human Kinetics.
5. Landers, D. M. (1999). The influence of exercise on mental health. In C. B. Corbin & R. P. Pangrazi (Eds.), *Toward a better understanding of physical fitness and activity*. Scottsdale, AZ: Holcomb Hathaway.
6. Krane, V., & Williams, J. (1994). Cognitive anxiety, somatic anxiety, and confidence in track and field athletics: The impact of gender, competitive level and task characteristics. *International Journal of Sport Psychology, 25*, 203–217.
7. Fazey, J., & Hardy, L. (1988). *The inverted-U hypothesis: a catastrophe for sport psychology? British association of sports sciences monograph 1*. Leeds: The National Coaching Foundation.
8. Ampofo-Boateng, K. (2009). *Understanding sport psychology*. Malaysia: UPENA.
9. Kamsi, M. A. (1997). Semangat Bertanding dan Kebimbangan Bertanding Di Kalangan Pemain Badminton Negara (Competitive anxiety on national badminton players). *Master Thesis*. University Kebangsaan Malaysia, Malaysia.
10. Aufenanger, S. J. (2005). Relationships between mental skills and competitive anxiety interpretation in open skill and close skill athletes. *Master Thesis of Miami University*, Oxford, Ohio.
11. Thatcher, J., Thatcher, R., & Dorling, D. (2004). Gender differences in the pre-competition temporal patterning of anxiety and hormonal responses. *Journal of Sports Medicine Physical Fitness, 44*, 300–308.
12. Martin, K. A., & Hall, C. R. (1997). Situational and intrapersonal moderators of sport competition state anxiety. *Journal of Sport Behavior, 20*, 435–446.
13. Hardy, L. (1999). Stress, anxiety and performance. *Journal of Science and Medicine in Sport, 2*(3), 227–233.
14. Raglin, J. S. (1992). Anxiety and sport performance. In J. O. Holloszy (Ed.), *Exercise and sport sciences reviews*. Baltimore: Williams & Wilkins.
15. Mahoney, M. J., & Meyers, A. W. (1989). Anxiety and athletic performance: Traditional and cognitive-development perspectives. In D. Hackfort & C. D. Spielberger (Eds.), *Anxiety in sports*. New York: Hemisphere.
16. Cashmore, E. (2002). *Sport psychology*. London: Routledge.
17. O'Neil, H. F., & Abedi, J. (1992). Japanese children's trait and state worry and emotionality in a high-stakes testing environment. *Anxiety, Stress, and Coping, 5*, 253–267.
18. O'Neil, H. F., Baker, E. L., & Matsuura, S. (1992). Reliability and validity of Japanese trait and state worry and emotionality scales. *Anxiety, Stress, and Coping, 5*, 225–239.
19. Caruso, C. M., Dzewaltowski, D. A., Gill, D. L., & McElroy, M. A. (1990). Psychological and physiological changes in competitive state anxiety during noncompetitive and competitive success and failure. *Journal of Sport and Exercise Psychology, 12*, 6–20.
20. Hanton, S., O'Brien, M., & Mellalieu, S. D. (2003). Individual differences, perceived control and competitive trait anxiety. *Journal of Sport Behavior, 26*, 39–55.
21. Jones, G. (2000). Stress and anxiety. In S. J. Bull (Ed.), *Sport psychology: A self-help guide*. Ramsbury, Marlborough: Crowood.
22. McNally, I. M. (2002). Contrasting concepts of competitive state-anxiety in sport: Multidimensional anxiety and catastrophe theories. *Athletic Journal: On line Journal of Sport Psychology*, http/www.Contrasting Concepts of Competitive State-Anxiety in Sport.htm.

23. Cerin, E., Szabo, A., Hunt, N., & Williams, C. (2001). Temporal patterning of competitive emotions: A criteria review. *Journal Sports Science, 18*, 605–626.
24. Aronson, E., Wilson, T. D., & Akert, R. M. (2010). *Social psychology*. New York: Harper Collins.
25. Smith, R. E., & Smoll, F. L. (1990). Sport performance anxiety. In H. Leitenberg (Ed.), *Handbook of social and evaluation anxiety*. New York: Plenum.
26. Hatzigeorgiadis, A., & Biddle, S. J. H. (2008). Negative self-talk during sport performance: Relationships with pre-competition anxiety and goal-performance discrepancies. *Journal of Sport Behavior, 31*(4), 389–412.
27. Swain, A., & Jones, G. (1993). Intensity and frequency dimensions of competitive state anxiety. *Journal of Sport Sciences, 11*, 533–542.
28. Hanton, S., & Connaughton, D. (2002). Perceived control of anxiety and its relationship to self-confidence and performance. *Research Quarterly for Exercise and Sport, 73*, 87–97.
29. Ntoumanis, N., & Biddle, S. (1998). The relationship between competitive anxiety, achievement goals, and motivational climates. *Research Quarterly for Exercise and Sport, 2*, 176–187.
30. Maynard, I. W., & Cotton, P. C. (1993). An investigation of two stress-management techniques in a field setting. *Sport Psychologist, 7*, 375–387.
31. Anshel, M. H., Kim, K. W., Kim, B. H., Chang, K. J., & Horn, H. J. (2001). A model for coping with stressful events in sport: Theory, application, and future directions. *International Journal of Sport Psychology, 32*, 43–75.
32. Skinner, E. A., Edge, K., Altman, J., & Sherwood, H. (2003). Searching for the structure of coping: A review and critique of category systems for classifying ways of coping. *Psychological Bulletin, 129*, 216–269.

The Effects of Rahim Training Model on Psychological Performance Among Malaysian Junior Tennis Players

Mohd Rahizam Abdul Rahim, Balbir Singh, Vincent Parnabas, Mazlan Ismail and Nagoor Meera Abdullah

Abstract This study investigated the effects of RAHIM Training Model as an intervention program on psychological performance among Malaysian junior tennis players. Twenty four (n = 24) nationally ranked junior tennis players between the age of 12–16 years old (18 males and 6 females, mean age 14.29 ± 1.65 years) were selected using purposive sampling method and were equally divided into training and control groups. The training group performed 4 sessions of intervention every week for eight weeks while control group performed a normal training sessions and match play for the same frequency and duration. All participants completed a 42 items Psychological Performance Inventory (PPI) Questionnaire one day before and after the intervention. The results of independent t-test analysis proved that most of the psychological performance variables showed significant difference between the training and the control groups after the intervention program. Therefore, it validates the positive effect of the RAHIM Training Model in enhancing the psychological performance of junior tennis players. The training group showed higher improvement results which dictate the optimistic effect of intervention training program developed in this study. Based on this study, RAHIM Training Model can be used as guideline for Malaysian coaches to formulate a more structured and effective training program.

Keywords Training · Psychology · Performance · Tennis · Training

M. R. Abdul Rahim (✉) · V. Parnabas · M. Ismail · N. M. Abdullah
Faculty of Sports Science and Recreation, UiTM, Pahang, Jengka, Pahang, Malaysia
e-mail: mrahizam@salam.uitm.edu.my; rahizamrahim@gmail.com

B. Singh
University of Malaya, Sports Center, UM, Kuala Lumpur, Malaysia

R. Adnan et al. (eds.), *Proceedings of the International Colloquium on Sports Science, Exercise, Engineering and Technology 2014 (ICoSSEET 2014)*,
DOI: 10.1007/978-981-287-107-7_33, © Springer Science+Business Media Singapore 2014

1 Introduction

It is an aspiration of any coach to train and to produce a successful athlete. Effective training is about providing not only quality programs but also measureable progress of athletes' performance. Athletes must also be able to demonstrate excellent levels of psychological abilities. A simple subjective view of how well the athletes perform is no longer applicable as a foundation of training and competing. The need to record, analyze and evaluate information on key performance areas such psychological skill has become fundamental to elite performance. The information on several characteristics of sports performance is the foundation for providing feedback on how the athlete is performing. This feedback will lead to the development of up to date coaching interventions or methods centered on evidence-based training programs. If athletes are to attain world class performances, information from continuous assessments of training and competition must be made available in order to assist in the evaluation of athletes' performance and progress. Therefore, it is important for coaches to possess a solid framework of sport science support system which is designed to help foster the skills of athletes and improve their performance.

It is commonly accepted that psychological preparation for athletic pursuits is critical for best performance, and that athletes generally perform better with knowledge and application of psychological principles and strategies. Sport psychology has been defined as "the subdiscipline of exercise science that seeks to understand the influence of behavioral processes on skilled movement" [1]. It can be a great help to sports coaches, sports leaders and those who want to understand the mental aspects of sport and how to enhance the performance of their athletes. Participation and success in any kind of activity, exercise or sport not only relies on physical ability but also requires some degree of mental strength. An athlete's performance can be affected by many factors such as their confidence level, ability to concentrate, other people such as spectators and competitors and the environment in which they are performing. The factor of mental toughness concludes that the athletes need to equip themselves with a package of a combined psychological skills starting in the early stage of involvement, training session, during and after the match [2]. This could be a factor that determines both their success and failure.

Mental training and coaching is aimed in supporting competitive athletes to perform at their most optimal level in athletic training during the regular season and play-offs, as well as at team selections, try-outs and major championships like the Olympics, Worlds, Nationals, and Provincials. Acknowledging the essential need of mental toughness, coaches have come to agree that mental toughness is a factor that could determine the success of an athlete [3] and it is also seen as an important element that could shape a successful athlete to be a champion [4]. Both athletes and coaches admitted that at least fifty percent of the success is influenced by psychological factors that are related to mental toughness [5]. Most of the times when two athletes of the same strength and skills are competing against each other, the determining factor of the winner depends on the mental strength.

The central involvement of the athlete in the training process is the forte that may increase their motivation and encourage adherence to any intervention strategies devised. The comprehensive monitoring of athletes' performance is necessary particularly during phases of training overload in order to allow a coach to make a decision regarding the effects and consequent planning of training. The principle of individualisation suggests that every athlete will react differently to any training stimulus [6].

The purpose of the study is to investigate the effectiveness of RAHIM Training Model as an intervention in enhancing the psychological performance among the existing national elite junior tennis players aged between 12 and 16 years old and to provide evidence based recommendations that can be adopted by players, parents, coaches and national associations in the training programs with the objective of upgrading the performance level and achievements of Malaysian junior tennis players.

The aim of this study is to seek a realistic and feasible solution in order to improve and to enhance the performance of Malaysian junior tennis players as well as to upgrade the standard of their achievement at international level.

The objectives of this study are:

1. To identify the psychological performance characteristics of junior tennis players.
2. To examine the effects of an intervention program on psychological performance of Malaysian junior tennis players.

2 Methodology

2.1 Sample

A total of 24 junior tennis players ($n = 24$) were selected for this study using the purposive sampling method where all the participants were the top tennis players in Malaysia between the age of 12 and 16 years old (18 males and 6 females, mean age 14.29 ± 1.65 years) and were ranked in the National Junior Ranking System. The ranking system was based on the accumulated points collected by the players in various national level tennis tournaments. The points collected determine the ranking achievement of each of the players who participated in the national tennis tournaments. All testing and training procedures were fully explained, and written parental consent was obtained for each participant, all of whom agreed to participate in the study.

2.2 *Measures (Procedures of Study)*

All subjects were tested on four different performance variables namely the physical, psychological, technical and tactical profile. These four performance components were tested utilizing established and proven reliable instruments as well as reliable procedures. Each of the performance profiling was assessed using a standard scoring methodology. These methods correlate with the objective of this study which focuses on effects of an intervention program on the four selected performance components. RAHIM Training Model was developed in the study and employed as an intervention. The participants were divided into two groups namely the Experimental Group and Control Group. The Experimental Group was inducted into the RAHIM Training Model which was employed for eight weeks while the Control Group did not participate in any specific form of training but engaged in a normal individual training and match play. Both the experimental and the control groups were tested on their psychological performance before and after the completion of the intervention training program. During the pre and post test sessions, all participants were asked to complete a Psychological Performance Inventory [5]. A 42 items self report inventory with seven subscales, designed to measure factors that reflect mental toughness in an athlete.

3 Statistical Analysis

Mean scores and standard deviation were calculated for each variable of the two groups according to each attributes of psychological performance. Independent t-test was carried out to examine the group differences based on psychological performance level. The statistical techniques provide the comparison of the two groups over time. An interaction effect between the performance level and measurement reveals differences between the experimental group and control group that changed as a function of time. An alpha of 0.05 was adopted for all the tests of significance.

4 Results

After running the t-test, results in Table 1 show that self-confidence, attention control, visual and imagery control, motivational level, positive energy, and attitude control proved to be significantly difference after the intervention program based on RAHIM Training Model between the experimental and control group. The results show that training group scores mean = 21.08 while control group has mean = 17.75 in self-confidence and statistically it is $t = 4.82$, $p < 0.001$, meaning that the 12 players who participated in intervention program based on RAHIM Training Model were significantly difference or possess a higher self-confidence as

Table 1 Comparison of psychological profiles between training and control groups

Psychological profiles		Mean	t-test	p-value
Self confidence	Pre Exp	18.33	4.03***	0.00
	Pre Con	15.92		
	Post Exp	21.08	4.82***	0.00
	Post Con	17.75		
Negative energy	Pre Exp	17.00	0.65	0.53
	Pre Con	16.33		
	Post Exp	17.92	0.62	0.54
	Post Con	17.42		
Attention control	Pre Exp	17.17	1.96*	0.06
	Pre Con	15.83		
	Post Exp	19.50	3.84***	0.00
	Post Con	17.25		
Visual and imagery control	Pre Exp	18.33	0.93	0.36
	Pre Con	17.50		
	Post Exp	20.83	2.77**	0.01
	Post Con	19.08		
Motivational level	Pre Exp	20.67	3.17***	0.00
	Pre Con	18.25		
	Post Exp	23.17	6.20***	0.00
	Post Con	19.00		
Positive energy	Pre Exp	17.42	1.74*	0.10
	Pre Con	16.42		
	Post Exp	19.67	3.71***	0.00
	Post Con	18.00		
Attitude control	Pre Exp	18.83	3.64***	0.00
	Pre Con	16.58		
	Post Exp	19.75	3.61***	0.00
	Post Con	17.75		

*($p < 0.1$), **($p < 0.05$), ***($p < 0.001$)

compared to the players who did not participate in the program. The training group showed higher results in attention control with mean = 19.50 while control group showed mean = 17.25 where it is statistically proven to have significant difference at $t = 3.84$, $p < 0.001$, meaning that the 12 players who participated in the intervention program based on RAHIM Training Model had significant difference or possess a good attention control compared to the players that did not participate in the program. The results in other psychological attributes showed that visual and imagery control with $t = 2.77$, $p < 0.05$, motivational level $t = 6.20$, $p < 0.001$, positive energy $t = 3.71$, $p < 0.001$ and lastly attitude control $t = 3.61$, $p < 0.01$. In short, most of the components in psychological profiles have significant difference effect except negative energy. This was proven by an intervention training program based on RAHIM Training Model which is a good predictor to foster the psychological profiling of the 12 tennis players as compared to the other 12 tennis players who had no intervention program.

5 Discussion

The findings showed that both experimental and control groups have shown significant difference in almost all psychological attributes except for energy control. However, the training group which was inducted into the training intervention still showed higher mean in negative energy as compared to the control group. The RAHIM Training Model is a training structure for accelerating the athlete's progress toward achieving peak performance and was developed in this study based on previous research findings and the coaches' experience that can be used as guideline for Malaysian coaches to formulate a more structured and effective training program. It has indicated five combinations of vital characteristics to enhance players' performance based on the acronym of RAHIM for Repetition, Attitude, High Intensity, Independence and Movement.

5.1 R for Repetition

Repetition is required in order to make technique an automatic process. The best tennis practice drills provide repetition, match experience and enjoyment. They may also be employed to improve other performance components. A tennis drill is designed to be performed over and over in a repetitive fashion. It has been proven that only practice drills performed with high intensity and purposeful will result in better performance in tennis. Repetition is important in mastery of skills as it is proven from the intervention and results of the study that repetition of drills is a key to consistency. As such repetition elements is an important characteristic that must be emphasized in any tennis intervention program as supported by previous research shows that it requires approximately 10,000 repetitions of a movement or stroke to make a change [7]. Practice volume is central to many prevalent perspectives on expertise, such as the 10,000 h rule, the power-law of practice, and deliberate practice.

5.2 A for Attitude

Tennis is a game of skill and heart. A professional tennis player needs to have ample amounts of both. Positive Attitude is directly related to self-confidence. Most sport psychologists agree that attitude plays a major role in the establishment of long- and short-term goals. The great majority of successful athletes have very positive attitudes. They enjoy what they do in both practice and games and they generally see potential in every circumstance [8]. In understanding the predicament of attitude, it is actually different with psychological control of the players. Although coaches definitely play a vital role in building mental toughness in a player such as self-confidence as previously proven in the research findings as one of the important attributes in psychological profile, thus a coach has to make efforts in creating the right atmosphere and attitude from the beginning in the training of a player for achieving maximum performance [9].

5.3 *H for High Intensity*

The RAHIM training model indicates the importance of high intensity as crucial factors in determining the right method of training as most of the endurance athletes use the high-intensity training to prepare for competitions [10]. A game of tennis is characterized by multiple high intensity efforts interspersed with variable periods of recovery. Under optimal conditions, the demands in sport are closely related to the athlete's physical capacity, which can be divided into the following categories: (i) the ability to perform prolonged exercise (endurance); (ii) the ability to exercise at high intensity; (iii) the ability to sprint; and (iv) the ability to develop a high power output (force) in single actions during competition such as kicking in soccer and jumping in basketball. This is in accordance with more studies show that performance can be maintained and improved by reducing the amount of low-intensity training and keeping a sufficient amount of high-intensity training [11, 12].

5.4 *I for Independence*

The development of independence in tennis players must be one of the primary goals in training. As in tennis, a player's ability to evaluate situations and make positive decisions is important for successful player development. A player's independence is highly inter-related to his or her level of self-confidence and it should be developed at the beginning stage of a player's involvement in the game. Self-confidence is the belief that one can successfully perform a desired behavior [13]. Coaches also have to foster an excellence atmosphere during training whereby the significance of high quality training is recognized by players. This will encourage players to develop their independence and increase their self confidence. This is a characteristic that coaches are trying to develop in the end, a strong and independent player with a clear vision of his or her goals, plans and abilities. By creating the culture of excellence which fosters the successful performance, coaches will help to develop the independence of players in making decision and the skill to handle difficult situations with confidence.

5.5 M for Movement

Tennis is a sport requiring explosive bursts of movement involving high forces, repeated over an extended period of time [14]. Whole movement potential of a tennis player is determined by the individual's conditioning and coordination abilities and that is why these abilities must be continually integrated into tennis technique [15]. It has been stated repeatedly in the tennis literature that tennis places demands on the ability of a player to move quickly in all directions, change directions often, stop and start, while maintaining balance and control to hit the ball effectively. The sprinting, stopping, starting and bending nature of tennis puts repetitive demands on the bones, ligaments and muscles to absorb the sheer forces [16]. Tennis is the ultimate multi-directional sport and players need to be trained accordingly. Research has found that players change direction between four and five times every point at the elite level, so when training for tennis, a player should focus on drills and specific movements that involve direction changes. Tennis is a sport requiring explosive bursts of movement involving high forces, repeated over an extended period of time [14].

In conclusion, the practicality of information gathered from this study can be applied in designing training programs for elite tennis players. The findings also revealed that training intervention programs play a significant role in enhancing the psychological performance of tennis players. As such, the RAHIM training intervention model is a viable tool in formulating a training program.

Acknowledgments The authors would like to thank Selangor Tennis Association, the coaches and players for the support and cooperation in completing this study.

References

1. Hatfield, B. D., & Brody, E. B. (1994). The psychology of athletic preparation and performance: The mental management of physical resources. In T. Baechle (Ed.), *Essentials of strength training principles* (pp. 163–187). Champaign: Human Kinetics Publishers.
2. Jones, G., Hanton, S., & Connaaughthon, D. (2007). A framework of mental toughness in the world's best performers. *Journal the Sport Psychologist. Human Kinetics, 21*, 243–264.
3. Gould, D., Hodge, K., & Petlichkoff, L. (1987). Psychological foundations of coaching: Similarities and differences among intercollegiate wrestling coaches. *Sport Psychologist., 1*, 293–308.
4. Norris, E. K. (1999). *Epistemologies of champions: a discursive analysis champions' retrospective attributions; looking back and looking within*. Michigan: Michigan University Microfilms International.
5. Loehr, J. E. (1986). *Mental toughness training for sport: achieving athletic excellence*. Lexington: Stephen Greene Press.
6. Bompa, T. (1999). *Periodization: Theory and methodology of training* (4th ed.). Champaign: Human Kinetics.

7. Ericsson, K. A. (1990). Peak performance and age: An examination of peak performance in sports. In P. Baltes & M. M. Baltes (Eds.), *Successful aging: Perspectives from the behavioral sciences* (pp. 164–195). Cambridge: Cambridge University Press.
8. McCord, J. R. (1998). *Attitude: The Mental Edge*. http://www.alliancegym.com/attitude-the-mental-edge/. Retrieved on, 10th June 2012.
9. Kawaljeet, S. (2009). *Mental Training—How as a Coach You Can Help*. http://www.itftennis.com/media/108084/108084.pdf. Retrieved on the 23rd November 2012.
10. Paton, C.D., Hopkins, W.G. (2004). Effects of high-intensity training on performance and physiology of endurance athletes. Sportscience, 8, pp. 25–40, (sportsci.org/jour/04/cdp.htm).
11. Mujika, I. (1998). The influence of training characteristics and tapering on the adaptations in highly trained individuals: A review. *International Journal of Sports Medicine, 19*, 439–446.
12. Shepley, B., MacDougall, J. D., Cipriano, N., Sutton, J. E., Tarnopolsky, M. A., & Coates, G. (1992). Physiological effect of tapering in highly trained athletes. *Journal of Applied Physiology, 72*, 706–711.
13. Lubbers, P., Grannito, G. (2004). How To Develop Independence In Tennis Players. ITF Coaching and Sport Science Review, 34.
14. Roetart, E. P., & Ellenbecker, T. (2001). Biomechanics of movement in tennis. *ITF, CSSR, 24*, 15–17.
15. Schönborn, R. (1998). *Advanced training techniques for competitive players*. Aachen: Meyer & Meyer Verlag.
16. Chandler, J. T. (1995). Exercise training for tennis. *Clinical of Sports Medicine, 14*, 33–46.

Part IV
Training Methodology and Technology Application

Breathing Pattern Influence to the Shooting Performance

Muhamad Noor Mohamed, Wan Mohd Norsyam Wan Norman, Adam Linoby, Mohd Hanifa Sariman and Mohd Zulkhairi Mohd Azam

Abstract Breathing pattern potentially affects aiming stability, and thus increases the chance of inconsistencies in shooting performance. The present study seeks to determine the relationship between breathing pattern and performance or score of shooting. Twelve archers, volunteered to participate in this study. Breathing pattern was assessed in terms of bits per minute, where larger values indicate larger amounts of air in the lung cavity and vice versa, using ZEPHYR Bio-Harness devices. Breathing pattern were analysed at the following three phases; (i) setup, (ii) aiming, and (iii) release. Participants shot 12 arrows to a 30-m target. The result showed a significant relationship between breathing pattern with shooting performance for both groups. During setup, both groups of archers were inhaling with the value of 1,538.93 and 1,668.42 bits respectively, and started to exhale during aiming (1,470.40 and 1,567.21 bits) towards the release phase (1273.81 bits) which worked only for the skilled group. The unskilled group showed a slightly higher value at the release phase (1,587.96 bits). The correlations between breathing pattern with shooting performances only reached the statistically significant state at the release phase for skilled archers ($r = 0.233$: $p < 0.05$), while for the unskilled all phases were noted to be significant ($r = -0.363$, -0.208, and -0.286: $p < 0.05$). Coefficients of breathing pattern during release phase for the skilled contributed 29.4 % towards the overall model, while for the unskilled, the setup phase contributed 44.4 % and release phase 24.3 % towards the overall model. In summary, these results showed that controlling the breathing

M. N. Mohamed (✉) · W. M. N. Wan Norman · A. Linoby · M. H. Sariman · M. Z. Mohd Azam
Faculty of Sport Science and Recreation, Universiti Teknologi MARA, Jengka, Pahang, Malaysia
e-mail: muhamad_noor@pahang.uitm.edu.my; muhamadnoor86@gmail.com

M. N. Mohamed
Universiti Teknologi MARA Kampus Seremban, Shah Alam, Malaysia

W. M. N. Wan Norman · A. Linoby · M. H. Sariman · M. Z. Mohd Azam
Universiti Teknologi MARA (Pahang), Bandar Tun Abdul Razak, Malaysia

R. Adnan et al. (eds.), *Proceedings of the International Colloquium on Sports Science, Exercise, Engineering and Technology 2014 (ICoSSEET 2014)*,
DOI: 10.1007/978-981-287-107-7_34, © Springer Science+Business Media Singapore 2014

process can help increase or decrease the overall shooting performance. There was a clear significant relationship between the breathing pattern and shooting performances of skilled and unskilled archers.

Keywords Breathing pattern · Shooting performance · Skilled archers · Unskilled archers

1 Introduction

Inconsistency in breathing pattern will impose a threat to the shot aiming stability [1, 2]. In archery, a type of non-contact, precision aiming task sport, it depends on the skill of the performer, as to whether the archers adopted a similar or rhythmic pattern of breathing. There are minute mechanical movements known as jerking actions that needs to be avoided if possible or needs to be minimized by the athlete by altering the time taken to take a shot. This jerking action is usually caused by the depolarization process of the heart ventricular function [1–5]. There are also certain sweet spots that expert archers tend to use at the release phase, which usually is in the early (10–50 %) part of the heart beat interval. Archers and shooters, where accuracy is the main priority, usually avoid the later part of the breathing interval (50–90 %). An increased and persistent control in breathing is needed in this kind of sport in order to avoid minute jerking actions that eventually will reduce the aiming stability [1, 3, 5, 6]. Increasing the jerking towards the end of the shooting performance will increase the chance of an unstable posture during the performance that cause the body to sway, thus decreasing the precision aiming stability, and resulting with a decreased chance to obtain the highest marks for each shot delivered [5, 7–11].

High performance athletes or skilled athletes usually execute the shot at a lower and under controlled heart rate value [6]. It appears probable that heart rates were parallel with the breathing capacities. When the breathing sequence is under control, the value of the heart rate also seems reduced. The lowering of heart rate causes a decrease in sway movement, thus showing an increase in postural stability. The decrease was likely due to the minimizing of muscular movements associated with the breathing or the respiration process such as the abdominal muscles and the pectoralis [6, 9, 12–15]. Previous researchers suggest that skilled athletes, tend to execute the shot at the early stage of the breathing interval, and that the placement of the shots were similar throughout the whole performances [1, 3–6]. The similarity of the placement was noted to yield similar shooting outcomes for the same situation. However, without consistency in repetitive performances, fluctuating results were displayed [3, 5, 6].

As for archery, it is widely debated as to whether during the initiation of the release phase archers should either exhale or completely hold their breath. However, a decrease in the heart rate value usually indicates that a control of breathing was present [1, 6, 9, 16]. Besides that, decreasing value of heart rate also acts as an indicator, which shows that the athletes are increasing their attention especially during the aiming task, and simultaneously the cardiac cycle starts to lengthen prior to the release phase [2, 16, 17]. Past researchers had noted that this was an indicator of high quality arrow shots [1, 4]. Simultaneously, mechanical movements, due to breathing and heartbeat, were minimized and postural movements were also minimized. This can be achieved by adequate control of the breathing process in order to reduce the heart rate volume [1, 4, 5].

1.1 Breathing Pattern and Postural Consistency

In precision aiming task sport, heart rate and breathing are common aspects that build a relationship with the performance. The theory usually used by researchers to explain the relationships between heart rate variations and focus channelled is the 'intake-rejection' hypothesis. This theory displays that when athletes channel their focus externally, their heart rate will simultaneously decrease, and when the focus was internal their heart rate will increase [2, 18]. External focus or intake tasks usually result in lowered heart rates. This normally considered as an indicator of the increasing attention demand level. In precision aiming tasks such as archery and shooting, athletes focus towards the target and the alignment of bow-sight-target are an example of intake task. Internal focus or rejection task is when the athletes channel their focus internally, such as trying to execute perfect shooting techniques and in the same time thinking of the consequences of imperfect techniques, which result in a rise of the heart rate. The raising and lowering of heart rate is often demonstrated by irregularity in breathing speed or inconsistent breathing pattern [1, 4, 5, 9, 16].

Inconsistency in breathing pattern also imposes effects toward the shot aiming stability. Usually archers adopt a similar breathing pattern during every shot, but this pattern was dependent on their performance level. Expert, experienced or skilled athletes usually have a sweet spot in the cardiac cycle to execute the shot in order to avoid or to minimize the unnecessary mechanical movements, also known as jerking action caused by the heartbeat and the breathing process. Shooters were more prominent toward this phenomenon compared to archers. However, as a part of precision aiming task sport, both of these sports require a consistent and adequate control of breathing or respiration process in order to avoid any unwanted movements. These minute jerking actions increase the chances of athletes' posture to be unstable and sway, and simultaneously decreases the capability of good aiming due to aiming stability being compromised. By decreasing the stability of aiming, the chances to obtain the highest possible mark for each shot was also diminished [9].

It is beneficial to identify correctly the breathing pattern during shooting phases, as it would impose an impact towards overall performance. By having a clear picture of the specific phases in which a proper control of breathing pattern should be implemented, it will help maximize training strategies and at the same time speed up the fundamental training process. A comparative study between skills level would provide further insight regarding the trainability of this variable. Later on, a specific training program can be formulated in precision aiming athletes training regime, where the main focus is on improving the consistency and persistency of breathing pattern and at the same time increase postural control [19].

2 Materials and Methods

2.1 Participants

Twelve ($N = 12$) athletes comprising of skilled ($n = 6$) and unskilled ($n = 6$) archers from Universiti Teknologi MARA Pahang participated voluntarily in this study. Skilled and unskilled groups comprised of both genders and aged between 18 and 22 years. There was no significant difference in age for the skilled ($M = 19.50$, $SD = 1.64$) and unskilled archers ($M = 19.67$, $SD = 1.86$); $t(10) = -0.164$, $p > 0.05$. The skilled group was considered skilled due to their qualification score of 1,150 upon 1,440 full FITA score in either national or international rank competitions. The unskilled were athletes that had the basic knowledge of shooting but had never participated in rank competitions either nationally or internationally.

2.2 Instrumentation

Breathing pattern was quantified using Zephyr Bio-Harness devices (Model PSM Research version 1.5, single transmitter and receiver, Annapolis, Maryland, United State of America: reliability 0.860–0.941). The transmitter was set to transmit live data feed as opposed to hard drive recording. Subsequently, the live data were transformed to graphs and figures in 10-s lengths per draw with 15 frames per second drawing feed. A laptop (model Toshiba Satellite L510, 3 Gb ram capacity, 4.60 Ghz processing capabilities, by Toshiba, assembled in Singapore) was used to compute all equations with software from Zephyr (version 2.3.0.5) that enabled comparison of multiple data and capture real time data transmission. A digital video recorder (model Sony Handy cam DCR-SR68E, Sony [Malaysia] Sdn. Bhd., assembled in Japan) was used for video recording purposes. Every participant's shooting performances were recorded for further analysis.

2.3 Procedures

The shooting area was constructed at the Universiti Teknologi MARA Pahang archery training range. The shooting area comprised of two target butts and target stands placed at 30 m from the shooting line for official target practice and official data collection. Multiple 10-m targets were set up for warming up purposes prior to target practice and the official data collection period (Fig. 1).

The purpose of present study and all the underlying procedures involved in it were brief to all participants, coaches and managers. Prior to the data collection period, sufficient warm up time was given to all participants involved. The warm up session was conducted at the 10-m target. The warming up session lasted for 30–45 min and was design to simulate actual conditions according to the rules and regulations. During target practice and actual data collection session, all participants were required to obey all the rules and regulations of archery competitions, which limited a long end shooting to 4-min per end. In one-end, participants were required to execute six shoots to the target provided, where if the arrow misses the target, it was considered as a miss, and no re-shoot was given. In the warming up sessions, participants were allowed to deliver as many shots as possible within the time limit, which was also a 4-min end. Shooting speed usually corresponds with the expertise level. Expert or skilled archers are usually able to shoot more arrows compared to unskilled in the same time limit. For a 30 min-warming up session, six-ends of shooting can be conducted. Warming up sessions only involved either rhythmic or fast shooting techniques at designated 10-m targets without the presence of target face. Warming up session was important to increase body temperature, and simultaneously promote blood flow to the limbs involved in order to minimize the possibility of injuries and reduce muscle soreness.

After completing the familiarization process, official target practice or sighters being carried out with only 12 arrows per archer or two-ends. During sighters, they were able to adjust the bow sight in order to get the perfect or near-to-perfect sighting line. No data recording occurs during this session. Participants individually paced their own shooting time accordingly.

Subsequently, the participants will complete 12 official shots at the 30-m target. Both breathing patterns and score of the shots will be recorded during this phase. Participants start their shooting after the cue given by the researcher. In order to obtain shooting performance, official target faces (FITA 80 cm 30-m face) were used which consist of five colours and 11-point rings that reflect the score ranging from 1 to 10 with the center ring marked as ‘X’. The ‘X’ ring or the inner most ring brings a score of ten and was considered as accurate compared to the outer ten rings (Fig. 2).

In order to increase the environmental validity, participants were required to use their own bow and arrows instead of standardized prepared equipments for performance measurement purposes. This was because archery equipments and settings are individualized rather than standard: thus, using same standard

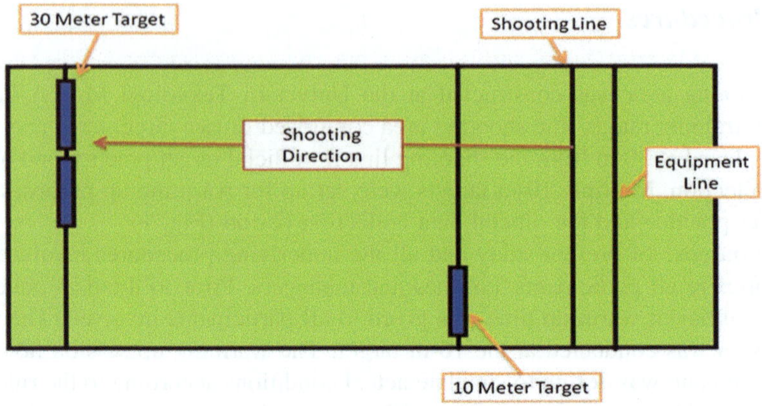

Fig. 1 Shooting area for data collection

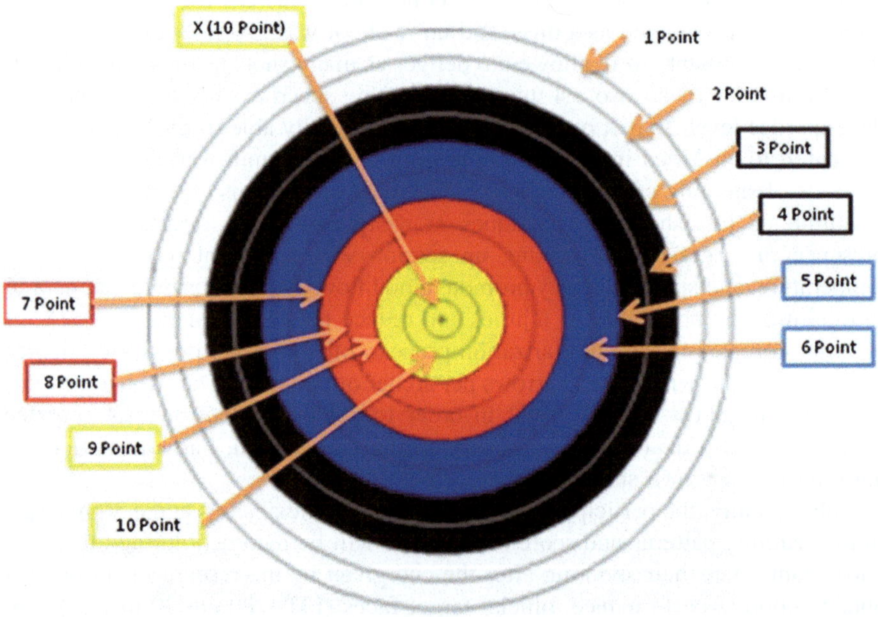

Fig. 2 Target face diagram (score ring)

equipment would not reflect participant's own shooting style and shooting performance would be affected due to differences in ergonomics and equipments characteristics.

All participants were equipped with Zephyr Bio-Harness devices, which were worn on the xyphoid process under the sternum region. The device belt or garment with transponder attached on it was placed under the participant's shirt in direct contact with the skin. A pilot study carried out, illustrated that the devices do not interfere with the bowstring pathway and do not affect participants original shooting characteristics. Hence, no modification to both equipment and technique was needed. All the participants used straight stance technique in order to minimize variations that would affect participant's stability. Whenever the participant was in the stance phase, they were given a "start" cue and the data were start being collected by recording of the live data transmitted by the device to the computer. The transmitting of the real-time data was stopped after the participants finished the release and follow-through phases of each end. The recording for a single end took about 3–4 min with roughly 10 s for each arrow intermittent with rest between shots. The present study utilized individual observations while at the same time the performances were digitally recorded.

2.4 Statistical Analysis

Descriptive statistics (Mean ± standard deviation) was used to describe the shooting performance and breathing pattern data. Regression analysis was used to seek relationships and contribution of each respective phase towards shooting performance for both groups. Significant level was set at $p < 0.05$.

3 Results

3.1 Shooting Performance Result

Shooting performances were determined based on shooting scores of the two-end shots by each participant. The shooting scores ranged from zero to bulls eye (X, or ten points, the highest mark). Zero point was awarded for arrows that missed the target mattress or target face during data collection, while the highest score was given to the participant that hits the bull's eye. The score distribution between skilled levels is as illustrated in Fig. 3.

The mean score for the skilled group was 7.51 ± 2.29 points. In this group, the highest score was seven points (27 %) followed by eight point (25 %). Four percent or a total of three arrows that missed the target, contributed to the lowest score in this group. The lowest percentage score obtained by the skilled participant group was one point (0 %). This showed that the score was positively skewed, which indicated that most of them shot a good score. As for the unskilled group, the mean score was 3.93 ± 3.54. The highest noted score was zero for missing the target completely or missing the target face (37 %) followed by eight points (14 %). The highest score was 10 points with only 1 % or one arrow. Although there were some arrows that hit high marks such as eight to ten, however most of them missed the target.

Fig. 3 Score distribution for skilled group

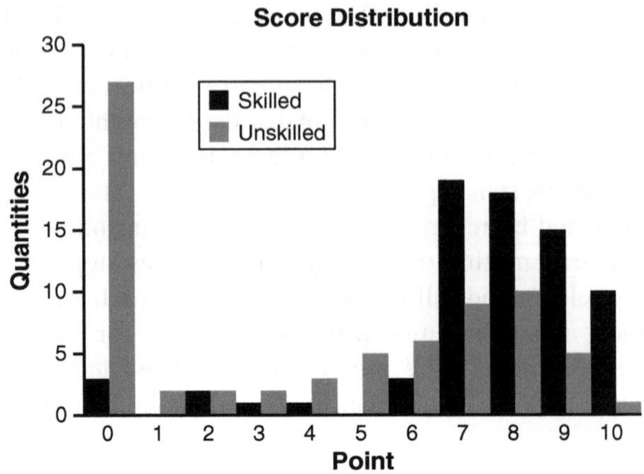

3.2 Breathing Pattern

The breathing pattern means across shooting phases of skilled and unskilled group archers are presented in Table 1. The figures simulate the volume of air inside the lung cavity measured by the expansion and deflation of the rib cage. Smaller figures indicate that the air inside the lung was fewer compared to the larger figures, which indicate the opposite condition. The smaller figures indicate lung-cavity deflation process, while larger figures indicate an expansion process.

The mean (±SD) value for breathing pattern during set up (1,538.93 ± 375.957) was higher compared to during aiming (1,470.41 ± 415.328) and release (1,273.81 ± 388.286) phase as showed in Table 1. However, for the unskilled, the (Mean ± SD) value were slightly higher for breathing pattern during set up, which showed a figure of (1,668.41 ± 364.316), as compared to aiming (1,567.21 ± 353.902) and release (1,587.96 ± 287.517) phases. The results suggest that for skilled archers, the air inside the lungs were decrease throughout the shooting process. It also acts as an indicator that the archers were exhaling throughout the shooting process. However, for unskilled archers, it showed a decrement from setup to aiming, but was slightly higher towards the release phase, which indicates inconsistencies in breathing. Figure 4 further illustrates the process.

3.3 Relationship Between Breathing Pattern with Shooting Performance

The correlations between breathing patterns and shooting performances of Universiti Teknologi MARA Pahang skilled and unskilled archers are presented in Table 2.

Table 1 Breathing pattern value for skilled and unskilled group (Mean ± SD)

Phase	Breathing pattern value (Mean ± SD) [skilled]	Breathing pattern value (Mean ± SD) [unskilled]
Phase 1 (set up)	1538.93 ± 375.957	1668.42 ± 364.316
Phase 2 (aiming)	1470.40 ± 415.328	1567.21 ± 353.902
Phase 3 (release)	1273.81 ± 388.286	1587.96 ± 287.517

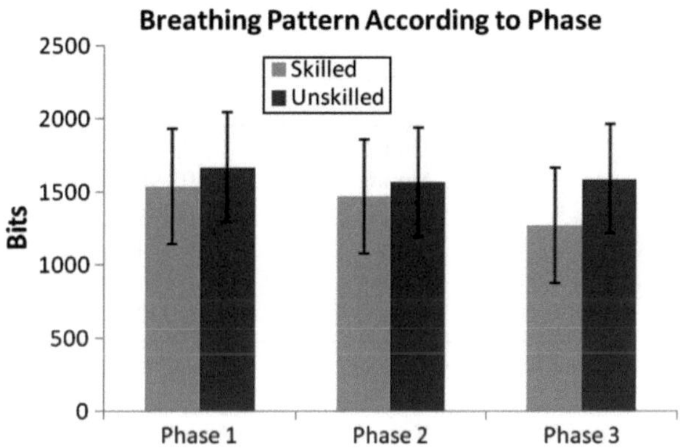

Fig. 4 Breathing pattern across shooting phases for skilled and unskilled group

Regression analysis indicated that a significant relationship occurred between breathing pattern and shooting performance for both parties. Skilled archers recorded a lower value ($r^2 = 0.118$, $p < 0.05$) compared to the unskilled ($r^2 = 0.201$, $p < 0.05$). From Table 2, the relationship between breathing pattern during the release phase was only significant for the skilled group, and the results indicated a positive relationship. This indicated that by increasing the air capacity simultaneously increases the shooting performance of the athletes. As for the unskilled group, the setup, aiming and release phases did show a significant relationship between breathing pattern and shooting performance.

According to Table 3, breathing during the release phase was the sole significant and highest contributor for the skilled group performance with a standardized coefficient value of 0.294, which indicates a partial correlation of 29.4 % towards the overall model. As for the unskilled group, the largest contributor towards the overall model was breathing pattern during the setup phase with a coefficient value of −0.444, which contributed 44.4 % to the total model. Breathing during release phase was noted to contribute 24.3 % to the total model with a coefficients value of −0.243. Breathing pattern from setup and release phases significantly contributed towards the unskilled group shooting performance: however, it showed a negative relationship. This indicates that increasing the amount of air inside the lung cavity will simultaneously give an inverse effect towards unskilled group shooting performance.

Table 2 Correlations between breathing pattern with shooting performances for skilled and unskilled group

Variables	Pearson correlation		
	Phase 1	Phase 2	Phase 3
Breathing pattern (skilled)	−0.157	−0.168	0.233*
Breathing pattern (unskilled)	−0.363*	−0.208*	−0.286*

* Significant level ($p < 0.05$)

Table 3 Coefficients of breathing pattern toward shooting performances for skilled and unskilled group

Group	Phases	Standerdized coefficients	*t*	Sig.
Skilled	Setup	−0.125	−1.007	0.318
	Aiming	−0.196	−1.550	0.126
	Release	0.294	2.484	0.015*
Unskilled	Setup	−0.444	−2.637	0.010*
	Aiming	0.141	0.844	0.402
	Release	−0.243	−2.227	0.029*

*Significant level ($p < 0.05$)

4 Discussion

According to previous studies, it appears probable that the breathing rate was related to the heart rate. Increasing the breathing rate or inconsistencies in breathing pattern will usually affect the heart rate [15, 20]. This is usually due to force breathing action and anxiety level. Continuation of breathing after breathlessness condition also contributes to an increase heartbeat [21].

Although past researchers suggest that the heart rate was shown to have a relationship with breathing, however, this was not tested in the present study. The purpose of the present study was to discuss the ability of controlling the breathing and the pattern of breathing between skill levels. This study was also designed to show the relationships between breathing pattern with the shooting outcomes.

Breathing in the archers was observed to show a significant relationship with the shooting performance. However, the amount of breathing or air in the lung cavity considered as optimal in archery performance for certain phases was still vague. Further studies are needed to determine the optimal range of breathing capacity and patterns that matched the phases involved in archery performance and archer's skill levels.

It is highly recommended that archers begin to inhale during the set up phase in order to obtain an additional force for the drawing phase (the push-pull action) [1, 9, 10]. Inhaling during the set up process helps in preparing the body rigidness for the shooting technique to be performed later. A rigid yet flexible body ensures a good shooting technique foundation. Besides, it also minimizes the chances for instability caused by muscular contraction as the same muscles are involved in maintaining core stability as well as breathing process [14, 15].

In continuing with the aiming phase, archers are either exhaling or holding their breath [1, 9]. This is to reduce the level of body rigidity while simultaneously preparing the body for the release phase where flexibility would be the main aspect [13]. It also enables the expansion of shoulder width. Although the expansion is not significantly visible, it will help in the execution of the release and follow through phases later.

The present condition was also in line with the intake-rejection hypothesis as proposed [2], which explains that when a person-focused attention were diverted to external factors known as the intake process, in this case towards the aiming phase, heart rate will be reduced. This is the state of vigilance, an attentive state of focus towards external stimulus events. The breathing rate also will be reduced and be under control [2, 17]. Increasing in breathing rate or inconsistencies in breathing pattern was also associated with the rejection process or internally channelled focus. Internal focus resulted in increasing of heart rate due to the personal focus being channelled more on how to actually perform the movement without flaws rather than make it autonomous. It also decreases the control of breathing rate and pattern of the particular person, which eventually increased postural movements due to muscle use and increasing of the ventricular depolarization phase, although minute. Other researchers have also suggested similar outcomes, where an increase in breathing rate or improper breathing pattern was observed to increase the heart rate correspondingly [1, 15, 16, 20].

During the release phase, the skilled group displayed a pattern of exhaling which was evident from the decreasing value of recorded data. However, with the unskilled group, it showed increasing figures during the release phase, which indicated the presence of the inhaling process, and simultaneously indicating inadequate control of breathing. The unskilled tend to hold their breath longer during the aiming phase, thus inducing the state of breathlessness, which force them to inhale during the later phase. The forced breathing process often causes minute jerking due to muscular contraction and an increase of ventricular depolarization [14, 15, 21].

Thus, it was likely that the skilled participants showed a significant increase in attentional demand in order to obtain a perfect shooting technique, which decreased undesired mechanical jerking. By minimizing all unwanted movements, shooting accuracy increased by ensuring stable aiming and uninterrupted arrow flight trajectory [9]. Increasing the attentional demand level enabled the archers to concentrate on the stability of the release phase. This also suggest that the skilled athletes were more focused on the target, rather than the movement of their own self. It also indicates that the movement executed by skilled archers were more autonomous, thus requiring less attention by the performer compared to the unskilled archers. The unskilled were noted to have irregularities in breathing pattern. This likely indicates that most of the focus was directed internally rather than towards the target itself.

An internally focused or rejection phase as suggested by the researcher [2], was usually carried out by unskilled athletes. They were more prone to this as they try to perform the skill as best as possible, thus making them control every movement of their body. The feeling of inferiority and surrounding conditions also cause them to increase their anxiety level. Without proper control of the anxiety can cause them to start feeling fear, thus their heart rate will increase and breathing pattern starts to be inconsistence, which decreases their aiming stability and disable them to attain good shooting outcome [2, 16, 20].

This was parallel with previous research [2], which indicated that the heart rate of skilled golfers' decreased during the last phase of putting. This indicates an increase in attention with decreasing heart rate. The attentional demand was channelled externally towards the target rather than focusing on how to perform the putting itself. The present study also showed that the unskilled participants were inhaling during the aiming to release phases causing the body to be rigid, thus compromising the aiming and release phase continuum and stability. In precision aiming task sport, decreased stability of the performer, result in a decrease in the aiming ability, and thus increases shooting inconsistencies.

5 Conclusion

In conclusion, the breathing pattern does impose an effect towards shooting performance. Irregular breathing patterns especially during the release phase partially affect the outcome of shooting in both skilled and unskilled groups. However, the intensity of the effect due to breathing pattern towards the shooting performance and the optimal level of breathing needed in particular phases is yet to be determined. Specialized training programs may be implemented to enhance breathing control and at the same time obtaining optimum breathing capacity and thus improving aiming and shooting stability.

References

1. Miyamoto, M. (1994). Changes of heart rate, somatic anxiety, and performance of Japanese archers during practice and matches. *The Japanese Journal of Experimental Social Psychology, 33*(3), 191–200.
2. Neumann, D. L., & Thomas, P. R. (2009). The relationship between skill level and patterns in cardiac and respiratory activity during golf putting. *International Journal of Psychophysiology, 72*, 276–282.
3. Fenici, R., Ruggieri, M. P., Brisinda, D., & Fenici, P. (1999). Cardiovascular adaptation during action pistol shooting. *Journal of Sports Medicine and Physical Fitness, 39*(3), 259–266.
4. Keast, D., & Elliot, B. (1990). Fine body movements and the cardiac cycle in archery. *Journal of Sport Sciences, 8*, 203–213.

5. Mets, T., Konttinen, N., & Lyytinen, H. (2007). Shot placement within cardiac cycle in junior elite rifle shooters. *Psychology of Sport and Exercise, 8*, 169–177.
6. Konttinen, N., Mets, T., Lyytinen, H., & Paananen, M. (2003). Timing of triggering in relation to the cardiac cycle in nonelite rifle shooters. *Research Quarterly for Exercise and Sport, 74*(4), 395–400.
7. Chi, K. H., & Chien, F. W. (2008). Analysis of the desired aiming adjustment based on the ARX part of ARMAX model. *Journal of Science and Engineering, 6*(4), 29–36.
8. Kuo, B. L., & Chi, K. H. (2005). Analyses of the relationship between the aiming time and the shot points in archery. *Journal of Physical Education in Higher Education, 7*(4), 161–173.
9. McKinney, R. (1996). *The simple art of winning*. Tokyo: Leo Planning Inc.
10. Sorrells, B. J. (2004). *Beginner's guide to traditional archery*. Mechanicsburg: Stockpole Books.
11. Tsung, Mh., Wen, T. T., & Tzyy, Y. S. (2009). A case study of integrated sport sciences for an olympic archer. *Journal of Medical and Biological Engineering, 29*(4), 164–171.
12. Barrett, K., Brooks, H., Boitano, S., & Barman, S. (2010). *Ganong's review of medical physiology*. United States of America: McGraw Hill.
13. McConnell, A. (2011). *Breathe strong perform better*. United States of America: Human Kinetics.
14. Netter, F. H. (2011). *Atlas of human anatomy* (5th ed.). Philadelphia: Saunders Elseviers.
15. Taylor, A. (1960). The contribution of the intercostal muscles to the effort of respiration in man. *Journal of Physiology, 151*, 390–402.
16. Robazza, C., Bortoli, L., & Nougier, V. (1999). Emotions, heart rate and performance in archery. *Journal of Sports Medicine and Physical Fitness, 39*(2), 169–176.
17. Tremayne, P., & Barry, R. J. (2001). Elite pistol shooters: Physiological patterning of best versus worst shots. *International Journal of Psychophysiology, 41*, 19–29.
18. Lacey, J. I. (1967). Somatic response patterning and stress: some revisions of activation theory. In M. H. Appley, & R. Trumball, Psychological stress: isues in research (pp. 14–42). New York: Appleton-Century-Crofts.
19. Era, P., Konttinen, N., Mehto, P., Saarela, P., & Lyytinen, H. (1996). Postural stability and skilled performance—A study on top level and naive rifle shooters. *Journal of Biomechanics, 29*(3), 301–306.
20. Shea, S. A. (1996). Behavioral and arousal-related influences on breathing in humans. *Experimental Physiology, 81*, 1–26.
21. Filho, L. G., Dajani, R. H., Leung, S. R., Floras, S. J., & Bradley, D. T. (1999). Entrainment of blood pressure and heart rate oscillations by periodic breathing. *American Journal of Respiratory and Critical Care Medicine, 159*, 1147–1154.

A Comparison of Periodization Models on Muscular Strength

Dina Asmadi Mansor, Zulkifli Abdul Kadir and Raja Firhad Raja Azidin

Abstract Long-term strength training program are commonly constructed based on periodization models. The purpose of this study was to determine the more superior periodization model between; linear periodization (LP, $n = 20$), daily undulating periodization (DUP, $n = 19$), and weekly undulating periodization (WUP, $n = 19$) in improving muscular strength. Fifty-eight recreationally trained college-aged male and female volunteers (18–25 years old) with a minimum of 6 months strength training experience were randomly assigned into three experimental groups. Pre- and posttest measures used the bench press to record upper body strength and leg press for lower body strength. A 9-week exercise program was implemented 3 days per week. Three sets training for each exercise was prescribed. Training loads at 90 % intensity of four repetitions was determined as heavy, 85 % intensity of six repetitions for medium, and 80 % of eight repetitions as light. All exercises were performed to the point of achieving momentary muscular fatigue. Training protocols subscribed to the guidelines by the National Strength and Conditioning Association of America (NSCA). All groups showed significant improvements ($p < 0.05$) in the bench press test; 24.31 % (LP), 38.23 % (DUP), and 37.46 % (WUP), and in the leg press test; 34.3 % for LP, DUP at 37 %, and WUP at 28.71 %. Although the DUP showed the highest strength increase in both the tests, statistical analysis, however, showed no significant differences existed between the groups ($p > 0.05$). In conclusion, all the periodization models were effective to elicit higher strength gain with the DUP showing a slight superiority over LP and WUP.

Keywords Linear periodization · Daily undulating · Weekly undulating · Bench press · Leg press · Strength training

D. A. Mansor (✉) · Z. Abdul Kadir · R. F. Raja Azidin
Faculty of Sports Science and Recreation, Universiti Teknologi MARA, Shah Alam, Malaysia
e-mail: dina.asmadi@gmail.com

R. Adnan et al. (eds.), *Proceedings of the International Colloquium on Sports Science, Exercise, Engineering and Technology 2014 (ICoSSEET 2014)*,
DOI: 10.1007/978-981-287-107-7_35,

1 Introduction

Approach in resistance training (RT) for high sport performance and fitness is very broad and varied, however, as knowledge and experience is gained in it, RT program is now advocated in a structured periodized manner and approach.

Designing strength training programs mainly aim to utilize and improve in many aspects especially in multiple and variety of fitness components focus on cardiovascular fitness, muscle strength, muscle mass, and body composition [1].

Traditional strength training programs will require a participant or athlete to lift or carry heavy loads (weight) with moderate interest recovery periods condition by using free weights equipments such as dumbbell, barbell and cable or resistive strength machines. Specifically, this increase in muscle mass and strength after one training period of close monitoring is very important, and the progression level is the output or result for improving in various aspects of athletic performance for competition and healthy lifestyle [2].

RT programs are widely regarded as an effective advance supplemental tool and program in the modern athlete's preparation and training for any sports and competition in the world [3].

The previous reliable research has enlarged, support, and discuss the advantages of performing strength training programs. The main goals are improving health status, enhance physical fitness level, and prolonged life span [4]. In brief, among the benefits to an individuals or athlete's performing strength training programs are improve in muscular strength (carry or lift loads), improve in muscular endurance (durations, sets and repetitions), increase power (explosive), and reduces fat free mass (fat percentage). These positive improvements and significant changes in physical fitness are achievable by planning and create more variations of exercise in program design or exercise prescription layout of training programs. Currently, no consensus exists as to the ideal strength training program design, but it typically involves a planning and monitoring resistance training regimen exercises, and it is widely advocated and accepted that some form of periodization is the most effective. To improve physical fitness components and to enhance sport performance in one field, resistance training has been implementing extensively [5].

The main objective and primary goals of periodization model is to reduce the potential to face with overtraining syndrome and reaching the peak level performance [6].

The aims of periodization programs are not only to maximize the principle of the overload but also it is also to understand and allowing a better relation between the two components of stress and recovery. Strength training periodization is a significant method and also a reliable concept of training program which can be design systematically and the exercise program can be prescribe suitable for recreationally trained group strength practitioners [4].

Another goal of periodization model is to prepare the individual or athletes a great maintenance program specific to the sports and to fulfill the high demand of competitive tournament especially for sports with a specific season. In additional, prescribing, and appropriately selecting exercises based on the nature of the sports and mimic to the sports specific will carry an advantage in the designed training program. There are three major distinct stages in a periodized training program or three different phases, which require different intensity and volume at each phase [6].

It becomes popular among coaches and researchers and it has been prove that most of the effectiveness strength training program conducted applied the periodization model method [6]. Monteiro et al. also agreed with the effectiveness of periodization system, whereby the concept of changes of volume and intensity in different phases over the time. The research above found that periodized training program with systematic and periodic alterations showed an improvement in optimizing physiological strain and greater in muscle strength improvement compare to a non-periodized training program which implement the constant load training paradigm. Great pattern of manipulation with ideal period of time to changes or switch with the correct sequencing of volume (repetitions per exercise, frequency of training in a week, microcycle, mesocycle, and macrocycle) and the total amount of intensity (loads) will determine and answer the result of the training program [6].

Studies have compared periodized and non-periodized training programs, and this research has been reviewed and showed that periodized program is superior to non-periodized program [7]. However, fewer studies have directly compared linear periodization (LP) model with (DUP) model and (WUP) model. Currently, there is debate as to which form of periodization either LP, DUP or WUP models is superior. One study compared LP model with DUP model over 12 weeks, which they found that DUP model elicited a greater percentage of strength gains [8]. Another study compared LP model with WUP model over 12 weeks, which resulted LP model was superior compared to WUP model over same duration of time, which is 12 weeks [9]. One study advocate 15 weeks of experimental design study period, by conducted and compared between non-periodized training, LP model, and undulating periodization intervention programs. That study was one of the first to use a training frequency of four times in a week rather than commonly used three times in a week. In brief, that study also used highly trained college football athletes as participants and result shown that, there is no significant difference in assessment of strength value in strength gains among this specific participants [10].

The main method for LP model is increasing the intensity (load) continuously or progressively while at the same time gradually decrease the training volume (repetitions) within and between the phase as the training progress [4]. DUP model consists of an increasing and decreasing the intensity (loads) and volume (repetitions) in the same week, whereby the alteration occurring quite frequent depends on the frequency of training in a week [4].

In WUP model, the strength training program will change from one week to another week or will alter by weekly basis. Usually these undulating periodization may place considerable stress on the neuromuscular system because of the rapid and continuous change in program available [9].

Ultimately, the purpose of this study was to determine which periodization is most superior in eliciting strength gains and to compare the training adaptations for nine weeks intervention of strength training LP, DUP, and WUP models.

2 Methodology

2.1 Research Design

The research design was pre- and post-experimental design that involves three groups with one set of sample data. The study was concerned with nine weeks intervention to develop cause-and-effect. This design was chosen, and the participants went through to their each intervention, which were LP, DUP, and WUP models. The participants went through a pretest before start the intervention and posttest after the intervention. The pretest on the participants worked as a baseline and to see whether there were any changes after following their each treatment.

The independent variables in this study were the training methods, which were resistance training using LP, DUP, and WUP models. Frequency of the exercise program was consistent for all three methods. All methods were conducted three times a week for 9 weeks continuously, which total of 27 sessions. Resistance training program was conducted at conditioning lab in which upper and lower body exercise were synchronized as closely as possible.

The sequences of exercises was maintained throughout the study, with the first (e.g., Monday) training sessions of each week focusing on bench press, leg press, seated row, lunges, preachers curls, and sit ups, and the second (e.g., Wednesday) training sessions focusing on bench press, leg press, lat pulls, leg extension, standing claves, and back extension, and third (e.g., Friday) training sessions focusing on bench press, leg press, upright rows, leg curls, triceps extension, and knee raises.

LP model trained 80 % (8 repetitions) from week 1 until 3, 85 % (6 repetitions) from week 4 until 6 and 90 % (4 repetitions) from week 7 until 9. DUP model trained 80 % (8 repetitions), 85 % (6 repetitions), and 90 % (4 repetitions) by alternated training session and WUP model trained 80 % (8 repetitions), 85 % (6 repetitions), and 90 % (4 repetitions) by alternated weekly basis. Participants performed warm up for seven minutes every time before starting the routine. Each exercise was performed with three sets and one minute rest interval in between.

2.2 Sample

Participants were referred to the strength conditioning students in Faculty Sports Sciences and Recreation, Shah Alam campus, Universiti Teknologi MARA, ranging in age from 18 to 25 years based on LP (Mean = 22.3, SD = 1.08),

DUP (Mean = 22.11, SD = 0.99), and WUP (Mean = 21.89, SD = 0.81). All participants have experiences more than 6 months in resistance training. Participants have no pending medical problems that could be affected by progressive physical testing and training, no pending medical problems that could influence testing outcomes, no ankle, knee, or back pathology within the preceding 9 weeks, sign statements, indicating they had no current or past use of anabolic steroids, have not ingested or currently be ingesting creatine, thermogenics, or other nutritional supplements (excluding multi vitamins), agree to follow the periodization resistance exercise program, criteria with and no medical contra indications to resistance exercise were selected to participate in the study. Participants were given and signed an informed consent letter.

2.3 Sampling Technique

The sampling technique used was purposive sampling in which participants were selected based on more than 6 months experience in resistance training. The gender was not fixed. Participants were grouped into three groups. All participants were required to undergo LP, DUP, and WUP resistance training program. There was no control group as this study was concerned with comparing the effects of the three types of training methods in all participants involved, which were LP, DUP, and WUP resistance training models.

After identifying the target participants (i.e., experience more than 6 months in resistance training), participants who agreed to the terms were assessed for eligibility. At pre-screening test session, participants filled out questionnaires evaluating their prior strength training experience, physical activity readiness questionnaire (PAR-Q) and were also asked to sign an informed consent form. Participant's recruitment and allocation are illustrated in Fig. 1.

2.4 Instrumentation

The instruments (including facilities) used in this study were the Weight Scale, Height Scale, Blood Pressure, Bench Press, and Leg Press. Participants were measured before the intervention and after the intervention. The procedure for conducting the test was in accordance with that described in American College of Sports Medicine (ACSM).

For upper body and lower body test protocol, the participant first should practice the bench press and leg press with a light warm up of 5–10 repetitions at 40–60 % of perceived maximum. After a 1 min rest, the participant does three to five repetitions at 60–80 % of perceived maximum. The participant should be close to the perceived maximum; a small amount of weight is added and 1-RM is attempted.

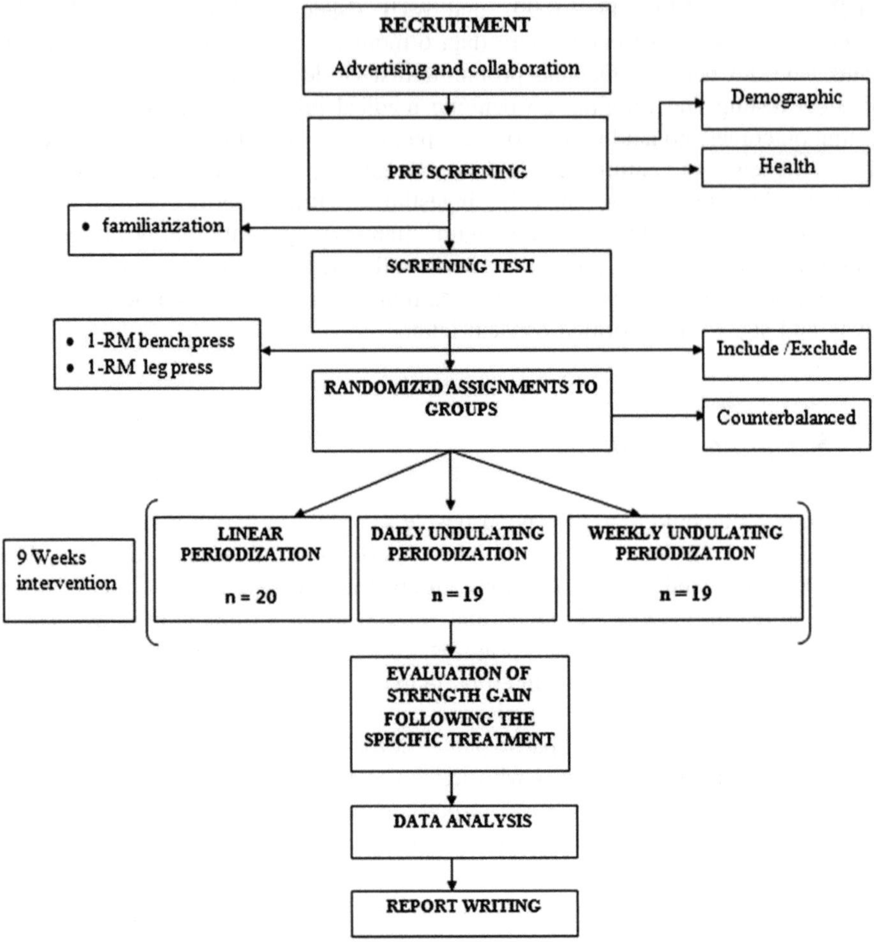

Fig. 1 Research design: LP, DUP and WUP models

If the lift is successful, a 3–5 min rest period is provided, and the test is continued until a failed lift occurs. The 1-RM should be found in three to five efforts.

2.5 Data Collection Procedures

The participants went through a pretest to determine the baseline of their strength level. The study began with a briefing of the participants on the objective, purpose, and significance of the study, the procedures, what the study requires them to do, and requirement as participants while the participants kept as double blinded. Data collection procedure was as outlined in Fig. 2.

2.6 Data Analysis

The statistical evaluation of the data will use paired sample t test to evaluate any changes between pre- and posttest.

3 Results

Results showed there is no significant different between age, weight, height, and BMI (Table 1).

The mean body weight (SD) for LP model was 60.09 ± 10.98 kg ($n = 20$), DUP model was 63.91 ± 13.03 kg ($n = 19$), and WUP model was 59.08 ± 13.37 kg ($n = 19$). Statistic showed that there is no significant different between three periodization models for body weight, $P > 0.05$.

The mean body height (SD) for LP model was 162.98 ± 7.40 cm ($n = 20$), DUP model was 163.32 ± 7.71 cm ($n = 19$), and WUP model was 161.71 ± 8.82 cm ($n = 19$). Statistic showed that there is no significant different between three periodization models for body height, $P > 0.05$.

The mean BMI (SD) for LP model was 22.51 ± 3.01 cm ($n = 20$), DUP model was 23.9 ± 4.39 cm ($n = 19$), and WUP model was 22.42 ± 3.85 ($n = 19$). Statistic showed that there is no significant different between three periodization models for BMI, $P > 0.05$.

3.1 Upper, Lower, and Total Body Strength Improvement for LP Model

Result of the paired sample t test from LP model bench press, leg press, and total body showed there is an improvement in mean score of strength level from pre-test compared to posttest. Mean score of the bench press post ($M = 49.7$ kg) was higher compared to mean score of the bench press pre ($M = 41.5$ kg) and mean score of the leg press post ($M = 257.25$) was higher than mean score of the leg press pre ($M = 170.75$). The mean score of the total post-LP model ($M = 306.95$) was higher compared to the mean score of the total pre-LP model ($M = 212.25$) in term of the strength level.

From the data, the posttest of LP model increase 8.2 kg (24.3 %) for bench press, 86.5 kg (34.3 %) for leg press, and 94.7 kg (33.1 %) for total body compared to the pretest of LP model. The result of t test yielded at t value of 10.21 which was statistically significant at $P < 0.05$ (Table 2).

Table 1 Participants demographic data

Variables	Independent variables	*N*	Mean	Standard deviation	Anova, *F*	*P*
Age	LP	20	22.30	1.08	0.85	0.43
	DUP	19	22.11	0.99		
	WUP	19	21.89	0.81		
Weight	LP	20	60.09	10.98	0.79	0.46
	DUP	19	63.91	13.03		
	WUP	19	59.08	13.37		
Height	LP	20	162.98	7.40	0.22	0.81
	DUP	19	163.32	7.71		
	WUP	19	161.71	8.82		
BMI	LP	20	22.51	3.01	0.92	0.40
	DUP	19	23.90	4.39		
	WUP	19	22.42	3.85		

$P > 0.05$

3.2 Upper, Lower and Total Body Strength Improvement for DUP Model

Result of the paired sample *t* test from DUP bench press, leg press, and total body showed there is an improvement in mean score of strength level from pretest compared to posttest. Mean score of the bench press post ($M = 57.11$ kg) was higher compared to mean score of the bench press pre ($M = 44.21$ kg), and mean score of the leg press post ($M = 280.79$) was higher than mean score of the leg press pre ($M = 187.68$). The mean score of the total post-DUP model ($M = 337.89$) was higher compared to the mean score of the total pre-DUP model ($M = 231.89$) in term of the strength level.

From the data, the posttest of DUP model increase 12.9 kg (38.23 %) for bench press, 93.11 kg (37 %) for leg press, and 106 kg (37.1 %) for total body compared to the pretest of DUP model. The result of t test yielded at *t* value of 8.81 which was statistically significant at $P < 0.05$ (Table 3).

3.3 Upper, Lower and Total Body Strength Improvement for WUP Model

Result of the paired sample t test from WUP bench press, leg press, and total body showed there is an improvement in mean score of strength level from pretest compared to posttest. Mean score of the bench press post ($M = 52.11$ kg) was higher compared to mean score of the bench press pre- ($M = 39.47$ kg), and mean score of the leg press post ($M = 248.68$) was higher than mean score of the leg press pre ($M = 176.32$). The mean score of the total post-WUP model

Table 2 Paired sample *t* test of strength level on upper, lower, and total body for LP model

Linear periodization (LP)	*N*	Mean	Standard deviation	*t*	*P*
Bench press pre	20	41.50	I8.99	−12.01	0.001*
Bench press post	20	49.70	19.14		
Plate loaded pre	20	170.75	75.61	−951	0.001**
Plate loaded post	20	257.25	99.01		
Total pre	20	212.25	92.03	−10.21	0.001**
Total post	20	306.95	115.41		

**Significant $P < 0.05$

Table 3 Paired sample *t* test of strength level on upper, lower, and total body for DUP model

Daily undulating periodization (DUP)	*N*	Mean	Standard deviation	*t*	*P*
Bench press pre	19	44.21	21.16	−6.67	0.001**
Bench press post	19	57.11	25.86		
Plate loaded pre	19	187.68	68.72	−8.39	0.001**
Plate loaded post	19	280.79	101.77		
Total pre	19	231.39	87.11	−8.81	0.001**
Total post	19	337.39	123.64		

**Significant $P < 0.05$

Table 4 Paired sample t test of strength level on upper. lower and total body for WUP model

Weekly undulating periodization (WUP)	*N*	Mean	Standard deviation	*t*	*P*
Bench press pre	19	39.47	19.00	−6.31	0.001**
Bench press post	19	52.11	23.94		
Plate loaded pre	19	176.32	89.39	−7.76	0.001**
Plate loaded post	19	248.68	98.32		
Total pre	19	215.79	106.63	−9.22	0.001**
Total post	19	300.79	119.40		

**Significant $P < 0.05$

($M = 300.79$) was higher compared to the mean score of the total pre-WUP model ($M = 215.79$) in term of the strength level.

From the data, the posttest of WUP model increase 12.64 kg (37.46 %) for bench press, 72.36 kg (28.71 %) for leg press, and 85 kg (29.8 %) for total body compared to the pretest of WUP model. The result of t test yielded at *t* value of 9.22 which was statistically significant at $P < 0.05$ (Table 4).

4 Discussion

The study was designed to identify the effects of pre- and post-intervention for LP, DUP, and WUP models on muscular strength.

4.1 Strength Gain for LP Model Between Pre- and Posttest

Based from the findings, it shows that the mean for strength gain in posttest is improved compared to the pretest after the 9 weeks of intervention. These results were accordance with the previous study where there is an improvement after training three times per week for 12 weeks intervention of strength training program among women aged between 20 and 35 years old with more than 6 months training experience [4].

Another study found that there is an improvement using LP model in total of 40 participants among men aged around 21 years old with more than 1-year experience in 12 weeks duration at four training session per week in 50 min [11]. However, result in that study found that DUP model is superior compared with LP model and no significant different was found between groups. That study reported similarity to the study done by the researcher. This might be due to the frequent changes of the volume and intensity in DUP model.

This is an important finding because LP model using progressively increase loads has previously been shown to promote and elicit strength higher than DUP model [12]. These results are also important because of its contrary with this study. Same study by Buford et al. which compare LP, DUP, and WUP model is contrary with the current study where previous study found that LP model is superior to DUP model after train three times per week for 9 weeks intervention of resistance training. They found no significant different between groups in total of 28 participants among men and women aged around 22 years old. However, results showed that all periodization models are workable improve and there is reduction in body fat.

One study applies LP model as a treatment for a total of 49 men participants between college age and middle age group. Result found that both group improve in strength, reduce in DXA lean mass, and no changes in 30 s Wingate test. In addition, result showed that middle age group is superior to college age group using LP model as a training program [13].

In this study, researcher found that even DUP model is superior to LP and WUP model for both upper and lower body, however, result showed that LP model is superior to WUP model for lower body strength. Based on the array of similarities and differences between the undulating periodized strength programs, caution must be used when comparing the results from these various studies.

4.2 Strength Gain for DUP Model Between Pre- and Posttest

Data analyzed shows that the mean for strength gain in posttest are improve compared to the pretest after the 9 weeks intervention of strength training program. Past study reported significant improvement in both LP and DUP models in 20 men participants aged 21 years old. In that study, participants train three times per week for 12 weeks programs in 40 min per training session. Even no significant for body composition in both groups, however, result showed that DUP model is superior to LP model [8]. The finding was consistent with the current study in term of strength gain.

Monteiro et al. found that, 27 men participants with strength training background and randomized into DUP model, LP model, and Non-Periodized (NP) which undergo the 12 weeks intervention of resistance training program resulted DUP model is superior to LP and NP models. This finding is similar to the current study by the researcher, which found that DUP model is superior to the LP model.

One they compare between DUP model and standard training control (LP) among 14 men participants from firefighters trainee background. Beside measured upper and lower body strength, they also measured lower body power output, speed, power, and anthropometric. The result showed that DUP model is superior compared to LP model after the 9 weeks strength training intervention at three sessions in a week. The major finding of the present research was that 9 weeks of resistance intervention program has been shown to elicit a strength gain in all three periodization model [14]. These important findings are accordance with current study by the researcher whereby DUP model is superior compare to LP model in term of strength.

4.3 Strength Gain for WUP Model Between Pre- and Posttest

Based from the findings, it shows that the mean for strength gain using WUP models in posttest are improve compared to the pretest after the 9 weeks of intervention. These results were accordance with the previous study where there is an improvement after training three times per week for 9 weeks intervention of strength training program among men and women aged around 22 years old [12].

Buford et al. found that, 28 men and women participants, which randomized into DUP model, LP model, and WUP model, which undergo the 9 weeks intervention of resistance training program, resulted WUP and LP model is superior to DUP models. This finding was contradicted with the study done by the researcher where the current study found that DUP model is superior to the WUP model.

Apel et al. found that LP and WUP models showed a significant increase in muscular strength (i.e., back squat, flat bench press, leg extension, lat pull down, and DB shoulder press) were noted from baseline to week 8 in both the LP and WUP models. Interestingly, the LP model showed a significant increase in muscular strength from week 8 to 12 for all exercises except leg extension. Moreover, the WUP model did not show a significant improvement in muscular strength at 8 until week 12 on any exercise over the same period. In summary, when pre-post comparison were conducted, it became evident that the LP model made many significant improvements from week 8–12, whereas the WUP model's improvements were marginal from week 8–12.

5 Conclusion

The findings of this study are as follows:

- There was a significant improvement in strength gain following LP model in more than 6 months resistance training experience participants after 9 weeks of intervention.
- There was a significant improvement in strength gain following DUP model in more than 6 months resistance training experience participants after 9 weeks of intervention.
- There was a significant improvement in strength gain following WUP model in more than 6 months resistance training experience participants after 9 weeks of intervention.

All periodization, LP, DUP, and WUP models with similar training frequency of 3 days per week for 9 weeks caused significant gains in the 1-RM bench press and leg press test. We conclude that a DUP program that adjusts to the day to day variation in performance capabilities would result in larger strength gains in previously strength trained individuals compared to a program employing the traditional LP method and WUP model.

References

1. Alcaraz, P. E., Gomez, J. P., Chavarrias, M., & Blazevich, A. J. (2011). Similarity in adaptations to high resistance circuit vs. traditional strength training in resistance-trained men. *Journal of Strength & Conditioning Research, 25*(9), 2519–2527.
2. Marx, J. O., Kraemer, W. J., Nindl, B. C., Gotshalk, L. A., Duncan, N. D., Volek, J. S., et al. (1998). The effect of periodization and volume of resistance training in women. *Journal Medicine & Science in Sports & Exercise, 30*(5), 164.
3. Murray, D. P., & Brown, L. E. (2006). Variable velocity training in the periodization model. *Journal of Strength & Conditioning Research, 28*(1), 88–92.
4. Prestes, J., Lima, C. D., Frollini, A. B., Donatto, F. F., & Conte, M. (2009). Comparison of linear and reverse linear periodization effects on maximal strength and body composition. *Journal of Strength & Conditioning Research, 23*(1), 266–274.
5. Monteiro, A. G., Aoki, M. S., Evangelista, A. L., Alveno, D. A., Monteiro, G. A., Picarro, Id C, et al. (2009). Nonlinear periodization maximizes strength gains in split resistance training routines. *Journal of Strength & Conditioning Research, 23*(4), 1321–1326.
6. Graham, J. (2002). Periodization research and an example application. *Journal of Strength & Conditioning Research, 24*(6), 62–70.
7. Rhea, M. R., & Alderman, B. L. (2004). A meta-analysis of periodized versus noneriodized strength and power training programs. *Research Quarterly for Exercise and Sport, 75*(413), 413–422.
8. Rhea, M. R., Ball, S. D., Philips, W. T., & Burkett, L. N. (2002). A comparison of linear and daily undulating periodized programs with equated volume and intensity for strength. *Journal of Strength & Conditioning Research, 16*, 250–255.

9. Apel, J. M., Lacey, R. M., & Kell, R. T. (2011). A comparison of traditional and weekly undulating periodized strength training programs with total volume and intensity equated. *Journal of Strength & Conditioning Research, 25*(3), 694–703.
10. Hartmann, H., Bob, A., Wirth, K., & Schmidtbleicher, D. (2009). Effects of different periodization models on rate of force development and power ability of the upper extremity. *Journal of Strength & Conditioning Research, 27*(3), 1921–1932.
11. Prestes, J., Frollini, A. B., Lima, C., Donatto, F. F., Foschini, D., Marqueti, R. C., et al. (2009). Comparison between linear and daily undulating periodized resistance training to increase strength. *Journal of Strength & Conditioning Research, 23*(9), 2437–2442.
12. Buford, T. W., Rossi, S. J., Smith, D. B., & Warren, A. J. (2007). A comparison of periodization models during nine weeks with equated volume and intensity for strength. *Journal of Strength & Conditioning Research, 21*(4), 1245–1250.
13. Kerksick, C. M., Wilborn, C. D., Campbell, B. I., Roberts, M. D., Rasmussen, C. J., Greenwood, M., et al. (2009). Early-phase adaptations to a split-body, linear periodization resistance training program in college-aged and middle-aged men. *Journal of Strength & Conditioning Research, 23*(3), 962–971.
14. Peterson, M. D., Dodd, D. J., Alvar, B. A., Rhea, M. R., & Favre, M. (2008). Undulating training for development of hierarchical fitness and improved firefighter job performance. *Journal of Strength & Conditioning Research, 22*(5), 1683–1695.

Monitoring of Rehabilitation Process via Gyro and Accelerometer Sensor

Safyzan Salim and M. Mahadi Abdul Jamil

Abstract This paper proposes a post-stroke rehabilitation system for hemiparetic arm based on the application of both gyro and accelerometer sensor. The study focuses on designing, developing, and simulating the results. The result is documented for the purpose of post-processing and progressive status tracking. The subject needs to wear a set of sensors over the wrist while performing a few basic arm movements. The data will be converted into series of readable data and saved into a microSD card and then carried out to a computer for analyzing the pattern. The experiment demonstrates the capabilities of the sensors to produce extended information regarding arm movement activities. It is believed that the system offers more information than conventional method and also the ability to improve training quality, results, and patients progress. For initial proof of concept, the system will be tested to a healthy normal subject.

Keywords Post-stroke · Hemiparetic arm · Rehabilitation · Accelerometer · Gyro · Monitoring

1 Introduction

Post-stroke patients in Malaysia always turn to traditional massage and medicine as the alternative treatment [1–3].

The easy access, relatively cheap, convenient, and less hassle compared to the one provided in the hospital makes them go for the traditional. The question is how

S. Salim (✉) · M. M. Abdul Jamil
Department of Electronic Engineering, Faculty of Electrical and Electronic Engineering, Modeling and Simulation Research Lab, UTHM, Batu Pahat, Malaysia
e-mail: he110275@siswa.uthm.edu.my; safyzan@bmi.unikl.edu.my

S. Salim
Communication Section Department, UniKL-BMI, Kuala Lumpur, Malaysia

R. Adnan et al. (eds.), *Proceedings of the International Colloquium on Sports Science, Exercise, Engineering and Technology 2014 (ICoSSEET 2014)*,
DOI: 10.1007/978-981-287-107-7_36,

to measure the success of the traditional treatment? Will there be any record or data logged for its progress? How about the potential side effects produced by the herbs that are yet to be scientifically proven? [4].

2 Literature Review

It is learned that studies have been made in rehabilitation of stroke patients using gyroscopes and accelerometer. They have discussed the better way to covey the data to the computer for post-processing, how to correspond effectively between the sensors and the microcontroller and so on.

A Web-based system for stroke patients has been developed that allow the result to be transmitted real time from the patient's home to their physicists over the Internet [5]. The chances of packet loss during the therapy may happen due to instability connection, resulting incorrect analysis by the therapists.

Another related contribution was done by Ambar et al. [6]. We notice that there are a few similarities between these systems with ours. However, it is not comfortable to wear since too many sensors to hook up and a tedious to wear.

A wearable sensor, which will be able to document the data during rehabilitation process, needs to be considered. The proposed system is secured to a hand glove. The wearable sensor consists of a low-cost single-board microcontroller, that is, to handle data flow from the sensor, a gyroscope sensor; measuring orientation and also microsecure digital memory card (microSD) for data logging. The data will be used as the evidence of the progress for rehabilitation process.

This paper reports the development of an affordable wearable device that is capable to capture the data produced during the post-stroke rehabilitation session for monitoring the progress of the patient.

3 Experimental Methods

The experimental methods have been composed of two exercises that are part of the rehabilitation treatment. The experiments are listed in Table 1. Each experiment needs to clock at least a minimum of 1,100 samples for the first experiment and 600 samples for raising hand activity. The reason behind this is that both samples are referred to the movement made by a healthy subject.

In this study, the exercises were done by healthy person. The measured person also tries to simulate post-stroke movements during the suggested exercises. In Fig. 1, the subject is in the motion of rising and stabilizing his hand.

Table 1 Selected exercises

Experiments	Execution of experiments
Palm movement	Able to move the palm clockwise and anticlockwise
Raising hand	Able to raise hand as high as possible

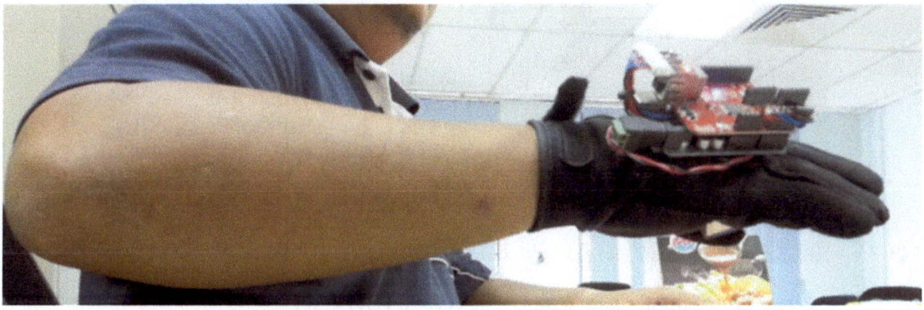

Fig. 1 Picture of a subject testing wearable data acquisition system (wDAQs)

4 Development of the Project

The elements of the system consist of two parts: the wearable sensor with integrated microSD card and accelerometer, and also a computer for data manipulation.

4.1 Arduino Uno

Arduino Uno is a microcontroller board based on the ATmega32. It has 14 digital input/output pins (of which 6 can be used as pulse width modulation outputs), 6 analog inputs, a 16-MHz ceramic resonator, a Universal Serial Bus (USB) connection, a power jack, an In-circuit Serial Programming (ICSP) header, and a reset button. It contains everything needed to support the microcontroller; simply connect it to a computer with a USB cable or power it with an AC-to-DC adapter or battery to get started.

4.2 Degrees of Freedom ITG3200/ADXL345

The details of accelerometer setup for this project are discussed in the previous research work [7]. The board comes with a full 6 degrees of freedom. The sensors communicate over I2C, and one INT output pin from each sensor is broken out.

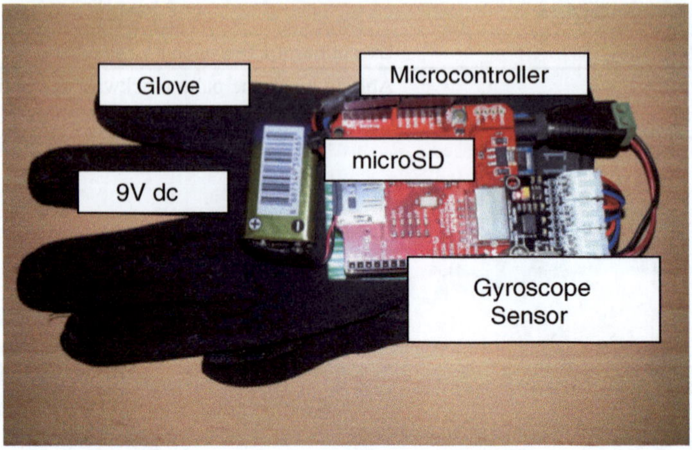

Fig. 2 The protoype wearable data acquisition system (wDAQs)

4.3 MicroSD Shield

Communication with microSD card is achieved over a serial port interface (SPI). The serial clock (SCK), digital input (DI), and digital output (DO) pins of the microSD socket are broken out to the ATmega168/328's standard SPI pins (digital 11–13), while the Chip Select (CS) pin is broken out to Arduino's D8 pin.

4.4 Prototype

The final prototype setup showed the device that was attached to a glove that contains a microcontroller with accelerometer connected. The data will be saved into the microSD and then transferred to a notebook for the results. The prototype can be seen in Fig. 2.

5 Results and Discussion

From the experimental works, the followings are the recorded measurement for the developed device that has been segregated into three parts.

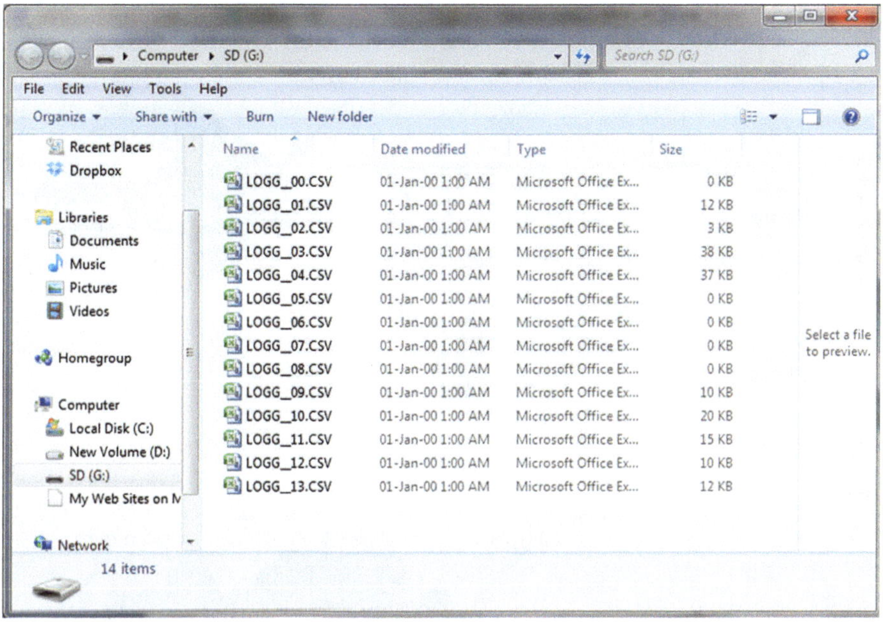

Fig. 3 Snapshot of files in the microSD

5.1 Managing the Files in the MicroSD

Figure 3 represents the files in the microSD. Do note that the file names are in order. It means that, every time the wDAQs is in use, it will not overwrite or amend the existing file; in fact, it will create a new file. The higher the number at the suffix, the new or latter the file will be. By having this system, the progress of every single treatment can be tracked and monitored easily.

5.2 Exercise 1: Palm Movement

During this modeling, the activities were resting right hand on top of a table, tilt to the right, then to the left, and finally back to idle position. Focus to Fig. 4a, it can be seen that the healthy subject is producing a smooth waveform for all axes that indicate ability to tilt the palm.

As for the simulated results for post-stroke patient in Fig. 4b, the pattern seemed distorted which indicated that the inability of subject to tilt neither to the right nor to the left.

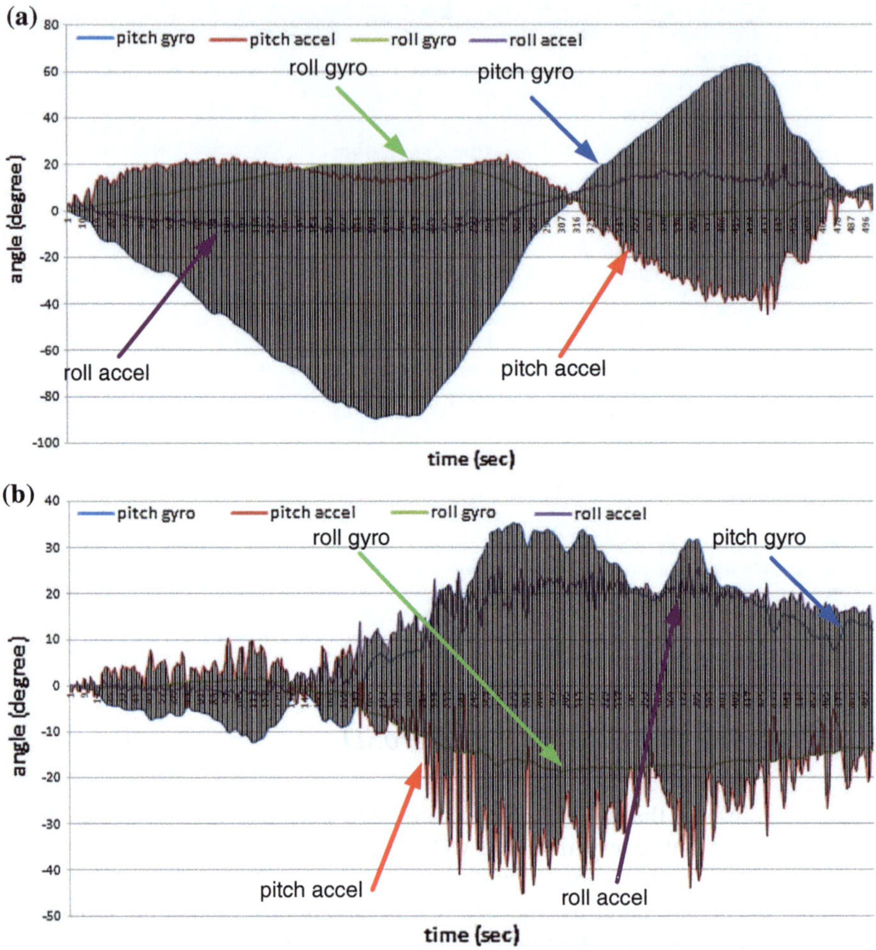

Fig. 4 Basic movement ability between **a** a healthy subject and **b** a simulated post-stroke subject

5.3 Exercise 2: Raising Hand

Figure 5 is the results for the second modeling, that is, raising the hand as high as possible. As for this experiment, only three segments involved; idle and resting the hand on top of a table, starts raising the hand, and halt at the highest point as possible as shown in Fig. 5a.

As expected, in Fig. 5b, it can be concluded that the post-stroke subject unable to complete the task. This is due to the inability of either moving or raising his hands.

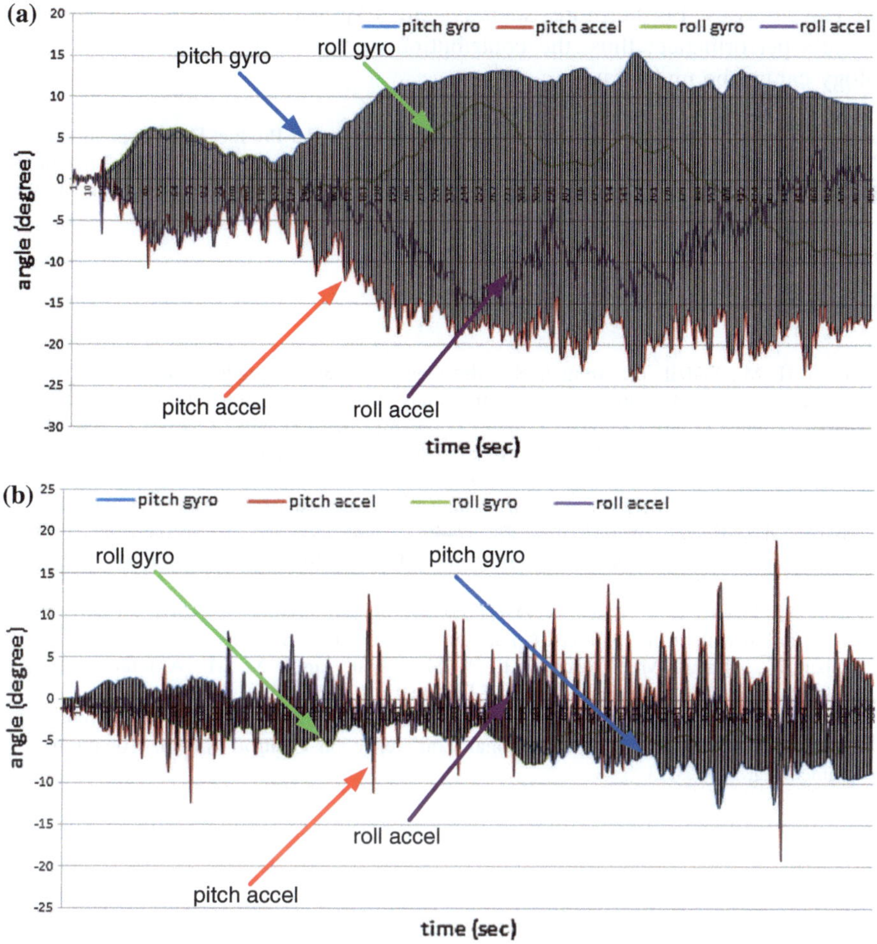

Fig. 5 Raising right-hand activity between **a** a healthy subject and **b** a simulated post-stroke subject

6 Conclusion

In this paper, we have presented and evaluated the potential of combining accelerometer and gyroscope sensor for rehabilitation process of post-stroke patients. Experimental result shows that the prototype system successfully documented the movement of one's hand in the microSD. In addition, the system is also able to save the data in different file names.

Based from the experiments, it can be seen that the application of this device can be extended to other area of physical rehabilitation such as physical movement and gait analysis. Furthermore, such system may also be used in evaluation of athlete's performance; thus, the contribution of this study toward Sports Technology cannot be understated.

Acknowledgments We acknowledge the financial support under Product Development Grant of UTHM (UTHM/PPI/600-5/1/14 Jld 3 (75)). The author would like to thank AP Dr Muhammad Mahadi Abdul Jamil for his contributions, guidance, and suggestions in completing this paper.

References

1. Anuar, H. M., Fadzil, F., Ahmad, N., Abd Ghani, N., & Sallehuddin, M. S. (2013). Urut Melayu for post-stroke patients: a qualitative study. National Committee for Clinical Research 2010. Available at http://www.nccr.gov.my/view_file.cfm?fileid=25. Accessed 7 Sept 2013.
2. Wu, B., Liu, M., Liu, H., et al. (2007). Meta-analysis of traditional Chinese patent medicine for ischemic stroke. *Stroke, 38*(6), 1973–1979.
3. Jamal, J. A. (2006). Malay traditional medicine an overview of scientific and technological progress. *Asia-Pacific Tech Monitor, 23*, 37–49.
4. Keat, W. L., & Siew, H. G. (2012). Burden of stroke in Malaysia. *International Journal of Stroke, 7*, 165–167.
5. Zheng, H., Davies, R. J., Black, N. D. (2005). Web-based monitoring system for home-based rehabilitation with stroke patients. Computer-Based Medical Systems, 419–424.
6. Ambar, R., Ahmad, M. S, Mohd Ali, A. M., Abdul Jamil, M. M., Arduino based arm rehabilitation assistive device. Journal of Engineering Technology, *7*, 5–13.
7. Safyzan, S., Badri, A. B., & Abdul Jamil, M. M. (2013). Exploitation of accelerometer for rehabilitation process. IEEE International Conference on Control System, Computing and Engineering, Nov 2013.

Effect of Stable Versus Unstable Exercises Among Chronic Low Back Pain Patients

Nursyuhada Zainal Abidin, Rahmat Adnan, Norasrudin Sulaiman, Shariman Ismadi Ismail and Amal Farah Abidin

Abstract This study aimed to investigate the effect of stability training in stable and unstable exercises among chronic low back pain (CLBP) patients. Method: A total of 38 subjects were involved in this study with 12 sessions over 6 weeks of intervention for the experiment group in which all of them were classified as CLBP patients. The subjects were divided into 3 groups, which were the stable group ($n = 15$), unstable group ($n = 15$) and the control group ($n = 8$). Results: The findings showed statistically significant differences in pain level between the stable and control groups ($p = 0.013$), while the stable group showed a significant decrease in pain level with 11.47 + 9.49 compared to the unstable group with a 14.27 + 11.03 pain level after the whole treatment. Moreover, both stable and unstable groups showed significant improvement in RROM with score of 4.52 + 0.41 and 4.24 + 0.53, respectively. Significant results were observed in the AROM in which the stable group recorded a value of 5.13 + 0.74, while the AROM for the unstable group recorded a slightly higher 5.40 + 1.73 number of people with disabilities. Conclusion: Hence, the results of this study suggest that stable exercises were superior in managing people with CLBP injury.

Keywords Stable exercises · Unstable exercises · Chronic low back pain · Rehabilitation · AROM · RROM

A. F. Abidin
Health Center, Universiti Teknologi MARA, Shah Alam, Malaysia

N. Zainal Abidin · R. Adnan (✉) · N. Sulaiman · S. I. Ismail
FSR Strength Training and Conditioning Laboratory, Universiti Teknologi MARA, Shah Alam, Malaysia
e-mail: alangr3@gmail.com

R. Adnan · N. Sulaiman · S. I. Ismail
Sports Science Center of Studies, Kuala Lumpur, Malaysia

R. Adnan et al. (eds.), *Proceedings of the International Colloquium on Sports Science, Exercise, Engineering and Technology 2014 (ICoSSEET 2014)*,
DOI: 10.1007/978-981-287-107-7_37,

1 Introduction

Low back pain is a common medical problem. Chronic low back pain (CLBP) is seen in 85 % of the population and up to 80 % of patients experience at least one recurrence [12]. It occurs among society members because of various reasons. CLBP can be categorized as symptoms frequently associated with pain for more than 3 months, and usually localized between the twelfth rib and the inferior gluteal folds with or without pain [7]. CLBP can be divided into specific or nonspecific forms. According to Sullivan [20], 85 % are classified as nonspecific because a definitive diagnosis cannot be achieved by the latest radiological methods. Many persons with CLBP will have a problem, especially during walking [13].

Persons with CLBP have a physical problem. It is already known that CLBP is associated with reduced walking speeds, and changes in coordination between trunk segments and stride lengths become shorter [13]. Graves et al. claimed that one factor related to the progression of CLBP has poor strength in the muscles that extend to the lumbar spine.

Rehabilitation is a superior treatment used by the health specialist when involved with musculoskeletal injury [4]. Basically, when a person suffers with CLBP, their daily activities will be limited because of the sensation of pain at the low back region. The muscular control of the trunk becomes unstable due to the lack of recruitment core muscles. Exercise is found to be one of the alternative treatments recommended, especially in managing people with CLBP [14]. One of the recommended exercises that should be started as the first step during the CLBP rehabilitation program is strengthening the stabilizer muscles via stability exercises. The stability of muscles was found to be effective as a tonic muscle group that enhanced the natural bracing mechanism and initiates the deep abdominal core muscles. These muscles increase the ability of the lumbo-pelvic region to provide a dynamic and functional response to the spinal column while in motion.

This is in agreement with Norris and Matthews [16], where a stability exercise program was integrated in a rehab program utilizing muscle isolation in musculoskeletal therapy [12]. In fact, the isometrics of specific stabilization exercise decreased the pain and disability of the person. However, there were many types of stable exercises suggested in the literature. The stability exercises can comprise of two different situations, referring to stable and unstable conditions. The difference in these two exercises is dependent on the tools that have been used to create the unstable environment during exercise.

To date, Marshall and Murphy [12] found in their pilot study that exercise in unstable environment gives significant benefit in CLBP rehabilitation particularly in erector spinae muscle. This gives clear information that unstable environment exercise may improved neuromuscular profile eventually enhance flexion–relaxation phenomenon to the people with CLBP injury. This study gives small information that unstable exercise may have benefits in exercise rehabilitation program [12].

According to Franca et al. [6], deep superficial muscles such as the transversus abdominis (TrA) become atrophied and decrease the anticipatory capacity and reduce the segmental function [5]. Thus, Ferreira et al. [4] stated that TrA exercising improves muscle activation for CLBP [3]. The weakness of the superficial trunk and abdominal muscles, and strengthening of these muscles via stability exercises is always associated with significant improvements in CLBP, as well as with decreased functional disability [5].

In addition, specific stability exercise directed at the local muscle system that consists of transversus abdominis, lumbar multifidus and internal oblique can enhance the dynamic stability of the lumbar spine as their functions are known to be tonically active during upright postures and during active spinal movements [13]. So far, there are tools that are famously used in the market such as BOSU balance, fitball and the like. All of these tools can provide an unstable environment when performing the exercise. In fact, the tools increase the number of tonic/synergist muscle groups and assist in the exercise [10]. As suggested by Marshall and Murphy [12], there are possibilities that comparing with different type of environment or protocol can give different outcome. Hence, this present study was designed to investigate the outcome between stable versus unstable exercises among CLBP patients in terms of three aspects, which are pain level, flexibility or active range of motion (AROM), and muscle strength or resistive range of motion (RROM).

2 Methods and Subjects

2.1 Research Design

This study used a quasi-experimental research design. Two experimental groups received different treatments, while the control group did not received any treatment.

As illustrated in the above Fig. 1, the 3 groups, which were stable, unstable and control groups, were tested for pain level and range of motion with AROM and RROM among CLBP patients. Stable and unstable exercise groups underwent 10 exercises based on their relative disabilities. The stable exercise group did the exercises using body weight, while the unstable exercise group used tools (BOSU ball and fitball) to assist them in doing the exercises. The control group did not receive any treatment.

2.2 Subject and Sampling Technique

A total of 38 subjects diagnosed with chronic back pain injury by the UiTM Health Center with their age range from 18 to 50 years old participated in this study. All of the subjects were diagnosed by the General Practitioner (GP) doctor in the

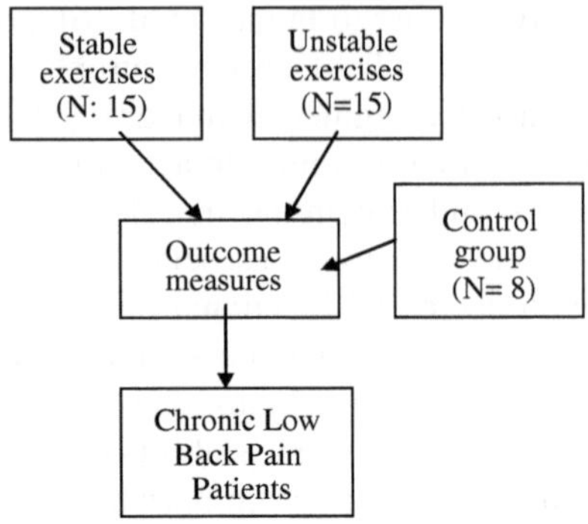

Fig. 1 Research design

Health Center UiTM before being referred into this research program. The inclusion criteria were set only for individuals with symptoms of localize back pain in the lumbar and hip region for more than 3 months and with symptoms located in the low back of the buttocks.

Subjects with severe cardiovascular or metabolic diseases, inflammatory disease, spinal tumor or fracture, nerve root compression/compromise, spinal cord irritation/cauda equina signs, osteoporosis, pregnancy and low back pain patients that involved sclerosis were excluded in this study [18].

All selected subjects were divided into three groups using a purposive randomized sampling technique. The three groups consisted of the stable group (N:15), unstable group (N:15) and control group without exercise (N:8). All participants had signed informed consent to voluntarily participate in this study as a subject.

2.3 Instrumentations and Tools

2.3.1 Functional Profile: Modified Oswestry Disability Questionnaire

This questionnaire was designed to identify how the back pain affects the daily life routine. This tool is calculated the each segment of questions from 0 to 5 point which can be converted into percentages. The 100 % percentage will give information that the pain is entirely affect person's daily life such as disability (Fairbank and Pysnet 2000).

2.3.2 Active Range of Motion

The ideal and valid way to measure the active joint range of motion is using goniometer [7]. This study use a plastic universal 360° E-Z with 31 cm size that to obtain joint AROM. Due to aim of this study, the focus areas of joint range of motion been measured were trunk and hip region only as the most affected area. The particular motions were measured such as flexion, extension, abduction and adduction, internal and external rotation.

2.3.3 Muscle Grading Scale

According Kendall method, this test is simplest and practical way to measure muscle function in injured people. Thus, all affected muscles were graded from 0 to 5 where the 0-lowest strength and 5 the strongest indicator of muscle function. The test was set the same particular areas of muscles that normally affected by this back injury. Hence, trunk and hip areas were tested based on this particular joint motion such as flexion, extension, abduction, adduction, external rotation and internal rotation [10].

2.3.4 Unstable Tool

A swiss ball which is an air-pressurized ball and BOSU balance which is a hemispherical physio-ball with an inflated dome side, and hard rubber flat side was used in the treatment of the unstable group.

2.3.5 Training Program

For the treatment, they were successfully completed for 12 sessions [19] at FSR gymnasium UiTM Shah Alam. They came based on their free time until this 12 session's program is completed. Selection of exercises for the subjects is started with 10- to 15-min walking on treadmill for warm up and followed by selected stability training based on their group segregation. All exercises were design same, but only stable and unstable surface were determine there are different groups. They did 10 types of exercises between 5 and 10 sets based on selected stability exercise with 10 s hold for every set. Rest interval period was set at 1 min in each session [8].

2.3.6 Data Collection Procedure

All 38 subjects were participated in this study. This study was participated by the subjects who passed the screening by General Practitioner in the University Health Center. Subjects did receive verbal and written information about how the research was conducted before signed the inform consent form. Next, the subjects required to answer the Modified Oswestry Disability Questionnaire (MODQ) which to determine their pain level that affect their daily life due to back pain. They were undergoing an assessment to determine their AROM for trunk and lower extremities joint angles. The data for AROM during pretest was recorded and analyze thoroughly and comparing with the posttest results.

2.3.7 Data Analysis

This study used the Statistical Package of Social Science (SPSS) 20.0 software to analyze the data. ANOVA method was used to compare mean between pre- and posttest. The significant level intended for this study will be set at $p < 0.05$.

3 Results

Out of thirty-eight participants, 39.5 % ($n = 15$) were in the stable group, 39.5 % ($n = 15$) were in the unstable group, and 21.1 % ($n = 8$) were in the control group. The control group served as a benchmark for the experimental groups.

The post hoc test showed statistically significant differences between the stable and control groups at the 5 % level of significance ($p = 0.013$).

There were significant differences in the mean pain level between all groups (Fig. 1). The bar graph above stable group showed a decrement in the level of pain by recording the lowest score with 11.47 + 9.49, and the unstable group had a score of 14.27 + 11.03. Thus, it can be concluded that stable exercises were effective in reducing level of pain among low back pain patients.

At session 8, at least two groups showed statistically significant differences (Table 1; $p = 0.002$). The post hoc test revealed statistically significant differences between the stable and unstable groups ($p = 0.029$) and between the stable and control group ($p = 0.003$).

3.1 Resistive Range Of Motion

There were differences in mean scores between the groups (Table 1). The stable group showed significant increment in improvement in muscular strength by recording the highest score with 4.52 + 0.41, while the unstable group also displayed some improvement, but was slightly lower than the stable group with a score of 4.25 + 0.35 (Table 1). Thus, it can be concluded that both stable and unstable exercises were effective in improving muscular strength among low back pain patients.

Table 1 Mean RROM at pre-session 8 and pre-session 12

Muscle strength (RROM)	Group	N	Mean	SD
Pre	Stable	15	3.06	0.91
	Unstable	15	2.86	0.48
	Control	8	2.98	1.24
S8*	Stable*	15	4.20	0.45
	Unstable*	15	3.45	0.62
	Control*	8	3.02	1.29
S12*	Stable*	15	4.52	0.41
	Unstable*	15	4.25	0.35
	Control*	8	3.04	1.36

**p* value < 0.05, **S* = session, ***mean RROM = muscle grading scale 0–5 T

3.2 Active Range of Motion

The comparison between mean frequencies of disabilities was obtained from the flexibility of the subjects using AROM for stable, unstable and control groups (Table 2). Data were collected at pre-assessment, and after session 8 and session 12 assessments. At pre-assessment, the results for all the groups showed relatively high number of mean disabilities with the stable group having 13.60 ± 2.85, unstable group 11.87 ± 2.92 and with 7.25 ± 2.51 for the control group. At session 8, however, the flexibility assessments from two of the groups showed a decrement in the frequency of disabilities in which the stable group had the lowest number of disabilities with 7.33 ± 2.51 and the unstable group was second with a decrement of 9.80 ± 2.90. The control group showed no significant improvement in terms of disability frequency with 4.75 ± 2.25. The most significant result was obtained with AROM test during session 12 where the stable group had a frequency of only 5.13 ± 0.74, and the unstable group displayed a slightly higher frequency of 5.40 ± 1.73. The control group remained at 5.50 ± 3.88. In a nutshell, the most appropriate exercise selections to improve flexibility or AROM among low back pain patients are stable.

4 Discussion

4.1 Level of Pain

Significant differences were obtaining between the stable, unstable and control groups on pain level. It is found that there was a significant difference between stable versus unstable exercises on pain among CLBP patients. The pretest results of the stable group recorded the highest pain level followed by the unstable and control groups (Table 3). Whereas, at session 8, the results showed a decrease in

Table 2 Frequency of disabilities from all types of AROM tests

Active range of motion (AROM)	Group	N	Mean	SD
Pre	Stable	15	13.60	2.85
	Unstable	15	11.87	2.92
	Control	8	7.25	2.51
S8	Stable	15	7.33	1.82
	Unstable	15	9.80	2.90
	Control	8	4.75	2.25
S12	Stable*	15	5.13	0.74
	Unstable	15	5.40	1.73
	Control	8	5.50	3.88

* S = session, mean = frequency of anomalies AROM

Table 3 Comparison of pre-session 8 and pre-session 12 mean pain levels among the subject

Pain level	Group	N	Mean	SD
Pre	Stable	15	33.33	14.16
	Unstable	15	33.47	13.26
	Control	8	29.25	18.05
S8	Stable	15	19.33	9.79
	Unstable	15	19.33	12.09
	Control	8	31.50	21.43
S12*	Stable*	15	11.47	9.49
	Unstable	15	14.27	11.03
	Control*	8	30.00	23.42

*p value < 0.05, S = session, Pain mean in percentage (%)

pain level for the stable compare to the unstable group. Both groups showed a decrease, but the decrease was more in the stable group.

In the final test after session 12, the results obtained in the stable group showed the lowest score for pain level followed by the unstable group and the control group displayed minimal decrement, but still recorded the highest score for pain level. Thus, it was concluded that stable exercises were most effective in reducing pain level among CLBP patients.

These results are in agreement with the findings by May and Johnson [15] who showed significant differences for pain favoring stabilization exercises [10]. The reasons that cause increment in pain level in the control group can be explained by the nature of chronic low back patients in that symptoms recur. A study on lumbar stabilization training in chronic low back patients using 'no intervention' as a control also found participants to stay the same or get significantly worse [12].

The unstable group did show a decrease in the pain level, but still the stable group showed the most significant decrement in terms of pain level. Hence, it was clear that stable exercises were the best option to reduce pain levels among CLBP patients. Norris and Matthews [17] also reported specific stabilization exercises are the most effective treatments to reduce pain level among CLBP patients [12].

4.2 Resistive Range Of Motion

This study found that there was a significant difference between stable versus unstable exercises on RROM among CLBP patients. According to Carpes et al. [2], stability training was effective in improving strength and body balance for CLBP [2]. The pre-resistive range of motion test (RROM) to measure subjects muscle strength showed that the stable group had slightly higher score, while the unstable group displayed the lowest score, whereas the control group showed higher results than the unstable group (Table 1). However, after session 8, the stable group recorded the highest score, while the unstable group was the second highest and the lowest score was recorded in the control group. The final muscular strength measurements showed that the stable group had the highest score and the unstable group showed an improved, but was slightly behind the stable group, while the control group had the lowest score.

Lee et al. [12] stated that an imbalance in trunk muscle strength in lower extensor muscle strength and flexor muscle strength might be a risk factor for low back pain [8]. One of the important risk factors for low back pain is weakness of superficial trunk and abdominal muscles, and strengthening of these muscles is often associated with significant improvement of chronic low back patients, as well as decreased functional disability [5].

Both stable and unstable groups showed significant improvements in muscle strength throughout this study. Thus, stable and unstable exercises were effective in improving muscular strength among low back pain patients. These findings are in agreement with Franca et. al. [6] who also observed that stabilization exercises effectively reduce pain and functional disability in individuals with CLBP in which stabilization improves TrA muscle activation capacity [5].

4.3 Active Range of Motion

This study also showed significant differences between stable versus unstable exercise on the AROM among CLBP patients. This was a clear comparison between frequencies of disabilities obtain from measuring flexibility of the subjects using AROM test for stable, unstable and control groups (Table 2).

The results of the stability program from kinematics assessment between pre and post in terms of range of motion in the selected movements were statistically significant [2]. The pre-assessment results obtained from the entire group showed relatively high frequency for disabilities in the stable group, while the unstable group showed lower disability than the stable group, while the control group recorded the least disability.

According to Magee [13], tissue damage would lead to decreased stability of spinal structure, increased challenges to the already inefficient muscles, and the perpetuation of a degenerative cascade [9]. However, flexibility assessment in session 8 showed decrement in the frequencies of disabilities in which the stable group had the lowest number of disabilities followed by the unstable group as the second lowest group.

The control group showed no significance decrement in terms of disability frequency. The AROM test in session 12 showed the stable group with the least disability frequency, while the unstable group displayed a slightly higher frequency. The results were similar to that conducted by Baard et. al. [1] who noted that stated spinal stabilization increased hamstring flexibility and had positive effects on the range of motion around the hip joint and pelvis [1]. This is because the exercises enhanced local blood flow through the muscles which generated additional heat, thereby enhancing muscle elasticity and facilitating an increase in range of motion in the hamstring muscles.

It was apparent that the most appropriate exercise selection to improve flexibility or AROM among low back pain patients are the stable exercises.

5 Conclusion

In summary, stability exercises were found to be more effective in managing people with CLBP. The frequency of training may influence the stability. Based on the results, 12 sessions of training was the most effective. However, 8 sessions were found to be sufficient for stable exercises compared to the unstable exercises. Overall, stable exercises were more effective compared to unstable exercises in reducing pain level, flexibility and muscle strength.

References

1. Baard, M. L., Pietersen, J., & van Rensburg, S. J. (2011). Interventions for chronic low back pain: whole body vibration and spinal stabilisation. *South African Journal of Sports Medicine, 23*(2), 35–39.
2. Carpes, P. F., Reinehr, B. F., & Mota, B. C. (2008). Effects of a program for trunk strength and stability on pain, low back and pelvis kinematics, and body balance: A pilot study. *Journal of Bodywork & Movement Therapies, 9*, 22–30.
3. Fairbank, J. C. T., & Pynsent, P. B. (2000). The Oswestry disability index. *Spine, 25*(22), 2940–2953.
4. Ferreira, P. H., Ferreira, M. L., Maher, C. G., Refshauge, K., Herbert, R., Hodges, P. W. (2009). Changes in recruitment of transversus abdominis correlate with disability in people with chronic low back pain. *British Journal of Sports Medicine.* doi:10.1136/bjsm.2009.061515 (published online 26 May 2009).
5. Field, J. (2009). Back pain: What is happening when it becomes chronic? *Clinical Chiropractic, 12*(1), 45–51.
6. Franca, R. F., Burke, N. T., Hanada, S. E., & Marques, P. A. (2010). Segmental stabilization and muscular strengthening in chronic low back pain—a comparative study. *Clinics, 65*, 1013–1017.
7. Heyward, V. H. (2006). Assessing flexibility and designing stretching program. In V. H. Heyward (Ed.), *Advance fitness assessment and exercise prescription* (Vol. 4). Champaign: Human Kinetics.

8. Koumantakis, A. G., Watson, J. P., & Oldham, A. J. (2005). Trunk muscle stabilization training plus general exercise only: Randomized controlled trial of patients with recurrent low back. *Pain, 85*, 209–225.
9. Krismer, M., & van Tulder, M. (2007). Low back pain (non-specific). *Best Practice & Research Clinical Rheumatology, 21*(1), 77–91.
10. Kendall, F. P., McCreary, E. K., & Provance, P. G. (1993). *Muscles, testing and functions* (4th ed.). Baltimore: William and Wilkens.
11. Lee, J. H., Hoshino, Y., Nakamura, K., Kariya, Y., Saita, K., & Ito, K. (1999). Trunk muscle weakness as a risk factor for low back pain: a 5-year prospective study. *Spine, 24*(1), 54–57.
12. Marshall, P. W., & Murphy, B. A. (2006). Evaluation of functional and neuromuscular changes after exercise rehabilitation for low back pain using a Swiss Ball: A pilot study. *Journal of Manipulative and Physiological Therapeutics, 29*, 5550–5556.
13. Magee, D. J. (2002). Lumbar spine. In *Orthopedic physical assessment,* 4th edn (pp. 467–566). Philadelphia: Elsevier Sciences.
14. May, S., & Johnson, R. (2008). Stabilisation exercises for low back pain: a systematic review. *Physiotherapy, 94*(3), 179–189.
15. Newell, D., & Laan, D. V. M. (2010). Measures of complexity during walking in chronic non-specific low back pain patients. *Clinical Chiropractic, 13*(1), 8–14.
16. Norris, C., & Matthews, M. (2008). The role of an integrated back stability program in patients with chronic low back pain. *Complementary Therapies in Clinical Practice, 14*(4), 255–263.
17. O' Sullivan, P. B., Phyty, M., Twomey, L. T., & Allison, G. T. (1997). Evaluation of Specific Stabilizing Muscle in the Treatment of Chronic Low Back with Radiologic Diagnosis of Spondylolysis or Spondylolisthesis. *Spine, 22*(24), 2959–2967.
18. Slade, S. C., & Keating, J. L. (2009). Effects of preferred-exercise prescription compared to usual exercise prescription on outcomes for people with non-specific low back pain: a randomized controlled trial [ACTRN12608000524392]. *BMC musculoskeletal disorders, 10*(1), 14.
19. Sokunbi, O., Cross, V., Watt, P., & Moore, A. (2010). Experiences of individuals with chronic low back pain during and after their participation in a spinal stabilisation exercise programme–A pilot qualitative study. *Manual therapy, 15*(2), 179–184.
20. Sullivan, O. P. (2005). Diagnosis and classification of chronic low back pain disorders: Maladaptive movement and motor control impairments as underlying mechanism. *Manual Therapy, 10*, 242–255.

Coaching Efficacy Level Among Individual and Team Sports Coaches in Malaysia

Raja Nurul Jannat and Kang Mea Kee

Abstract Coaches can be considered as individuals who can influence athletes in their lives and sports performance. Coaches have the responsibility in guiding the athletes to be successful in their sports performance by helping them to improve their skills. One of the factors that may influence athletes to perform at their optimal level is their belief in their coaches' ability to guide them during training and competition. Therefore, the purpose of this study was to determine the level of coaching efficacy among individual and team sports coaches. A total of 154 coaches who coached in SUKMA 2012 (a sports event that involved young athletes below 21 years of age) were selected through purposive sampling to participate in this study. They were categorized into individual ($N = 77$) and team ($N = 77$) sports coaches. Coaching efficacy scale (CES) questionnaire was used to measure the coaches coaching efficacy. Overall, Malaysian coaches who coached individual sports showed that their level of coaching efficacy was higher ($M = 7.86$, SD = 0.55) for all the subscales, namely character building ($M = 7.91$, SD = 0.62), motivation ($M = 7.86$, SD = 0.58), technique ($M = 7.90$, SD = 0.63), and game strategy ($M = 7.78$, SD = 0.61) compared with Malaysian coaches who coached team sports during SUKMA 2012. In conclusion, coaches who coached individual sports have higher level of coaching efficacy which indicates that under the guidance of efficacious coaches, athletes can gain knowledge on sport's skills, develop leadership qualities, become more confident in their physical and mental abilities and also work hard toward achieving goals.

Keywords Coaching efficacy · SUKMA · CES

R. N. Jannat (✉) · K. M. Kee (✉)
Faculty of Sports Science, Universiti Teknologi MARA,
Shah Alam, Malaysia
e-mail: raja_master11@yahoo.com

K. M. Kee
e-mail: kee@salam.uitm.edu.my

R. Adnan et al. (eds.), *Proceedings of the International Colloquium on Sports Science, Exercise, Engineering and Technology 2014 (ICoSSEET 2014)*,
DOI: 10.1007/978-981-287-107-7_38,

1 Introduction

Currently, participation in sports has become a major activity for young and adult athletes. Their participation provides many opportunities such as enhancing their sports skills, gaining new experiences, and also developing social skills among themselves. The most important factor in helping them to be successful in sports activities is the individuals who serve as coaches. Coaches can be considered as individuals who can influence athletes in their lives and sports performance [9]. Furthermore, they also can positively affect the behavior and psychological state of an athlete. Becoming a coach is an important and a challenging role, as their responsibility is not only teaching but also guiding athletes to succeed in their sports performance.

One of the key elements to be a successful coach is through their perception of their ability to lead the athletes through the competitive experience, and the belief in one's personal ability is known as self-efficacy [7]. If a coach has a low level of self-efficacy, it may have a detrimental effect on athletes' performance. Conversely, if the coach has high self-efficacy, it may help in influencing his or her own performance, as well as the athletes' performance during training and competition. Past studies have also shown that athletes were also more confident in themselves and their teams when they believed their coach was a confident leader. There are numerous studies on coaches' level of coaching efficacy in various sports environments [7]. Throughout these studies, there were many factors that were revealed to have an effect on the coaches' level of coaching efficacy.

Lack of numerous factors, such as how to teach technical skills and various methods of motivation, may impair the coach' beliefs in their ability to guide their athletes, and hence, it will affect the leadership abilities of current and future athletes [11]. In the sports context, understanding the coaches' beliefs in their competence to lead and guide individual athletes as well as their teams during competition may continue to be superior [4]. Coaching efficacy can be divided into four types of subscales: motivation efficacy, technique efficacy, game strategy efficacy, and character-building efficacy [14]. Motivation efficacy can be defined as the coaches' confidence in their ability to influence his or her athletes' psychological state and skills. Technique efficacy refers to the coach's belief on his or her ability to teach skills and to diagnose and correct the errors made by the athletes. The third subscale, which is game strategy efficacy, represents the coach's belief in leading the athletes to perform successfully during competition. The last subscale, which is character-building efficacy, involves the coach's ability to affect the athlete's attitudes such as sportsmanship and positive attitude.

Many factors affect coaches' coaching efficacy, including their playing experience and coaching courses attended. In Malaysia, there are three levels (level 1: beginner; level 2: intermediate; level 3: advance) of coaching courses conducted by the National Coaching Licensing Board that can be attended by all sports coaches. Besides serving as a national standard for the recognition of coaching qualifications, the program provides systematic coaching education with the purpose of improving the knowledge and skills required by sports coaches. The program has three basic components: sports science, specific sports coaching, and practical components. Every coach who attends this course must go through all the components before they can be certified. Although there are many studies that measure coaching efficacy, most of these studies were based on a Western context. Limited information is available on the coaching efficacy among Malaysian coaches, particularly in youth sport. Therefore, this study was conducted to gather more information with regard to Malaysian youth coaches and their coaching efficacy. The purpose of this study was to determine significant differences in the level of coaching efficacy among individual and team sports coaches in Malaysia.

2 Methodology

2.1 Participants

A total of 154 coaches from 14 states who coached in Sukan Malaysia (SUKMA) 2012 volunteered to serve as participants in the study. These participants were selected through a purposive sampling comprising of both individual and team sports (athletics, aquatic, weight lifting, badminton, gymnastic, hockey, lawn ball, archery, tenpin bowling, sepak takraw, volleyball, equestrian, and boxing). They were categorized into individual ($N = 77$) and team ($N = 77$) sports coaches.

2.2 Outcomes Measure

The coaching efficacy scale (CES) [7] questionnaire was used to measure their coaching efficacy. The reliability of the CES questionnaire in this study was 0.91.

3 Results

The following analyses were based on the result from social science statistical package version 17. Descriptive analysis such as frequencies was calculated to demonstrate the characteristics of coaches. Table 1 shows that the majority of respondents for both groups are male (individual = 49, team = 56). Furthermore, the majority of coaches in both groups were in the range of 41–45 years of age (individual = 23, team = 22). In terms of education level, the majority of coaches in both groups have a diploma (individual = 37, team = 32).

Majority of coaches had experience playing at the state level, regardless of their category of coaching (individual = 47, team = 39), while only eight individual

Table 1 Gender, age, and education level among individual and team sports SUKMA 2012 coaches

Characteristics	Coaches		
	Individual	Team	Total
Gender			
Male	49	56	105
Female	28	21	49
Age (years)			
Below 30	9	13	22
31–35	15	6	21
36–40	20	20	40
41–45	23	22	45
Above 45	10	16	26
Education level			
SPM	15	25	40
Diploma	37	32	69
Degree	22	8	30
Postgraduate	3	12	15

coaches had experience playing at the national level and 26 team coaches had experience playing at the national level. Furthermore, the findings showed that 81.8 % ($N = 126$) of the respondents have attended the coaching courses organized by the Malaysian Sports Council, while 18.2 % ($N = 28$) of them did not attend any coaching courses organized by the Malaysian Sports Council. The majority of coaches who had attended the coaching courses had attained the intermediate level of coaching (individual = 28, team = 26) (Table 2).

The means and standard deviations of each subscale in the CES questionnaire, which includes motivation, technique, game strategy, and character-building efficacy, are presented in Table 3. Malaysian SUKMA coaches who coached individual sports had higher mean scores in four efficacy subscales, which are motivation efficacy ($M = 7.86$, SD = 0.58), strategy efficacy ($M = 7.78$, SD = 0.61), technique efficacy ($M = 7.90$, SD = 0.63), and character-building efficacy ($M = 7.91$, SD = 0.62), compared with coaches who coached team sports. Coaches who coached team sports showed lower mean scores in motivation efficacy ($M = 7.69$, SD = 0.51), strategy efficacy ($M = 7.66$, SD = 0.57), technique efficacy ($M = 7.61$, SD = 0.65), and character-building efficacy ($M = 7.78$, SD = 0.57). The results indicate that Malaysian SUKMA 2012 coaches who coached individual sports were more confident in handling tasks such as motivating their athletes, detecting the strengths and weaknesses of the opposing teams, carrying out the instructional aspects of coaching, and developing their athletes' characteristics.

An independent-samples *t* test was conducted to compare the level of coaching efficacy and the subscales (motivation, game strategy, technique, and character building). There was a significant difference in scores for coaches who coach individual sports technique efficacy ($M = 7.90$, SD = 0.63) and team sports ($M = 7.61$, SD = 0.65; $t\ (154) = 2.78$ $p = 0.006$, two-tailed). The magnitude of the differences in the means (mean difference = 0.29, 95 %CI 0.08–0.49) was

Table 2 Level of playing experience and coaching course attended among individual and team sports SUKMA 2012 coaches

Characteristics	Coaches		
	Individual	Team	Total
Level of playing experience			
School	22	12	34
State	47	39	86
National	8	26	34
Level of coaching course			
Did not attend	16	12	28
Beginner	27	22	49
Intermediate	28	26	54
Advance	6	17	23

Table 3 Coaching efficacy among individual and team sports SUKMA 2012 coaches

Characteristics	Coaches			
	Individual		Team	
	M	SD	M	SD
Motivation	7.86	0.58	7.69	0.51
Game strategy	7.78	0.61	7.66	0.57
Technique	7.90	0.63	7.61	0.65
Character building	7.91	0.62	7.78	0.57
Overall CES	7.86	0.55	7.68	0.51

small (eta-squared = 0.05). There was also a significant difference in overall coaching efficacy scores for coaches who coach individual sports ($M = 7.86$, SD = 0.55) and team sports ($M = 7.68$, SD = 0.51; t (154) = 2.08 $p = 0.04$, two-tailed). The magnitude of the differences in the means (mean difference = 0.18, 95 %CI 0.01–0.35) was very small (eta-squared = 0.01). This indicates that coaches who coached individual sports, in contrast to coaches who coached team sports, are hypothesized to be more effective in correcting and providing skills and techniques to the athletes. Furthermore, in overall coaching efficacy, coaches who coach individual sports tend to have higher efficacy in coaching compared with coaches who coach team sports.

4 Discussion

Coach acts as the most important person who has responsibility to ensure their athletes' success, influencing the athlete's mental development, self-esteem, and skill satisfaction in sports performance satisfaction. The CES measurement developed by Feltz has been widely used in past studies to measure the confidence or efficacy level among school and collegiate coaches [7]. It measures the coaches' efficacy in carrying out duties based on the four subscales of coaching tasks, which are motivation, technique, game strategy, and character building. Coaches who have a high degree of coaching efficacy do give a more positive feedback. In addition, when compared to past studies [8, 10], the overall coaching efficacy score for both groups of coaches was high (Malaysia {individual 7.86, team 7.68}, [8]: 6.72). It was evident that Malaysian youth coaches showed similar capabilities in coaching and had the same potential as others to be successful coaches, regardless of the type of sports they coached. The Malaysian coaches believed that self-confidence in coaching is important to inspire athletes' performance during competition. The findings of the present study are supported by past studies, which indicate that Malaysian coaches were confident in their ability to handle coaching tasks [2, 12]. The level of coaching efficacy produced positive outcomes, and this was considered to be aligned with the difficulty of tasks [14].

The findings also showed that coaches who coached individual sports had higher scores in motivation, game strategy, character-building, and technique subscales compared with coaches who coached team sports during the SUKMA 2012 event. It was also mentioned that getting a group of athletes to perform as a team is not always easy [1, 5]. Rather, a group of athletes becomes a team when they all possess a common identity, have shared goals and objectives, exhibit structured patterns of interaction and communication, and most importantly consider themselves to be a 'team.' These issues raise difficulties for coaches who coached team sports and try to make them united and perform well as a team. The role of the team coach is very different compared with one-to-one coaching, and when a team coach is also part of the team, they have to take the role of coach and player simultaneously. This requires the ability to handle complex relationships and a high degree of self-awareness to make the coaching effective [1, 5].

It has been mentioned that coaches who have higher coaching efficacy were significantly linked to technique, game strategy, motivation, and character-building efficacy. Other studies have also shown that highly efficacious coaches often displayed more positive coaching behaviors, offered more positive reinforcement and instruction, displayed more commitment to coaching, and were more likely to increase player satisfaction levels, performance, and team's winning percentages. Positive reinforcement, for example, can influence the athlete's confidence in their ability to achieve their goals, which can in turn influence the level of effort expended to achieve those goals [7]. It has been highlighted that coaches who were less educated in pedagogy, kinesiology, or exercise physiology that specifically addresses the training requirements of children and adolescents make them less confident in coaching athletes. Therefore, these elements need to be mastered by the coaches of team sports to help boost their efficacy level in motivation, technique, game strategy, and character building.

Coaches of individual sports with a higher coaching efficacy in character-building subscale suggest that they were more inclined toward carrying out character-building duties and were more efficacious in handling tasks of instilling positive attitudes such as respect for others, fair play during competition, and also instill good moral character [8], because the coaches believed that inculcating a positive attitude toward sports in their athletes will result in successful performance. Furthermore, they also reflect efficacy in carrying out tasks such as instilling an attitude of fair play among athletes, instilling an attitude of respect for others, promoting good sportsmanship, demonstrating the skills of the sports, and instilling an attitude of good moral character. The coaches who coached individual sports have to deal with individual athletes that vary in attitudes. Therefore, they need to convey various types of character-building strategies to athletes with different attitudes.

One possible reason that Malaysian SUKMA coaches who coached team sports had low mean scores in game strategy efficacy compared with coaches who coached individual sports might be due to limited games and matches being organized that expose coaches to the real competition environment and use their ability in making decisions during competition. Furthermore, the SUKMA event is organized only once every 2 years, which gives little opportunity of exposure for coaches to guide their athletes in real game situations [13]. In addition, most of the youth athletes involved in team sports came from schools where most of the time they were being supervised by school coaches [6]. The Malaysian SUKMA coaches were only able to be with their athletes during centralized training just before the event. Due to the short duration of centralized training, there is limited contact between coaches and their athletes. Therefore, coaches have to look for other ways to improve their game strategy efficacy. For example, coaches may have to organize more competitions or friendly matches in increasing their team performance so that they can practice their game strategies [3]. Further research is paramount to expanding the understanding of coaching efficacy in this environment. However, this study has provided additional knowledge of coaching efficacy related to Malaysian youth coaches.

It was evident that the coaches who coached individual sports demonstrated a higher coaching efficacy level compared with coaches who coached team sports. Under the guidance of efficacious coaches, young athletes can learn the technical and tactical skills of a sport, gain confidence in their physical abilities, develop leadership qualities, and work toward a common goal. This study only examined the antecedents of Malaysian SUKMA coaches and did not investigate the impact of the coaching efficacy dimensions. Past studies had already noted that coaches who had high coaching efficacy used more positive coaching styles, had more players who were satisfied with their playing experiences, had higher winning percentages, and had higher efficacy levels among athletes and teams [2, 9]. Therefore, it would be interesting to examine these variables on Malaysian coaches to determine whether the outcomes were similar.

5 Conclusion

This study has successfully demonstrated that Malaysian SUKMA coaches have a high level of coaching efficacy to coach youth athletes. Their level of coaching efficacy was primarily determined by their playing experiences and coaching courses attended. Measuring coaching efficacy is important because it does not only provide a direct impact on coaching behavior, but also include positive and negative influences that a coach may have on character and athletic development as well as performance of individuals and teams. This study has provided a foundation to build a body of knowledge on coaching efficacy that is related to the Malaysian environment.

In conclusion, coaches who coached individual sports demonstrated a higher coaching efficacy level and technique efficacy subscale compared with coaches who coached team sports. Under the guidance of efficacious coaches, young athletes can learn the technical and tactical skills of a sport, gain confidence in their physical abilities, develop leadership qualities, and work toward a common goal.

Past studies conducted elsewhere had noted that coaches who had high coaching efficacy used more positive coaching styles, had more players who were satisfied with their playing experiences, had higher winning percentages, and had higher efficacy levels among athletes and teams [2, 9]. Therefore, it would be interesting to examine these variables on Malaysian coaches to see whether the outcomes were similar.

References

1. Carron, V. A., & Hausenblas, H. A. (1998). *Group dynamics in sport*. WV: Fitness Information Technology Morgantown.
2. Chiu, L. K., Mahat, N. I., & Hua, K. P. (2013). R.B. Radzuwan, Student-athletes' perceptions of coaches' coaching competency at the malaysian public institution of higher learning. World. *World Journal of Education, 3*(1), 13.
3. Christmas, V. H. (2009). The newsletter of the psychometrics forum unpublished 2009.
4. Chu, R. D., & Tingzon, C. J. (2009). The relationship of coaching competency on the athlete's self-efficacy and hope. *The International Journal of Research and Review, 1*, 84–121.
5. Clerk Maxwell, J. (1892). *A treatise on electricity and magnetism* (3rd ed., Vol. 2, pp. 68–73). Oxford: Clarendon.
6. Delano, J. (2009). Communicating across differences: the case for becoming a cosmopolitan coach, (2), 92–103.
7. Feltz, D. L., Chase, M. A., Moritz, S. E., & Sullivan, P. J. (1999). A conceptual model of coaching efficacy: Preliminary investigation and instrument development. *Journal of Educational Psychology, 91*(4), 765–776.
8. Fung, L. (2003). Coaching efficacy as indicators of coach education program needs. *Athletic Insight*, *5*(1), 12–18.
9. Gilbert, W., Gallimore, R., & Trudel, P. (2009). A learning community approach to coach development in youth sport. *Journal of Coaching Education, 2*(2), 1–21.
10. Kuan, G., & Roy, J. (2007). Goal profiles, mental toughness and its influence on performance outcomes among Wushu athletes. *Journal of Sports Science and Medicine, 6*(2), 28–33.
11. Majzub, R., & Muhammad, T. A. (2011). Goal orientation and achievement of junior in golfers Malaysia. *Procedia-Social and Behavioral Sciences, 15*, 1644–1649.
12. Malete, L., & Sullivan, P. (2009). Sources of coaching efficacy in coaches in Botswana. *International Journal of Coaching Science, 3*(1), 17–27.
13. Marback, T. L., Short, S., Short, M., & Sullivan, P. J. (2005). Coaching confidence: An exploratory investigation of sources and gender differences. *Journal of Sport Behavior, 28*(1), 18–34.
14. Myers, N. D., Vargas-Tonsing, T. M., & Feltz, D. L. (2005). Coaching efficacy in intercollegiate coaches: Sources, coaching behavior, and team variables. *Psychology of Sport and Exercise, 6*(1), 129–143.

A Pilot Study: Effects of Aquatic and Land Spinal Stabilisation Training on the Management of Back Pain

Bashtiah Nahrul Khair, Rahmat Adnan, Hamid Ahmad, Norasrudin Sulaiman and Shariman Ismadi Ismail

Abstract The objective of this study was to determine the effects of aquatic spinal stabilisation training and land spinal stabilisation training on the management of back pain. This study use quasi-experimental crossover design which all subjects underwent both intervention treatments with one week of washout period. The patients managed to complete 12 sessions of both water- and land-based exercises in 6 weeks of intervention period. The pain level, functional status, deep abdominal muscle function, trunk flexibility, balance, and centre of vertical force of 4 chronic low back pain patients were assessed to achieve the stated objective. There was a significant difference in the pain level between post-land and post-aquatic spinal stabilisation training ($z = -2, p < 0.05$). The findings of this study indicated that aquatic spinal stabilisation training and land spinal stabilisation training decreases pain level and improves functional status, deep abdominal muscle function, and centre of vertical force. Both aquatic and land spinal stabilisation training demonstrated positive results in pain level, functional status, deep abdominal muscle function, and centre of vertical force.

Keywords Chronic back pain · Aquatic spinal stabilisation training · Land spinal stabilisation training · Pain level · Trunk flexibility · Deep abdominal muscle function

H. Ahmad
Physiotherapy Department, Terendak Military Hospital, Malacca, Malaysia

B. Nahrul Khair · R. Adnan (✉) · N. Sulaiman · S. I. Ismail
Faculty of Sports Science and Recreation, Universiti Teknologi MARA, Shah Alam, Malaysia
e-mail: rahmatadnan@salam.uitm.edu.my; alangr3@gmail.com

R. Adnan · N. Sulaiman · S. I. Ismail
Sports Science Center of Studies, Universiti Teknologi MARA, Shah Alam, Malaysia

R. Adnan et al. (eds.), *Proceedings of the International Colloquium on Sports Science, Exercise, Engineering and Technology 2014 (ICoSSEET 2014)*,
DOI: 10.1007/978-981-287-107-7_39, © Springer Science+Business Media Singapore 2014

1 Introduction

Low back pain is a global health problem which could affect the patient's quality of life, social status, productivity, and happiness [1]. Chilibeck et al. [2] believed that people with non-specific chronic low back pain can safely perform a variety of physical activities that are progressive. Musculoskeletal problems, including low back pain have been treated with aquatic therapy for quite some time [3]. Water is a safe medium of exercise because the water buoyancy and hydrostatic pressure supports the spine and pelvis [3]. Intveid et al. [4] also suggest aquatic physiotherapy as one of the treatments for low back pain. Koumantakis et al. [5] agreed no specific exercise on the management of low back pain amongst non-specific low back pain patients; however, they suggested exercise safety is the most important aspect for a better low back pain management.

The most widespread method to deal with low back pain includes strengthening the trunk and spine [6]. Carter et al. [7] stated that stabilisation exercises are exercises in which the spine is held in a static position. Spinal stabilisation helps to maintain control over the position of the spine for a longer time with lower force output [7]. It is important to maintain segmental control within the trunk because it plays a part in spinal stability and is therefore able to decrease the risk of back pain in sedentary people [7]. Danneels et al. [8] also believed that although rigorous physical therapy improved mobility and back pain, stabilisation training could still be essential in active chronic low back pain patients.

Transversus abdominis and other local muscles play a big role in maintaining the position of the spine and are able to increase trunk endurance [7]. Deep abdominal muscles such as transversus abdominis and multifidus provide local stability [9]. Normal muscle activation is vital to maintain spine stability and prevent injury [10]. The deep muscles ability to actively stabilise the lumbar spine can be assessed using a pressure biofeedback unit [11].

The functional instability of the lumbar spine may also be the result of the pain from unsuccessful back surgery patents, because mainly the computed tomography (CT) or magnetic resonance imaging (MRI) showed only normal post-operative differences [12]. Bouche et al. [12] also noted that the quality of functional spinal stability is difficult to assess because it involves different components. Hence, Bouche et al. [12] suggested that the specific conditions in which functional stability is challenged involve balance or postural control, activities.

Head-out (the body is submerge in the water) aquatic exercises are one of the most important physical activities conducted in primary and tertiary health prevention systems [13]. Not all head-out aquatics have the same intensity, and water temperature must be varied to ensure the patient stays comfortable whilst preventing thermo-regulation stress [13]. According to the Aquatic Exercise Association, water temperatures near or above 32 °C are suitable for therapeutic-type activities such as water massage or rehabilitation exercises for musculoskeletal injuries, and water temperatures between 28 and 30 °C allow the body to respond naturally and gain the benefits of physical fitness without the associated problems of conserving heat or overheating [14].

When choosing aquatic exercises as a treatment method, the intervention characteristics that have to be taken into account are fluid mechanics, temperature, type of exercise, the professional's experience, and cost [15]. Verhagen et al. [15] noted that aquatic exercises are as effective as other interventions in treating patients with chronic low back pain. Besides that, it was also observed that water provides natural resistance which assists in muscle strengthening.

Exercise programmes to decrease pain and increase function can be developed by utilising hydrodynamic principles and natural water physical properties learned through water immersion [4]. However, Intveid et al. [4] did not determine how long the immediate reduction on low back pain would be sustained.

2 Methodology

2.1 Research Design

The research design was a quasi-experimental crossover design that involves one group with two sets of sample data. The study was concerned with treatment manipulations intended to develop cause-and-effect relationships [16]. This design was chosen as the patients went through two treatments, which were aquatic spinal stabilisation training and land spinal stabilisation training. The patients went through a pretest and washout method for first post-test and second post-test after water-based exercise. The pretest on the patients served as a baseline as to whether there were any differences following each treatment. The research design is illustrated in Fig. 1.

The independent variables in this study were the training methods, which were aquatic spinal stabilisation training and land spinal stabilisation training. Frequency of the exercise programme was similar for both methods. Both methods were conducted twice a week for 3 weeks, which sums up to 6 sessions. Aquatic spinal stabilisation training was conducted in a hydrotherapy pool (32–34 °C) in which land spinal stabilisation exercises and aquatic spinal stabilisation exercises were synchronised as closely as possible.

2.2 Sample

Participants were patients referred to the Physiotherapy Department in Terendak Camp Military Hospital, Terendak Camp in Malacca ranging in age from 36 to 55 years (Mean = 42.75, SD = 8.38). All patients had complained of back pain occurring for more than 3 months. Patients were free from chronic pulmonary or

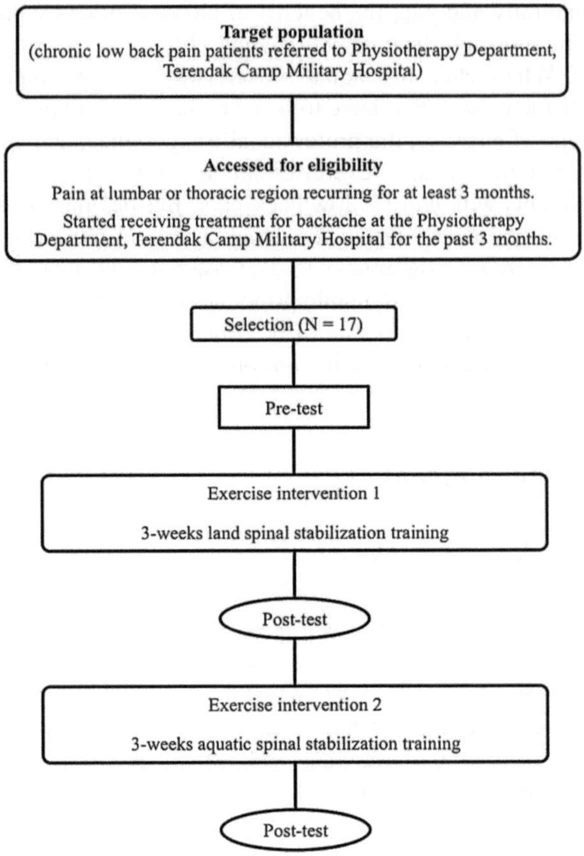

Fig. 1 *Crossover research design* washout method

metabolic disease, inflammatory diseases, vertebral fracture, and osteoporosis. Participants were selected from the Physiotherapy Department patient attendance records. Patients who had visited the Physiotherapy Department for treatment for the past 3 months were selected to participate in the study. Patients were given an informed consent letter and a subject information sheet. 17 patients agreed to participate. However, 5 patients defaulted after baseline assessment and another 8 patients defaulted after 3 weeks into the treatment. Therefore, only 4 patients complied with the study.

2.3 Sampling Technique

The sampling technique used was purposive sampling in which participants were selected based on chronic back pain symptoms. The gender was not fixed. The inclusion and exclusion criteria were as stated in Table 1. Participants were

Table 1 Descriptive statistics of age, weight, and height of subjects

Variable	*N*	Minimum	Maximum	Mean	SD
Age (years)	4	36	55	42.75	8.38
Weight (kg)	4	52.2	90	75.05	16.11
Height (cm)	4	143	178	162.25	15.24
BMI[a] (kg m^{-2})	4	25.25	36.05	28.45	5.12

[a] *BMI* body mass index

grouped into only one group. All participants were required to undergo aquatic spinal stabilisation training and land spinal stabilisation training. There was no control group as this study was concerned with comparing the effects of the two types of training methods in all subject involved, which were aquatic spinal stabilisation training and land spinal stabilisation training.

After identifying the target population (i.e. chronic low back pain patients referred to the Physiotherapy Department at Terendak Camp Military Hospital), patients who agreed to the terms were assessed for eligibility. They were asked to fill in a form containing details of their background and were also asked to sign an informed consent form. Patients referred to the Physiotherapy Department at Terendak Camp Military Hospital were chosen because the facilities to conduct the treatments were available at the Physiotherapy Department.

2.4 Instrumentation

The instruments (including facilities) used in this study were the EWAC Medical hydrotherapy pool (8,025 × 4,475 mm), visual analogue scale (VAS), Oswestry Disability Questionnaire (ODQ), pressure biofeedback unit, measuring tape, and the Zebris Static Force Plate 1504. Subjects were assessed, before the intervention, after the end of the land spinal stabilisation training, and after the end of the aquatic spinal stabilisation training, on their pain level, functional status, deep abdominal muscle function, trunk flexibility, balance, and centre of vertical force. Pain level was assessed using the VAS. The subjects were asked to rate their level of pain from a scale of 0 to 10. Older subjects might not illustrate pain the same way as younger subjects; thus, more questions have to be asked in clarifying the level of pain [17]. Functional status was calculated as a percentage using the ODQ. ODQ contains 10 questions on pain and pain-related disabilities in activities of daily living and social participation [18]. The ODQ evaluated various aspects of disability that affect daily life functions [19]. Deep abdominal muscle function was assessed using the pressure biofeedback unit. The procedure for conducting the test was in accordance with that described in Cynn et al. [11], where the subjects were required to maintain the required pressure and constant lumbar position whilst conducting lower extremity movements. All the subjects were comfortable with the standard position and usage of the pressure biofeedback unit. Trunk flexibility was determined by measuring trunk active flexion, extension, and lateral flexion using a measuring tape. For trunk flexion and extension, the measuring tape was placed proximately at the spinous process of C7 and distally to S1 during the starting position. The ending position followed the flexion of the vertebrae using the same landmark. The difference between the starting and ending position was calculated. Side lateral flexion was measured using the finger-to-floor method. The patient was instructed to stand heels together and knees straight with the hands hanging on the side [20]. The distance from the tip of the middle finger until the floor was measured. Then, the subject was asked to bend to the side, whilst the fingers were still facing down and laid by the side. The distance from the tip of the middle finger was again measured, and the difference was calculated. Balance and centre of vertical force were assessed using the Zebris Static Force Plate 1504. The subjects were required to stand in a bilateral stance on the force plate for 10 s whilst facing forward.

2.5 Data Collection Procedures

The participants went through a pretest to determine the baseline of their pain level, functional status, deep abdominal muscle function, trunk flexibility, balance, and centre of vertical force. The study began with a briefing of the subjects on the objective, purpose, and significance of the study, the procedures, what the study requires them to do, and why they were chosen as subjects.

2.6 Data Analysis

The statistical technique used was inferential statistics. Inferential statistics was applied because generalisations were being made from a smaller group to a bigger group [21]. Wilcoxon's matched-pairs signed-rank-sum test was used to compare the scores within treatments in order to determine the effects of both treatments. This was because the same subjects went through two types of treatments, starting with land spinal stabilisation training and followed by aquatic spinal stabilisation training. The pain level, functional status, deep abdominal muscle function, trunk flexibility, balance, and centre of vertical force of the subjects were measured.

Data were summarised using means and standard deviations (SD). The means of the pre-intervention and post-intervention of the two training methods were calculated.

3 Results

The weight range of the subjects was between 52.2 and 90 kg, with a mean weight (SD) of 75.05 (16.11). The height range of the subjects was between 143 and 178 cm, with a mean height (SD) of 162.25 (15.24) cm. The range of BMI for the subjects was between 25.5 and 36.05 kg m^{-2}, with a mean BMI (SD) of 28.45 (5.12). Results of demographic data of the subjects are displayed in Table 1.

3.1 Effects of Aquatic Spinal Stabilisation Training and Land Spinal Stabilisation Training

Results of the Wilcoxon's matched-pairs signed-rank-sum test analyses showed that there was no significant difference in the pain level between baseline and post-land spinal stabilisation training ($z = -1.890$, $p > 0.05$). However, there was a significant difference in the pain level between post-land and post-aquatic spinal stabilisation training ($z = -2$, $p < 0.05$).

There was no significant difference in functional status, deep abdominal muscle function, trunk flexibility, balance, and centre of vertical force between baseline and post-land spinal stabilisation training ($p > 0.05$). There was also no significant difference in functional status, deep abdominal muscle function, trunk flexibility, balance, and centre of vertical force between post-land and post-aquatic spinal stabilisation training ($p > 0.05$) (Table 2).

3.2 Comparison on the Effects of Aquatic Spinal Stabilisation Training and Land Spinal Stabilisation Training

Mann–Whitney test was used to compare the effects between aquatic spinal stabilisation training and land spinal stabilisation training. The results indicate that there was a significant difference in VAS scores between land (Mean rank = 3) and aquatic (Mean rank = 12) spinal stabilisation training ($p = 0.040$). However, there were no significant differences in functional status, flexibility, deep abdominal muscle function, balance, and centre of vertical force between land and aquatic spinal stabilisation training (Table 3).

4 Discussion

The study was designed to identify the effects of land and aquatic spinal stabilisation training on pain level, functional status, trunk flexibility, deep abdominal muscle function, balance, and centre of vertical force in chronic back pain patients.

Table 2 Wilcoxon's matched-pairs signed-rank-sum test at baseline, post-land spinal stabilisation training, and at post-aquatic spinal stabilisation training

Outcome	Variable	Z	p-value
Baseline–post-land	VAS	−1.89	0.059
Post-land–post-aquatic		−2	0.046*
Baseline–post-land	ODQ	−1.826	0.068
Post-land–post-aquatic		−1.826	0.068
Baseline–post-land	Pressure biofeedback unit	−1.826	0.068
Post-land–post-aquatic		−1.841	0.066
Baseline–post-land	Trunk flexibility	−0.365	0.715
Post-land–post-aquatic		−1.461	0.144
Baseline–post-land	Balance	−1.461	0.144
Post-land–post-aquatic		−0.73	0.465
Baseline–post-land	Centre of vertical force	−1.095	0.273
Post-land–post-aquatic		−1.826	0.068

*$p \leq 0.05$, reject H_0

Table 3 Mann–Whitney test comparison between land spinal stabilisation training and aquatic spinal stabilisation training

Outcome		Mean rank	Sum of ranks	z	Asymp. sig (2-tailed)
VAS	Land	3	6	−2.049	0.040*
	Aquatic	12	24		
ODQ	Land	4.13	16.5	−0.436	0.663
	Aquatic	4.88	19.5		
Flexibility	Land	5.5	22	−1.155	0.248
	Aquatic	3.5	14		
Pressure biofeedback unit	Land	4.63	18.5	−0.146	0.884
	Aquatic	4.38	17.5		
Balance	Land	3	12	−1.732	0.083
	Aquatic	6	24		
Centre of vertical force	Land	3.5	14	−1.155	0.248
	Aquatic	5.5	22		

*$p < 0.05$, reject H_0

4.1 Pain Level

This study showed a significant difference in VAS scores between aquatic and land spinal stabilisation training ($p < 0.05$). There was a reduction in the mean VAS scores between baseline ($M = 5.50$), post-land aquatic spinal stabilisation training ($M = 4.25$), and post-aquatic spinal stabilisation training ($M = 2.25$). These results were in agreement with previous findings by Bello et al. [22] who reported lower VAS scores on chronic low back pain patients after 6 weeks of land-based exercises and hydrotherapy. However, in terms of differences between the types of training, the results from this study were contrary to the report by Bello et al. [22], who found no significant differences in the mean post-treatment VAS scores between land-based exercises and hydrotherapy.

4.2 Functional Status

The ODQ was used to determine functional status of chronic low back pain patients. Keller et al. [18] had used the same questionnaire on patients with chronic low back pain. Results of the present study showed a reduction in the mean ODQ percentages from baseline ($M = 44.6675$), following post-land spinal stabilisation training ($M = 32.09$), and post-aquatic stabilisation training ($M = 15.39$). These results showed that both types of training had positive effects on the functional status of chronic low back pain patients. However, Keller et al. [18] did question the results of ODQ as they believed that the ordinal scales in ODQ should not be measured by adding or subtracting the values since each category in the questionnaire are distinct from each other.

4.3 Deep Abdominal Muscle Function

The pressure biofeedback unit was used to assess ability of the deep abdominal muscles to actively stabilise the lumbar spine. Cynn et al. [11] had also analysed deep abdominal muscle function using the pressure biofeedback unit. In the present study, the minimum pressure biofeedback unit reading difference for post-aquatic spinal stabilisation training was zero, which means that the pressure biofeedback unit reading remained at 40 mm Hg. Thus, the results indicate that after undergoing aquatic spinal stabilisation training, the subjects had greater lumbo-pelvic control.

4.4 Trunk Flexibility

Trunk flexibility was determined by calculating trunk flexion, trunk extension, and trunk lateral flexion. Trunk flexibility was measured using a measuring tape, and the data are presented in centimetres. Results of the analysis on trunk flexibility was somewhat controversial as there was a reduction in the mean scores between post-land spinal stabilisation training ($M = 43.75$) and post-aquatic spinal stabilisation training ($M = 38.675$). This may be due to the fact that the method of measuring trunk flexibility was not in accordance with the modified Schober flexion technique (MSFT) as used in the research conducted by Bello et al. [22].

4.5 Balance

Difference in balance between the left and right leg was measured using the Zebris Static Force Plate 1504 (32 × 47 cm). Based on the results, the land spinal stabilisation training illustrated more positive effects compared to the aquatic spinal stabilisation training. Improvements in balance will benefit those who are functionally impaired [23].

4.6 Centre of Vertical Force

Centre of vertical force which was the force exerted by the feet against the support surface was determined using the Zebris Static Force Plate (32 × 47 cm). Results displayed positive effects in the mean centre of vertical force from baseline ($M = 3.975$), compared to post-land spinal stabilisation training ($M = 4.6$), and post-aquatic spinal stabilisation training ($M = 6.55$).

5 Conclusion

The findings of this study are as follows:

- There was a significant difference in pain level following aquatic spinal stabilisation training and land spinal stabilisation training in chronic back pain patients.
- There was a significant difference in functional status following aquatic spinal stabilisation training and land spinal stabilisation training in chronic back pain patients.
- There was a significant difference in deep abdominal function following aquatic spinal stabilisation training and land spinal stabilisation training in chronic back pain patients.
- There was a significant difference in centre of vertical force following aquatic spinal stabilisation training and land spinal stabilisation training in chronic back pain patients.

The results above provide evidence that aquatic spinal stabilisation training and land spinal stabilisation training decreased the pain level and improved functional status, deep abdominal muscle function, and centre of vertical force. The reduction in pain level and improvements in functional status, deep abdominal muscle function, and centre of vertical force between baseline and post-land treatment (post-test 1) and also between post-test 1 and post-aquatic treatment (post-test 2) verified the improvement after 3 weeks of land spinal stabilisation training and 3 weeks of aquatic spinal stabilisation training. In conclusion, both aquatic and land spinal stabilisation training demonstrated positive results in four out of six measures, which were pain level, functional status, deep abdominal muscle function, and centre of vertical force.

References

1. Demirel, R., Ucok, K., Kavuncu, V., Gecici, O., Evcik, D., Dundar, U., et al. (2008). Effects of balneotherapy with exercise in patients with low back pain. *Journal of back and musculoskeletal rehabilitation, 21*(4), 263–272.
2. Chilibeck, P. D., Vatanparast, H., Cornish, S. M., Abeysekara, S., & Charlesworth, S. (2011). Evidence-based risk assessment and recommendations for physical activity: Arthritis, osteoporosis, and low back pain. *Applied Physiology, Nutrition, and Metabolism.*
3. Waller, B., Lambeck, J., & Daly, D. (2009). Therapeutic aquatic exercise in the treatment of low back pain: A systematic review. *Clinical Rehabilitation, 23*(1), 3–14.
4. Intveid, E., Cooper, S., Kessel, G. v. (2010). The effect of aquatic physiotherapy on low back pain in pregnant women. *International Journal of Aquatic Research and Education.*
5. Koumantakis, G. A., Watson, P. J., & Oldham, J. A. (2005). Trunk muscle stabilization training plus general exercise versus general exercise only: Randomized controlled trial of patients with recurrent low back pain. *Journal of the American Physical Therapy Association, 85*(3), 209–225.
6. Souza, G. M., Baker, L. L., & Powers, C. M. (2001). Electromyographic activity of selected trunk muscles during dynamic spine stabilisation exercises. *Physical Medicine and Rehabilitation, 82*(11), 1551–1557.
7. Carter, J. M., Beam, W. C., McMahan, S. G., Barr, M. L., & Brown, L. E. (2006). The effects of stability ball training on spinal stability in sedentary individuals. *Journal of Strength and Conditioning Research, 20*(2), 429–435.
8. Danneels, L. A., Vanderstraeten, G. G., Cambier, D. C., Witvrouw, E. E., & Cuyper, H. J. D. (2000). Ct imaging of trunk muscles in chronic low back pain patients and healthy control subjects. *European Spine Journal, 9*(4), 266–272.
9. Marshall, P. W., & Murphy, B. A. (2005). Core stability exercises on and off a swiss ball. *Physical Medicine and Rehabilitation, 86*(2), 242–249.
10. Newcomer, K. L., Jacobson, T. D., Gabriel, D. A., Larson, D. R., Brey, R. H., & An, K.-N. (2002). Muscle activation patterns in subjects with and without low back pain. *Physical Medicine and Rehabilitation, 83*(6), 816–821.
11. Cynn, H.-S., Oh, J.-S., Kwon, O.-Y., & Yi, C.-H. (2006). Effects of lumbar stabilizationstabilisation using a pressure biofeedback unit on muscle activity and lateral pelvic tilt during hip abduction in sidelying. *Physical Medicine and Rehabilitation, 87*(11), 1454–1458.
12. Bouche, K., Stevens, V., Cambier, D., Caemaert, J., & Danneels, L. (2006). Comparison of postural control in unilateral stance between healthy controls and lumbar discectomy patients with and without pain. *European Spine Journal, 15*(4), 423–432.
13. Barbosa, T. M., Marinho, D. A., Reis, V. M., Silva, A. J., & Bragada, J. A. (2009). Physiological assessment of head-out aquatic exercises in healthy subjects: A qualitative review. *Journal of Sports Science and Medicine, 8*(2), 179.
14. *Aquatic fitness professional manual.* (2010). (6th ed.). United States of America: Human Kinetics.
15. Verhagen, A. P., Cardoso, J. R., & Bierma-Zeinstra, S. M. A. (2012). Aquatic exercise and balneotherapy in musculoskeletal conditions. *Best Practice & Research Clinical Rheumatology, 26*(3), 335–343.

16. Thomas, J. R., Nelson, J. K., Silverman, S. J. (2011). *Research methods in physical activity* (6th ed.): Human Kinetics.
17. Bruckenthal, P., Reid, M. C., & Reisner, L. (2009). Special issues in the management of chronic pain in older adults. *Pain Medicine, 10*(S2), S67–S78.
18. Keller, A., Brox, J. I., & Reikeras, O. (2008). Predictors of change in trunk muscle strength for patients with chronic low back pain randomized to lumbar fusion or cognitive intervention and exercises. *Pain Medicine, 9*(6), 680–687.
19. Gronblad, M., Hupli, M., Wennerstrand, P., Jarvinen, E., Lukinmaa, A., Kouri, J.-P., et al. (1993). Intercorrelation and test retest reliability of the Pain Disability Index (PDI) and the Oswestry Disability Questionnaire (ODQ) and their correlation with pain intensity in low back pain patients. *The clinical journal of pain, 9*(3), 189–195.
20. Frost, M., Stuckey, S., Smalley, L. A., & Dorman, G. (1982). Reliability of measuring trunk motions in centimeters. *Journal of the American Physical Therapy Association.*
21. Berg, K. E., & Latin, R. W. (2008). *Research methods in health, physical education, exercise science, and recreation* (3rd ed.). Philadelphia: Lippincott Williams & Wilkins.
22. Bello, A. I., Kalu, N. H., Adegoke, B. O. A., Agyepong-Badu, S. (2010). Hydrotherapy versus land-based exercises in the management of chronic low back pain: A comparative study. *Journal of Musculoskeletal Research.*
23. Willardson, J. M. (2007). Core stability training: Applications to sports conditioning programs. *Journal of Strength and Conditioning Research, 21*(3), 979–985.

Differences in Game Statistics Between Winning and Losing Football Teams in Malaysia Super League: A Pilot Study

Norasrudin Sulaiman, Muhammad Sufyan Mohamad Zaki, Mubin Ali, Rahmat Adnan and Shariman Ismadi Ismail

Abstract The aim of this study was to identify specific performance indicators to distinguish winning and losing football teams in thc Malaysia Super League (MSL). Eight selected games from the 2012 season were analyzed. The studied variables were divided into three groups, which were related to goals scored (shot on goal, shot off goal, total shots), related to offenses (fouls committed, offside received, crosses, corners), and related to defenses (fouls received, offside committed, yellow cards, and red cards received). Data were analyzed using the Sportcode gamebreaker software. The results showed shots on goal had a significant difference between winners and losers ($Z = -2.375$, $p = 0.018$, $p < 0.05$), while no significant differences were found with the other study variables. The paper presents values that might be utilized as references to plan and assess projects and likewise for creating future groupings.

Keywords Performance indicators · Match analysis · Football

1 Introduction

The important of performance analysis of team sports, especially in technical aspects are crucial in improving team performance. Knowledge on characteristics of the winning team can be transformed in training sessions. Coach preservers give extra focus on significant indicators that are able to contribute to the betterment of the team's performance. Identifying reasons why some individuals and groups

N. Sulaiman (✉) · M. S. Mohamad Zaki · M. Ali · R. Adnan · S. I. Ismail
FSR Performance Analysis Laboratory, Universiti Teknologi MARA, Shah Alam, Malaysia
e-mail: noras878@salam.uitm.edu.my

N. Sulaiman · M. S. Mohamad Zaki · R. Adnan · S. I. Ismail
Sports Science Center of Studies, Universiti Teknologi MARA, Shah Alam, Malaysia

R. Adnan et al. (eds.), *Proceedings of the International Colloquium on Sports Science, Exercise, Engineering and Technology 2014 (ICoSSEET 2014)*,
DOI: 10.1007/978-981-287-107-7_40,

achieve success is a common aim in the performance psychology literature. Hence, methodologies have been used to study the performance at work and sport, and have been implemented to compare cognitions, decisions, and actions of successful and less successful groups and individuals. Lago et al. [1] proposed that scientists ought to center upon the improvement and use of performance indicators. The training procedure is about upgrading execution and by giving criticism about the execution to a player or any group. Past studies have indicated that human perception and memory are not dependable enough to give exact and objective data on perplexing games, for example, soccer, ball, and handball. Hence, target measuring apparatus are important to empower and encourage the sentiment process. The methods connected with measuring games execution are frequently alluded to as execution dissection and typically take the manifestation of feature investigation, utilizing either hand or modernized frameworks both throughout and post-occasion, from either a specialized, strategic, or development examination point of view.

A similar analysis remains in field hockey, where video analysis and hand notation systems were used to record individual actions and players positions to characterize successful patterns of play [2]. Performance indicators are characterized by the determination and mixture of variables that characterize some part of execution and assistance accomplished physical triumph [2]. Lago et al. [3] compared the performance of successful and unsuccessful teams in the 1990 World Cup. They found differences between the two in respect of the latter converting possession into shots on goal, with the successful team having the better ratio. However, Lago et al. [3] analyzed the play of great and unsuccessful groups prompting shots and objectives throughout the Copa America Tournament of 2001. Lago et al. [1] indicated that fruitful groups in the English Premier association regularly had longer belongings than unsuccessful groups regardless of the match status.

Based on these finding, it can be concluded that performance indicators have been used widely in past studies. However, until now it is unclear whether the performance indicators were used to evaluate past match performance or rather depends on individual characteristics that were played at their own team. Previous findings have suggested the use of performance indicators as key factors to predict future performance and improve their skill level through the development process.

To date, only a few studies have endeavored to give pointers of group execution through the correlation of winning and losing groups. As stated by Lago and Dellal [4], a large number of dubious discoveries of past studies looking at the ownership methodologies of fruitful and unsuccessful groups may start from reasonable and methodological issues. Ortega et al. [5] expressed that the investigation of amusement detail concerning singular and group aptitudes is one of the instruments that could be used to portray and screen conduct when in rivalry. One of the questions that need to be answered in this study is whether we can differentiate the successful team through the performance indicators and what are the key factors that contribute toward a successful team? In order to provide better understanding to the concept of this study, a review of literature on several related comparisons between successful and unsuccessful teams was carried out. From the articles evaluated, we found that explorations were seldom done into the development of group execution pointers and profiles in affiliation football. The point of this study was to recognize particular execution markers that recognize winning and losing football groups in the Malaysia Super League (MSL).

2 Method

2.1 Sample

A total of 10 matches from the Malaysian Super League season 2012 were randomly selected and analyzed in this study. All games were recorded using a video camera (model JVC HD Everio GZ-HM400). All games were analyzed using the Sportcode elite software (Sportstec Limited, Australia).

2.2 Procedures

All matches recorded were transferred into the Sport Code elite® (Sportstec limited, Australia) software for analysis. All the 11 performance indicators were divided into three groups of variables (Table 1). All the data gathered were transferred into the Statistical Package for the Social Sciences (SPSS) version 17.0 for analysis.

3 Statistical Analysis

Descriptive analysis was performed to determine the mean and standard deviation of each variable. Wilcoxon Signed Ranks Test was performed to distinguish between winning and losing teams. The level of significance was set at $p < 0.05$.

4 Results

The Wilcoxon signed ranks test results are summarized in Table 2. The first group of variables which were related to goals scored and shots on goal showed high significant differences between the winning and losing teams ($Z = -2.375$, $p = 0.018$). Shot off goal ($Z = -0.594$, $p = 0.553$) and total shots ($Z = -1.405$, $p = 0.160$) showed no significant differences between the winning and losing

Table 1 Variables studied in Malaysia Super League

Group of variables	Variables or game statistics or performance indicators
Variables related to goals scored	Shots on goal; shots off goal; total shots
Variables related to offense	Offside committed; fouls received; crosses; corners
Variables related to defense	Offside received; fouls committed; yellow card; red card

Table 2 Differences in game statistics between winning and losing football teams in Malaysia Super League

Group of variables		Winners		Losers		*P*
		Mean	SD	Mean	SD	
Related to goals scored	Shot on goal	6.88	1.73	4.38	1.85	0.018
	Shot off goal	6.00	2.73	5.5	3.5	0.553
	Total shots	12.88	2.95	9.88	4.7	0.16
Related to offense	Offside committed	2.5	2.00	3.88	2.1	0.228
	Fouls received	12.00	3.82	13.5	4.93	0.574
	Crosses	17.5	5.98	21.38	4.5	0.123
	Corners	4.5	2.2	3.88	1.64	0.552
Related to defense	Fouls committed	13.5	4.93	12	3.82	0.574
	Offside received	3.88	2.1	2.5	2.00	0.228
	Yellow card	1.38	1.3	0.75	0.71	0.236
	Red card	0.13	0.35	0.00	0.00	0.317

teams. The second group of variables which were related to offenses, showed that there were no significant differences between winning and losing teams for offside committed ($Z = -1.205$, $p = 0.228$), fouls received ($Z = -0.561$, $p = 574$), crosses ($Z = -1.542$, $p = 0.123$), and corners ($Z = -0.595$, $p = 0.552$). For group of variables related to defense, all variables showed no significant differences between winning and losing teams. These variables include fouls committed ($Z = -0.561$, $p = 0.574$), offside received ($Z = -1.205$, $p = 0.228$), yellow card ($Z = -1.186$, $p = 0.236$), and red card received ($Z = -1.000$, $p = 0.317$).

5 Discussion

The point of this study was to distinguish particular execution markers that recognize winning and losing football groups in the Malaysian Super League. The present study indicates that winning teams made more shots on goals compared to losing teams. This is because by shooting on goals, they probably can score more goals and win the match. Past studies [2] have shown that top teams made more shots and shots on goals than middle and bottom teams. Lago et al. [1] demonstrated that winning groups made a bigger number of shots than losing and drawing groups. This indicates how important this variable. In order to win the match with the possibility of creating goals and shots on goal are needed to win the match. The results of the present study support the idea that the winning teams are stronger in the variables related to goals scored. Future studies should emphasize more on causes related to shots on goal. According to Lago et al. [3], there were differences between successful and unsuccessful teams in converting possession into shots on goal.

Concerning the performance related to offenses, there were no significant differences between winning and losing teams in offsides committed, fouls received, crosses, and corners. Previous studies showed different results, and there were differences between winners, drawers, and losers teams related to offenses [3]. However, Lago et al. [3] also found different results compared to Hughes et al. and Low et al. Regarding performance related to defenses, there were no significant differences between the winning and losing teams. The articles reviewed for this study, also showed similar results [2]. From the articles reviewed, there were no studies that analyzed the relationship between indicators related to goals scored and playing pattern. Future research should concentrate on this area.

6 Conclusion

This study provides general values that help to comprehend football diversion and may be seen as standardizing values in outlining preparation sessions. The data can be used as reference or as guidelines for Malaysian footballers in order to form a better understanding and developing a training program that will emphasize more on game strategies.

Acknowledgments This work was supported in part by the Fundamental Research Grant Scheme (FRGS), Ministry of Higher Education of Malaysia [FRGS/1/2011/SKK/UITM/03/02].

References

1. Lago, C., Lago, J., & Rey, E. (2011). Differences in performance indicators between winning and losing teams in the UEFA champions league. *Journal of Human Kinetics, 27*, 137–148.
2. Lago, J., & Lago, C. (2010). Performance in team sports: Identifying the keys to success in soccer. *Journal of Human Kinetics, 25*, 85–91.
3. Lago, C., Lago, J., Dellal, A., & Gomez, M. (2009). Game-related statistics that discriminated winning, drawing and losing teams from the Spanish soccer league. *Journal of Sports Science and Medicine, 9*, 288–293.
4. Lago, C., & Dellal, A. (2010). Ball possession strategies in elite soccer according to the evolution of the match-score: The influence of situational variables. *Journal of Human Kinetics, 25*, 93–100.
5. Ortega, E., Villarejo, D., & Palao, J. M. (2009). Differences in game statistics between winning and losing rugby teams in the six nations tournament. *Journal of Sports Science and Medicine, 8*, 523–527.

Validity of YYIR1 and MST in Estimating VO_{2max} Among U-15 National Football Players

Norasrudin Sulaiman, Adde Shah Naddra Din, Rahmat Adnan, Shariman Ismadi Ismail and Rezian-na Muhamad Kasim

Abstract The purpose of this study was to determine the criterion validity of Yo-Yo intermittent recovery level 1 test (YYIR1) and 20-m multistage shuttle run test in estimating VO_{2max} among U-15 national football players. A total of 30 players participated in this research (mean ± SD; height, 154 ± 6.1 cm; weight, 46.35 ± 8.1 kg; and BMI 19.4 ± 3.1 kg). Field-based tests selected for this research were YYIR1 and multistage shuttle run test (MST), and the criterion test was the Bruce running protocol test. Subjects had to undergo the entire selected test, and 7 days was given as an interval time between the tests. The descriptive and inferential statistics were implemented in data analysis using the Statistical Package for Social Sciences version 2.0 (SPSS ver. 2.0). The results showed significant correlations between these variables (MST vs. criterion test, $r = 0.69$; YYIR1 vs. criterion test, $r = 0.55$). Thus, the results confirm that both tests can be used to measure the cardiovascular fitness among under-15 football players. MST had a better correlation with Bruce protocol compared with YYIR1. Hence, coaches are recommended to use MST to measure the cardiovascular fitness among young football players.

Keywords VO_{2max} · YYIR1 · MST

N. Sulaiman (✉) · R. Adnan · S. I. Ismail
FSR Performance Analysis Laboratory, Universiti Teknologi MARA, Shah Alam, Malaysia
e-mail: noras878@salam.uitm.edu.my

N. Sulaiman · R. Adnan · S. I. Ismail · R. Muhamad Kasim
Sports Science Center of Studies, Universiti Teknologi MARA, Shah Alam, Malaysia

A. S. N. Din
UiTM Sport Center, Universiti Teknologi MARA, Shah Alam, Malaysia

R. Adnan et al. (eds.), *Proceedings of the International Colloquium on Sports Science, Exercise, Engineering and Technology 2014 (ICoSSEET 2014)*,
DOI: 10.1007/978-981-287-107-7_41,

1 Introduction

It is essential to determine the cardiovascular ability among football players since they have different roles in the team. Rahmat et al. [1] had found that there was a difference in cardiovascular ability among football players based on their position. Hence, coaches have to look into a multifaceted approach in developing good football players [2]. Football is also known as an intermittent sport that usually uses repeated movements that include anaerobic and aerobic capacity, agility, strength, and power. The ability to use all the movements will give an impact on their physiological and technical performance in football. Based on the above scenario, football player must be well developed especially in aerobic capacities before they compete in competitive games. For football players, there is a 90-min demand for high physiological adaptation and maximal effort to perform during the game.

Aerobic capacity directly measures and involves a large number of muscles for physiological adaptation of an individual [3]. Aerobic capacity can be measured by laboratory tests or field-based tests. This study used field-based tests, namely Yo-Yo intermittent recovery level 1 test (YYIR1) and the 20-m multistage shuttle run test (MST). Football is a multifaceted sport that involves various skills and movements such as walking, jogging, sprinting, jumping, and acceleration.

The physiological aspect in football was assessed through laboratory tests and field-based tests. Laboratory tests are valid as compared to field-based tests, but laboratory tests involve high-technical tools and only one player can be assessed at one time. Coaches usually use field-based tests because this form of testing is easy to administer to a large number of subjects. According to Leger and Lambert [4], the capacity of an athlete can be measured using field-based tests such as the 12-min run, the Leger shuttle run test, and the maximal oxygen uptake (VO_{2max}). Based on the Leger multistage fitness test, previous researchers had developed a different form of the test for football players, such as YYIR1 and MST. The validity of the test among adult football players has been established [5], but is it valid for use among elite under-15 football players? Therefore, the objectives of the current research were to investigate the criterion validity of YYIR1 and MST in estimating VO_{2max} among U-15 national football players.

2 Method

2.1 Sampling Technique

The current research used purposive sampling by choosing the entire under-15-year-old players who represent the national squad. Players who were involved with injuries were excluded from this research. A permission letter was given to the coaching team, as they are responsible to take care of all the players during the training at Bukit Jalil, Kuala Lumpur, Malaysia.

2.2 Inclusion and Exclusion Criteria

A total of 30 players participated in this study. The inclusion criterion was males, age 14–15 years, and represented at the state level at least once. All the participants played different positions such as goalkeeper, midfielder, defender, and striker. In addition, all participants were required to undergo the pretesting field test procedure. The important inclusion criteria were that all participants were current students of Bukit Jalil Sport School (BJSS) and that they are representing the Malaysian under-15 national squad. In this study, the exclusion criterions for the subjects were those on medication and exposed to any form of injury.

2.3 Data Collection Procedures

Test selection was based on the tests that had been used to measure cardiovascular fitness among the football players. Two tests had been selected to be investigated in this research, while the Bruce protocol acted as a criterion test. Subjects were required to fill the PAR-Q before the physical fitness test as a screening tool. This research was approved by the Faculty Sport Science and Recreation Ethical Committee.

All subjects had to undergo all the tests, namely YYIR1, MST, and Bruce protocol test, and the rest interval between all the tests was 1 week [6]. All the field-based data (YYIR1 and MST) were compared with the criterion test (Bruce protocol) to determine the strength of the relationship between field-based tests and laboratory tests.

2.4 Data Analysis Procedures

The data were analyzed using the Statistical Package for Social Sciences version 19.0 (SPSS ver. 19). The descriptive statistics (mean ± SD) was used to analyze demographic data, and inferential statistics (Pearson's correlation) was used to answer the research hypothesis.

3 Research Findings

3.1 Demographic Data

A total of 30 under-15 national football players participated in this study, and their demographic data are presented in Table 1.

Table 1 Demographic data of subjects

	Min	Max	Mean	SD
Height (cm)	154	177	167.7	6.1
Weight (kg)	46.35	78.8	59.5	8.1
BMI (kg)	19.5	24	21.8	3.1

Table 2 Correlation between YYIR1 and criterion tests

		YYIR1	Criterion test
Distance covered	Pearson's correlation		0.550**
	Sig. (two-tailed)		0.002
	N	30	30

**Significant correlation, $p < 0.05$, YYIR1

The results indicate that the minimum height of the subjects was 154.0 cm and the maximum height was 177.0 cm with a mean of 167.7 cm and a standard deviation of 6.1 cm (Table 1). The lightest subject was 46.35 kg, and the heaviest subject was 78.8 kg with a mean of 59.5 kg and a deviation among the subjects of 8.1 kg. The players' mean BMI was 21.8 kg with a deviation of 3.1 kg and a minimum of 19.5 kg and a maximum of 21.8 kg.

3.2 Hypothesis Testing

Pearson's correlation was used to answer the research hypothesis. The details of the analysis between the YYIR1 and criterion tests are presented in Table 2.

Based on the findings, there was a significant correlation between YYIR1 and the criterion tests ($r = 0.55$, $p = 0.002$, $p < 0.05$). The correlation between the MST and the criterion test is presented in Table 3.

The results showed that there was a significant correlation between MST and criterion test ($r = 0.69$, $p = 0.002$, $p < 0.05$). MST showed a higher correlation with the criterion test compared with YYIR1 test.

4 Discussion

Based on the results, it was evident that all the selected tests showed significant correlations with the criterion test. Similar results were reported in a previous study which stated that there was a high correlation ($r = 0.91$) between MST and criterion test [4]. However, the subjects in this previous research were adults compared with the subjects in the present study.

Table 3 Correlation between MST and criterion test

		MST	Criterion test
Total shuttle	Pearson's correlation		0.69**
	Sig. (two-tailed)		0.000
	N	30	30

**Significant correlation, $p < 0.05$, YYIR1

The findings from this study are also supported by previous research on investigations into the validation of two running tests to estimate maximal aerobic power among children [6]. They used 82 participants aged between 12 and 14 years. They found that there was a high correlation ($r = 0.76$) between the two running tests in estimating aerobic power among children.

Another previous research had stated that there was a significant correlation between MST and VO_{2max} ($r = 0.95$, $p < 0.01$) among children football players [7]. Researcher used 35 male football players in the age range of between 14 and 16 years as their participants. The study developed a new equation for prediction of the aerobic capacity, but it may different from the current study because of differences in altitude level.

The present research found that there was a significant correlation between distance covered in YYIR1 and that covered in the criterion test in estimating VO2max among U-15 national football players. According to Bangsbo et al. [8] and Krustrup et al. [9], YYIR1 is a valid tool to measure aerobic capacity for adult football players. Atkins [10] showed that YYIR1 was an appropriate test to determine near-maximal performance of elite game players. It also consisted of the total distance covered to perform in the game. A valid field-based test to predict VO2max is important because the results will be useful input for coaches to develop individual training programs. Improvements in a specific individual training program will improve team performance.

The finding of the current study was similar to previous research, but the strength of the relationship was different from previous research, probably due to differences in the age of maturity of the subjects. The construct validity of the YYIR1 as a measure of match-related physical performance has been clearly demonstrated, showing significant correlation between the YYIR1 and the distance covered during the match.

5 Conclusion

Based on the current findings, it can be concluded that MST and YYIR1 are valid tools to measure VO2max among young football players. Coaches can use either MST or YYIR1 to measure VO2max of their players; however, MST is a better tool to measure VO2max among young football players compared with YYIR1 [11]. Future research needs to be done to compare other types of cardiovascular test among under-15 football athletes.

Acknowledgments We would like to express our gratitude to coaches and the under-15 national football players who gave excellent cooperation in the data collection process. Special thanks to the Faculty of Sports Science and Recreation, Universiti Teknologi MARA (UiTM) Shah Alam, Selangor, Malaysia, for the support throughout the whole process of this study.

References

1. Rahmat, A., Sharifudin, S.A., Tumijan, W., & Norasrudin, S. (2011). The VO_{2max} differences based on the position roles amongst Malaysia leagues soccer athletes. In *IEEE Colloquium on Humanities, Science and Engineering (CHUSER), 2011* (pp. 290–293). New York: IEEE.
2. Stolen, T., Chamari, K., Castagna, C., & Wisloff, U. (2005). Physiology of soccer. *Sport medicine*, *35*(6)501–536.
3. Astrand, P. O., & Rodahl, K. (1970). *Text book of work physiology*. New York: McGraw Hill Book Company.
4. Leger, L., & Lambert, J. (1982). A maximal multistage shuttle run test to predict VO2max. *European Journal of Applied physiology*, *49*(1), 1–12.
5. Norasrudin, S., Nasir, M., Akmal, M., Adnan, R., Misdan, M., & Ahmad, H. (2011). Criterion validity of Yo-Yo intermittent endurance test in estimating maximal oxygen consumption among malaysian elite football players. In *IEEE Colloquium on Humanities, Science and Engineering (CHUSER), 2011* (pp. 281–284). New York: IEEE.
6. Van Mechelen, W., Hlobil, H., & Kemper, H. C. G. (1986). Validation of two running tests as estimates of maximal aerobic power in children. *European Journal of Applied Physiology and Occupational Physiology, 55*(5), 503–506.
7. Cheterjee, P., Banarjee, A.K., Das, P., & Debnath, P. (2008). A regression equation to predict VO2 max of youn football players of Nepal. *International Journal of Applied Sport Sciences*, 21(2), 113–121.
8. Bangsbo, J., Iaia, F.M., & Krustrup, P. (2008). The yo-yo intermittent recovery test: A useful tool for evaluation of physical performance in intermittent sports. *Sports medicine*, *38*(1), 37–51.
9. Krustrup, P.M., Amstrup, T., Rysgaard, T., Johansen, J., Steensberg, A., Pedersen, P.K., & Bangsbo. J. (2003). The yo-yo intermittent recovery test: physiological response, reliability and validity. *Medicine and Science in Sports and Exercise*, 697–705.
10. Atkins, S.J. (2006). Performance of the yo-yo intermittent recovery test by elite professional and semiprofessional rugby league players. *Journal Strength and Conditioning Research,* 222–225.
11. Krustrup, P., Mohr, M., Ellingsgaard, H., & Bangsbo, J. (2005). Physical demands during an elite female soccer game: importance of training status. *Medicine and Science in Sports and Exercise, 37*, 1242–1248.

The Effects of High-Intensity Interval Training and Continuous Training on Weight Loss and Body Composition in Overweight Females

Syazwani Airin, Adam Linoby, Muhammad Sufyan Mohamad Zaki, Hafizuddin Baki, Hanifa Sariman, Badli Esham, Mohd Zulkhairi Mohd Azam and Muhamad Noor Mohamed

Abstract It is now known that aerobic capacity and weight reduction can be induced by either continuous or interval training. However, for many years, researchers have been debating the superiority of one training regime (continuous vs. interval training) over another. Furthermore, there are limited studies investigating the effect of both types of training towards the overweight population, particularly among women. The aim of this study was to investigate the comparative effects between high-intensity interval training and continuous training in inducing the improvement of body weight and body composition among overweight females. The subjects were randomly separated into two groups: high-intensity interval training group (HIIT) and continuous training group (CONT). Each group performed the activities three times per week for six weeks. HIIT ($n = 16$) required 8 s of sprint at a cadence range of 120–130 rpm within 80–95 % of heart rate maximum (HRmax) interspersed with 12 s of recovery period at cadence of 40 rpm for approximately 20 min. The CONT group ($n = 16$) performed at 60–70 % of HRmax continuously for 30 min. The HIIT group showed significant improvement in body fat percentage (2.2 % vs. 0.3 %), lean body mass (−0.5 kg vs. 0.8 kg) and waist-to-hip ratio. Nevertheless, there were no apparent differences in weight and body mass index (BMI) between the two groups. But there was a significant decrease in body weight in HIIT and CONT groups, respectively, whereas BMI was significantly reduced by 0.5 kg/m^2 for both groups. The HIIT group showed greater decrease in body fat percentage as well as the improvement of overall anthropometric indices in overweight females.

Keywords Aerobic · Intensity · Anthropometry · Women · Overweight

---

S. Airin · A. Linoby (✉) · M. S. Mohamad Zaki · H. Baki · H. Sariman · B. Esham · M. Z. Mohd Azam
Faculty of Sports Science and Recreation, Universiti Teknologi MARA, Pahang, Malaysia
e-mail: linoby@pahang.uitm.edu.my

M. N. Mohamed
Faculty of Sports Science and Recreation, Universiti Teknologi MARA, Negeri Sembilan, Malaysia

R. Adnan et al. (eds.), *Proceedings of the International Colloquium on Sports Science, Exercise, Engineering and Technology 2014 (ICoSSEET 2014)*,
DOI: 10.1007/978-981-287-107-7_42,

1 Introduction

Aerobic exercise is well known for its benefits in reducing the risks of chronic diseases, weight management and maintenance of fitness level [1, 2]. It is widely acknowledged that obesity has emerged as an epidemic problem on a worldwide scale. Studies conducted in 2005 by the World Health Organization found that 400 million people (9.8 % of the world's population) were classified as obese [3]. In addition, in Malaysia, the number of obese adult females was rising dramatically higher than males (7.6 % in 1996 to 17.4 % in 2006), while male obesity increased from approximately 20.1 % in 1996 to 29.7 % in 2006 [4]. Knowing this, people are now becoming more aware of the importance of endurance-type activities in maintaining their overall fitness.

Generally, the effectiveness of aerobic exercise has been shown to improve the anthropometrics indices of the body. Nowadays, aerobic exercise interventions are designed to reduce body weight and body composition. Many of the widely used programmes still focus on traditional continuous training protocols, for example jogging or walking in a steady state [5]. Meanwhile, more studies have been able to show that high-intensity interval training (HIIT) has the potential to be an alternative approach to gain aerobic cardiovascular and anaerobic fitness [6] and also the improvement of body composition in fewer workout durations [7, 8]. However, in an attempt to reassure the public on the superiority of one training programme over another (interval and continuous) in improving body composition, several studies have shown conflicting results [9, 10]. Despite large number of studies showing the benefit of HIIT in relation to weight loss, there were limited studies conducted on the effects of both types of training on the overweight population, particularly women. Hence, the purpose of this study is to look into the outcome of both training modality (interval and continuous) inducing weight loss among overweight females.

2 Methodology

2.1 Study Population

A total of 32 overweight females (20.5 ± 1.9 years, 25.2 ± 1.4 kg/m^2) were recruited from the Contours Express gymnasium members. This included those who are accustomed to exercises before but are not trained in any sport. Participants were randomly assigned into two groups: high-intensity interval training (HIIT; $n = 16$) and continuous training (CONT; $n = 16$).

2.2 Procedures

In order to avoid different changes to the result, the subjects included are those who are not taking any medication or supplements, non-smokers and those who have a stable weight. Participants who reported injuries or suffered from any medical complications were excluded from this study. The selected subjects were advised to avoid strenuous activity and caffeine consumption for 24 h prior to the testing. In order to determine the level of activity, they completed the physical activity readiness questionnaire (PAR-Q), and all subjects provided a written and informed consent letter prior to the study. Subjects were informed about the benefits and also possible injuries which might result from the study. The selected subjects were asked to keep a record of their calorie intake throughout the duration of the study, and they were not allowed to change their eating habits as that can give negative changes to the data.

2.3 Anthropometric Measurements

Trained assistants measured the subject's height and weight. Subject's bodyweight such as body fat percentage was measured using BIA (Body Impedance Analysis) method that helps by HBF-82 352 body fat monitor (Omron Healthcare Co, Kyoto, Japan) tool. Height of subjects was taken as a mean of three measurements with a Holtain stadiometer (Holtain Ltd., Crymych, Dyfed, UK) of the subject's standing height without shoes and was read to the nearest 0.5 cm. Waist and hip circumferences were measured by measuring tape to the nearest 0.1 cm.

2.4 Training Protocols

Both groups completed 3 sessions of training programmes a week, continuously for 6 weeks. Every training session consisted of a 5-min warm-up and cool down within 30–35 % of heart rate maximum (HRmax) and performed using Kettler Giro M bicycles. The HIIT subjects completed supervised training periods consisting of 20 min per session. Each training session included an 8-s sprint at a cadence range from 120 to 130 rpm within 80–95 % of HRmax interspersed with 12 s of recovery period at a cadence of 40 rpm for approximately 20 min. The subjects also performed at the specific level of intensity and were instructed to maintain that level in order to achieve the 85–95 % of HRmax. The CONT group (n = 16) performed at 60–70 % of HRmax continuously for 30 min. Their heart rate was monitored at every interval. The total training period per day is approximately 30 min including warming-up and cooling down. The power output was similarly set between two groups.

2.5 Statistical Analysis

Data were analysed using the Statistical Package for Social Sciences (version 20.0 for Windows, SPSS Inc., Chicago, IL, USA). Independent-samples t tests were used to ensure consistency of both groups prior to training. Paired t tests were used to assess any differences and comparisons between the two groups within inviolate variables of paired t test assumptions. The statistical analysis was considered significant at $p < 0.05$.

3 Results

All subjects successfully completed the training protocols. There was no significant difference in body weight, body mass index (BMI) and age before starting the training programme (Table 1).

3.1 Weight and Body Mass Index

There were no apparent differences in weight and BMI between the two groups. But there was a significant decrease in their body weight ($p < 0.05$) by 1.2 kg (1.9 %) and 1.3 kg (2 %) in HIIT and CONT groups, respectively, whereas BMI was significantly reduced by 0.5 kg/m^2 for both groups (Table 1, Fig. 1).

3.2 Body Fat Percentage and Lean Body Mass

When compared to the CONT group, the HIIT group induced a significantly higher decrease in body fat percentage (2.2 % vs. 0.2 %, $p < 0.05$) (Fig. 2). The HIIT group also resulted in increased calculated lean body mass more than the CONT group (Table 2).

3.3 Waist/Hip Circumference and Waist-to-Hip Ratio

After the intervention, waist-to-hip ratio was significantly decreased in the HIIT group ($p = 0.001$); waist and hip circumference also decreased (from 82.3 $\pm$ 5.5 to 80.9 $\pm$ 5.7 and from 101.9 $\pm$ 5.0 to 100 $\pm$ 5.1) compared with the CONT group (from 84.3 $\pm$ 5.8 to 84.0 $\pm$ 5.7 and from 100.5 $\pm$ 5.4 to 100.3 $\pm$ 5.1) (Table 2).

Table 1 The mean (SD) values of both groups prior to training programme

	HIIT Pre*	CONT Pre*
Age (years)	20.6 ± 1.9	20.5 ± 1.9
Height (cm)	159.5 ± 5.1	160.3 ± 5.0
Weight (kg)	63.8 ± 5.8	65.1 ± 5.1
BMI (kg/m^2)	25.1 ± 1.4	25.3 ± 1.3

*$p < 0.05$

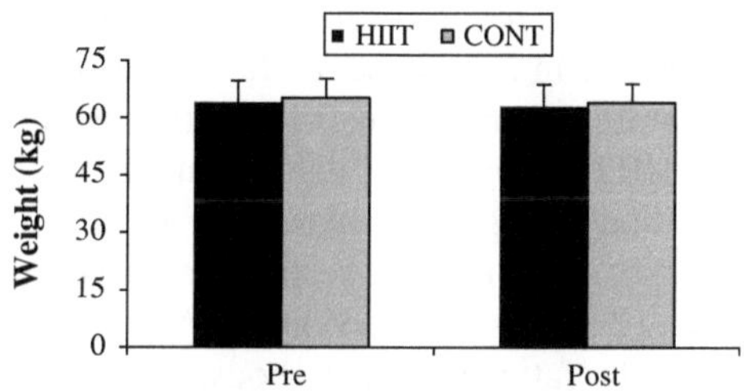

Fig. 1 Within-group and between-group changes in weight (mean and standard deviation)

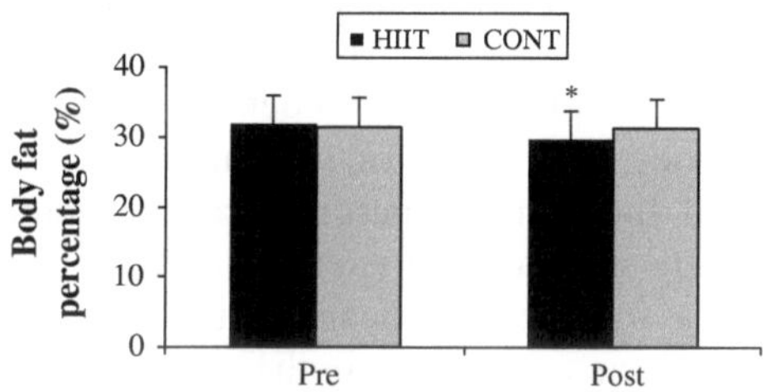

Fig. 2 Within-group and between-group changes in body fat percentage (mean and standard deviation)

Table 2 The mean (SD) values of anthropometric changes for both groups prior to and after the training programme

	HIIT		CONT	
	Pre	Post	Pre	Post
Weight (kg)	63.8 ± 5.8	62.9 ± 6.0	65.1 ± 5.1	63.8 ± 4.9
BMI (kg/m^2)	25.1 ± 1.4	24.5 ± 1.6	25.3 ± 1.3	24.8 ± 1.3
Body fat percentage (%)	31.9 ± 4.1	29.7 ± 4.0	31.5 ± 4.2	31.3 ± 4.3
Lean body mass (kg)	43.3 ± 2.9	43.8 ± 3.0	44.6 ± 3.7	43.8 ± 3.7
Waist circumference (cm)	82.3 ± 5.7	81.0 ± 5.5	84.3 ± 5.8	84.0 ± 5.7
Hip circumference (cm)	101.2 ± 5.0	100.0 ± 5.1	100.5 ± 5.3	100.3 ± 5.1
WHR	0.814 ± 0.04	0.810 ± 0.04	0.840 ± 0.06	0.839 ± 0.06

4 Discussion

In general, the main finding of this study was that six weeks of HIIT led to significant improvement in body weight, BMI, body fat percentage, lean body mass and waist-to-hip ratio (WHR) of overweight females (Table 2).

Researchers [11] conducted a 15-week study of HIIT programme comprised of 8/12 s of work–rest-ratio cycling protocol in approximately 20 min per session study. The other group in the study performed traditional aerobic cycling of 40 min per session. The HIIT group shows a decrease of 2.5 kg in fat, but no change was found in the second group. Reference [12] also reported a significant reduction in subcutaneous fat by 2.6 kg (8 %), total body mass by 1.9 kg (3 %), abdominal trunk fat by 0.12 kg (6 %) and waist circumference by 3.5 cm (4 %) in the HIIT group when using similar protocols. In addition, researchers [7] found that body weight, fat mass and waist circumference were significantly decreased by 2, 6.7 and 11.7 %, respectively, whereas the control group remains unchanged after 12 weeks of intervention. A recent study demonstrates an increase in changes of body fat percentage, body mass index and waist-to-hip ratio among HIIT subjects [13]. Researchers [13] conducted research in which sedentary women performed 4–6 repeated maximal efforts of sprint running in a specific measured area interspersed by 20–30 s of recovery rest. They concluded that HIIT may be a good training regime in order to reduce subcutaneous fat and also the improvement of body anthropometric measurements including body mass index and waist-to-hip ratio. Interestingly, past researchers examined the comparative effects of HIIT and steady-state aerobic exercise regimes for individuals with type 2 diabetes [14]. They found a significant reduction in visceral fat of 48 % compared with the steady-state group where there was only an 18 % decrease. Study by [15] in which 8-week HIIT intervention was implemented noted that abdominal trunk fat also significantly decreased by 44 %. Furthermore, past researchers utilized longer duration, in terms of training programmes in order to discover the effect of HIIT on the reduction of fat loss [16]. In their study, lasting 24 weeks, they compared both exercise regimes and found that the HIIT group lost more subcutaneous fat than did the steady-state group.

However, in this study, there was no significant difference in body weight and BMI between the HIIT and CONT groups. These findings are supported by previous researchers, who discovered that HIIT induced significant fat loss, but resulted in an increase of 0.3 % of body mass in a 12-week training intervention [9]. The subjects underwent a 1:1 work–rest ratio of four intervals. Study by [17] investigated the effects of exercise intensity performed by 27 women with obesity, 3 days a week for 8 weeks. They expressed that the changes in fat loss between the independent groups were low (40–50 % of VO_{2max}) and high intensity (80–90 % of VO_{2max}) produced no significant difference. Researchers [18] claim that there are no apparent differences in weight changes and body fat for untrained women who had performed either a low or high intensity of exercise in 12 weeks. They recruited 12 obese women (25–40 years) and divided them into two groups:

low-intensity group (LI; women with 31.0 ± 4.8 % body fat; $n = 6$) and high-intensity group (HI; women with 31.1 ± 3.8 % body fat; $n = 6$). They utilized the differences between the low-intensity group, which is 50 % of VO_{2max}, and the high-intensity group, which is 80 % of VO_{2max}, to discover the changes in body composition in a 12-week training programme. In their study, researchers [19] also found no significant differences between interval training and continuous training with regard to total body weight. In their study, thirteen subjects participated in a training programme with a 1:2 work–rest ratio. The HIIT group performed at the level of 90–105 % of their VO_{2max} and also low intensity of continuous training at 30–45 % of VO_{2max}. All trainings were completed within 30 min. In the case of [20], there were no significant changes of subjects' body composition either in the HIIT or in the continuous training group among pre-menopausal women. The programme included a 2-min sprint interval at 95 % VO_{2max} separated by 3 min of recovery period at 1.2 mph of speed with a 0 % grade; it was an 8-week programme with three sessions per week. The authors concluded that significant changes could be achieved for a longer period of training, with a larger sample size; thus, increasing in the volume of training was not necessary.

For instance, the improvement of anaerobic capacity is closely related to the production of skeletal muscle or muscle mass resulting in the increase in resting metabolic rate (RMR) [20]. Hence, this training regime will generate greater calorie burning while at rest and facilitate more fat loss. So, these situations may be due to the increase in muscle mass as well as muscle oxidative capacity which are associated with greater fat oxidation and decrease the insulin resistance among HIIT subjects in the current study. Aerobic and anaerobic fitness of chronic responses is well achieved by high-intensity interval training in relation to the original HIIT protocols. The result in getting an increase in aerobic and anaerobic capacity was readily approved with high-intensity training exercises [6]. Traditionally, according to the past study [6], the ratio of the original protocol, also known as the IE1 protocol, is set at 2:1 of work and recovery times, respectively. Exercise session begin with warm-up, follow by 8 sets of bout interval running with all out effort in 20 second and 10 second resting, and complete the test with cooling down. The high-intensity interval training group in the study resulted in significant improvements in aerobic and anaerobic capacity among male subjects (mean age 23 ± 1 years of age). However, there is only a significant improvement in aerobic capacity and not anaerobic capacity. The authors concluded that HIIT not only provides a potential training regime for the aerobic system but also gives significant benefit to the anaerobic system. Researchers [1] used the short-term HIIT protocols and performed the Wingate test, which consisted of 30 s of maximum effort cycling against high resistance. Within 2 weeks, HIIT effectively shows higher muscle oxidative capacity than traditional endurance training programmes. In addition, HIIT has the potential to increase rapid skeletal muscle remodelling in order to elicit the level of the recruitment of muscle fibre [1]. Previous studies revealed significant increases in trunk muscles in young women after HIIT interventions and also in the tight muscles by using MRI of older males and females [15]. Another studies also demonstrated the effects of HIIT to increase whole-body and skeletal muscle capacities in either trained or untrained individuals [21, 22]. Similarly, past study conducted found that the capacity of efficient oxidation only took 2 weeks of HIIT intervention [23].

5 Conclusion

In conclusion, HIIT may be a time-efficient exercise that provides greater decrease in body fat percentage as well as the improvement of anthropometric indices in overweight females. Consequently, HIIT performed for 20 min per day may have greater potential in increasing the aerobic and anaerobic fitness than traditional continuous training. As the HIIT exercise regime required minimal time commitment, the impact still produces benefits for the body, including reduction in the subcutaneous and visceral fat in order to prevent the risk factors for chronic disease, mainly related to obesity and cardiovascular disease [24].

Acknowledgments This work was supported in part by the Fundamental Research Grant Scheme (FRGS), Ministry of Higher Education of Malaysia FRGS/1/2011/SKK/UITM/03/30.

References

1. Gibala, M. J. (2007). High-intensity interval training: a time-efficient strategy for health promotion? *Current Sports Medicine Reports, 6*, 211–213.
2. Gill, J. M., & Malkova, D. (2006). Physical activity, fitness and cardiovascular disease risk in adults: interactions with insulin resistance and obesity. *Clinical Science, 110*, 409–425.
3. World Health Organization (WHO) (2006). Obesity and overweight. Retrieved 8 Sept 2013, from http://www.who.int/mediacentre/factsheets/fs311/en/index.html.
4. The National Health and Morbidity Survey (NHMS) (2006). Ministry of Health, Malaysia. Retrieved 20 Aug 2013.
5. Jacobsen, D., Donnelly, J., Snyder-Heelan, K., & Livingston, K. (2003). Adherence and attrition with intermittent and continuous exercise in overweight women. *International Journal of Sports Medicine, 24*(6), 459–464.
6. Tabata, I., Nishimura, K., Kouzaki, M., et al. (1996). Effects of moderate-intensity endurance and high-intensity intermittent training on anaerobic capacity and VO_{2max}. *Medicine and Science in Sports and Exercise, 28*, 1327–1330.
7. Heydari, M., Freund, J., & Boutcher, S.H. (2012). The effect of high-intensity intermittent exercise on body composition of overweight young males. *Journal of Obesity*, 1–8.
8. Boutcher, S.H. (2010). High-intensity intermittent exercise and fat loss. *Journal of Obesity*, 1–10.
9. Tjonna, A. E., Stolen, T. O., Bye, A., et al. (2009). Aerobic interval training reduces cardiovascular risk factors more than a multitreatment approach in overweight adolescents. *Clinical Science, 116*(4), 317–326.
10. Brian, A. I., Christopher, K. D., David, W. B., Judy, Y. W., et al. (2008). Effect of exercise training intensity on abdominal visceral fat and body composition. *Medicine and Science in Sports and Exercise, 40*(11), 1863–1872.

11. Trapp, E. G., Chisholm, D. J., Freund, J., & Boutcher, S. H. (2008). The effects of high-intensity intermittent exercise training on fat loss and fasting insulin levels of young women. *International Journal of Obesity, 32*(4), 684–691.
12. Dunn, S.L. (2009). Effects of exercise and dietary intervention on metabolic syndrome markers of inactive premenopausal women (Doctoral dissertation, University of New South Wales).
13. Kordi, M. R., Choopani, S., Hemmatinafar, M., & Choopani, Z. (2013). The effects of the six week high-intensity interval training on resting plasma levels of adiponectin and fat loss in sedentary young women. *Journal of Jahrom University of Medical Sciences, 11*(1), 20–27.
14. Mourier, A., Gautier, J.-F., Kerviler, E. De et al. (1991). Mobilization of visceral adipose tissue related to the improvement in insulin sensitivity in response to physical training in NIDDM: effects of branched-chain amino acid supplements. *Diabetes Care, 20*(3), 385–391.
15. Boudou, P., Sobngwi, E., Mauvais-Jarvis, F., Vexiau, P., & Gautier, J. F. (2003). Absence of exercise-induced variations in adiponectin levels despite decreased abdominal adiposity and improved insulin sensitivity in type 2 diabetic men. *European Journal of Endocrinology, 149*(5), 421–424.
16. Tremblay, A., Simoneau, J.-A., & Bouchard, C. (1994). Impact of exercise intensity on body fatness and skeletal muscle metabolism. *Journal of Metabolism, 43*(7), 814–818.
17. Ballor, D. L., McCarthy, J. P., & Wilterdink, E. J. (1990). Exercise intensity does not affect the composition of diet- and exercise-induced body mass loss. *The American journal of Clinical Nutrition, 51*, 142–146.
18. Grediagin, M. A., Cody, Mildred, Rupp, Jeffrey, Benardot, D. A. N., & Shern, Robin. (1995). Exercise intensity does not effect body composition change in untrained, moderately overfat women. *Journal of the American Dietetic Association, 95*(6), 661–665.
19. Wallman, K., Dawson, B., & De La Motte, P. (2006). Effects of interval and continuous aerobic training on body composition and metabolism in overweight/obese person. *Journal of Science and Medicine in Sport, 12*, 109.
20. King, J., Panton, L., Broeder, C., Browder, K., & Panton, L. (2001). A comparison of high intensity vs. low intensity exercise on body composition in pre-menopausal women. *Medicine and Science in Sports and Exercise, 33.* 228.
21. Kohn, T. A., Essen-Gustavsson, B., & Myburgh, K. H. (2011). Specific muscle adaptations in type 2 fibers after high-intensity interval training of well-trained runners. *Scandinavian Journal of Medicine and Science in Sports, 21*, 765–772.
22. Christoper, G. R., Heigenhauser, J. F., Bonen, A., & Lawrence, L. S. (2008). High-intensity aerobic interval training increases fat and carbohydrate metabolic capacities in human skeletal muscle. *Applied Physiology, Nutrition and Metabolism, 33*, 1112–1123.
23. Talanian, J. L., Galloway, S. D., Heigenhauser, G. J., Bonen, A., & Lawrence, L. S. (2007). Two weeks of high-intensity aerobic interval training increases the capacity for fat oxidation during exercise in women. *Journal of Applied Physiology, 102*, 1439–1447.
24. Drigny, J., Gayda, M., Nigam, A., Guilbeault, V., et al. (2011). Long-term lifestyle intervention and optimized high intensity interval training program improve body composition, cardiovascular risk and exercise capacity in obese patients with or without metabolic syndrome. *Journal of Rehabilitation, 7*, 532.

Relationship Between Mental Toughness and Sport Performance Among Contact and Non-contact Sport Athletes

Mardiana Mazaulan and Mohamad Rahizam Abdul Rahim

Abstract The purpose of this study was to determine the relationship between mental toughness and sport performance among contact sport and non-contact sport athletes. In addition, differences in mental toughness dimensions between contact sport and non-contact sport athletes and difference in mental toughness between genders were compared. It was hypothesized that there was a relationship between mental toughness and the level of achievement among the athletes. A total of 285 athletes who participated in the Sukan Kolej-Kolej Bersekutu (SUKOB) competition (males = 147, females = 138) aged 18–26 years participated in this study. The method involved in this study was self-evaluation by participants themselves. The psychological performance inventory (PPI) questionnaire was distributed to the athletes who participated in the event during that time. After finishing the match, the researcher evaluated the sport performance based on the results of the match, either a win or a loss. Findings from this study showed a significant, weak, and negative relationship between overall mental toughness and sport performance among contact and non-contact sport athletes ($r = -0.398$, $p = 0.000$). Contact sport athletes scored higher in overall mental toughness compared with non-contact sport athletes ($t = 2.99$, $p = 0.003$). It can be concluded that mental toughness is an important component of psychological skills that should be included in training session in order to achieve excellent performance in competitions.

Keywords Mental toughness · Sport performance · Contact sport athletes · Non-contact sport athletes

M. Mazaulan (✉) · M. R. Abdul Rahim
Faculty of Sport Science and Recreation, Universiti Teknologi MARA, Shah Alam, Malaysia
e-mail: mardiana9840@salam.uitm.edu.my; dianazz2328@yahoo.com

R. Adnan et al. (eds.), *Proceedings of the International Colloquium on Sports Science, Exercise, Engineering and Technology 2014 (ICoSSEET 2014)*,
DOI: 10.1007/978-981-287-107-7_43, © Springer Science+Business Media Singapore 2014

1 Introduction

Mental toughness is one of the common important elements of athlete performance in a game. It comes together with physical ability, expertise growth, work mentality, and respect of high-achieving athletes and coaches and describes the main factors that contribute to their success [1].

Coaches and athletes usually know that success in sport performance cannot be achieved without basic mental skills. Mental toughness can be described as the preparations that athletes need to make in terms of psychological skills training starting early during participation in training sessions and during and after the games [2]. This factor could determine an athletes' success or failure in competitions. Mentally tough performers are disciplined thinkers and respond to stress by staying comfortable, relaxed, and energized [3].

Sport performance in a competition is typically assessed through outcome measures. The outcome measure in competitions is determined using performance satisfaction questionnaire or by winning the competition [4]. A previous study revealed that athletes achieved good performance in their game when they were at an advanced stage of mental toughness [5]. In addition, athletes from team sports and contact sports revealed that when they scored advanced stage of mental toughness, they got better results in their game compared with individual and non-contact sport athletes [6].

Sports can be categorized into contact sports and non-contact sports. Nicholls et al. [7] noted that there were no differences in sport performance and mental toughness among athletes who participated in team or individual sports and athletes who participated in contact or non-contact sports. However, an earlier study had suggested that mental toughness might be specific to certain sports [8]. The result could be differences between team and individual athletes and those competing in contact and non-contact sports with the suggestion that those athletes participating in contact sports being more mentally tough during the competition.

In Malaysia, studies that were carried out specifically on the relationship between mental toughness and the performance of athletes were only on *Wushu* participants [4] and football players [9]. There is still a lack of understanding among athletes about the real meaning and importance of mental toughness to their performance, especially in contact and non-contact sport athletes. An earlier study reported that mental toughness influenced performance and achievement of athletes in competitions [1]. However, scientific support for the relationship between mental toughness and performance satisfaction has yet to be fully established [1].

Determining the level of mental toughness and its relation to sport performance may help athletes in their competition. There is a need to conduct a systematic study on a larger scale in Malaysian athletes on the level of mental toughness, especially in contact and non-contact sport athletes. Therefore, the purpose of this study was to investigate the relationship between mental toughness and sport performance among contact and non-contact sport athletes.

2 Methodology

2.1 Participants

Participants for this study were selected from athletes that were involved in Sukan Kolej-Kolej Bersekutu (SUKOB, UiTM). All the participants were athletes from different local Private Institutions of Higher Learning (IPTS) and Polytech MARA College (UiTM) at Jengka, Pahang. The criteria for selection of the participants were that they were SUKOB athletes, male and female, and 18 years of age and above. Based on the population of 1,100, the sample size needed for this study was 285 participants [10].

2.2 Instruments

2.2.1 Demographic Questionnaire

The demographic questionnaire was used to obtain respondents' data on personal information including gender, age group, academic background, and years of experience.

2.2.2 Psychological Performance Inventory

The PPI developed by Loehr [11] was used to measure mental toughness. This 42-item scale yields an overall mental toughness score, as well as seven (7)-item subscale scores, including self-confidence (items 1, 8, 15, 22, 29, 36), negative energy control (items 2, 9, 16, 23, 30, 37), attention control (items 3, 10, 17, 24, 31, 38), visualization and imagery control (items 4, 11, 18, 25, 32, 39), motivation (items 5, 12, 19, 26, 33, 40), positive energy control (6, 13, 20, 27, 34, 41), and attitude control (7, 14, 21, 28, 35, 42). Each subscale of mental toughness had 6 items. Mental toughness subscale dimension scores ranged from a low of six to a desirable high of 30, and total scores ranged from 42 to 210. Scores were recorded on a five-point Likert scale anchored by 'almost always' and 'almost never.' The PPI was found to be internally consistent with Cronbach's alphas for the seven subscales indicating high reliability (self-confidence—0.70, negative energy control—0.62, attention control—0.75, visualization and imagery control—0.82, motivation—0.70, positive energy control—0.71, attitude control—0.71) [9].

2.2.3 Sport Performance Evaluation

The sport performance evaluation was based on the results of the game, either win or lose. The scale for sport performance was 1 = win and 2 = lose.

2.3 Data Collection Procedure

The questionnaire was distributed to the athletes who participated in the event. The demographic questionnaire was also used to obtained data including gender, age group, academic background, level of achievement, years of experience, and types of sports. Participants were asked to complete the PPI questionnaire in 2 h before the match begins. This time frame was appropriate, and comfortable conditions were provided for the athletes to answer the questionnaire. The time frame chosen was similar to that employed by other researchers who utilized the PPI [5, 12]. Upon the completion of the questionnaire, the researcher collected the questionnaire from the participants. After the match was completed, the researcher evaluated the sport performance based on the result of the match, either win or lose.

2.4 Statistical Analysis

Descriptive statistics (percentage, mean, and standard deviation) were used to describe the categorical and demographic variables of respondents.

Biserial correlation was used to analyze and identify the relationship between mental toughness to sport performance among contact and non-contact sport athletes. Independent-samples t test was used to analyze and identify the differences in mental toughness dimension between contact and non-contact sport athletes. The statistical significance was set at an alpha level of $p < 0.05$.

3 Results

Positive energy (22.07 $\pm$ 2.55) showed the highest mean score among the dimensions of mental toughness, followed by motivation (21.33 $\pm$ 2.57), attitude control (21.22 $\pm$ 2.51), visual and imagery (20.43 $\pm$ 2.84), self-confidence (19.74 $\pm$ 2.03), attentional control (18.54 $\pm$ 2.36) and negative energy (18.14 $\pm$ 2.89).

The relationship between mental toughness and sport performance was determined using Pearson's correlation analysis. The results showed a significant, weak, and negative relationship between overall mental toughness and sport performance among contact and non-contact sport athletes ($r = -0.398, p < 0.05$). Dimensions of mental toughness for self-confidence ($r = -0.14$) and negative energy ($r = -0.15$) also showed a significant, weak, and negative correlation with sport performance among contact and non-contact sport athletes ($p < 0.05$). In contrast, the results found that there was no significant relationship between attentional control ($r = 0.00$), visual and imagery control ($r = 0.01$), motivation ($r = -0.12$), positive energy ($r = -0.02$), and attitude control and sport performance ($r = 0.01$) ($p > 0.05$) (Tables 1, 2, 3).

Table 1 Descriptive analysis of mental toughness dimension

Mental toughness	M	SD
Self-confidence	19.74	2.03
Negative energy	18.14	2.89
Attentional control	18.54	2.36
Visual and imagery	20.43	2.84
Motivational	21.33	2.57
Positive energy	22.07	2.55
Attitude control	21.22	2.51

Table 2 Correlation between performance and mental toughness

	C	NE	AC	VI	M	PE	ATT	O
Performance	−0.14*	−0.15*	0	0.01	−0.12	−0.02	−0.01	−3.98*

* $p < 0.005$

Table 3 Differences of mental toughness dimension between contact and non-contact sport athletes

	Types sport	N	Mean	T	P
Self-confidence	Contact sport	171	20.00	2.64	0.008*
	Non-contact sport	114	19.36		
Negative energy	Contact sport	171	18.65	3.70	0.000**
	Non-contact sport	114	17.37		
Attentional control	Contact sport	171	18.46	−0.67	0.504
	Non-contact sport	114	18.66		
Visual and imagery	Contact sport	171	20.52	0.64	0.521
	Non-contact sport	114	20.30		
Motivational level	Contact sport	171	21.49	1.27	0.205
	Non-contact sport	114	21.10		
Positive energy	Contact sport	171	22.04	−0.21	0.833
	Non-contact sport	114	22.11		
Attitude control	Contact sport	171	21.20	−0.17	0.868
	Non-contact sport	114	21.25		
Overall	Contact sport	171	21.10	2.99	0.003*
	Non-contact sport	114	20.32		

**$p < 0.001$, *$p < 0.05$

In conclusion, the results showed that the independent-samples t test was significant in overall mental toughness scores for contact and non-contact sports ($t = 2.99$, $p < 0.05$).

For mental toughness dimension, there was a significant difference in self-confidence and negative energy scores for contact and non-contact sports ($t = 2.64$, $p = 0.008$, $t = 3.70$, $p < 0.05$).

4 Discussion

The importance of mental toughness in sporting success has been highlighted in previous studies [5, 13]. Moreover, these characteristics including self-belief, focus, motivation, positive and tough attitude, and handling pressure have been repeatedly associated with mental toughness. The different findings suggest the continued choice of researchers to focus on specific sports. This study emphasizes the gap in the literature when examining intervention training on mental toughness in contact and non-contact sport athletes.

4.1 Relationship Between Mental Toughness and Sport Performance

As postulated, the present study showed a significant weak and negative relationship between mental toughness and sport performance among contact and non-contact sport athletes ($r = -0.398$, $p < 0.005$). In the present study, significant relationships were observed in mental toughness when winners and non-winners were analyzed. The winners showed better self-confidence and better energy control than the non-winners. Athletes who have a high level of mental toughness show excellent achievement in their game or competition. A previous study pointed out that psychological skill was one of the factors that affect sport performance in competitions [4].

In addition, the results of the current study demonstrated some similarities with previous studies [8], which reported that those athletes who have high mental toughness have positive relationship with sport performance. This was because athletes who were mentally tough were able to remain calm and relaxed; they are competitive in most situations and have lower anxiety levels than others. The current study also showed negative relationship between mental toughness and performance. Although these results showed significant relationships, it does not address the winners of the competition and the extent to which mental toughness is related to their performance success [2]. Anizu et al. [13] suggested that success in performance is relative, rather than absolute, and theoretically, mental toughness is a concept applicable to athletes of all skills and ability levels.

4.2 Differences of Mental Toughness Dimension Between Contact and Non-contact Sport Athletes

The current findings showed that the dimension of mental toughness was significant between contact and non-contact sports ($t = 2.99$, $p < 0.05$). Contact sport athletes scored a higher level of mental toughness in overall mental toughness and in other four dimensions of mental toughness, including self-confidence, negative energy control, visual and imagery control, and motivational level. These results have similarities with a previous study where the level of mental toughness among footballer players was excellent and the same thing goes to players' mental toughness dimensions where it was found that all dimensions were at an excellent level [9]. The differences in mental toughness dimensions between contact and non-contact sport athletes were not obvious, but despite that, contact sport athletes still showed a higher level of mental toughness in overall mental toughness and four dimensions of mental toughness including self-confidence, negative energy control, visual and imagery control, and motivation. Rationally, this implied that contact sport athletes can create pictures or images in mind which is one of the most powerful techniques to develop mental toughness as it is the connecting link between the mind and body. Furthermore, contact sport athletes were also able to control negative emotions such as fear, anger, and temper, increase awareness, and stimulate competitive situations and high level of motivation, which is essential for effective learning by which an individual is inspired to do something.

Findings obtained by Bull et al. [8] have suggested that mental toughness may be specific to certain sports. This could be the reason for the different results between contact and non-contact sports, and this suggests that contact sport athletes have a high mental toughness level compared with non-contact sport athletes. However, there was evidence that mental toughness was not different from situation to situation like individual versus team sports or contact versus non-contact sports and thus was acting like a personality trait [7].

4.3 Recommendations

Based on the current study, a lot of knowledge, experience, and information have been gained. It has been suggested that future research should increase the sample size, so that the range of the highest level achievements leads to findings that are more representative. Besides, future research should focus on high-standard competitions such as MASUM or SUKMA, in order to get more reliable results at the highest level of athletes' achievement.

As this study used the PPI questionnaire by Loehr [11], it has some limitations. Several dimensions of the questionnaire should be modified to enhance its efficacy. By using PPI, participants rated themselves on all items using the scale from 1 (almost never) to 5 (almost always), but two negatively worded items were designed to be reverse-scored and stand out differently than others. Instead of going from 1 to 5, the response choices on these items should go from 5 to 1.

Furthermore, physiological indicators should be added in future research on mental toughness. It would seem important to consider physiological toughness, given that researchers consider physical toughness as a component of mental toughness [1]. Assessing the impact of mental toughness training on biomarkers of stress like cortisol would help to understand exactly how mental toughness operates [14].

5 Conclusion

Three main conclusions were reached in this study. First, it was concluded that the mental toughness was one of the indicators in athletes' performance. The overall findings showed that mental toughness had a relationship with sport performance and influenced athletes' performance achievement in competitions. Athletes who have higher level of mental toughness showed better performance. Second, it was evident that contact sport athletes have higher level of mental toughness and achieved good results in their competition. Overall, contact sport athletes were mentally tough competitors. This implies that they are self-motivated and self-directed and do not need to be pushed as they are capable of controlling themselves from within. They are mentally alert, focused, confident, and responsible for their actions. Lastly, male athletes were more mentally tough compared with female athletes. Male athletes were more positive and realistic about their goals and ready for action and were usually energetic and determined. Overall, from this current study, it can be concluded that mental toughness is an important psychological skill that should be included in training sessions in order to improve and enhance athletes' performance in competitions.

Acknowledgments I would like to express my deepest gratitude and thanks to my supervisor Encik Mohamad Rahizam Bin Abdul Rahim for his unending support and help throughout this research. My thanks also go to all the lecturers of the Sport Science Faculty, UiTM Shah Alam, for their many hours of explanation and support. Special thanks to the SUKOB organizers and athletes who gave me their full support and cooperation during this study.

Last but not least, to my family, who is always there for me in any situation. Not forgetting all my friends who have helped me throughout this study. Thank you to all who had contributed directly or indirectly with their sweat and tears toward the completion of this study.

References

1. Newland, A., Newton, M., Finch, L., Harbke, C.R., & Podlog, L. (2012). Moderating variables in the relationship between mental toughness and performance in basketball. *Journal of Sport and Health Science*, 1–9.
2. Jones, G., Hanton, S., & Connaaughthon, D. (2007). A framework of mental toughness in the world's best performers. *Journal of the Sport Psychologist, 21*, 243–264.
3. Loehr, J. E. (1995). *The New Mental Toughness Training for Sport*. New York: Plume.
4. Kuan, G., & Roy, J. (2007). Goal profiles, mental toughness and its influence on performance outcomes among Wushu athletes. *Journal of Sports Science and Medicine, 6*, 28–33.
5. Bhambri, E., Dhillon, P. K., & Sahni, S. P. (2005). Effects of psychological intervention in enhancing mental toughness dimensions of sport persons. *Journal of the Indian Academy of Applied Psychology, 31*, 65–70.
6. Sohrabi, F., Atashak, S., & Aliloo, M. M. (2011). Psychological profile of athletes in contact and non-contact sports. *Journal of Scientific Research, 50*, 638–644.
7. Nicholls, A.R., Polman, R.C.J., Levy, A.R., & Backhouse. S.H. (2009). Mental toughness in sport: achievement level, gender, age, experience and sport type's differences. *Personality and Individual Differences*, *47*, 73–75.

8. Bull, S. J., Shambrook, C. J., James, W., & Brooks, J. E. (2005). Towards understanding of mental toughness in elite English cricketers. *Journal of Applied Sport Psychology, 17*, 209–227.
9. Mohamad Nizam, A., Omar Fauzee, M. S., & Abu Samah, B. (2009). The affects of higher score of mental toughness in the early stage of the league towards winning among Malaysian football players. *Journal of International Studies, 12*, 67–78.
10. Krejcie, R. V., & Morgan, D. W. (1970). Determining sample size for research activities. *Educational and Psychological Measurements, 30*, 607–610.
11. Loehr, J. E. (1986). *Mental toughness training for sports: achieving athletic excellence*. Lexington, MA: Stephen Greene Press.
12. Golby, J., & Sheard, M. (2004). Mental toughness and hardiness a different levels of rugby league. *Journal of Personality and Individual Differences, 37*, 933–942.
13. Anizu, M., Kumaraswamy, N., Singh, R., & Rusli, M. (2006). Mental toughness profile as one of psychological predictors of injuries among Malaysian professional football players. *Journal of International Studies, 12*, 128–138.
14. Crust, L. (2008). A review and conceptual re-examination of mental toughness: Implication for future researchers. *Personality and Individual Difference, 45*, 576–583.

Development of Heart Rate Monitor Using Colour-Coding System to Communicate Exercise Intensity

Adam Linoby, Fauzan Khairi and Fadzil Kamaruddin

Abstract Currently, many heart rate monitors or HRM have been developed with numerical digital displays providing indications of target heart rate. Such displays do not provide for ease of reading under most conditions of use, particularly when the user is exercising vigorously. This paper presents a prototype of a bracelet-like HRM which responds to the problem by integrating a colour-coding system. The HRM displays a uniform colour homogenously, enabling the user to tell at a brief glance the optimal exercise intensity range. The transmission medium of ANT+ was used to transfer heart rate data from the HRM chest strap to a microcontroller which then processed to display the appropriate colour according to project specifications. The LED was programmed using Arduino to automatically suggest to the user their optimum heart rate zones in accordance with their personal data and pre-set exercise goal. Analysis of the prototype demonstrated that the device is effective in communicating optimal training heart rate zone and would certainly help many athletes, coaches and the public to monitor and regulate their exercise training regime.

Keywords Heart rate monitor · Colour coded · Exercise intensity · Heart rate zone · Wireless sensor

A. Linoby (✉)
Faculty of Sports Science and Recreation, Universiti Teknologi MARA, Jengka, Pahang, Malaysia
e-mail: linoby@pahang.uitm.edu.my

F. Khairi
Faculty of Electrical Engineering, UTM, Johor Bahru, Malaysia

F. Kamaruddin
INSTEDT, Johor Bahru, Johor, Malaysia

R. Adnan et al. (eds.), *Proceedings of the International Colloquium on Sports Science, Exercise, Engineering and Technology 2014 (ICoSSEET 2014)*,
DOI: 10.1007/978-981-287-107-7_44,

1 Introduction

The measurement of heart rate is commonly used by medical and health professionals to assist in the diagnosis of a person's health condition. It is also increasingly used by athletes, coaches and members of the public interested in monitoring their heart rate in order to gain optimal benefits from their exercise training. The first wireless electrocardiogram (ECG) heart rate monitor available for sport and exercise use was developed by Polar Electro in 1977 and was used as a training aid for the Finnish National Cross Country Ski team [1]. Since then, with the increasing popularity of various fitness training and exercise programmes, more athletes, coaches and the public are paying greater attention to specific heart rates achieved during exercise. This phenomenon has also helped accelerate the development of a more practical and efficient heart rate monitoring device.

The heart rate monitor or HRM is a popular tool for evaluating exercise training status. The device has the ability to constantly monitor the heart rate and allow a user to be in their desired 'training zone'. Heart rate zone training allows an individual to exercise within a specific target heart rate to achieve the desired exercise objective [2]. Studies have shown that the optimal benefit from an exercise routine can be achieved by elevating the heart rate to a specific predetermined range for the duration of the exercise [3–5]. HRM can really ease the process by communicating to the user the optimal exercise heart rate intensity.

The intensity of exercise is difficult to control. Exercise at a too low exercise intensity and the desired training effect may not be reached, whereas overtraining may result from high volume of high training intensity [6]. Hence, having a practical and effective means of monitoring exercise intensity is essential to gain the optimal outcome of training. However, many HRMs have been developed with numerical digital displays and such displays do not provide for ease of reading under most conditions of use, particularly when the user is exercising vigorously [1, 7, 8]. This paper presents a prototype of a bracelet-like HRM which responds to the problem by integrating a colour-coding system. The HRM displays a uniform colour homogenously, enabling the user to tell at a brief glance the optimal exercise intensity range. The work presented in this paper will highlight the development of the prototype application in exercise and sport training context and extend the work in [2, 7, 9].

2 Related Research and Invention

Heart rate is the measurement of the number of contractions of the ventricles and typically determined as the number of heartbeats per unit of time. Commercial heart rate monitors were invented for individuals based on electrocardiograph heart monitoring, which consist of a chest strap with electrodes. Heart rate monitors allow accurate measurements to be taken continuously and can be used when manual measurement would be difficult or impossible to make, especially during vigorous activities [8].

One of the related prior arts is [10] which disclosed a heart rate monitoring device including two separated devices. One of the devices is a display unit to be attached at the wrist to display heart rate, and another one is a transmitter unit to transmit the heart rate to the display unit. The cited invention used a percentage of 60–80 % from the maximum heart rate as a guide to determine the minimum and maximum heart rates of the user, regardless of the age. There has not been any formula disclosed in the invention to determine the user's specific heart rate. Unlike the current invention, the cited invention is only based on the reading display for the result, and there is no other alternative for the user to check on the heart rate during vigorous exercise.

Another related prior art is US patent [11]; the patent has disclosed an electronic, wrist-worn device which has three display units on the device, and the device can be a heart rate monitor, a sportsman's watch or a diving computer. The cited invention has a light-emitting diode to only one side of the device for better indicator of the heart rate display. However, there are no different settings for different age and gender to determine the minimum and maximum heart rates in the cited invention, as the current project proposed.

In another US patent [12], the cited patent disclosed an apparatus and method for non-invasive monitoring of heart performance parameters and comprises at least one sensor adapted to continuously sense factors correlated with blood flow and collect data related to the flow of blood. The sensor is adapted to be positioned adjacent to the peripheral blood and preferably worn on a wrist. However, the maximum heart rate is not based on the age and gender as the features were not disclosed in the cited invention.

In 2010, Adidas introduced Micoach: the Interactive Personal Coaching and Training System which also uses colour to indicate heart rate zone [8]. This invention is considerably different from the proposed invention, in that the Micoach system colour display comprises a very small area of single LED lights, rather than the entire body of the monitor as proposed in the current prototype. The Finnish company Spektikor also developed a disposable heart rate indicator in 2011 which consists of a heart rate sensor together with two EKG electrodes and an LED indicator [8]. However, this device needs a connection of wire 35 cm long which will cause inconvenience when moving. Moreover, unlike current inventions that focus on exercise and sport setting, the cited invention is intended to be used for medical and hospital settings.

It is an objective of the present prototype invention to provide an apparatus for monitoring heart rate and displaying heart rate information in a more convenient display, especially during vigorous activities, by providing information not only on the display panel, but together with multicolour light sources for indicating different ranges of heart rate information.

3 Methodology

In this methodology section, the process sequence in relation to hardware involved in designing the colour-coded HRM system is described. This project prototype can continuously monitor heart rate and send feedback to user via specific colour of RGB LED lights. Figure 1 is a process flowchart of the prototype from setting user's information to the displayed result and colour LED output. The process begins with user options displayed on the screen of μOLED, either to enter self-determined maximum heart rate or to select to use the predicted maximum heart rate determined from user's information. If the user selects to use the predicted maximum heart rate, the user is then prompted to enter their age and gender. Then, the microcontroller determines the user's predicted maximum heart rate based on the equation [13]:

Male: Heart rate maximum $= 202 - (0.55 \times \text{age})$
Female: Heart rate maximum $= 216 - (1.09 \times \text{age})$

The user is then prompted to be in a resting state for the system to capture his/her resting heart rate. The Karvonen method equation [14] was used to calculate the target heart rate (THR) zone as stated below:

$$\text{THR} = ((\text{HRmax} - \text{HRrest}) \times \%\ \text{intensity}) + \text{HRrest}$$

The user is also prompted to select a mode of exercise (e.g. fat burning mode, cardiovascular mode, anaerobic mode and maximum mode) on the μOLED display panel. Then, the microcontroller, which is Atmega 328, determines the user's detected heart rate information during the selected exercise mode. The transmission of data from the chest strap to the heart rate monitor starts after the user has worn the chest strap properly. There is a transmitter that resides in the chest strap (Garmin) that is placed closely to the chest to detect the heart rate of the user. This detected heart rate data will then be transmitted to the ANT+ receiver. The ANT+ wireless protocol medium was chosen because of a small volume, high expansion, low power consumption, stylization and two-way transmission system [9]. Taking into consideration the maximum heart rate input and the mode of exercise selected, the microcontroller then controls the plurality of coloured light sources that correspond to the intensity of the user's heart rate as shown in Table 1.

4 Result and Discussion

4.1 Overall System Performance

The colour-coded heart rate monitoring system prototype shows excellent performance in transmitting data from the user's (tester) chest strap to the colour-coded HRM device. The prototype also displays the correct colour for

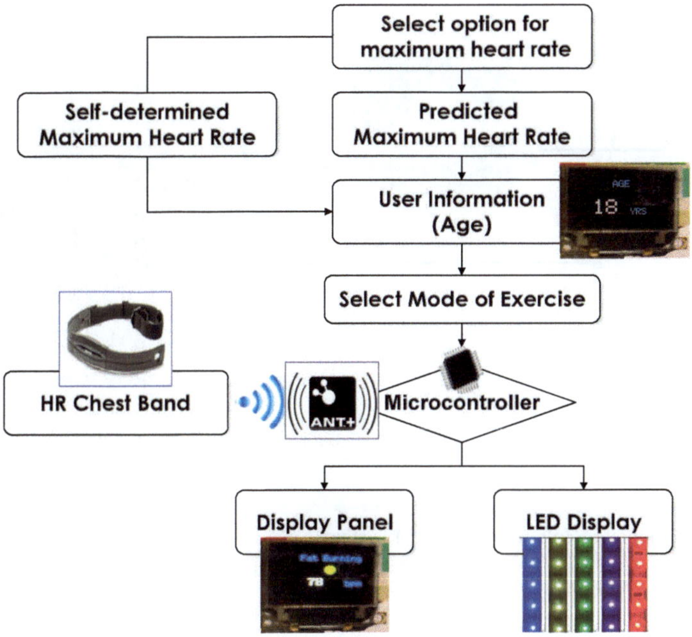

Fig. 1 Process flowchart of the prototype

Table 1 Relationship of the exercise mode selection, colour indication and corresponding percentage of heart rate ranges [8]

Exercise mode	Fat burning	Cardio	Anaerobic	Max
Colour (intensity status)	Exercise intensity (%)	Exercise intensity (%)	Exercise intensity (%)	Exercise intensity (%)
No colour (resting)	N/A	N/A	N/A	N/A
Yellow (very low) 	<60	<70 %	<80	<90
Green (optimal) 	60–69	70–79	80–89	90–100
Orange (high) 	70–79	80–89	90–100	N/A
Red (very high) 	80–89	90–100	N/A	N/A

corresponding heart rate range/zone as programmed. Figure 2 shows the completed colour-coded HRM system and the user placement for chest strap and the colour-coded HRM.

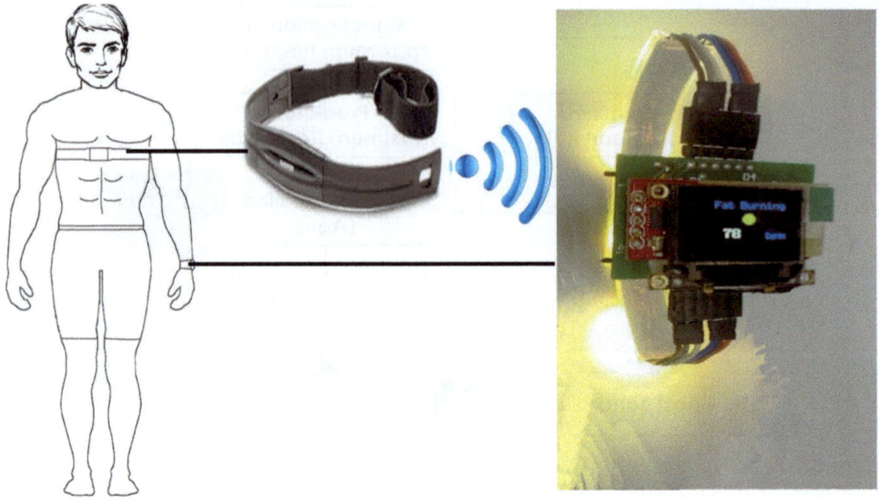

Fig. 2 The completed colour-coded HRM system

Fig. 3 Heart rate data collected using colour-coded HRM system and pulse oximeter

4.2 Data Accuracy

A test was conducted to determine the data accuracy of the colour-coded HRM prototype. Comparison of active heart rate data collected was made using the prototype and pulse oximeter on four levels of intensity as shown in Fig. 3. The percentage error for each intensity level was calculated by taking the heart rate data collected on both devices and using the standard pulse oximeter as reference.

Figure 4 presents very small percentage of error between data collected on the current prototype and pulse oximeter (reference data) ranging from 0.12 to 0.83 %

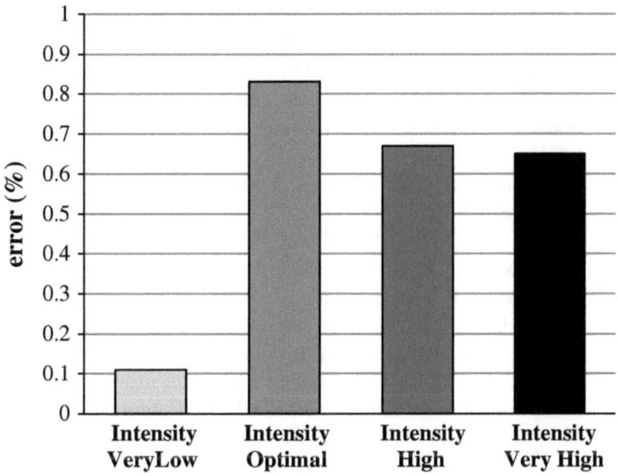

Fig. 4 Percentage error of colour-coded HRM (using pulse oximeter as reference) on four different levels of exercise intensity

(less than 1 %). Therefore, it can be inferred that heart rate data collected by the colour-coded HRM prototype system are reliable when using standard pulse oximeter as the reference data.

5 Conclusion

In conclusion, the colour-coded HRM prototype system is designed to help eliminate the need of a user to monitor numerical display (as in conventional HRM) which does not provide for ease of reading under most exercising conditions. The large LED colour surface of the current prototype enables an exercising person to determine the level of their target heart rate intensity at a glance, without having to slow or stop the exercise activity. The present heart rate monitor will also be beneficial in enabling users to make progress towards achieving their objective of a specific fitness domain. As the primary information required of an exercising user is their heart rate range and the knowledge of appropriate exercise exertion level, the colour-coded HRM system will prove to be beneficial in a range of populations including athletes, coaches and the general public to monitor and regulate their exercise training regime in a more effective, time-efficient and safer manner.

References

1. Laukkanen, R. M. T., & Virtanen, P. K. (1998). Heart rate monitors: state of the art. *Journal of Sports Sciences, 16*(supl), 3–7.
2. Che Harun, F.K., Zulkarnain, N., Ab Aziz, M.F., Mahmood, N.H., Kamarudin, M.F., Linoby, A. (2011). Pulse oximetry color coded heart rate monitoring system using zigbee. In 5th Kuala Lumpur international conference on biomedical engineering (pp. 348–351). Berlin: Springer.

3. Karvonen, J., & Timo, V. (1988). Heart rate and exercise intensity during sports activities. *Sports Medicine, 5*(5), 303–311.
4. Gutin, B., Paule, B., Scott, O., Lemmon, C. R., Mara, B., Jerry, A., et al. (2002). Effects of exercise intensity on cardiovascular fitness, total body composition, and visceral adiposity of obese adolescents. *The American journal of clinical nutrition, 75*(5), 818–826.
5. Lucia, A., Hoyos, J., Perez, M., & Chicharro, J. L. (2000). Heart rate and performance parameters in elite cyclists: a longitudinal study. *Medicine and Science in Sports and Exercise, 32*(10), 1777–1782.
6. Bompa, T.O., Haff, G. (2009). Periodization: theory and methodology of training. *Human Kinetics.*
7. Che Harun, F.K., Uyop, N, Ab Aziz, M.F, Mahmood, N.H, Kamarudin, M.F, Linoby, A. (2011) Color coded heart rate monitoring system using ANT + . In 5th Kuala Lumpur international conference on biomedical engineering (pp. 283–286). Berlin: Springer.
8. Linoby, A. (2009). Color-coded heart rate monitoring bracelets (COBRA) to guide specific exercise training intensity. RMI-UiTM (Patent). pp. 1–12.
9. Mahmood, N.H., Uyop, N., Zulkarnain, N., Che Harun, F.K., Kamarudin, M. F., Linoby, A. (2011). LED indicator for heart rate monitoring system in sport application. In IEEE 7th international colloquium on signal processing and its applications (CSPA) (pp. 64–66).
10. Pulkkinen, I., Kiuru, E., Lähdesmäki, T. (2006). Heart rate monitor. U.S. Patent 7,076,291, issued July 11, 2006.
11. Nissil, S., Puolakanaho, P. (2004). Electronic wrist-worn device and method of controlling the same. U.S. Patent 6,745,069, issued June 1, 2004.
12. Gorenberg, M., Gorenberg, N.V., Marmor, A., Naroditsky, M., Rotstein, H. (2002). Apparatus and method for non-invasive monitoring of heart performance. U.S. Patent No. WO2002000094 A2, issued January 3, 2002.
13. Whyte, G. P., George, K., Shave, R., Middleton, N., & Nevill, A. M. (2008). Training induced changes in maximum heart rate. *International Journal of Sports Medicine, 29*(02), 129–133.
14. Karvonen, M. J. (1957). The effects of training on heart rate, A longitudinal study. *Ann Ned Exp Biol Fenn, 35*, 307–315.

Part V
Sports Science and Performance

Efficacy of Core Stability Exercise and Muscular Stretching on Chronic Low-Back Pain

Ebby Waqqash, Rahmat Adnan, Sarina Md Yusof, Norasrudin Sulaiman and Shariman Ismadi Ismail

Abstract Chronic low-back pain (CLBP) affects most adults at some point in their lives. The purpose of the review was to search and analyze contemporary studies on muscular stretching and core stability (CS) exercise in rehabilitation of CLBP. Science Direct, Springer, and Google Scholar database were searched for related articles published from 2008–2013. A total of 57 articles were identified; however, only 10 papers fulfilled the criteria for this review. The results showed that both CS exercise and muscular stretching improve pain level and functional disability among CLBP patients. However, there is modest evidence from the relevant reviews that CS exercise is superior to muscular stretching in CLBP rehabilitation since CS exercise improves the deep core muscles, whereas muscular stretching has no effect on the deep core muscles. The effects of specific types of stretching techniques (e.g., effect of dynamic stretching and PNF stretching) are unclear because most studies reviewed utilize only basic stretching techniques. Hence, there is a need for further studies to examine the effects of specific stretching techniques on CLBP.

Keywords Low-back pain · Core exercise · Stretching · Rehabilitation

1 Introduction

Chronic low-back pain (CLBP) is one of the most prevalent causes of disability and morbidity in our society [1, 2]. One of the primary causes of CLBP is weakness and/or tightness in the core muscles [3]. A weak core is the fundamental cause of many inefficient movements that lead to injury. The tightness of core

E. Waqqash · R. Adnan (✉) · S. Md Yusof · N. Sulaiman · S. I. Ismail
Faculty of Sports Science and Recreation, Sports Science Center of Studies,
Universiti Teknologi MARA,
Shah Alam, Malaysia
e-mail: rahmatadnan@salam.uitm.edu.my

R. Adnan et al. (eds.), *Proceedings of the International Colloquium on Sports Science, Exercise, Engineering and Technology 2014 (ICoSSEET 2014)*,
DOI: 10.1007/978-981-287-107-7_45,

muscles also leads to CLBP, as it may influence the entire muscle kinetic chain of body movement. The core also known as lumbo-pelvic-hip complex is attached to the local (postural, tonic) muscles, i.e., transverse abdominis, multifidi, and quadratus lumborum [4]. These local muscles are responsible for providing segmental stability and directly controlling the lumbar segments during movement.

Rehabilitation therapy for CLBP has emerged over time. Core stabilization (CS) exercise is the basic and fundamental component of all comprehensive functional rehabilitation programs [3]. CS exercise program is designed to help an individual to gain functional strength, neuromuscular control, and muscle endurance of the core muscles [3, 5]. Muscular stretching is another physical therapy increasingly used in rehabilitation technique programs in CLBP [6]. Muscular stretching for CLBP patients is designed progressively stretching the muscle groups which are assumed to be too short [3].

Commonly, the low-back ache is only due to muscle tightness and eventually leads to soft tissue injuries. Therefore, a combination treatment of stretching and CS exercise was advocated by many physical and exercise therapists. However, combination treatment of both exercises thoroughly is time-consuming, unclear, and anecdotal by many studies and may not be necessary as muscle tightness can be alleviated with stretching alone. Hence, this article was aimed to review the empirical evidence related to muscular stretching and core stability exercises in rehabilitation of CLBP.

2 Methodology

2.1 Search Strategy

Searches were conducted in Springer, Science Direct, and Google Scholar databases. Keywords are as follows: low-back pain, core exercise, stretching, and rehabilitation.

2.2 Inclusion Criteria

A study was included if it met the following criteria:

1. Written in English.
2. The target population was people with CLBP.
3. Stretching or core stability exercise as intervention.
4. Date of publication 2008–2013.

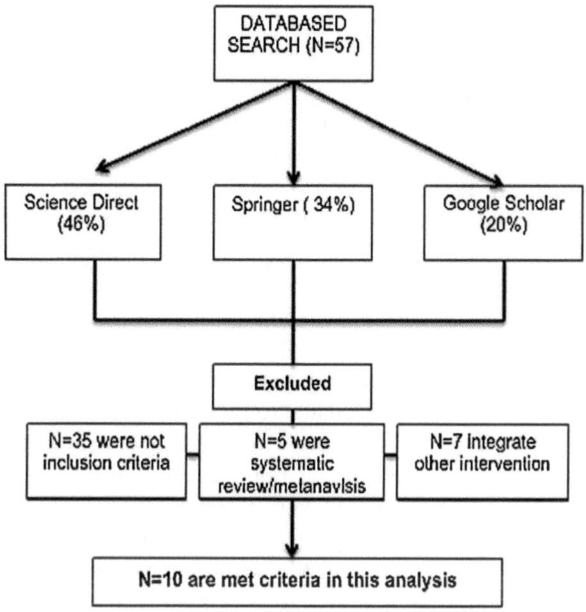

Fig. 1 Data collection process

2.3 Exclusion Criteria

A study was excluded if

1. It was a systematic review or meta-analysis.
2. Other interventions apart from stretching and core stability exercises (examples include massage, strength training, and locomotion training).

3 Results

Fifty-seven papers were identified, but only ten papers fulfilled the inclusion and exclusion criteria for this review. The papers were extracted from the three different databases: Science Direct (46 %), Springer (34 %), and Google Scholar (20 %). In ten published papers (2008–2013), it was discovered that the primary intervention for CLBP core strengthening was core stability exercise. Muscular stretching was identified as the secondary approach for rehabilitation of CLBP. The outcome measures most frequently used by the studies were thoracolumbar ROM, functional disability, pain, and trunk muscle strength and endurance (Fig. 1).

3.1 Core Stability Exercise

Five studies on core stability exercise were identified. The reviewed studies showed that core stability exercise helped to improve pain level, functional disability, and activation of tonic core muscle groups.

In the first CS exercise study reviewed, the author [7] studied the effects of integrated back stability (IBS) program in the CLBP population. The IBS program helps improve the level of pain and functional impairment in 89 % of patients suffering from CLBP. In the second CS exercise study reviewed, efficacy of supervised core exercise, spinal manipulation, and home exercise for the treatment of CLBP were evaluated [8]. The researchers concluded that supervised core exercise was significantly better than chiropractic spinal manipulation and home exercise in terms of satisfaction with treatment and trunk muscle endurance. In the third CS exercise study reviewed, the effects of CLBP on the trunk muscle activity, body balance, and lumbar range of motion while performing core stability exercises were investigated [9]. Their findings showed that CLBP patients inhibit greater muscle activity levels than control CLBP participants. In the fourth CS exercise study reviewed, Mohseni-Bandpei and associates [2] studied the effects of pelvic floor muscle exercise on women with chronic back pain. The overall finding showed that the pelvic floor muscle exercise combined with routine treatment was not more superior than routine treatment alone in terms of improving pain and functional disability, but was more efficient in improving core muscle strength and endurance. Finally, in the last CS exercise study reviewed, Chung, Lee, and Yoon [10] compared the effects of lumbar stabilization exercises on unstable surface versus a stable surface. The researchers concluded that both lumbar stabilization exercises on stable and unstable surfaces improve cross-sectional areas of the multifidus segment, weight bearing, pain relief, and functional disability. They further mentioned that the lumbar stabilization exercise on an unstable surface was superior to that on a stable surface.

3.2 Muscular Stretching

Four studies on muscular stretching were identified. The reviewed studies show that muscular stretching helps to improve pain level, functional disability, and joint range of motions (ROM).

In the first muscular stretching study reviewed, Rajaratnam et al. [11] conducted a biomechanical study on the effects of performing stretching during prolong sitting. The researchers discovered that stretching exercise in between prolong sitting does not have any significant effect on the postural alignment (spinal posture). Nevertheless, the stretches were efficient in reducing the participants' level of discomfort. In the second muscular stretching study reviewed, Sherman et al. [12] conducted a randomized controlled trial (RCT) comparing yoga, stretching, and a self-care book for CLBP. The overall finding was yoga sessions, and conventional stretching was equally efficient than the self-care book in improving function and pain level. In the third muscular stretching study reviewed, Chen and associates [13] examined the effectiveness of stretching exercise program on low-back pain (LBP) and exercise self-efficacy among nurses in Taiwan. The researchers reported that after 6 months of undergoing the stretching program, 81 % of participants reported a moderate-to-high level of LBP relief. Furthermore, they also improved on their exercise self-efficacy. Finally, in the last muscular stretching study reviewed, Sairyo et al. [14] discovered that active-static stretching (jackknife technique) was efficient in increasing flexibility of tight hamstrings, which was reported to be one of the main causes of CLBP (Table 1).

3.3 Core Stability Exercise Versus Muscular Stretching

França et al. [15] compared muscular stretching versus segmental stabilization in patients with CLBP. They discovered both interventions improved pain and functional disability although they found out that segmental stabilization was superior to muscular stretching for improving transverse abdominis (TA) muscle activation capacity among CLBP patients.

4 Discussion

The purpose of this review was to evaluate the effectiveness of CS exercises versus muscular stretching as a therapeutic exercise intervention on patients with backache. Ten related studies were identified and reviewed.

4.1 Core Stability Exercise Rationales

Core stability exercise has become an integral component in the management of patients with low-back pain for the past decade [5]. Core stability exercise is defined as exercises which activate deep core muscles such as the TA and multifidi [16]. The authors' reviewed the existing evidence and suggested that core stability exercise was effective in improving pain, function, and activation of tonic core muscle groups among heterogeneous group of patients with CLBP [2, 7–10].

The hypothetical rationale for the CLBP improvement after CS exercises includes the following; a CS exercise program is designed to help an individual to gain functional strength, neuromuscular control, and muscle endurance of the lumbo-pelvic-hip complex [3, 5]. This approach offers a biomechanically efficient

Table 1 Reviews on the efficacy of core stability exercise in rehabilitation of CLBP

Authors	Intervention	Outcomes
Norris and Matthews [7]	6 week treatment • Integrated back stability (IBS)	Pain and functional impairment improved
Bronfort et al. [8]	6 week treatment • Supervised trunk exe (STE) • chiropractic spinal therapy • home exercise	STE improves treatment satisfaction and trunk muscle endurance
Desai and Marshall [9]	2 sessions • exercise on labile (unstable) surface	CLBP patients > muscle activity
Mohseni-Bandpei et al. [2]	Control group • Routine therapy Experimental group • Routine therapy • Pelvic floor muscle (PFM) exercise	PFM is efficient in improving core muscles strength and endurance
Chung et al. [10]	8 week program (3x/week) • Stabilization program (SP) + balls • Stabilization program (SP) only	SP + ball were superior to SP only

mechanism for the entire kinetic chain of the lumbo-pelvic-hip complex, thus allowing the body to decelerate gravity, ground reaction forces, and momentum at the right joint, in the right plane and at the right time [3]. The increased stability of the trunk will enable the individual to maintain the spine and pelvis in the most comfortable and acceptable mechanical position that control the forces of repetitive microtrauma and protect the structures of the back from further damage (Table 2).

4.2 Muscular Stretching Rationales

Stretching is another physical therapy increasingly used in rehabilitation technique programs in CLBP [6]. Any motion of body parts that causes movement of a joint can be called stretching [17]. After extensive searching and reading on studies of muscular stretching on CLBP, the authors' concluded that there are very few studies on the effects of muscular stretching for CLBP rehabilitation. The stretching protocols used in previous studies were very general and not specific to recognized stretching protocols (for example, dynamic and PNF stretching). Despite the limited studies of stretching on CLBP, the authors managed to identify 4 studies related to the field. Muscular stretching is effective in improving pain and function, but does not improve muscular strength and endurance of tonic core muscle groups [10–12, 14]. What is the relationship between flexibility and low-back pain?

Table 2 Reviews on the efficacy of muscular stretching in rehabilitation of CLBP

Authors	Intervention	Outcomes
Rajaratnam et al. [11]	2 h to watch a movie 1. Control: sat continuously 2. Experiment: performed stretch exercise when sat	Prolong sitting no effect on postural alignment Stretches efficient reduce discomfort
Sherman et al. [12]	12 weeks, 1x a week 1. Yoga group 2. Stretching group 3. Self-book care group	Yoga and stretching equally efficient in improving function and pain level
Chen et al. [13]	6 Months; 3x/wk, 50 min per session Experiment: stretching	81 % of experimental: moderate-to-high level of LBP relief; improved on exercise self-efficacy
Sairyo et al. [14]	4 weeks, every day, 2 sets of jackknife stretch	Jackknife stretch > flexibility of tight hamstrings

According to Brukner and Khan [4], specific muscle tightness or shortening is commonly found in association with low-back pain. Commonly, shortened muscles include erector spinae, psoas, iliotibial band, hip external rotators, hamstrings, rectus femoris, and gastrocnemius. Tightness of these muscles affects the biomechanics of the lumbar spine. Alter [18], based on a theoretical model, explained decreased lumbar flexion in those with low-back pain; tight hip extensors decrease the lordotic curve that naturally exists in the lumbar spine which can exaggerate the posterior tilt of the pelvis. This condition will diminish the shock-absorbing capacity of the lumbar segments and increases compression forces on the lumbar spine. The thoracolumbar often maintains chronic muscular tension, which can be greatly alleviated through effective stretching [19]. Stretching for low-back pain patients is designed to progressively stretch the muscle groups which are assumed to be too short, especially the lumbar spinal muscle and the hip flexors and extensors [3].

4.3 Core Stability Exercise Versus Muscular Stretching

In the beginning, it appears that CS exercise and muscular stretching have equal effects in rehabilitation of CLBP. Nonetheless, there is modest evidence that core stability exercise is more beneficial than muscular stretching in CLBP rehabilitation. It was discovered that core stability exercise promotes improvement in tonic core muscle strength and endurance, whereas muscular stretching does not [15].

This review has several limitations. First, the literature searching scope is small as only 3 databases (Springer, Science Direct, and Google Scholar) were used to search the related articles; hence, the authors may have overlook other recent findings. Second, despite thorough and extensive searching using the common keywords, the search may not have identified all articles. This is because some of the articles may have been termed differently and thus were hard to find.

5 Conclusion

In summary, CLBP has been one of the most common complaints in modern society. The core also known as lumbo-pelvic-hip complex is attached to local (postural, tonic) muscles which are responsible for providing segmental stability and for directly controlling the lumbar segments during movement. Weakness and/or tightness of these groups of muscles will cause many disorganized movements that lead to low-back injury. Core stability exercise has become a fundamental component in the management of patients with CLBP as it helps individuals to regain functional strength, neuromuscular control, and muscle endurance of the lumbo-pelvic-hip complex. Stretching is another intervention widely used in physical therapy, which helps alleviates CLBP by progressively stretching the muscle groups which are assumed to be too short, especially the lumbar spinal muscle and the hip flexors and extensor. Core stability exercise is more beneficial than muscular stretching in CLBP rehabilitation as CS exercise develops strong deep core muscles to provide segmental stability and directly controlling the lumbar segments during movement.

Acknowledgments Researcher would like to express upmost gratitude to RMI—RIF Grant for sponsoring this study. Last but not least, thanks to everyone involved in preparing this paper directly or non-directly.

References

1. Ferreira, M. L., et al. (2007). Comparison of general exercise, motor control exercise and spinal manipulative therapy for chronic low back pain: a randomized trial. *Pain, 131*(1–2), 31–37.
2. Mohseni-Bandpei, M. A., et al. (2011). The effect of pelvic floor muscle exercise on women with chronic non-specific low back pain. *Journal of Bodywork and Movement Therapies, 15*(1), 75–81.
3. Prentice, E. W. (2011). *Rehabilitation techniques for sports medicine and athletic training* (5th ed., p. 707). New York: McGraw-Hill.
4. Brukner, P., & Khan, K. (2007). *Clinical sports medicine* (3rd ed.). Sydney: McGraw-Hill.
5. Standaert, C. J., Weinstein, S. M., & Rumpeltes, J. (2008). Evidence-informed management of chronic low back pain with lumbar stabilization exercises. *The Spine Journal, 8*(1), 114–120.
6. Poiraudeau, S., & Revel, M. (2000). Rehabilitation therapy in chronic low back pain. *Joint Bone Spine, 67*(6), 582–587.

7. Norris, C., & Matthews, M. (2008). The role of an integrated back stability program in patients with chronic low back pain. *Complementary Therapies in Clinical Practice, 14*(4), 255–263.
8. Bronfort, G., et al. (2011). Supervised exercise, spinal manipulation, and home exercise for chronic low back pain: a randomized clinical trial. *The Spine Journal, 11*(7), 585–598.
9. Desai, I., & Marshall, M. W. P. (2010). Acute effect of labile surfaces during core stability exercises in people with and without low back pain. *Journal of Electromyography and Kinesiology, 20*, 1155–1162.
10. Chung, S., Lee, J., & Yoon, J. (2013). Effects of Stabilization Exercise Using a Ball on Multifidus Cross-Sectional Area in Patients with Chronic Low Back Pain. *Journal of Sports Science and Medicine, 12*, 1–9.
11. Rajaratnam, B.S., et al. (2009). Can Stretching Exercises Reduce Your Risks of Experiencing Low Back Pain?. In Lim, C., Goh J.H. (Eds.), *13th international conference on biomedical engineering* (pp. 2114–2117) Berlin: Springer.
12. Sherman, K. J., et al. (2011). A randomized trial comparing yoga, stretching, and a self-care book for chronic low back pain. *Archives of Internal Medicine, 171*(22), 2019–2026.
13. Chen, H.-M., et al. (2012). Effectiveness of a stretching exercise program on low back pain and exercise self-efficacy among nurses in Taiwan: a randomized clinical trial. *Pain Management Nursing*.
14. Sairyo, K., et al. (2013). Jack-knife stretching promotes flexibility of tight hamstrings after 4 weeks: a pilot study. *European Journal of Orthopaedic Surgery & Traumatology, 23*(6), 657–663.
15. França, F. R., et al. (2012). Effects of muscular stretching and segmental stabilization on functional disability and pain in patients with chronic low back pain: a randomized, controlled trial. *Journal of Manipulative and Physiological Therapeutics, 35*(4), 279–285.
16. May, S., & Johnson, R. (2008). Stabilisation exercises for low back pain: a systematic review. *Physiotherapy, 94*(3), 179–189.
17. Nelson, G. A., & Kokkonen, J. (2007). *Stretching Anatomy* (p. 147). Chicago: Human Kinetics.
18. Alter, J. M. (2004). *Science of flexibility* (3rd ed.). Chicago: Human Kinetics.
19. McAtee, E. R., & Charland, J. (2007). *Facilitated Stretching* (3rd ed., p. 183). Chicago: Human kinetics.

Nutritional Status and Activity Level of Children in Kuala Tahan National Park, Malaysia

Hafizuddin Baki, Adam Linoby, Sarina Md Yusof, Anuar Suun, Muhammad Sufyan Mohamad Zaki, Hanifa Sariman, Badli Esham and Muhamad Safiq Saiful Annur

Abstract A cross-sectional study related to anthropometric indices and activity level among Batek and Malay children was carried out in Taman Negara, Pahang, Malaysia. The study included a total of 165 children, 128 Batek and 37 rural Malay. Participants were children between the ages of 9 and 12 years old. Anthropometric measurements such as height, weight and waist-to-hip ratio were taken. Body mass index (BMI) was used as a measure of obesity and to describe nutritional status. Children were classified into four body mass categories (underweight, normal weight, overweight and obese) based on the International Obesity Task Force's age and gender specification definitions. Nutritional status was determined by the distribution of the z-score of weight for age (WAZ), height for age (HAZ) and weight for height (WHZ), from the WHO growth chart of reference. Pedometers worn over three days measured children's activity level. Malay children have a higher prevalence of obesity compared with its Batek counterpart; 20.4 % are overweight and 28.8 % are obese. Prevalence of malnutrition (stunting and wasting) among Batek children for both genders is high. Due to inadequate or imbalanced energy consumption among the children, a comprehensive effort in the health care program is essential for both populations to improve their nutritional status.

Keywords Anthropometry · Activity level · Batek · Children · Malaysia

---

This work was supported in part by the Fundamental Research Grant Scheme (FRGS), Ministry of Higher Education of Malaysia FRGS/1/2011/SKK/UITM/03/30.

H. Baki · S. Md Yusof · A. Suun
Faculty of Sports Science and Recreation, Universiti Teknologi MARA,
Shah Alam, Selangor, Malaysia

A. Linoby (✉) · M. S. Mohamad Zaki · H. Sariman · B. Esham · M. S. Saiful Annur
Faculty of Sports Science and Recreation, Universiti Teknologi MARA,
Jengka, Pahang, Malaysia
e-mail: linoby@pahang.uitm.edu.my

R. Adnan et al. (eds.), *Proceedings of the International Colloquium on Sports Science, Exercise, Engineering and Technology 2014 (ICoSSEET 2014)*,
DOI: 10.1007/978-981-287-107-7_46,

1 Introduction

Worldwide rapid socioeconomic development has brought marked changes in lifestyle of the Orang Asli including diet and activity level [1, 2]. For the past two decades, Malaysia has recorded a marked increase in the prevalence of lifestyle non-communicable diseases which includes diabetes mellitus type II, coronary heart disease and hypertension [3, 4]. Similarly, this phenomenon is also affecting the population of indigenous Orang Asli in Malaysia [5]. Although some of the indigenous Orang Asli still maintains their isolation from the urban population, many of them are going through the westernization of diet and lifestyle [6]. This may partially explain the increase in the prevalence of chronic diseases among the population. In addition, the original active lifestyle of fishing, hunting and herding among the indigenous people is gradually shifting to a sedentary lifestyle [7]. However, unlike the area of diet and nutrition, fewer studies have been conducted on the activity level of the Orang Asli population.

Although malnutrition remains a major health issue among the Orang Asli population, there is accumulating evidence to show that the level of adiposity and obesity is also on the rise in this 'at risk' population, particularly among their children [6].

One related factor may due to resettlement activity by the government to the special land scheme. This change transformed their dietary and lifestyle practice, from a nomadic culture of hunting and gathering to a permanent cash crop agriculture. Usually, this resettlement was located in close proximity to Malay villages [3]. However, the current nutritional status and lifestyle of the aboriginal Batek tribe in Taman Negara Kuala Tahan are still unknown. Studies also report a high prevalence of obesity among children, which has been directly related to low activity levels [1, 8]. Given the link between activity level and obesity, it is of interest for this study to assess and provide a nutritional profile among the Batek children and compare it with the Malay children living in the area of Taman Negara Kuala Tahan, Pahang.

2 Methodology

2.1 Study Population

This study was carried out in Kuala Tahan, Taman Negara, Pahang, which covers more than 4,343 km^2 of land area, and has approximately 2,909 total resident, which include the Malays and Batek among others. This study consists of two sample groups, the Batek children residing in Kuala Tahan, Taman Negara (National Park), Pahang, and Malay children living in a nearby village. The Batek children, aged 9–12, were selected using a purposive sampling method from the tribe's villages located in Taman Negara. Information sheets and consent forms were given to the subjects. Data reported here were collected during the months of September and October 2012.

2.2 Anthropometric Measurements

Trained assistants measured participant's height and weight. Children's weights were measured using a portable mechanical scale (SECA 762; Hamburg, Germany) to the nearest 0.5 kg, with subjects wearing light indoor attire and no shoes. A correction of 0.5 kg was made for the weight of the clothes. The height of the children was measured by using a wall-mounted stadiometer (SECA 206; Hamburg, Germany) with the subject standing height without shoes against a Frankfurt plane placed horizontally; the height was read to the nearest 0.1 cm. Weight-for-age z-score was used to denote underweight as an overall indicator for malnutrition, while height-for-age z-score was used as an indicator for stunting. Weight-for-height z-score was used as an indicator for wasting (acute malnutrition). The z-scores were calculated based on the median values of the National Center for Health Statistics United States (NCHS) reference population.

2.3 Physical Activity Level

Omron HJ-113 piezoelectric pedometers (Healthcare Ltd., UK) were worn by the children for three days. The device has a 7-day memory, which avoids the need for participants to record their own daily step counts. Pedometers were attached to the children's belt at a position directly in line with the patella. All pedometers were sealed with a cable tie and clear plastic wrap to prevent accidental resetting and to discourage behavior modification due to access to the pedometer's step count. The subjects were asked to not tamper with the seal. Researchers collected the pedometer each morning, recorded the pedometer score, then reset, resealed and reattached the pedometer to the subjects. From the pedometer data, the first day was excluded because of partial data collection and an average steps/day variable was computed.

2.4 Data Analysis

Data were analyzed using the Statistical Package for Social Sciences (version 19.0 for Windows, SPSS Inc., Chicago, IL, US). *T* tests were used to assess any differences, and comparisons of proportion were tested with the *z*-test. Nonparametric test (Mann—Whitney *U* test) to determine whether the distribution was not normal. The significance level was set at $p < 0.05$.

Table 1 Mean (SD) values for anthropometric characteristics

	Male		Female	
	Malay n = 19	Batek n = 80	Malay n = 18	Batek n = 48
Age (year)	10.57 (1.1)	9.28 (0.84)	10.44 (1.0)	9.71 (0.88)
Weight (kg)	32.2 (1.1)	30.3 (4.5)	32.8 (1.1)	28.8 (5.0)
Height (m)	1.34 (0.1)	1.33 (6.6)	1.38 (0.1)	1.33 (8.1)

2.5 *Ethical Approval*

Approval to conduct the study was granted by the Department of Orang Asli Development (JAKOA), Kuala Lumpur, Malaysia. Ethical approval was obtained from the The Research Ethics Committee (REC) of Universiti Teknologi MARA (UiTM), Malaysia.

3 Results

3.1 *Physical Characteristic*

This study demonstrates that mean body weight for both genders in the Batek group was lower compared with the rural Malay group (30.3 kg for males and 28.8 kg for females). In general, Batek children are lighter compared with the Malay population at Kampung Pagi, whereby their mean body weights were 32.2 and 32.8 kg for males and females, respectively. Table 1 presents the physical characteristic for each group including: age, height and weight. Based on the results obtained, both groups have almost similar characteristic in terms of number of age, height and weight.

3.2 *Nutritional Status*

Table 2 shows the percentage of children classified as normal or stunting (low height for age), wasting (low weight for height) and underweight (low weight for age). There was overall prevalence of underweight among Batek and Malay children, 57.3 and 7.8 %, respectively. As in underweight, a similar pattern of prevalence was also seen in stunting. The overall prevalence of stunting among Batek and Malay children was 69.6 and 11.3 %, respectively. The prevalence of malnutrition based on these measurements revealed that more Batek children were significantly stunted and underweight than the Malay children. However, there was no significant difference found in wasting measurement for the populations.

Table 2 Distribution of Batek and Malay children according to weight for age, height for age and weight for height

	Batek (n = 128) N (%)	Malay (n = 37) N (%)
Weight for age (*WAZ*)		
Normal	54 (42.7)	34 (92.2)
Underweight	73 (57.3)	3 (7.8)*
Height for age (*HAZ*)		
Normal	39 (30.4)	33 (88.7)
Stunting	89 (69.6)	4 (11.3)*
Weight for height (*WHZ*)		
Normal	113 (88.4)	35 (99.6)
Wasting	15 (11.6)	1 (0.4)

* z-test, $p < 0.05$

Table 3 BMI percentile according to race and gender

	Malay children		Batek children	
	Male (n = 19) N (%)	Female (n = 18) N (%)	Male (n = 56) N (%)	Female (n = 36) N (%)
Underweight (<5th % ile)	2 (11)	5 (28)	0 (0)	2 (6)
Normal BMI (5th–85th % ile)	14 (74)	9 (50)	49 (88)	34 (94)
Overweight (≥85th % ile)	3 (16)	4 (22)	7 (13)	0 (0)
Obese (≥95th % ile)	1 (6)	1(6)	1 (1)	0 (0)

3.3 Body Mass Index

Table 3 shows body mass index (BMI) percentile according to race and gender. Malay children showed higher BMI percentile of obesity. There was a female Batek child who was overweight, and no male Batek child was underweight. Batek male children showed one person were obese, and nine children were overweight. These results were also consistent with general knowledge that men, in comparison with women, had higher mean body weight. Some of the reasons for underweight among Batek children and Malay children could be poor dietary intake, poor education, early marriage and high morbidity caused by unhygienic practices, environmental and economic factors.

Table 4 Mean (SD) values for step counts

Parameters	Groups			
	Malay		Batek	
	Male ($n = 19$)	Female ($n = 18$)	Male ($n = 56$)	Female ($n = 36$)
Pedometer score	21,377 (4,035)	17,387 (3,663)	21,281 (3,823)	18,404 (3,049)

3.4 Physical Activity Level

Table 4 shows the mean values for step counts between the Batek group and rural Malay children. The lowest number of steps shown by Batek male students was 12,434, and the highest number of steps was 27,860; Batek female children showed the lowest number of steps as 16,344, and the highest number of steps was 27,452. The lowest steps showed by Malay male children were 13,982, and the highest number of steps was 27,905; the lowest number of steps shown by Malay male children was 11,087, and the highest number of steps was 26,154. Both Malay and Batek children were exceeding the steps recommendation for one day, which is 16,000 steps per day for male children and 12,000 steps per day for female children.

4 Discussion

Worldwide rapid socioeconomic body composition measurement (waist-to-hip ratio, body mass index, nutritional status and percentage body fat) yielded non-significant results. Due to their lifestyle, fewer Batek children are overweight. Most of the Batek children in the study were normal weight, and rural Malay children (30 %) were overweight and obese when body mass index was compared with age- and sex-specific cutoffs developed by the International Obesity Task Force of World Health Organization [9]. These prevalence estimates are similar to those reported by [10] in a nationally representative sample of seven- to 13-years-old Canadian children. The lack of statistical significance in body composition variables between groups may be partially explained by the inherent problems common to skin fold measurements and body mass index in small samples. For example, the triceps skin fold is one of the most commonly used sites for the assessment of subcutaneous fat distribution in growth studies [10]. Unfortunately, intra-individual and inter-individual reliability of skin fold measurements is difficult to establish with reliability decreasing as body fat increases for a given individual [11].

Body mass index is a frequently used index of body composition [11]. However, body mass index is limited because of its derivation from body dimension. Ideally, the definition of adiposity should be based on fat [9, 11]. Additionally, BMI has been found to have variable sensitivity, which has limited utility in detecting those who are at increased risk for becoming overweight [12]. Nevertheless, in field-based settings and epidemiological studies, skin folds and BMI are preferable because of their simplicity of computation and low cost [11].

To explain the lack of difference in BMI between groups, it should be observed that the difference in physical activity between groups may not have been biologically meaningful and therefore inefficacious in eliciting a more lean body mass among Batek children. The benefits of lifestyle related to activity level on body composition in Batek children were masked by a high fat diet. However, the likely explanation for the non-significance of the body composition variables in this showed no real difference.

From this study, there was a significant difference in body weight status of Batek children determined according to BMI. Being underweight was reported to be lower when BMI was used as the criterion (2 %) than when body fat classification was used (no under nutrition). In addition, body fat percentage shows a higher prevalence of normal subjects than BMI. BMI is unable to differentiate between body fat and fat-free mass.

Thus, the reason for the difference in the percentage of overweight/obese subjects determines using different criteria and may be necessary in a study population that has a high lean body mass relative to their height. This might be due to a high level of physical activity. Thus, these findings were also consistent with the idea that females accumulate fat subcutaneously, whereas males do not and that age factors specifically determine the linear correlation with BMI. An insufficient food supply can also contribute to poor dietary intake.

Previous studies indicate the poor health status of Orang Asli community and infections are still a common burden [13, 14]. In this study, there was a low prevalence of underweight and stunting among Batek children compared with rural Malay children living in same area. When genders were compared, male Malay children have a higher prevalence of being underweight, stunting and wasting compared with Batek children. This is contradict to other studies [15].

Based on three main nutritional problems (stunting, wasting and underweight), stunting or the slowing of skeletal muscle growth of children is stated to be more prevalent. This result reflects the poor overall economic conditions and inadequate energy consumption. The decreasing sources of food from hunting and gathering due to logging activities in the Batek settlement area also contribute to this finding. Stunting also may result from impaired growth in the uterus when the fetus is deprived of essential substances during pregnancy [16].

In this research, Batek children had a high prevalence of being underweight 57.3 %, stunting 69.6 % and wasting 10 %. However, among Malay children, the prevalence of being underweight was 3 %, stunting 4 % and wasting 1 %. This represents an excess of wasting, stunting and being underweight among the Batek children compared with the Malay children living in the same area. According to [15], an early study on nutritional status of Orang Asli children between 0 and 10 years old found that the prevalence of being underweight was between 32 and 65 % and stunting from 50 to 81 %. Compared with the prevalence found in this study, this indicates that there is relatively no improvement in the nutritional status of Batek children.

Wasting also showed a high prevalence among Batek children compared with rural Malay children; wasting leads to weight loss. But the number of children wasting is less than the number of stunting children. Wasted children need to have immediate intervention from improvised problems because wasting is related to a high level of morbidity and mortality.

Nutritional programs were strongly suggested in comprehensive primary health care programs [17, 18]. This program can help them to improve body weight and height in most infants and children. Nutritional programs need to be more organized and provide continuous information about nutritional status of the population or groups of children. The program must also include family planning, regular ante-helminthic medication and supplementary vitamins. This program may help to prevent malnutrition and improve their growth and demonstrate an ability to reduce mortality and morbidity in children.

All the findings may apply to other Batek settlement villages. However, this community is of unique culture and belief, they choose to refuse resettlement and are still performing a nomadic lifestyle. Although the aim of resettling these unique communities is to improve the culture of a nomadic lifestyle to permanent residence, the process is slow and adequate support must be provided to maintain this unique community satisfaction.

Batek children took 20,242 ± 3,562.3 (steps ± standard deviation) steps of moderate to vigorous activity level in two conservative days, which was significantly more than (19,436 ± 4,308.9 steps) rural Malay children took each day. Table 4 summarizes the studies that have reported mean steps stated by Batek children and rural Malay children. In terms of physical activity, most time was spent engaged in manual agricultural work and collecting jungle resources for survival. Therefore, these activities are the main daily physical activities for males, whereas females are involved in routine family responsibilities, such as cooking and washing clothes. In addition, females also help their family in farming, but hunting is usually performed by males as a headman of the family.

5 Conclusion

In conclusion, Batek children have a higher prevalence of being underweight, stunting and wasting than their Malay counterparts. It was also found that the nutritional status of the Batek children was poor compared with the Malay children. Batek children also prove to have a higher activity level. These results may indicate secular deterioration in the activity level of Batek children with nutritional inadequacy and still practicing nomadic lifestyle. Further research is necessary to investigate the extent of the impact of the nutritional problem and activity level in relationships to physical fitness in Batek children. Due to inadequate or imbalanced energy consumption among the children, a comprehensive effort in the health care program is essential in order for both populations studied to improve their nutritional status.

Acknowledgments We would like to express our gratitude to the community of Batek, all the children who participated in this study and the Department of Orang Asli Development (JAKOA) for their permission to carry out this study. The authors gratefully acknowledge En. Alias of JAKOA and Miss Nazratul Husna of UiTM for their help in this study.

References

1. Anderson, K. D., Baxter-Jones, A. D. G., Faulkner, R. A., Muhajarine, N., Henry, C. J., & Chad, K. E. (2010). Assessment of total and central adiposity in Canadian Aboriginal children and their Caucasian peers. *International Journal of Pediatric Obesity, 5*(4), 342–350.
2. Coble, J. D. (2010). Physical activity behaviors and motivations in an adult First Nation population.
3. Khambalia, A., & Seen, L. (2010). Trends in overweight and obese adults in Malaysia (1996–2009): a systematic review. *Obesity Reviews, 11*(6), 403–412.
4. The Third National Health and Morbidity Survey (NHMS III). (2006). Ministry of Health, Malaysia.
5. Haemamalar, K., Zalilah, M., Neng Azhanie, A. (2010). Nutritional Status of Orang Asli (Che Wong Tribe) Adults in Krau Wildlife Reserve, Pahang. *Malaysian Journal of Nutrition, 16*(1).
6. Khor, G. L., & Shariff, Z. M. (2008). The ecology of health and nutrition of "Orang Asli"(indigenous people) women and children in peninsular Malaysia. *Tribes and Tribals (Special Volume), 2*, 67–77.
7. Saibul, N., Shariff, Z. M., Lin, K. G., Kandiah, M., Ghani, N. A., & Rahman, H. A. (2009). Food variety score is associated with dual burden of malnutrition in Orang Asli (Malaysian indigenous peoples) households: implications for health promotion. *Asia Pac J Clin Nutr, 18*(3), 412–422.
8. Downs, S. M., Marshall, D., Ng, C., & Willows, N. D. (2008). Central adiposity and associated lifestyle factors in Cree children. *Applied Physiology, Nutrition and Metabolism, 33*(3), 476–482.
9. Cole, T. J., Belizzi, M. C., Flegal, K. M., & Dietz, W. H. (2000). Establishing a standard definition for child overweight and obesity worldwide: international survey. *British Medical Journal, 320*, 1240–1246.
10. Dietz, W. H., & Robinson, T. N. (1998). Use of the body mass index (BMI) as a measure of overweight in children and adolescents. *Journal of Pediatrics, 132*, 191–193.
11. Tyrrell, V. J., Richards, G. E., Hofman, P., Gillies, G. F., Robinson, E., & Cutfield, W. S. (2001). Obesity in Auckland school children: a comparison of the body mass index and percentage body fat as the diagnostic criteria. *International Journal of Obesity, 25*(2), 164–169.
12. Malina, R. M., & Katzmarzyk, P. T. (1999). Validity of the body mass index as an indicator of the risk and presence of overweight in adolescents. *American Journal of Clinical Nutrition, 70*, 131S–136S.
13. Ariff, T., Nazi, M. Z., Rizam, M. Z., et al. (1997). Health status of aborginal children in Post Brooke, Kelantan. *Maladies Journal Children Health, 9*, 60–65.
14. Norhayati, M., Norhayati, M. I., Fariza, N. N., et al. (1998). Health status of Orang Ash i (aborigine) community in PosPiah, Sungai Siput, Perak, Malaysia. *Southeast Asia Journal Tropicale Medicine Public Health, 29*, 58–61.

15. Sham, M. K., Zulkifli, I., & Lailanor, I. (1989). Nutritional status of various Orang Asli communities in Peninsular Malaysia. *Alidemilca, 35*, 69–74.
16. Martorelli, R., Keulel Khan, L., & Scroeder, D. C. (1994). Reversibility of stunting; epidemiological findings in children from developing countries. *Europe Journal Clinical Nutrition, 48*(suppl), S45–S57.
17. Chee, H. L. (1995). Health and nutrition of the Orang Ash. The need for primary health acre amidst economic transformation In: R. Rashid (ed.), *Indigenous minorities of Peninsular Malaysia: Selected Issues and Ethnographies*. INAS Kuala Lumpur.
18. Shephard, R. J. (2003). Limits to the measurement of habitual physical activity by questionnaires. *British Journal of Sports Medicine, 37*(3), 197–206.

Comparison of Handgrip Strength Among Winning and Non-winning Male Boxers

Hanifa Sariman, Adam Linoby, Muhammad Sufyan Mohamad Zaki, Mohd Zulkhairi Mohd Azam, Muhamad Noor Mohamed, Nadiah Diyana and Azizul Afandi

Abstract Boxing is a combat sport in which the boxers collect points by punching in a target area. It is also an individual sport in which courage, speed, strength, power, accuracy, and passion are of the utmost importance. The purpose of this study was to compare the strengths of winning and non-winning male boxers. An ex post facto design was employed. Thirty-five male boxers [$N = 35$; mean height 168.00 cm (±6.03 cm), weight 59.49 kg (±10.12 kg), and BMI 21.01 kg/m^2 (±2.93 kg/m^2)] who were either winners or non-winners were selected during the Malaysia Youth Boxing Tournament. The tests and measurements that were conducted were as follows: (1) measurements of weight and height and (2) a handgrip strength test using a handgrip dynamometer. The strengths of winning and non-winning boxers were significantly different ($p < 0.05$, $p = 0.01$). It was concluded that, in boxing, strength has a definite influence on whether a boxer will win or not, with winners being stronger than non-winners. In addition, excellent boxers were fit, as demanded by boxing.

Keywords Boxing · Strength · Winner and non-winner · Handgrip · Comparison

This work was supported in part by the Fundamental Research Grant Scheme (FRGS), Ministry of Higher Education of Malaysia FRGS/1/2011/SKK/UITM/03/30.

H. Sariman (✉) · A. Linoby · M. S. Mohamad Zaki · M. Z. Mohd Azam · A. Afandi
Faculty of Sports Science and Recreation, Universiti Teknologi MARA,
Jengka, Pahang, Malaysia
e-mail: hanifa4191@pahang.uitm.edu.my; hanifasariman@gmail.com

M. N. Mohamed · N. Diyana
Faculty of Sports Science and Recreation, Universiti Teknologi MARA, Negeri Sembilan,
Kuala Pilah, Negeri Sembilan, Malaysia

R. Adnan et al. (eds.), *Proceedings of the International Colloquium on Sports Science, Exercise, Engineering and Technology 2014 (ICoSSEET 2014)*,
DOI: 10.1007/978-981-287-107-7_47,

1 Introduction

Boxing is an ancient sport. The earliest evidence is from Egypt and dates back to around 3000 BC. Boxing was introduced to the Olympic Games by the Greeks in the late 7th century BC. Greek boxers wrapped their hands and forearms in soft leather for protection purposes. Ideal body fat percentages, speed, strength, and power are of significant importance for the development of boxers' performance. The relationship between these characteristics and the performance of the athletes is crucial to finding out which strengths and weaknesses affect the athletes' performance [1]. Data obtained from a specific test that suits the nature of boxing could lead to good development in boxing where the potential for success in the sport is high. According to Katralli [2], a coach who is sensitive in terms of anthropometric and physiological characteristics could develop an easy path to success. The development of the athlete needs to be inline with the demands created by the nature of the sport. Additionally, strength is a determinant for improving arm strength to be used for the purpose of punching. Furthermore, identifying the profiles of the arm strength of Malaysian boxers could be a significant indicator of champion boxers. As are results, the data can be a indicator in recruiting a new boxers. Moreover, the data can be used as a reference in designing sound training program in order to improve the arm strength. Development program consist the periodization process of strength training to ensure the boxers obtain the good arm strength. Arm strength is crucial to boxers, especially for the punching impact, prevention of injuries and the use of arm shields during matches.

Strength increases the impact of the punch and at the same time trains the body to accept the opponent's attack. Since boxing is an individual sport, there is nobody to help the boxer in the ring if he is not fit. Reference [3] a boxer has to develop wrist and forearm strength through a proper training program. Strength increases the impact of punching and this might lead to concussion when the punch is right on target on the head. Moreover, arm strength assists boxers to have a solid shield during a match. This is an advantage to the boxer to help him win the match. Reference [4] handgrip strength is significant in a boxer's performance. In boxing, strength tends to be a factor in the boxer's performance. Hand strength for a boxer is of benefit in terms of punching and shielding. A firm shield prevents the opponent from gaining points by punching. Besides this, a harder punch tends to break through the shield, and the boxer then gains points from punching. A lack of strength might lead to less impact in punching. Reference [4] also found that the strengths of junior and senior boxers show significant differences, so that strength has a significant impact on performance. An ideal boxer should have good strength in order to deliver a greater impact, especially in punching. Reference [5] a boxer's ranking is highly related to his handgrip strength. A high ranking boxer tends to have greater handgrip strength, which might be due to the level of resistance training he undergoes during training sessions.

2 Methodology

2.1 Research Design

An ex post facto design was employed to investigate the data to discover whether it showed any differences within the group. The subjects performed handgrip strength test: this was a quantitative part of the study and allowed us to develop an understanding of the strength levels of winning and non-winning boxers.

2.2 Study Population

A purposive sampling method was employed for this research. All the boxers who participate in this competition, which has become the platform for the Persatuan Tinju Amatur Malaysia (PTAM) to identify and recruit new boxers to represent Malaysia and compete at the international level. The population for this study came from these boxers. Eighteen ($N = 18$) gold and silver medalists, categorized as winners, and seventeen ($N = 17$) bronze medalists, categorized as non-winners, were selected to participate in this study. The boxers competed in Malaysia Youth Boxing Tournament and had undergone the training program with their appointed state coaches.

2.3 Data Collection Procedures

The researcher gave an explanation of the purpose of the study, and a short description of the tests that would be conducted, and the state boxing organization and the researcher then came to an agreement on the terms and conditions applying to each party. To begin the collection of data, the researcher again gave a brief explanation to the subjects of the purpose of the study and the tests that he would conduct before the consent form and the physical activity readiness questionnaire (PAR-Q) Malay version (S. M. Yusof, The effect of weight training and *Eurycoma longifolia* supplementation on physical and functional performance among middle-aged woman, unpublished) were distributed to the subjects. The participation of the subjects in this study was on a voluntary basis, and a subject had the right to withdraw at any time he wished, by informing the researcher and the coach of his intention to withdraw.

The subjects were divided into two groups, the 17 non-winning boxers and the 18 winning boxers. All the boxers who successfully reached the semifinals were tested. The subjects were weighed and measured and completed the handgrip dynamometer test early in the morning straight after the weighing session conducted by the tournament organizer. The weighing session for the boxers determined the category in which the boxers competed in the tournament. The test was held in a hall which was deemed to be an appropriate area for the tests mentioned earlier.

The anthropometric measurements, consisting of the weight and height of the boxers, were measured using a SECA electronic scale 7802321134, made in Germany, and a SECA 206 wall mounted measure, made in Germany. The handgrip strength was measured using a Takei A5401 dynamometer that was made in Japan. The boxer needed to stand erect and to hold the handgrip parallel to the side of his body. The grip of the dominant hand was measured when the boxer gripped the handgrip as hard as possible without moving his arms. Three trials were conducted with a 1-min rest and recovery between trials.

2.4 Data Analysis

Data were analyzed using the Statistical Package for Social Sciences (version 19.0 for Windows, SPSS Inc., Chicago, IL, US). Independent t test was used to assess comparisons between winner and non-winner boxers. Nonparametric test (Mann - Whitney U test) to determine whether the distribution was not normal. The significance level was set at $p < 0.05$.

2.5 Ethical Approval

The data collection procedure for the study began once approval had been received from the Universiti Teknologi MARA's ethics committee. Upon receipt of the approval from the ethics committee, the researcher embarked on his study by approaching each state boxing organization with an official letter to seek permission to conduct the study with that organization's boxers. A letter of approval from the faculty of Sports Science and Recreation was obtained before the researcher approached the boxing organizations.

3 Results

3.1 Physical Characteristic

The mean age of the winners was 19.22 years, higher than the mean age for the non-winners, of 19.06 years. The mean weight of the winners was 60.62 kg, and the mean weight of the non-winners was slightly lower, at 59.08 kg. The mean height of the winners was 168.67 cm, which was higher than the mean height of

Table 1 Descriptive statistics for age, weight, height, and BMI of winning and non-winning boxers

Variables	Winner $n = 18$	Non-winner $n = 17$
Age (year)	19.22	19.06
Weight (kg)	60.62	59.08
Height (cm)	168.67	168.23
BMI (kg/m^2)	21.18	20.83

Table 2 Independent sample *t* test of strength for winning and non-winning boxers

Group	N	Mean	SD	*t*-value	*p*	df
Winner	18	42.68	5.12	3.033	0.005	33
Non-winner	17	37.71	4.53			

168.23 cm for the non-winners. The mean body mass index (BMI) of the winners was 21.18 kg/m^2, which is higher than the mean BMI for the non-winners of 20.83 kg/m^2. Table 1 shows descriptive statistics for age, weight, height, and BMI of winning and non-winning boxers.

3.2 Handgrip Strength

As shown in Table 2, the mean strength score for the winners ($M = 42.68$) is higher than the mean for the non-winners ($M = 37.71$). The results of the *t* test yielded a *t*-value of 3.033 which was statistically significant at $p < 0.05$. The result indicated that there was a statistically significant difference in strength between the winning and the non-winning boxers. Therefore, the null hypothesis for the strength component was rejected.

4 Discussion

Strength was measured using the handgrip dynamometer device. The data analysis shows that there was a significant difference in handgrip strength between the winning and the non-winning boxers who participated in the Malaysia Youth Boxing Tournament. This study only focusing on dominant hand of the boxers. Study conducted by Ramirez Garcia [6] found that dominant hand of the boxers have greater strength compared nondominant hand. This is probably resulted by the boxer's improper strength training regime. Besides that, priority given to the dominant hand could affect the arm strength. A previous study reported by Khanna and Manna [4] found a significant difference between the handgrip strength of senior and junior boxers. The finding was consistent in terms of strength as a measurement of fitness for the boxers currently participating in the Malaysia Youth Boxing Tournament, because the winners have greater handgrip strength than the non-winners. In addition, the boxers would be excel in the resistance training part of their training program if the winning boxers train harder than the non-winners. In boxing, different types of punching involve different extend utilization of muscles such as trapezius, triceps, and deltoid [5]. Proper strength training regime could enhance the upper limb strength. Study conducted by Ramirez Garcia [6] found that middleweight boxers had significantly increased their strength according to the training phases. Furthermore, boxers who were well trained would have less tendency of injury since the boxers have strengthen their upper limb bones and soft tissue.

Strength might refer to the impact of punching in boxing, where a harder impact is better in order to knock the opponent out. It would be an advantage to a boxer to become the champion through inflicting a complete knockout, particularly when the boxer has to fight for a short period. Reference [5] reported that handgrip strength was highly related to boxing ranking. They showed that differences in handgrip were significant between the high ranking and low ranking boxers. The finding was similar to the present study. Winning boxers, like high ranking boxers, have greater handgrip strength, since both are superior in their category. Study conducted by Ramirez Garcia [7] stated that arm strength is also related with the boxer's physical characteristics where larger-sized boxers are normally stronger and have greater impact of punching. A harder punch toward the face would probably cause concussion to the opponent.

Reference [8] found that differences in arm strength between two martial arts were not significant. The martial arts in the study were judo and taekwondo. This finding was contrary to the results of the present study conducted by the researcher. This scenario might be due to the differences in the martial arts of boxing, judo, and taekwondo. Judo and taekwondo are very similar in terms of movement, when compared to boxing. A previous study has also found that silat athletes have lower handgrip strength than taekwondo and judo athletes [9]. This might be because of the conditions of the sport, in that arm strength might not be too crucial for some martial arts. It is likely that boxers would get a better score than silat, taekwondo, and judo practitioners since boxing concentrates more on hand strength for punching.

Basically, handgrip strength reflects arm strength, because the boxer has to grip the device as strongly as possible, so that a boxer with greater arm strength will get a better result in the handgrip strength test. Proper strength training should be done without violating the training principles such as intensity, progression, and volume. According to [8] each individual martial arts practitioner needs to make a different effort in training in order to fulfill his own goals, and should not be stressed about his or her capabilities. The elite or winning martial arts athletes might have great strength because they have good resistance training and enough rest after the training to allow their muscles to recover. Moreover, nutrition also plays an important role in muscle strength.

5 Conclusion

In conclusion, fitness components such as strength are crucial in recruitment and development programs when the strengths and weaknesses of new boxers are identified. This study reported that handgrip strength of winning boxers was significantly different from that of non-winning boxers. The development of this fitness component was basically related to the nature of the sport.

References

1. Zar, A., et al. (2008). A survey of the physical fitness of the male taekwondo athletes of the Iranian national team. *Facta Universitatis-Series: Physical Education and Sport, 6*, 21–29.
2. Katralli, J., & Goudar, S. S. (2012). Anthropometric Profile and special Judo fitness levels of Indian Judo players. *Asian Journal of Sports Medicine, 3*, 113.
3. Schick, M. G., et al. (2010). Physiological profile of mixed martial artists. *Medicina Sportiva, 14*, 182–187.
4. Khanna, G. L., & Manna, I. (2006). Study of physiological profile of Indian boxers. *Journal of Sports Science & Medicine, 5*, 90.
5. Guidetti, L., et al. (2002). Physiological factors in middleweight boxing performance. *Journal of Sports Medicine and Physical Fitness, 42*, 309–314.
6. Ramirez Garcia, C. M., et al. (2010). Assessment of hand grip strength in Mexican boxers by training phase. *Archives of Budo, 6*, 1–6.
7. Blazevic, S., et al. (2009). Factor structure of boxer's morphological characteristics. *Acta Kinesologica, 3*(1), 94–98.
8. Babalola, J. F., & Nabofa, O. E. (2011). Influence of martial arts on selected performance variables. *Research Journal of international studies* (22), 147–154. ISSN: 1453-212X.
9. Aziz, A. R., et al. (2002). Physiological responses during matches and profile of elite pencak silat exponents. *Journal of Sports Science and Medicine, 1*, 147–155.

Design of Automated Physiotherapy Device for Knee Rehabilitation Using TRIZ

Velatchi Hema Palaniappan and Shahrizat Shaik Mohamed

Abstract Automated physiotherapy device for knee rehabilitation (APK) is a portable and weightless device for knee rehabilitation at home. The new design is improved using TRIZ, as one of the systematic innovation tools where the focus is to reduce the weight and size of the existing product. There are several elegant solutions have been established for the initial design of this APK where the device can be fit into any type of chair, thinner belts for leg positioning, hanging remote control, stack up LCD and keypad, retractable knee placement and worm gear to increase the torque. In addition, the device will allow the user to set the speed of movement and the timing for the exercise. The device has manual and auto mode for the user feasibility to lift up and lift down the leg based on their desire. It also consists of muscle and motion sensor to detect the movement of user's muscle and the knee. These sensors will trigger and send signal to deactivate the motor if the user is able to move their leg and vice versa. Graphical user interface (GUI) will be generated to monitor the patient's recovery by the physiotherapist via universal serial bus (USB) port.

Keywords Knee rehabilitation · TRIZ · Muscle sensor · Motion sensor

1 Introduction

The design is to develop an automated physiotherapy device for knee rehabilitation (APK) also known as APK which improved in terms of product weight. The general idea of the device is to help the user to carry out their daily physiotherapy exercise at home without any assistant from a professional physiotherapist. Knee

V. H. Palaniappan (✉) · S. S. Mohamed
Engineering Department, KDU College, George Town, Penang, Malaysia
e-mail: hema.p@live.com

R. Adnan et al. (eds.), *Proceedings of the International Colloquium on Sports Science, Exercise, Engineering and Technology 2014 (ICoSSEET 2014)*,
DOI: 10.1007/978-981-287-107-7_48, © Springer Science+Business Media Singapore 2014

rehabilitation is define as a treatment to help the user to recover from an injury or knee surgery [1]. The main intention to do knee rehabilitation is to allow the patient's physical movement be restored back to normal and to compensate for deficit that cannot be done through medicine [1].

The drawback of the current products is that they are inconvenient and not affordable where the users need to rent it if they would like to use it at home. Besides that, the current products available are not very user-friendly, as the user needs special training to assemble the device.

The elegant solution has been generated using TRIZ [2] as an innovation problem-solving tool that speed up the entire process of design improvement. There are six elegant solutions have been considered when design to improve knee rehabilitation device that will be discussed in detail later.

APK is a new and improved innovation compared to the other products in the market with the focus to reduce the weight and its complexity.

The new design will minimize the space needed to place and store the device beside its portability. Besides that, it can be attached to any type of chair. Another imaginative idea develops using the problem-solving tool is a slim LCD screen with an optional keyboard which can be slide in and out when desired. The main innovative feature of this product is its capability to detect muscle and leg movement for motor activation that will help to reduce the power consumption.

2 Research Background

2.1 Knee Anatomy

Knee is the most complicated and largest joint in human body. Knee is also important because it bears the weight and pressure of the body [1]. About 1.5 times of the body mass is supported by the knee while walking, and three-fourths of the body weight are supported by the knee when climbing, while squatting supports 8 times of the body weight [1]. The four major types of bones are tibia, patella, fibula and femur. The main types of injuries that need physiotherapy are anterior cruciate ligament overstretched, medical collateral ligament, lateral collateral ligament, lateral collateral ligament, posterior cruciate ligament, meniscal tears and patella tendonitis [3].

2.2 Existing Product in the Market

Kinetic prima advance knee CPM [4] machine is a manual device, whereas kinetic spectra data capture (DC) [5], Djo Artomot K4 [6] and Fisiotek 2000 GS [7] are automated devices. Moreover, except kinetic prima advance knee CPM machine, all the other devices have LCD display, numeric keypad and DC motor.

When comparing the features, the main difference between the products available in the market with the new design is the muscle sensor and motion sensor components, which makes it an intelligent device. Furthermore, kinetic spectra DC knee, Fisiotek 2000GS and APK are able to store and memorize the patients' data, whereas Djo Artomot K4 and kinetic prima advance knee CPM do not have that feature. All the existing devices are able to control the speed, time, stop and start as the basic features. Kinetic prima advance knee CPM Machine [4] is the lightest among all the devices with 11 kg [7]. Lastly, the common mode available among the devices is manual, auto and warm modes which allows the user to heat up the joint before the user start the session which is available in kinetic spectra DC knee and Fisiotek 2000 GS only. Based on the research conducted, all the existing products cost more than RM7000.00 [8].

3 Methodology

TRIZ pronounced as "treez" and is a Russian word, and the acronym is "Theory of Inventive Problem Solving" [2]. The typical problem-solving process includes defining the problem, identifying the root cause, generating solution for the problem, implementing the solution, evaluating to ensure problem pagination been fixed and lastly refining the solution if needed [2]. The TRIZ way problem-solving is much faster and easier where it starts off by specifying the problem, finds out the cause-and-effect chain analysis, applies trimming, determines the engineering contradiction and finally applies contradiction matrix to find the specific inventive solution as shown in Fig. 1 [2].

3.1 Flow Chart

Figure 2 explains how the device operates for different types of inputs. Firstly, the user is prompt to press the start button; the LCD displays the main menu, which displays manual mode, auto mode and data transfer. Figure 3 shows the flow when the user press the manual mode. The LCD displays another two options, which is UP and DOWN. The user's leg will lift up when the user press the UP option and lift down when the DOWN option is pressed.

Figure 4 shows the flow of auto mode option where the number of cycle and speed has to be entered. If the user selects option 2, the Arduino will check whether the muscle sensor is triggered, muscle sensor will be triggered if there is movement in the user's muscle, if the muscle sensor is triggered, then the Arduino next will check whether the potentiometer is triggered, potentiometer will be triggered if the user move their leg manually. If both sensors are triggered, the motor will be deactivated and the user has to do the exercise manually. If both sensors are not triggered, then the motor will be activated to help the user to do the

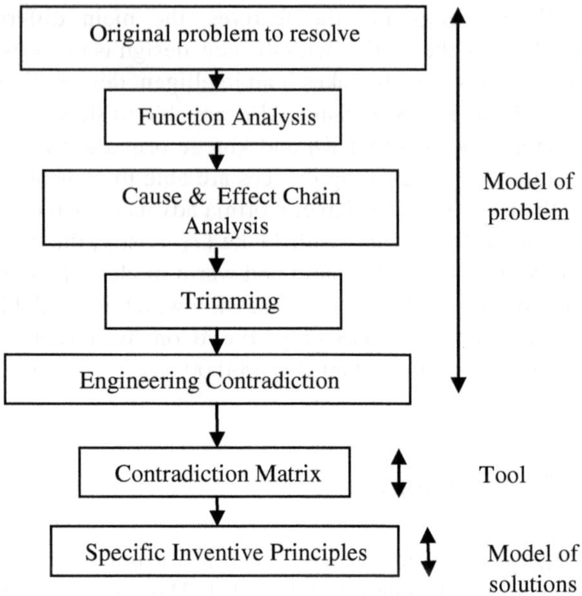

Fig. 1 TRIZ models and tools [2]

exercise automatically. If both sensors are not trigger, then the motor will activate to help the user to do the exercise automatically.

In Fig. 5, the Arduino first checks for the USB connection between the Laptop and the APK. If the connection is established, data will transfer to the laptop. The process will end if all the data are transfer or will continue until all the data transfer accordingly. If the user enter end button in between the process, the device will stop working.

3.2 Function Analysis

Figure 6 shows the super system and system products of the knee rehabilitation device. S denotes as support, H denotes hold, whereas A for access, U for use and T for trimmed. The black colour arrow represents the sufficient systems to the product, red arrow denotes that the system is insufficient and lastly blue arrow shows that the system is harmful to the system product. The floor supports the user's bed, the bed supports the base and the base supports the frame of the device. The frame supports the motor, force sensor, comfort pad, belt, extension, flexion, adjustable footplate, USB connection and a place for remote control storage. The force sensor may be harmful to the user because it might force the user's knee. Besides that, the frame supports the screw, whereas the screw can be loose at any time and is not sufficient enough for the adjustable footplate, flexion, extension and also the frame. The place for remote control storage holds the remote control and

Fig. 2 Flowchart for the three different inputs

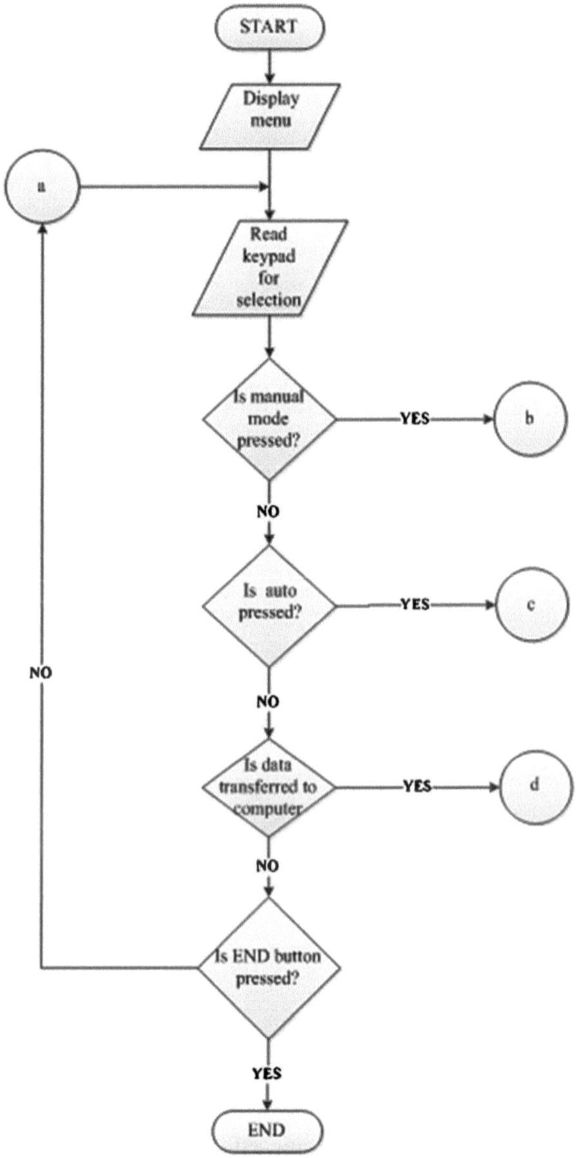

the remote control extension. The user accesses the remote control to set the setting and adjusts the footplate, flexion, extension and belt according to the leg size. Moreover, the user also can adjust the comfort pad according to their comfortless.

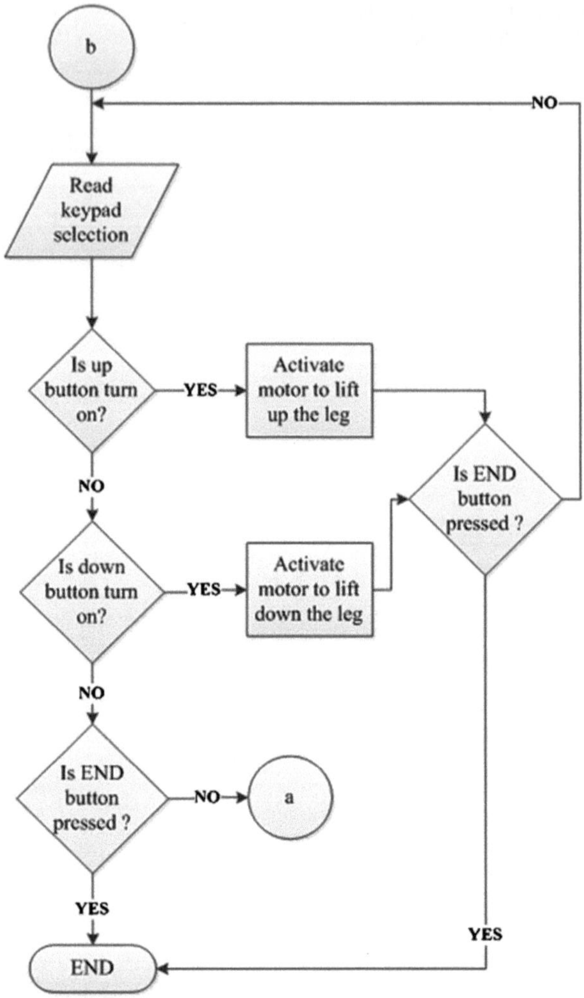

Fig. 3 Flow of the manual mode option

3.3 Cause–Effect Chain Analysis

Table 1 shows the cause-and-effect chain analysis method using 5 why's concept. Firstly, the main problem is listed in the first degree root which is current knee rehabilitation device in the market is bulky and complex.

The remote control storage makes the device bulky; remote control storage is needed to store the remote control to avoid misplacement and also to store remote control extension for high flexibility. The second reason for the device being big is due to bulky remote control which has big screen for clear view, large numeric keypad for easy selection and large ROM for high number of programming to be stored to cater different level of injuries. Besides that, the thick knee gripper to

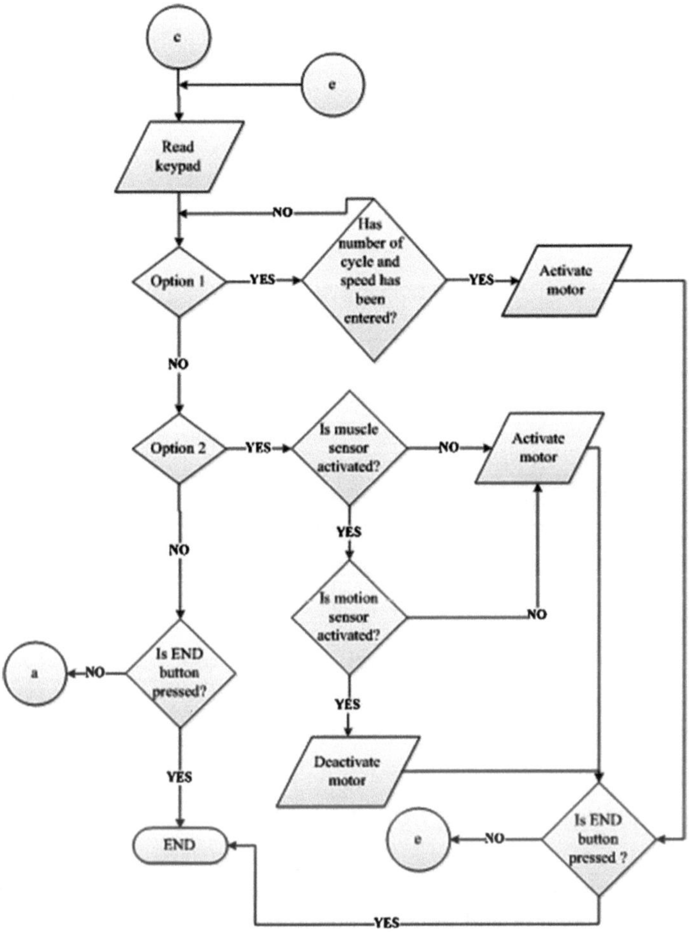

Fig. 4 Flow of the auto mode option

hold the leg tightly for correct leg positioning also makes the device bulky. Moreover, the thick comfort pad also increases the size of the device, but it will increase the comfort and relaxation for the user. The big size of DC motor can produce high torque to lift up the leg, but it increases the weight and size of the device. Lastly, the frame becomes bulky to allow extension, flexion and adjustment of the footplate according to the user size.

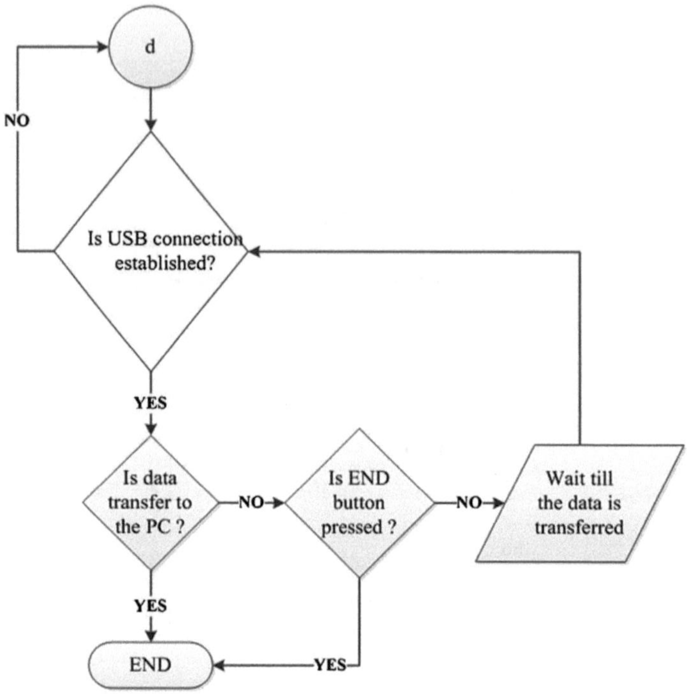

Fig. 5 Flow of the USB connection

3.4 Engineering Contradiction and Matrix

There are three types of contradiction, which are administrative contradiction, engineering contradiction and physical contradiction [2]. An engineering contradiction is a situation in which an attempt to improve one characteristic of the engineering system results in the worsening of another characteristics [2]. There are 39 parameters, which used to do the engineering contradiction. From the 39 parameters, contradiction matrix is use to generate the inventive principle solution. The engineering contradiction, the parameters, matrix and the inventive solution are shown in Table 2.

4 Results and Simulation

4.1 Elegant Solution

From the TRIZ implementation, there are several improvements have been identified as per discussed in the methodology section under engineering contract

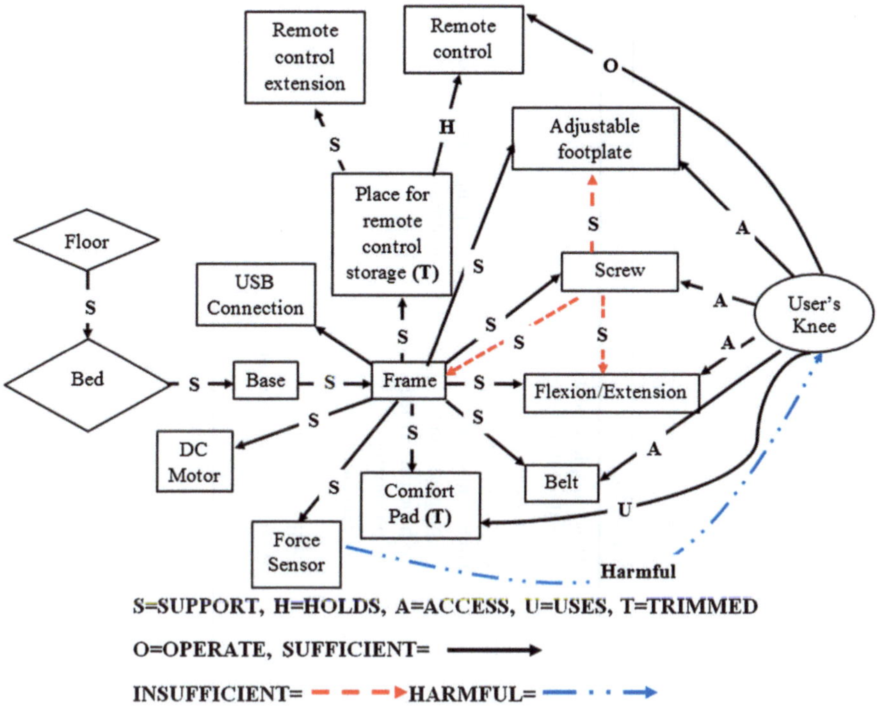

Fig. 6 Function analysis of the APK

diction matrix. There are five elegant solutions, which will be implemented and incorporated to the APK design, are listed as below:

(a) There are number of solutions that can be generated but one of the more elegant solutions would be to use principle #14 that is curvature. The worm gear will be implemented which uses rotary motion concept. Worm gear is chose for this project as a way to increase the torque because the cost of fabrication is cheaper.
(b) Principle #7 that is nested doll will be used as the most elegant solution. Different humans have different size; therefore, retractable surface can be used which will retract to the required length.
(c) There are number of solutions that can be generated but one of the more elegant solutions would be to use principle #17 that is another dimension. A stacking up design where numeric keypad is inserted below the LCD and whenever user needs to use the keypad they slide in and slide out can reduce the remote control size.
(d) The best solution is by selecting principle #3 which is local quality where the device can be attached to any type of chair.
(e) The elegant solution will be principle #13 that is the other way around where shoe straps will be used to replace the belts.

Table 1 Cause–effect chain (*CEC*) analysis

No	Number of degrees				
	First degree	Second degree	Third degree	Fourth degree	Fifth degree
1	Bulky and complex knee rehabilitation device	Bulky remote control storage	Store remote control	Avoid misplacement	
			Long and thick extension	High flexibility	
2		Bulky remote control	Big screen	Clear view	
			Large numeric keypad	Easy selection	
			Large room	High number of programming	To cater different level of injuries
3		Thick knee gripper	Hold the leg tightly	To have correct leg positioning	
4		Bulky comfort pad	Increase comfortness to patient	For relaxation	
5		Big DC motor	High torque	To lift up the leg	
6		Bulky frame	Extension/flexion	To suit different size of people	
			Adjustable footplate		

Table 2 Engineering contradiction, parameters, contradiction and elegant solution

Engineering contraction	Parameters	Contradiction	Elegant solution
If the remote control storage is bulky, then it store the remote control and the remote control extension, but this can make the device bulky	IP: 33 (ease of operation) WP: 2 (weight of stationary object), 4 (length of stationary object), 6 (area of stationary object), 8 (volume of stationary object), 12 (shape)	33, 2= 6, 13 1, 25 33, 4= – 33, 6= 18, 16 15, 39 33, 8= 4, 18 39, 31 33, 12= 1, 13 2, 4	Use principle #2 that is taking out
If the remote control is lightweight and small, it will be easy to hold by the patient, but this will increase the device complexity	IP: 1 (weight of moving object), 3 (length of moving object), 7 (volume of moving object), 12(Shape) WP: 36 (device complexity), 32 (ease of manufacture)	36, 1= 26, 30 36, 34 36, 3= 1, 19 26, 24 36, 7= 26, 1 36, 12= 16, 29 1, 2 32, 1= 27, 28 1, 36 32,3= 1, 29 17 32, 12= 13, 2 17, 28 32,7= 29, 1 40	The best solution is by selecting principle #17 another dimension
If the knee gripper belt is thick, then it can hold the leg tightly, but this make the size of the device big	IP: 13 (stability of the objects composition) WP: 2 (weight of stationary object), 4 (length of stationary object), 6 (area of stationary object), 8 (volume of stationary object), 12 (shape)	13, 2= 21, 35 2, 39 13, 4= 13, 15 1, 28 13, 6= 2, 11 13 13, 8= 3, 10 14, 24	Principle #13 that is the other way around will be used

(continued)

Table 2 (continued)

Engineering contraction	Parameters	Contradiction	Elegant solution
If the comfort pad is thick, then it will increase the comfort level, but it will make the knee rehabilitation device bulky	IP: 27 (reliability) WP: 2 (weight of stationary object), 4 (length of stationary object), 6 (area of stationary object), 8 (volume of stationary object), 12 (shape)	27, 2= 3, 10 8, 28 27, 4= 15, 19 28, 11 27, 6= 32, 35 40, 4 27, 8= 2, 35 24 27, 12= 35, 1 16,11	The best solution is by selecting principle #3 which is local quality
If the DC motor is big then it will increase the torque, but ultimately make the knee rehabilitation device bulky	IP: 9 (speed) WP: 1 (weight of moving object), 3 (length of moving object), 5 (area of moving object), 7 (volume of moving object), 12 (shape)	9, 1= 2, 28 13, 38 9, 3= 13, 14 8 9, 5= 2, 28 13, 38 9, 7= 7, 29 34 9, 12= 35, 15\ 18, 34	Use principle #14 that is curvature
If flexion, extension and adjustable footplate is universal shape, the more users can use, but this will indirectly increase the size of the product	IP: 3(length of moving object), 12 (shape) WP: 5(area of moving object), 7(volume of moving object)	5, 3= 15, 17 4 5, 12= 1, 7 4, 17 7, 3= 1, 7 4, 35 7, 12= 7, 17 4, 35	Principle #7 that is nested doll will be used

There are number of solutions that can be generated but one of the more elegant solutions would be to use principle #2 that is taking out. The place for remote control will be extracted out from the device, and the remote control can be hand on the device.

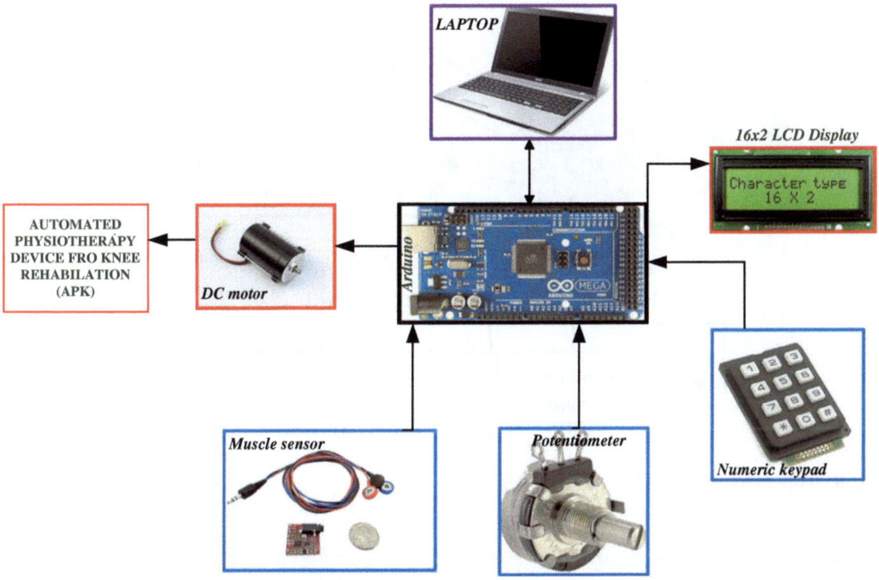

Fig. 7 Block diagram of APK device

Table 3 Block diagram components and its function

Component	Function
DC motor	• To drive the mechanical part of the product
Muscle sensor	• To detect the muscle movement of the user
Potentiometer	• To detect the movement of the user's leg
Numeric keypad	• To allow the user to enter the option
LCD display	• To display the option
Laptop	• To retrieve the data of the user's recovery through a pen drive and display GUI
Pen drive	• To save the data of the user's recovery
Arduino mega	• Whole brain of the product • Receive data from numeric keypad and display it in the LCD display • Receive sensor trigger from the muscle sensor and motion sensor • Rotate DC motor according to the programming code • Send and receive data from laptop • Display the graphic user interface (GUI)

4.2 Block Diagram

Based on the elegant solution, the above block diagram as shown in Fig. 7 has been developed and details of each function has been explained in Table 3.

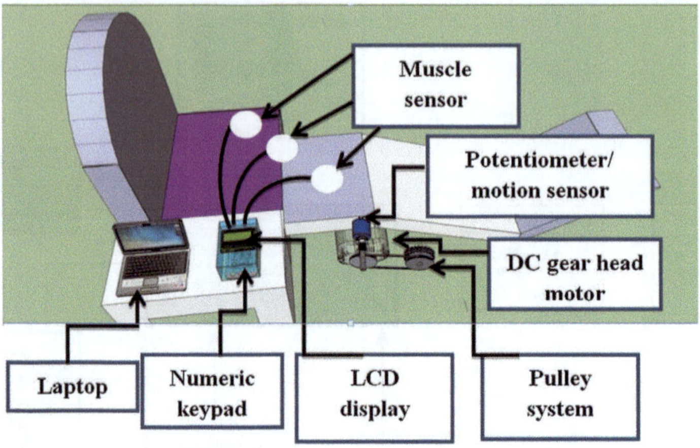

Fig. 8 Top view of the APK product design

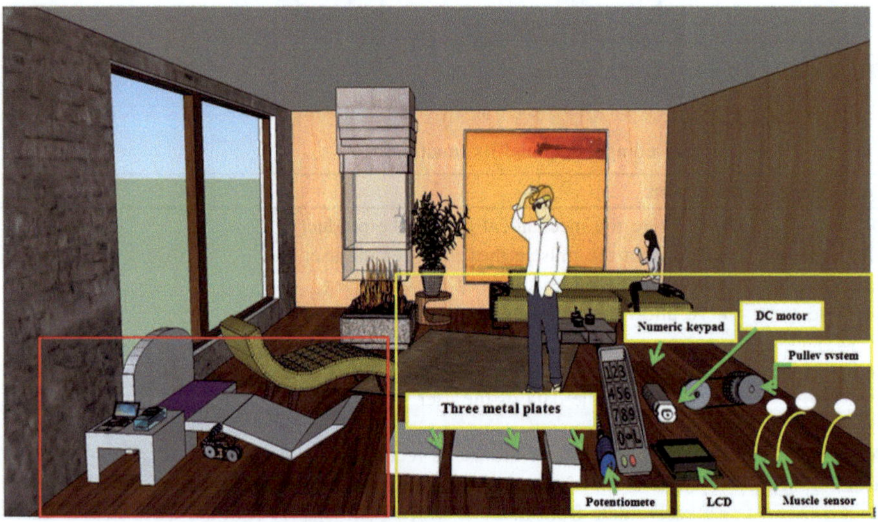

Fig. 9 Full overview of the APK product design

4.3 Product Design

Figure 8 shows the top view of the device, while Fig. 9 shows the full overview of product design, which has two boxes with different colours, which are red and yellow. The yellow colour box consists of the components provided for the user for assembled which are potentiometer, DC gearhead motor, pulley system, LCD display, numeric keypad, muscle sensor and three metal plates. The red box shows the device after assembling. The chair and laptop will not provide.

5 Recommendation and Conclusion

In conclusion, there are six solutions has been identified using TRIZ as the problem-solving method such as using taking out, another dimension, the other way around, local quality, curvature and nested doll.

The prototype design of APK is an affordable, portable and weightless device for knee rehabilitation at home. However, several future works need to be continue to develop the prototype of this product as listed below:

(1) *Develop a muscle sensor to detect the muscle movement.*
(2) *Develop a motion sensor to detect the movement of the leg.*
(3) *Build Graphic User Interface (GUI) using visual basic.*
(4) *Build H-Bridge to drive DC Motor.*
(5) *Interface LCD and numeric keypad with arduino in order to built the remote control.*

References

1. Knee Anatomy, Function: Bones, Ligaments, Muscles, Tendons, Images. [Online]. Available: http://healthpages.org/anatomy-function/knee-joint-structure-function-problems/. Accessed: 21 Sept 2013.
2. Li, S. C., San, Y. T., & Jin, Y. T. (2011). *TRIZ systematic innovation in manufacturing* (2nd ed., p. 175). Singapore: Firstfruits Publishing.
3. Knee Ligament Repair|Johns Hopkins Medicine Health Library. [Online]. Available: http://www.hopkinsmedicine.org/healthlibrary/test_procedures/orthopaedic/knee_ligament_repair_92,P07675/. Accessed 11 Mar 2014.
4. Untitled—Catalog Kinetec.pdf. [Online]. Available: http://www.ortovit.eu/ortopedie/kinetec/CatalogKinetec.pdf. Accessed 22 Sept 2013.
5. CPM 4 KINETEC SPECTRA DC (DATA CAPTURE) KNEE. [Online]. Available: http://neotech.ro/cpm-4-kinetec-spectra-dc-data-capture-knee/86.htm. Accessed 8 Sept 2013.
6. A-MOT GesPros e DV 705—ARTROMOT_product_range.pdf. [Online]. Available: http://storage.djoglobal.eu/en_US/Documents/Production_documents/ARTROMOT_product_range.pdf. Accessed 19 Sept 2013.
7. Passive mobilization, electromedical: RIMEC Srl (Bologna—Emilia Romagna—ITALY). [Online]. Available: http://www.rimec.it/IT/mobilizzazione-passiva-fisiotek-2000-e.php. Accessed 19 Sept 2013.
8. Kinetec Prima™ Advance Knee CPM Machine|1800wheelchair.com. [Online]. Available: http://www.1800wheelchair.com/product/4944/kinetec-prima-adnvace-knee-cpm-machine. Accessed 22 Sep 2013.

Relationship of Anthropometrics and Fitness Level Between Elite and University Male Rowers

Azreeany Abdul Rahim and Norasrudin Sulaiman

Abstract Successful competitive performance is particularly associated with four main characteristics which are skills, anthropometrics characteristics, appropriate psychological attitude, and physical fitness of individual athletes. Prior to that, selection of athletes for a particular sport should focus on those traits and abilities which have the most significant impact on sport performance. Hence, the purpose of this study was to determine the relationship between anthropometrics and fitness level among rowers. It consisted of one hypothesis as there was no statistical significant relationship on anthropometrics and fitness level among elite and university male rowers. Thirty male rowers (elite = 15, university = 15), aged 20–30 years, participated in this study. The design of this study was an ex post facto design. The method involved quantitative assessment of anthropometrics (height, weight, body fat percentage, body length, breadth, and girth) and fitness level tests (flexibility, power, relative strength, muscular endurance, and aerobic fitness). The findings showed significant positive relationship between nine of the anthropometrics measurements, which were height ($r = 0.76$), sitting height ($r = 0.65$), arm span ($r = 0.64$), arm length ($r = 0.73$), forearm length ($r = 0.54$), thigh length ($r = 0.74$), leg length ($r = 0.63$), shoulder breadth ($r = 0.43$), and calf girth ($r = 0.55$) with all fitness level variables among the elite and university rowers in this study. It can be concluded that anthropometrics and fitness level have a strong relationship toward rowers' performance.

Keywords Anthropometrics · Fitness level · Elite rowers · University rowers

A. Abdul Rahim (✉) · N. Sulaiman
Sports Science Centre of Studies, Universiti Teknologi MARA, Shah Alam, Malaysia
e-mail: azrean87@yahoo.com

A. Abdul Rahim · N. Sulaiman
Faculty of Sports Science and Recreation, Universiti Teknologi MARA, Shah Alam, Malaysia

R. Adnan et al. (eds.), *Proceedings of the International Colloquium on Sports Science, Exercise, Engineering and Technology 2014 (ICoSSEET 2014)*,
DOI: 10.1007/978-981-287-107-7_49, © Springer Science+Business Media Singapore 2014

1 Introduction

Rowing has recently become a sport for talent identification. In selection of athletes for a particular sport, biomechanical and kinesiological investigations into the technical performance of rowers are rare and coaches depend on the overall mentality of technique [15]. It has been established that no single parameter measures physical fitness and that variable composite factor was involved in each sport. Thus, physical fitness may contain a number of different elements including maximal oxygen uptake, local muscle endurance, muscle strength, and body composition [8].

Elite athletes in sports such as rowing differ in physical and physiological characteristics. The researcher may expect the elite athlete to represent an expression of genetic, physical training, nutrition, and sociocultural factors. As an Olympic sport, rowing requires technique, tactics, and a high level of physical fitness and body size which are undoubtedly performance-related factors [6, 20, 21, 23]. Some of the physical fitness and anthropometrical variables are considered as requisites for high performance in rowing competition.

The performance achievements of Olympic athletes come from a unique mixture of features and capabilities developed by way of training. Determining factors such as physical size and framework which may result in the best performance can assist the exercise researcher and physical trainer in choosing and creating skilled athletes [1].

In selection of athletes for a particular sport, the focus should be on those attributes and capabilities which have the most significant impact on sports performance, such as physiological and anthropometric characteristics [15]. In the sport of rowing, skill recognition programs have attempted to identify potential young athletes using various performance factors, including several anthropometric data gathered at Olympic Games level to accomplish this skills recognition strategy. The latest extensive anthropometric study of rowers was conducted at the Sydney Olympic Games in 2000 [1].

Physical structure is an important factor that contributes to success in rowing [23], and evidences indicate that anthropometric characteristics have influenced over rowing performance [7]. Indeed, rowing is a type of strength–endurance game, challenging high stages of both aerobic and anaerobic capabilities for effective performance by approximately 70–80 % aerobic and 20–30 % anaerobic [11].

Instead of having a good fitness level, the rowing motion is also reinforced by the sliding seat so that the drive phase is sequentially conducted by extension of the legs and extension of the trunk with simultaneous flexion of the arm [12, 21]. The rower has to attain a mixture of high action power and maximum stroke length to produce an effective stroke [12]. Strength enhancement not only has an impact on speed and power but also provides the basis for strength endurance [14]. Having excellent anaerobic and aerobic capacity, strength, and power are the most important elements needed to achieve great results in rowing competitions [14, 21].

Previous research had investigated some of the anthropometric features such as height, lean body weight, and limb lengths in rowing athletes [4, 13, 17, 24]. These studies showed that elite's rowers are usually taller and heavier than athlete in other sports. However, if rowers' anthropometric details are only used as talent identification at all stages, past details provide only restricted details and more knowledge is required on the anthropometric characteristics of the elite rowers from the other country. Therefore, the aim of the present research was to determine the preliminary database of anthropometric and physiological profiles between Malaysian elite and university male rowers.

A review of the literature showed that most studies on rowing profiling have been conducted abroad [3, 4]. These studies also showed that elite rowers are usually taller and heavier as compared to other athletes. Since rowing has been played at the Olympic level, athletes must have an excellent fitness level such as flexibility, power, strength, muscular endurance, and a good aerobic capacity to win in tournaments. A strong and specific training program based on the individual needs to be designed in order for the athletes to excel in the games. However, in Malaysia, there is lack of adequate profiling data among the elite and local university male rowing athletes.

Thus, it is important for the athletes and coaches to know anthropometric characteristics and fitness level of elite and university rowers in order to design a training program or for talent identification. Hence, the aim of the present study was to get baseline data of local elite and university male rowers. This profiling will enable us to relate the elite and university male rowers' performances as a track record in view of the fact that we need to prepare subelite athletes to ensure continuous success.

2 Methodology

2.1 Participants

The sample consisted of 30 male rowers. Malaysian elite male rowers and university male rowers aged 20–30 years were selected for this study. The participants of elite and university rowers only consisted of 15 people in each group. Prior to the intervention, all participants completed their Prior to the intervention, all volunteers completed their screening. The study was approved by the Ethics Committee of MARA Technology University (600-RMI: 5/1/6/01).

2.2 Outcome Measures

Precisely, the variables measured were anthropometrics measurements (height, weight, body fat percentage, body length, breadth, and girth) and fitness level (flexibility (Flx), power (Pwr), relative strength (R.S), muscular endurance (M.E), and aerobic fitness (A.F)) to determine the relationship between anthropometric and fitness levels among the participants. The anthropometrics measurements were performed according to the ISAK manual [12] by two testers who had taken part in using Harpenden Anthropometer, USA. Meanwhile, eight sites of skindfold to determine the body fat percentage by using the Harpenden Clipper equipment.

The flexibility was measured by sit and reach test, while power was measured by vertical jump test. Moving on to strength, it was measured by handgrip and leg dynamometer test. Next is muscular endurance which was estimated using the maximal number of half-sit-up that can be conducted in 1-min test.

Finally, after the rowers finish an incremental maximal test on the rowing ergometer Concept II model B, Morrisville, VT, USA, the aerobic fitness measurement shall be recorded. The test started with 5–10-min rowing session with "all-out" effort to ensure they achieved maximal levels of aerobic capacity. Heart rate was monitored using the Polar Heart Rate RS 100 Germany at the end of each moment and row until a steady state was obtained (5–10 min). Oxygen consumption will be measured using manual Eq. (1), to which we will connect the 2 points between submaximal heart rate (/min) and power (watt) due to record VO_2 max (L/min) level.

$$\mathrm{VO}_{2\mathrm{max}} = \frac{x \times 1,000}{\text{Body weight}} = x\,\mathrm{ml/kg/min} \tag{1}$$

3 Results

The demographic descriptive statistics for mean ± standard deviation (SD) of anthropometrics and fitness levels among the participants are presented in Table 1. Most of the athletes indicated normality of the data distribution on anthropometrics and fitness level. The present study indicated that elite rowers had higher values for most of the measurements, including height, length of the arm and leg, power, relative strength, muscular endurance, and also aerobic fitness.

The correlations between anthropometric characters and fitness level are presented in Table 2. The results showed that there were positive significant correlations ($p < 0.05$) between anthropometric variables (height, sitting height, arm span, arm length, forearm length, thigh length, leg length, and chest girth) with flexibility, power, relative strength, muscular endurance, and aerobic fitness.

4 Discussion

According to the present study, the findings showed that height, weight, and body length are the vital factors in rowing performance. It means the lighter a rower, the faster movement during rowing can be produced. In addition, identical findings were reported by several researchers [12, 21] where they discovered that body

Table 1 Anthropometrics and fitness level characteristics between rowers

Variables	Elite Mean ± SD	University Mean ± SD
Age	22.03 ± 1.06	21.07 ± 0.70
Height (cm)	177.12 ± 3.72	169.13 ± 3.19
Weight (kg)	71.67 ± 2.08	73.69 ± 2.86
Body fat (%)	6.13 ± 0.19	6.531 ± 0.26
Sitting height (cm)	132.47 ± 2.38	128.00 ± 3.48
Arm span (cm)	186.15 ± 3.98	180.39 ± 3.17
Arm length (cm)	31.86 ± 0.67	30.01 ± 0.88
Forearm length (cm)	25.37 ± 0.94	24.31 ± 0.79
Flexed arm girth (cm)	31.87 ± 0.99	31.33 ± 1.23
Femur breadth (cm)	9.22 ± 0.42	9.04 ± 0.43
Chest girth (cm)	94.88 ± 3.22	94.07 ± 2.86
Waist girth (cm)	74.28 ± 2.42	74.60 ± 3.20
Shoulder breadth (cm)	36.19 ± 1.85	34.73 ± 0.87
A-P chest depth (cm)	18.37 ± 1.64	16.40 ± 0.64
Humerus breadth (cm)	6.65 ± 0.37	6.72 ± 0.32
Thigh length(cm)	37.29 ± 2.03	32.99 ± 1.29
Leg length (cm)	48.18 ± 1.89	46.12 ± 1.19
Hip girth (cm)	92.96 ± 2.61	94.87 ± 3.98
Thigh girth (cm)	56.39 ± 1.56	55.63 ± 2.59
Calf girth (cm)	37.51 ± 0.94	34.80 ± 2.18
Flexibility (cm)	34.56 ± 3.55	30.73 ± 4.13
Power (cm)	47.67 ± 2.31	44.67 ± 2.52
Relative strength (kg)	5.06 ± 0.20	4.73 ± 0.18
Muscular endurance	76.60 ± 6.03	72.93 ± 2.84
Aerobic fitness (ml/kg/min)	51.91 ± 1.52	46.69 ± 1.58

weight was supported by a sliding seat in the boat; thus, they can afford to carry a greater mass and possess an advantage and it is believed rowing performance is significantly related to height [11]. Height and body length had a positive significant relationship for the entire fitness test variable. Higher rowers are able to make long rowing stroke length are thoroughly recognized with high-level rowing performance [18]. Upper limb and lower limb lengths are factors related to the stroke rate [7]. The length of feet improved upon the generate stage of the rowing stroke, during the catch and generate action including all arms and legs, leading to produce power stroke length [22]. The height and leg strength demonstrated the importance of leg and trunk length which could expand the driving phase. Though it is not a factor that can be trained, this could be handy to identify success in potential rowers.

Research on elite rowers have proven that size is not the only important factor, but also the ratio of height to arm span, such that proportionately longer arms provide a greater movement of stroke length. Thus, these factors might affect rowing performance, while the selection of competitive rowers might well focus upon these genetically determined aspects [7].

Table 2 Correlations between anthropometrics and fitness level among rowers

Anthropometric	Flx $n = 30$ (r)	Pwr $n = 30$ (r)	R.S $n = 30$ (r)	M.E = 30 (r)	A. F $n = 30$ (r)
Height	0.76**	0.74**	0.42*	0.45*	0.74**
Weight	0.29	0.18	0.43*	−0.12	−0.21
B. fat (%)	0.05	−0.21	−0.29	−0.26	−0.39*
Sitting height	0.60**	0.65**	0.29	0.45*	0.57**
Arm span	0.31**	0.58**	0.45*	0.48**	0.64**
Arm length	0.50*	0.68**	0.44*	0.30	0.73**
Forearm length	0.54*	0.22	0.32	0.37	0.45*
Thigh length	0.54*	0.57**	0.46**	0.43*	0.74**
Leg length	0.63**	0.26	0.29	0.38*	0.53**
Shoulder breadth	−0.03	0.32	0.24	0.43*	0.40*
A-P chest depth	0.16	0.48**	0.32	0.32	0.55**
Humerus breadth	0.16	0.07	−0.15	−0.06	−0.15
Femur breadth	0.10	0.34	0.28	0.09	0.25
Flexed arm girth	0.28	0.21	0.24	0.09	0.09
Chest girth	0.39*	0.48**	0.02	0.28	0.10
Waist girth	0.32	0.32	−0.17	0.04	−0.12
Hip girth	0.28	0.10	−0.32	−0.10	−0.32
Thigh girth	0.29	0.30	0.08	0.20	−0.04
Calf girth	0.22	0.55**	0.37*	0.39*	0.48**

*Correlation is significant at the 0.05 level (1 tailed)
**Correlation is significant at the 0.01 level (2 tailed)

Meanwhile, shoulder breadth had a positive significant relationship on muscular endurance and aerobic fitness. Meanwhile, chest girth had a positive significant relationship with flexibility and power. Calf girth had a positive significant relationship with power, relative strength, muscular endurance, and aerobic fitness. As the studies made by [22], informed rowing training for at least a year can improve some breadth and girth (arm and chest) among junior rowers. It was due to the years of training that affected the development of body structure which elite rowers were capable to develop large muscle mass (breadth and girth) which lead in increasing a higher muscle mass cross-sectional area, and therefore, a higher power and force output can be produced [24].

However, body fat percentage had negative significant relationship with aerobic fitness variables. In parallel with previous studies by [19], found that, body fat percentage was significantly negatively correlated with VO_2 max ($r = -0.534$, $p = 0.002$). An individual with a greater aerobic capacity most likely engages in enough physical activity to alter their body composition in a favorable manner. Engaging in exercise of a more anaerobic nature probably induces positive adaptations in a person's body composition. Therefore, having a low body fat percentage can be an advantage to succeed in rowing competitions [16]. Studies of male and female international rowers [10] have noted that the range of body fat percentage was from 6 to 10% and 11 to 15 % for male and female rowers, respectively.

Furthermore, the entire anthropometric variable (height, body length, and girth) had a positive significant relationship with flexibility. The elite rowers had a higher score in flexibility probably due to the contribution from body length (upper and lower limb). The reason for elite rowers to have a higher score of flexibility compared to university rowers was based on the years of training. Several exercises were included in the additional training which aimed at improving flexibility of the low back region and hamstrings [4].

The entire anthropometric variable (height, body length, breadth, and girth) had a positive significant relationship with power. Based on the result from the previous study made by Jurimae et al. [12], it indicates that strength and power assessment are highly predictive of traditional rowing performance.

Strength enhancement has an influence on speed and power besides providing the foundation for strength endurance [14]. In fact, the greater stage of maximal strength and muscular endurance and power in the elite rowers provide a benefit to maintain a more powerful stroke during the oar cycle [9, 12, 14]. Efficient elite rowers produce about 75–80 % of their power with their legs and 20–25 % with their arms during the rowing stroke [5].

Besides, relative strength had a positive significant relationship with height, weight, body length, and girth. Strength is the important criteria that must be developed among rowers. Strength is proportional to muscle size where this can be an indicator of a higher muscle mass cross-sectional area, and therefore, a higher power and force output may be produced during oar cycle and also stroke length [22]. But then, in rowing, the relative strength is most significantly variables that need to be implemented because it is closely related to the nature of rowing event.

The result was also supported by Mikulie and Ruzic [17], which claimed that relative strength is isometric vital to a rower than absolute strength due to the amount of body weight in a boat does affect the drag through water. An increase in absolute strength is of no benefit if the excess of bodyweight offsets the strength gain by increasing resistance through water when rowers increased the relative strength; on the other hand, they find it easier to speed up the boat because they have increased strength without increasing drag.

Muscular endurance had a positive significant relationship with height, body length, breadth, and girth. The result is quite similar to the previous study made by [3], which stated 1-min half-sit-up was used to measure the muscular endurance among elite Iranian Junior Rowers. Hence, it can be indicated that strength enhancement has an impact on speed and power. It also provides the basis for strength endurance. The truth is the greater levels of maximal strength, muscular endurance, and power in the elite rowers provide an advantage to maintain more highly effective stroke during the oar cycle.

Aerobic fitness had a positive significant relationship with height, body length, breadth, and girth among rowers. To gain a higher aerobic capacity, elite rowers are required to have the capability to maintain very high oxygen consumption, based on the fact that during a race, most oarsmen usually execute close to their maximum aerobic capacity for the entire length of the race [20].

Height and body length have a relationship with the aerobic fitness performance. It is because taller people have larger lungs. In this case, the vital capacity is typically increased in taller people especially by undergoing the incremental exercise [2]. Thus, establishing the exact levels of these parameters is required to succeed in this competition.

5 Conclusion

It can be concluded that anthropometrics (height, weight, and body length) and fitness level (power, relative strength, muscular endurance, and aerobic fitness) have a strong relationship toward rowers' performance. The results of the present study provides new profiling data on anthropometric and fitness levels of Malaysian rowers that can be used by sport scientists, coaches, and athletes. The data are useful for constructing anthropometric profiles for individual rowers to improve individual training programs in the future. The anthropometric and fitness profiles of elite and university male rowers can contribute toward skills selection and recognition and could also be of importance for maximal development of strength, power, and endurance training programs for talent identification.

References

1. Ackland, T., Kerr, D., Hume, P., Norton, K., Ridge, B., Clark, S. & Ross, W. (2001). Anthropometric normative data for Olympic rowers and paddlers. In *Australian conference of science and medicine in sport*. Perth: Sports Medicine Australia:157.
2. Adams, G. M., & Beam, W. C. (2007). *Exercise physiology: Laboratory manual* (5th ed.). New York: McGraw-Hill.
3. Arazi, H., Faraji, H., & Mohammadi, S. M. (2011). Anthropometric and physiological profiles of elite Iranian junior rowers. *Middle-East Journal of Scientific Research, 9*(2), 162–166.
4. Bourgois, J., Claessens, A. L., Vrijens, J., Philippaerts, R., Renterghem, B. V., Thomis, M., et al. (2000). Anthropometric characteristic of Elite male junior rowers. *Journal Sports Medicine, 34*, 213–217.
5. Cosgrove, M. J., Wilson, J., Watt, D., & Grant, S. F. (1999). The relationship between selected physiological variables of rowers and rowing performance as determined by a 2000 m ergometer test. *Journal of Sports Sciences, 17*(11), 845–852.
6. De Garay, A. L., Levine, L., & Carter, J. E. L. (1974). Genetic and anthropological studies of Olympic athletes. Waltham: Academic Press.
7. Dimakopoulou, E., Blazevich, A. J., Kaloupsis Diafas, S. V., & Bachev, V. (2007). Prediction of stroking characteristics of elite rowers from anthropometrics variables. *Journal of Sports Science, 1*(3), 89–96.

8. Grant, S. F. (1999). The relationship between selected physiological variable of rowers and rowing performance as determined by a 20,000 m ergometer test. *Journal of Sports Sciences, 17*(11), 845–852.
9. Gabarren, M. I., Exposito, G. D. T., & Izquierdo, M. (2010). Physiological factors to predict on traditional rowing performance. *Journal of Applied Physiology, 10*, 83–92.
10. Hagerman, F., & Toma, K. (1997). Physiological evolution of the rowing Athlete: A 25-year Study.
11. Ingham, S., Whyte, G., Jones, K., & Nevill, A. (2002). Determinants of 2,000 m rowing ergometer performance in elite rowers. *Journal of Applied Physiology, 88*(3), 243–246.
12. Jurimae, T., Turpin, J. A. P., Tormo, J. M. C., Mira, I. J. C., Anta, R. C., Maestu, J., et al. (2010). Relationship between rowing ergometer performance and physiological responses to upper and lower body exercises in rowers. *Journal of Science and Sport Medicine, 13*(4), 434–437.
13. Kaloupsis, S., Bogdanis, G. C., Dimakopoulou, E., & Maridaki, M. (2008). Anthropometric characteristics and somatotype of young Greek rowers. *Journal of Biology Sports, 25*(1), 57–69.
14. Maestu, J., Jurimae, J., & Jurimae, T. (2005). Monitoring of performance and training in rowing. *Journal of Sport Medicine, 35*(7), 597–617.
15 Markovic, G., Dizdar, D., Jukic, I., & Cardinale, M. (2004). Reliability and factorial validity of squat and countermovement jump tests. *Journal Strength and Conditioning, 18*(3), 551–555.
16. McLester, J. R., Green, J. M., Wickwire, P. J., & Crews, T. R. (2008). *Relationship of VO2 peak, body fat percentage, and power output measured during bouts of a wingate protocol.* DigitalCommons@ Kennesaw State University.
17. Mikulie, P., & Ruzic, L. (2008). Predicting the 1,000 m rowing ergometer performance in 12 and 13 year old rowers: The basis for selection process. *Journal of Science and Medicine in Sport, 11*(2), 218–226.
18. Muller, G., Hille, E., & Szpalski, M. (1994). Function of the trunk musculature in elite rowers. *Sportverletzung Sportschaden, 8*(3), 134–142.
19. Nolte, V. (2011). How to build rowing strength. 2nd edition. *Human Kinetics* (2nd ed.). Retrieved from http://www.humankinetics.com/all-outdoor-activities-articles/all-outdoor-activities-articles/how-to-build-rowing-strength.
20. Secher, N. H., & Vaage, O. (1990). Rowing performance, a mathematical model based on analysis of body dimensions as exemplified by body weight. *Journal of Applied Physiology, 52*, 88–93.
21. Secher, N. H., Vaage, O., & Jackson, R. C. (1983). Rowing performance and maximal aerobic power of oarsmen. *Journal of Sport Science, 4*, 9–11.
22. Singh, R., Singh, H. J., & Sirisinghe, R. G. (1995). Physical and physiological profile of Malaysian dragon boat rowers. *Journal of Sport Medicine, 29*(1), 13–15.
23. Slater, G. J., Rice, A. J., Mujika, I., Hahn, A. G., Sharpe, K., & Jenkins, D. G. (2005). Physique traits of lightweight rowers and their relationship to competitive success. *Journal of Sport Medicine, 39*, 736–741.
24. Piotrowski, J., Sklad, M., Krawczyk, B., & Majle, B. (1992). Somatic indices of junior rowers as related to their athletic experience. *Journal of Biology of Sport, 9*, 117–124.

Ricebag Induced Skin Interface Temperature in Shoulder Injury Patients

Mohamad Hisyam Izzuddin Mohamed Ishak, Rahmat Adnan, Norasrudin Sulaiman, Shariman Ismadi Ismail and Ridzuan Azmi

Abstract Ricebag has been found to be an alternative therapy in the management of musculoskeletal injury. The objectives of the present study were: (1) to determine the efficacy of skin heat transfer in ricebag treatments among shoulder injury patients, and (2) to determine the effect of active range of motions (AROM) on shoulder injury patients. This study used the repeated measure design to achieve its objectives. A total of 30 samples comprising of males ($n = 19$) and females ($n = 11$) (20 experiments, 10 controls) were recruited into the study, where the mean age for males was 43.21 years old (SD = 15.393), and the mean age for females was 37.00 years old (SD = 11.653). The mean weight of males was 76.53 kg (SD = 9.930) and the mean weight of females was 61.36 kg (SD = 5.679). The mean height of males was 166.05 cm (SD = 6.892) and the mean height of females was 161.76 cm (SD = 6.087). There was a significant difference in temperatures between the ricebag and Hydrocollator ($p < 0.05$), but there was no significant differences in temperatures on the skin interface between the Ricebags and Hydrocollator ($p = 0.512$). There were no significant differences on AROM (Flexion, Extension and Abduction) for both the Ricebag and Hydrocollator ($p = 0.300$, $= 0.548$, $= 0.305$). However, there was a significant difference between the ROM and pre- and posttests after 20 min of treatment ($p < 0.05$). In conclusion, Ricebags gave a similar effect as the Hydrocollator on skin interface temperatures and showed a modest improvement on AROM. Hence, Ricebags can be considered as one of the treatment tools in back pain management.

M. H. I. Mohamed Ishak · R. Azmi
Rehabilitation Department, Tuanku Mizan Military Hospital, Kuala Lumpur, Malaysia

M. H. I. Mohamed Ishak · R. Adnan (✉) · N. Sulaiman · S. I. Ismail
Sports Science Center of Studies, Universiti Teknologi MARA, Shah Alam, Malaysia
e-mail: rahmatadnan@salam.uitm.edu.my

R. Adnan · N. Sulaiman · S. I. Ismail
Faculty of Sports Science and Recreation, Universiti Teknologi MARA,
Shah Alam, Malaysia

R. Adnan et al. (eds.), *Proceedings of the International Colloquium on Sports Science, Exercise, Engineering and Technology 2014 (ICoSSEET 2014)*,
DOI: 10.1007/978-981-287-107-7_50,

Keywords Ricebag · Hydrocollator · Shoulder injury · Skin interface temperature · Active range of motion

1 Introduction

The shoulder has a complex joint which allows for a high range of motions (ROM). However, this high ROM may affect the joint instability which in turn could result in injury. Conditions causing shoulder pain are common and contribute substantially to the musculoskeletal morbidity of the community [1]. Injuries can also occur during everyday activities such as washing walls, hanging curtains, and gardening, but may also occur from accidents like motorcycle or car crash.

The treatments for shoulder injury are often through medication, surgery, and physical therapy. However, physical therapy claims to be the safest method using heat application in shoulder injury. Tomkins et al. [2] noted that most physical therapist respondents agreed (either strongly or somewhat) that physical therapy is effective in decreasing pain (88 %) and improving function (94 %), especially in treating shoulder injury [2]. Besides, it reduces the risk factor compared to medication and surgery.

Hydrocollators are widely used and considered as one of the best tools in the market and one of the best methods for treating chronic soft tissue injury [3]. However, this tool is impractical because it requires qualified personnel, needs space, and is time consuming. Thus, it is not user friendly for immediate treatment management.

Rahmat et al. [4] found an alternative method of using Ricebags to give a similar function as hydrocollators that may be effective in shoulder injury management. Ricebag theoretically can maintain prolonged heat for recommended durations. However, there was still a lack of information on the skin heat transfer, especially in shoulder injury patients.

Hence, this study sets out a twofold objective to answer the hypothesis:

1. To determine the efficacy of skin heat transfer in ricebag treatments among shoulder injury patients.
2. To determine the effect of active range of motions (AROM) on shoulder injury patients.

2 Methodology

2.1 Research Design

In this study, a quasi experimental research design was used to test the hypothesis. The research had pre- and post treatments for durations of 20-min period for every treatment. The experimental group was subjected to this treatment by strictly

Fig. 1 Research framework

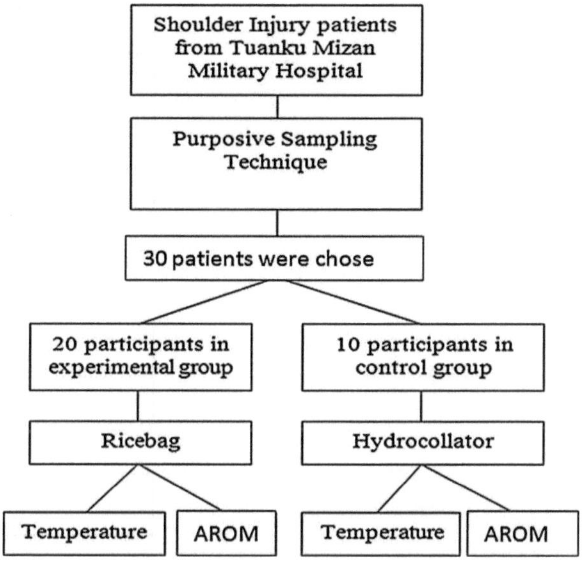

monitoring the skin heat transfer in each minute. The control group used the Hydrocollator treatment, and the results were compared to the heat transfer with the Ricebags. The independent variable (IV) was the application of Ricebag. The dependent variable (DV) was the efficacy of the skin heat transfer among shoulder injury patients (Fig. 1).

2.2 Sample

There were 30 shoulder injury patients comprising of males ($N = 19$) and females ($N = 11$) (20 experiment groups and 10 control groups) that were tested in this research. The participants were selected through the purposive sampling technique. The participants were taken from the Tuanku Mizan Military Hospital (TMMH) and the research was conducted at the same place.

2.3 Tools

The equipment used for this study was Ricebags, Hydrocollator, Digital Thermometer (Thermometer Hanna Instrument P/N: HI935005), a thermocouple probe (Hanna K-Type Waterproof Thermocouple), Goniometer, and a Hygrothermometer.

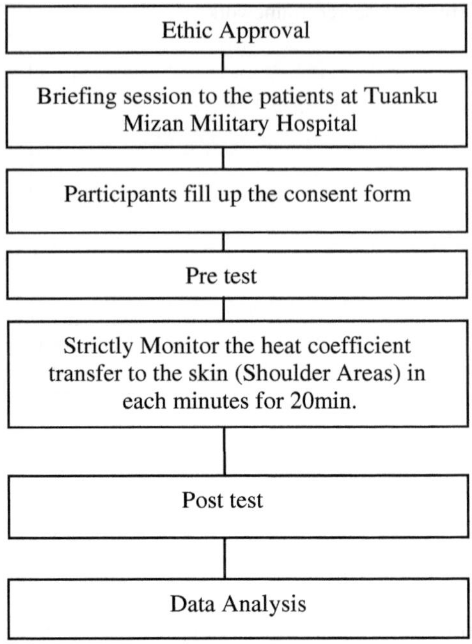

Fig. 2 Data collection procedure

2.4 Procedure

In order to conduct this research, firstly, the researcher had to submit an ethics form to the faculty for approval of the research project. Following the approval, a briefing session was conducted to inform participants about the purpose of this research and to make sure that they were clear about what they had to do. Subsequently, the participants filled the inform consent forms prior to the treatment procedures. The study was conducted in a room in the rehabilitation department at TMMH. This study had pre- and posttreatments and the application of Ricebags was strictly monitored each minutes. All subjects were tested individually with a thorough analysis and observations by the researcher (Fig. 2).

2.5 Statistical Analysis

The parameters were statistically analyzed using the Statistical Package for Social Sciences Software (SPSS) for windows version 19.0. The independent t test and repeated measure ANOVA was used to determine the differences between groups of ROM and temperature. The significance level was set at $p \leq 0.05$ for all statistical tests.

Table 1 Demographic data of the samples

Variable	Male (n = 19)		Female (n = 11)	
	Mean	SD	Mean	SD
Age	43.21	15.393	37.00	11.653
Weight	76.53	9.930	61.36	5.679
Height	166.05	6.892	161.76	6.087

Table 2 Comparison of temperature between groups

Measure	df	Sig.
Treatment*	2.809	0.045
Skin	2.102	0.512

Sig value set ($p \leq 0.05$)
*Treatment represents Ricebag and Hydrocollator

3 Results

3.1 Demographic Profile of Subjects

A total of 30 samples comprising of males (n = 19) and females (n = 11) (20 experiment groups and 10 control groups) were tested in this research. The mean age for males was 43.21 years old (SD = 15.393), and the mean age for females was 37.00 years old (SD = 11.653); the mean weight of males was 76.53 kg (SD = 9.930), and the mean weight of females was 61.36 kg (SD = 5.679); while the mean height of males was 166.05 cm (SD = 6.892), and the mean height of females was 161.76 cm (SD = 6.087) (Table 1).

3.2 Demographic Profile of Ricebag

The Ricebag and Hydrocollator were significantly different in retaining the heat during the 20-min period of treatment (Table 2). However, the skin interface temperature was not significantly different in both treatments, which are Ricebag and Hydrocollator, during the 20 min of application onto the skin.

3.3 Demographic Profile of Shoulder AROM

Based on the results presented in Table 3, there was no significant difference between the Ricebag and Hydrocollator on shoulder AROM.

Table 3 Demographic data of AROM

Group	Variable	Pre			Post		
		N	Mean	SD	N	Mean	SD
Ricebag	Flexion	20	110.35	35.538	20	122.20	35.416
	Extension	20	43.60	11.227	20	50.10	13.404
	Abduction	20	81.55	30.202	20	93.00	33.680
Hydrocollator	Flexion	10	111.60	29.923	10	111.90	27.962
	Extension	10	41.20	9.682	10	44.50	8.017
	Abduction	10	91.70	36.031	10	94.40	34.677

Table 4 Comparison of ricebag and hydrocollator on AROM

	Df	Sig.
Flexion	1.000	0.300
Extension	1.000	0.548
Abduction	1.000	0.305

The results also showed that the Ricebag and Hydrocollator were not significantly different on shoulder AROM after 20 min (Table 4).

However, the results showed that the AROM pre- and posttests were significantly different after 20 min of treatment (Table 4).

4 Discussion

The purpose of our study was to determine the effect of skin heat transfer in Ricebag among shoulder injury patients. Therapeutic heat has been widely used in clinical applications involving physical therapy and is especially relevant in resolving musculoskeletal injury [5]. We conducted the study based on the objectives that we wanted to achieve.

Based on the data that we had collected and analyzed, it was evident that the Ricebag can reach a higher temperature compared to the Hydrocollator. There was a significant difference between the Ricebag and Hydrocollator in retaining the heat temperature during the 20-min period of treatment (Fig. 3). This finding was in agreement with Rahmat et al. [4] who stated that rice was more capable of holding heat compared to other grains like barley and mung. However, when we comparing the skin interface, there was no significant difference between both treatments (Ricebag and Hydrocollator) during the 20 min of treatment. The results, however, showed that even if there was no significant difference on skin

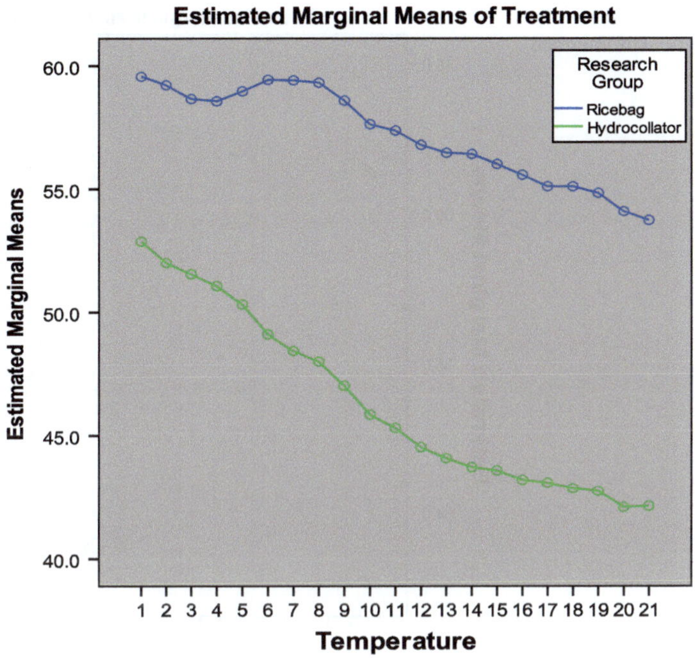

Fig. 3 Descriptive means for temperature of treatment

interface temperature between the Ricebag and Hydrocollator, but the Ricebag transferred more heat to the skin compared to Hydrocollator (Fig. 4). Therefore, the Ricebag was not only holding more heat, but also it could transfer more heat compared to Hydrocollator.

Another objective of this study was to measure the AROM. The ROM of the joints of the upper extremities, especially the shoulder joint, is important for normal physical function [6]. Impairment in joint ROMs may be due to stiffness caused from the protective mechanisms that respond to the injury [7].

Hence, from the results, there was no significant difference between the Ricebag and Hydrocollator (Table 4). The results presented in Table 3 imply that the mean of each variable of Ricebag was not much different from the Hydrocollator. However, there was an improvement on AROM between the pre- and posttests (Table 5), which was significantly different. Thus, heat therapy helps relax the muscles and reduce pain [5].

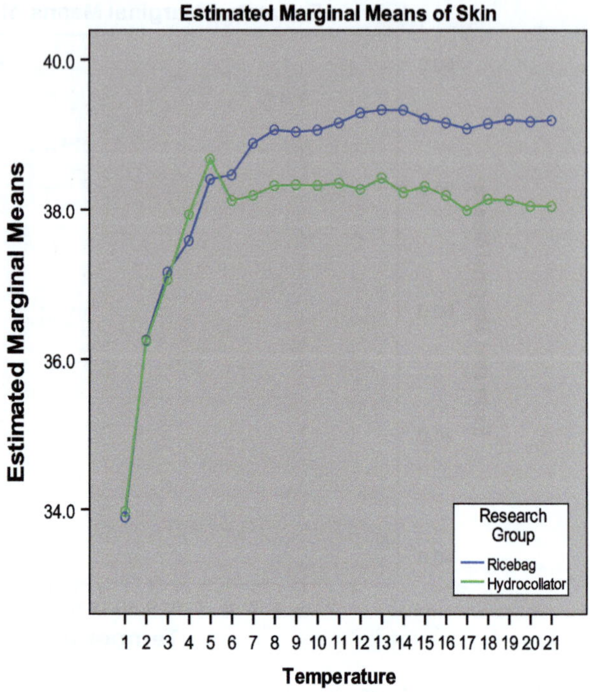

Fig. 4 Descriptive means for temperature of skin interface

Table 5 Comparison on AROM of pre- and posttest

	Pre		Post	
	df	Sig.	df	Sig.
ROM	1.985	0.000	1.999	0.000

5 Conclusion

In conclusion, this study found that there were significant differences in the AROM and also the heat transfer. The treatment provides benefits from the application of Ricebag as a heat therapy for chronic injury. It is also recommended that further research be conducted to provide evidence on the efficacy of the Ricebag.

Acknowledgments Researcher would like to express upmost gratitude to RAGS-RMI Grant for sponsoring this study. Thanks to Tuanku Mizan Military Hospital in giving opportunity to recruit participants and also to the Faculty of Sport Science and Recreation for approving our study. Last but not least thanks to everyone involved in producing this paper directly or nondirectly.

References

1. Green, S., Buchbinder, R., & Hetrick, S. (2010). *Physiotherapy interventions for shoulder pain (Review)*.
2. Tomkins, C. C., et al. (2010). Physical therapy treatment options for lumbar spinal stenosis. *Journal of Back and Musculoskeletal Rehabilitation, 23*(1), 31–37.
3. Kaiser, D. A., et al. (2004). Hot-pack warming in 4- and 8-pack hydrocollator® units. *Journal of Sport Rehabilitation, 13*(2), 103–113.
4. Adnan, R., et al. (2011). The development of beany/grainy pack to treat soft tissue injury. In *Humanities, science and engineering (CHUSER), 2011 IEEE Colloquium on.*
5. Lin, Y. H. (2003). Effects of thermal therapy in improving the passive range of knee motion: comparison of cold and superficial heat applications. *Clinical Rehabilitation, 17*(6), 618.
6. Escalante, A., Lichtenstein, M. J., & Hazuda, H. P. (1999). Determinants of shoulder and elbow flexion range: results from the San Antonio longitudinal study of aging. *Arthritis Care and Research, 12*(4), 277–286.
7. Kenia, J. L., et al. (2009) The effects of visual and verbal feedback upon gait biomechanics and functionality. In A. McGoron., C.-Z. Li., & W.-C. Lin (Eds.), *25th southern biomedical engineering conference 2009, 15–17 May 2009, Miami, Florida, USA* (pp. 39–44). Berlin Heidelberg: Springer.

Analysis of Goals Scored that Discriminated Between Winning and Losing Teams in EURO 2012

Muhamad Safiq Saiful Annur, Adam Linoby and Norasrudin Sulaiman

Abstract The objective of current study was to investigate the differences of winning and losing teams in terms of goal scoring and other selected performance indicators. A total number of 31 matches from UEFA-EURO 2012 were analyzed, and 5 matches were excluded from analysis due to matches end up drawn. There are five group of variables that was selected as performance indicators which include the possession (how possession gained and where possession gained), touches, types of shot used and also the directions of the shots in scoring goal. Data were analyzed using Wilcoxon's signed-rank test with significant value was set at $p < 0.05$. The intercept ($Z = -2.859$, $p < 0.05$), corner ($Z = -2.126$, $p < 0.05$), poor passing ($Z = -2.126$, $p < 0.05$), poor control ($Z = -3.000$, $p < 0.05$) and the goalkeeper distribution ($Z = -2.236$, $p < 0.05$) resulted in significant difference in how the possession was gained and the possession was significantly gained at first quarter ($Z = -2.449$, $p < 0.05$), second quarter ($Z = -2.309$, $p < 0.05$), third quarter ($Z = -2.804$, $p < 0.05$), and fourth quarter ($Z = -2.739$, $p < 0.05$). The winning teams preferred to use 2 touches or less, and they preferred to use driven shot, and inside shot in scoring goal ($p < 0.05$). Regarding the directions of

This work was supported in part by the Fundamental Research Grant Scheme (FRGS), Ministry of Higher Education of Malaysia FRGS/1/2011/SKK/UITM/03/30.

M. S. Saiful Annur (✉) · A. Linoby
Faculty of Sports Science and Recreation, Universiti Teknologi MARA, Pahang, Malaysia
e-mail: msafiq@pahang.uitm.edu.my; msafiq3927@yahoo.com

N. Sulaiman
Universiti Teknologi MARA, Shah Alam, Malaysia

R. Adnan et al. (eds.), *Proceedings of the International Colloquium on Sports Science, Exercise, Engineering and Technology 2014 (ICoSSEET 2014)*,
DOI: 10.1007/978-981-287-107-7_51, © Springer Science+Business Media Singapore 2014

the shot, more goals were scored at lower sections of the goals, either left or right. It can be concluded that the winning teams have significant higher performance indicators in goal scoring compared to losing teams.

Keywords Soccer · Possession · Shots · Directions · Goal

1 Introduction

Coaches nowadays usually analyzed their team's or player's performance using the method of performance analysis or match analysis. Match analysis is defined as the objective of recording and examining the behavioral event that occur during competition [1]. Performance analysis or match analysis is the systems or methods used by the coaches to gather information about the athlete's performance to plan their strategies such as tactics, technique, movement pattern and the rate of work during a game [2]. In the last decade, many analyst have applied the notational or computerized analysis to understand the different factors in the individuals or team performances. They used it for multiple purposes such as the technical and tactical evaluation, developing the norms, movement analysis and for feedback provision [3].

Without notational analysis, coaches, managers and the players will be subjective in their judgments and decision and cannot recall what their mistakes [4]. The difficulty to store and retrieve the information was due to the personal biases and emotional state during the game. The subjective observations by the coaches may or may not be accurate, and it is also unreliable. This was supported by one study where the percentages of coaches are correct in their post-game assessment was less than 45 % during the 45 min of a soccer game [2]. It was due to the human memory system has the limitations whereby we cannot remember the whole event during the entire soccer game. The poor viewing environment during the competition event, mind sets and prejudices of the coaches (what coaches want or expect to see) and the emotions during the competition (such as anger and stress) also affect the feedback collection by the coaches. The coaches often unable to view and see the action that occur during an event, and it is impossible for them to cover all the areas where the action are taking place. If the coaches fail to give the information (knowledge) or giving the inaccurate feedback, the process of learning cannot take place.

Hence, with the advancement of technology in computers and video viewing, the coaches are more preferred in using the video in providing the feedback to the players and teams [5]. The video feedback was given by the coaches in helping the players and the coaches themselves to clarify and understand what they are presenting about the team's performances. The main objective of doing match analysis is to analyze and identify the team's strength and weakness, which then can be improve and also to analyze the opponent's strength and weakness, and the data can be used as the way to encounter their opponent during competition [6]. Besides, match analysis also allows for the improvement of the physical, which related to physiological–biological and also the technical and tactical enhancement [7]. The information (feedback) from match analysis can be obtained during and also after the competition itself. The information gathered from the sports competition then is used during a sport practice to prepare the athlete for the next competition [8]. Usually the video feedback was accompanied by statistics to identify the areas that need improvement by the teams and also the areas that required more attention during the team's training session [9].

In order to bring success to a soccer team, coaches need to depend on various unanticipated event factors that they need to control [10]. Most of the empirical researches of match analysis are more focused and limited on the playing pattern or physiological work rate of the individual players, and it was suggested that match analysis should focus on the development and more utilizing the successful performance indicators [11]. Performance indicators were defined as the selection, or combination, of action variables that was aimed to define some or all of the aspects of performance. The performance indicators that are going to be evaluated need to be related to successful outcome. It has been suggested for coaches to be focus in developing the performance indicators and utilize them [12]. From the indicators, the coaches can establish a profile of ideal performance that can be included in the athlete's training activity to achieve the performance target and also can be used to predict the future behavior of the sporting activity [13].

Until now, there are limited number of studies that investigate the performance indicators by distinguishing between the winning teams and losing teams [11, 14–18]. Hughes has studied the differences between winning, drawing and losing teams in the Spanish soccer league [19]. From total of 380 matches ranged from the 2008 to 2009 season of Spanish Men's Professional League, they found that the of total shots, shots on goal, crosses, crosses against, ball possession and venue were the indicators that differentiate between winning, drawing and losing teams. Hughes and Churchill have compared the pattern of play between the successful and unsuccessful teams during the Copa America Tournament 2001 by looking at the shots and goals [15]. They have found out that the there was no significant difference between the successful and unsuccessful teams in shots to goal. Study by Hughes and Franks [17] have found that there are differences between successful and unsuccessful teams in converting the possession into shots to goal, with the successful teams have higher ratios in the 1990 World Cup. Total of 24 matches have been studied which focused on the possession of the ball between successful and unsuccessful teams in the English Premier League 2001–2002 season [16] .They found that the successful teams have longer possession than the unsuccessful teams regardless of the match status (evolving score). During the soccer World Cup 2002, the differences on passing between successful and unsuccessful teams in the event were found to be not significant difference [18]. This may be resulted because of the small sample size (6 teams, 3 teams each on successful and unsuccessful teams).

The successful play and shooting techniques in scoring goals are important in all level of football. The study is needed because football will continue to change and evolve as the mode of the game is concerned, although there are lots of studies concerning football [20]. Current study was carried out to examine the how the possession affected the goal scoring patterns and other selected performance indicators that was applied by the top level European football team which can be used to differentiate between winning and losing teams.

2 Method

2.1 Sample

All 16 teams involved in the UEFA-EURO 2012 Championship were observed which comprised a total of 31 matches. The reason of this championship was selected is the involvement of top European football teams (Poland, Greece, Russia, Czech Republic, Netherlands, Denmark, Germany, Portugal, Spain, Italy, Ireland, Croatia, Ukraine, Sweden, France, and England).

2.2 Instruments

Matches were recorded using a Astro B.yond™ PVR service which enables the researcher to play the video in normal speed, fast forward and also in slow motion speed. The Astro B.yond™ PVR service was used with the Panasonic TX14c3 T/S color television set. In the study, the software that was used in current study was SportsCode Elite software provided by Sportstec South East Asia (M) Sdn. Bhd. The function of the software was used to design and create a code window (Fig. 1) in the computer. The performer will click the performance indicators in the code window, and the data will be stored in the computer, and the results can be shown immediately upon entering the data.

2.3 Data Collection Procedure

A total number of 31 matches were recorded using the Astro B.yond™ PVR. The Astro B.yond™ PVR service allowed the user to record up to two channels at one time, and the video was downloaded into the MacBook Pro notebook for further process of analyzing. The process of analyzing was done by playing the video and at the same time, clicking the indicators in the code window to key in the frequency of actions happened throughout the 90-min match. Only 26 matches were analyzed as the remaining 5 matches resulted in drawn. The winning teams were defined as team that wins in a match while losing teams is the team that loses in a match. The variables being studied was shown in Table 1.

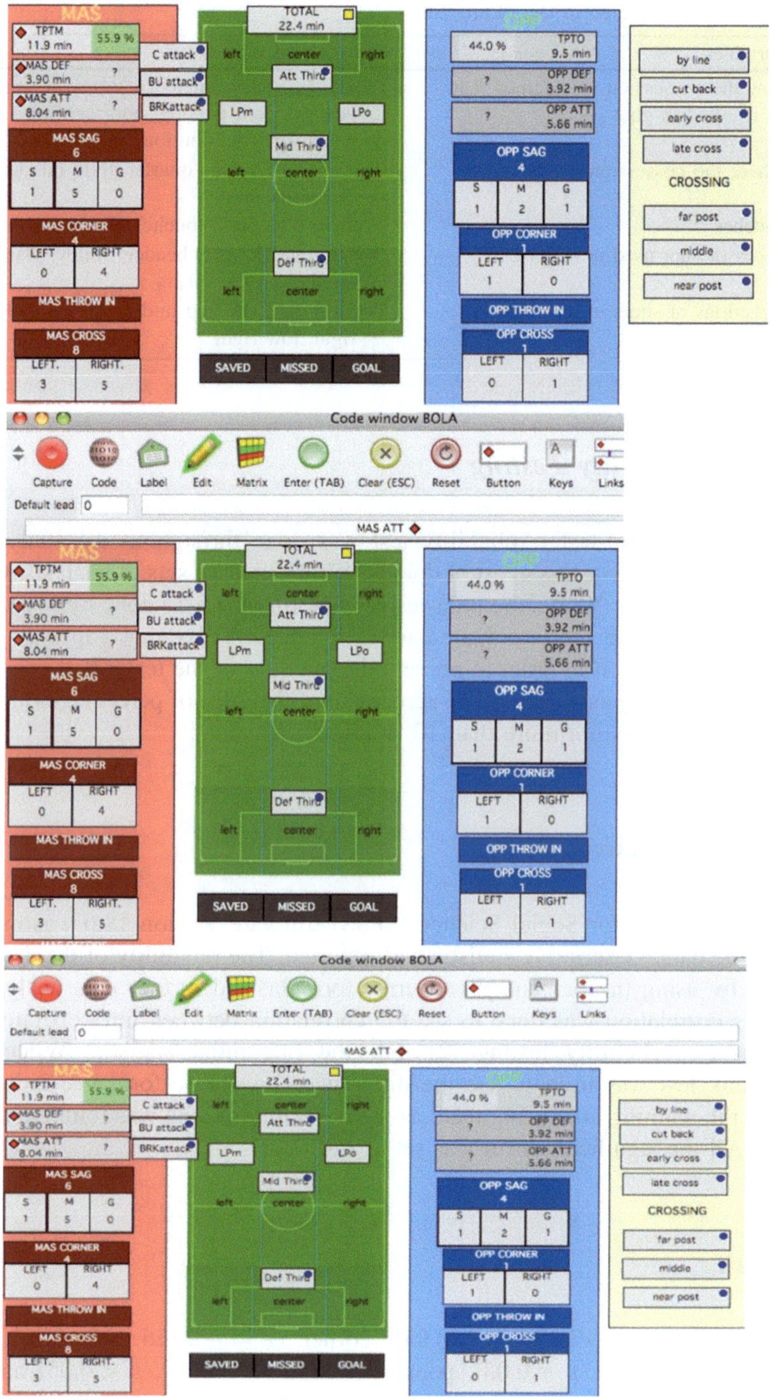

Fig. 1 Code window

Table 1 Variables studied in Uefa-Euro 2012

No.	Variables	Performance indicators
1	How the possession was gained that resulted in goal	Tackle, free kick, intercept, throw in, corner, penalty, poor passing, poor control, goalkeeper distribution, and others
2	Where the possession was gained	First quarter, second quarter, third quarter, final or fourth quarter
3	Touches	0 touch, 1 touch, 2 touches, 3 touches or more
4	Types of shot used in scoring	Driven, curl, lobbed, header, volley, place, overhead kick, others
5	Directions of shots in scoring goals	Top left, low left, top middle, low middle, top right, low right

2.4 Inter-reliability Testing

Inter-tester and intra-tester reliability test was done throughout the study. The intra-tester reliability test is a test of doing a number of two sets of analysis, which was done by a tester. While, the inter-tester reliability test was done by two testers and the results were then compared to each other. A five-day gap was given between each notation in order to prevent the memory of the tester that can affect the results and analysis process. The agreement of the error of percentages allowed in present study was not more than 10 % [30].

2.5 Data Analysis

Statistical Package for Social Science (SPSS) software version 18.0 was used to analyze the data. Descriptive statistic was reported. The reliability of the data was checked by using the percentages of error formula, and further data analysis of Pearson's correlation was done to see the correlation between intra-operator and inter-operator reliability testing [2]. For the inferential statistic, Wilcoxon's signed-rank test was used to see the differences on the performance indicators between the winning and losing team as the normality of data was found to be not normal. All the significant level was set at $p < 0.05$.

3 Result

The intercept ($Z = -2.859$, $p < 0.05$), corner ($Z = -2.126$, $p < 0.05$), poor passing ($Z = -2.126$, $p < 0.05$), poor control ($Z = -3.000$, $p < 0.05$) and the goalkeeper distribution ($Z = -2.236$, $p < 0.05$) were found to have significant

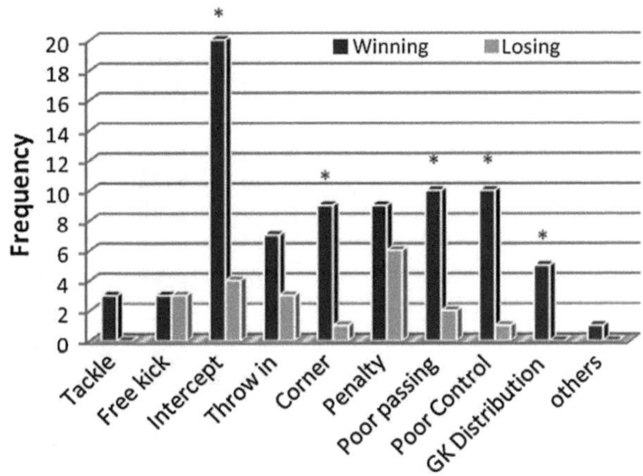

Fig. 2 How possession gained

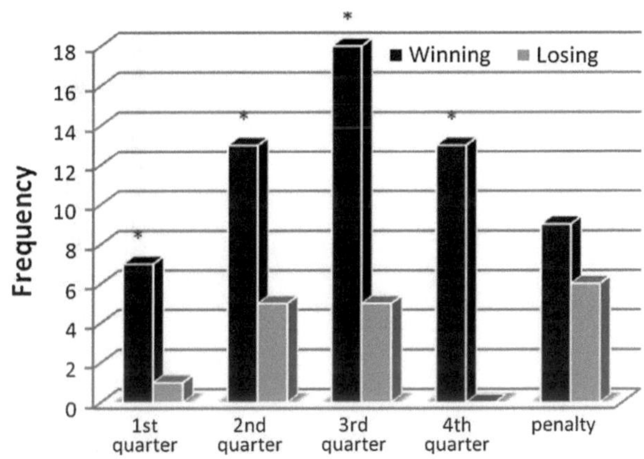

Fig. 3 Where possession gained

difference between both winning and losing teams. The tackle, $Z = -1.342$, $p > 0.05$), free kick ($Z = 0.000$, $p > 0.05$) throw in ($Z = -1.265$, $p > 0.05$), penalty ($Z = -0.966$, $p>0.05$) and others ($Z = -1.000$, $p > 0.05$) were not significant.

The winning and losing teams were also found to be significant difference in the variable where the possession was gained that resulted in goal which are the first quarter ($Z = -2.449$, $p < 0.05$), second quarter ($Z = -2.309$, $p > 0.05$) third quarter ($Z = -2.804$, $p < 0.05$), and fourth quarter ($Z = -2.739$, $p < 0.05$), while only one performance indicator, penalty ($Z = -0.966$, $p > 0.05$) was found to be not significant between the winning and losing teams (Figs. 2, 3, 4, 5, 6).

In the next variable which related to the number of touches before scoring the goal, 3 touches or more was the only variables that have no significant difference ($Z = 0.000$, $p > 0.05$), while 0 touch ($Z = -3.660$, $p < 0.05$), 1 touch $Z = -3.000$, $p < 0.05$) and 2 touches ($Z = -2.121$, $p < 0.05$) were found to be significant difference between both teams.

Fig. 4 Touches before scoring goal

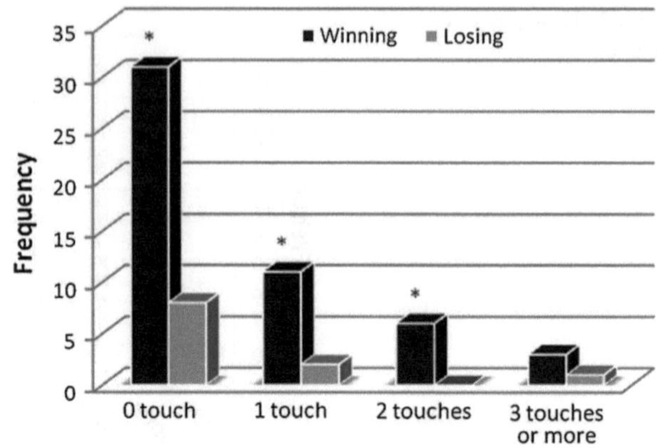

Fig. 5 Type of shot used to score goal

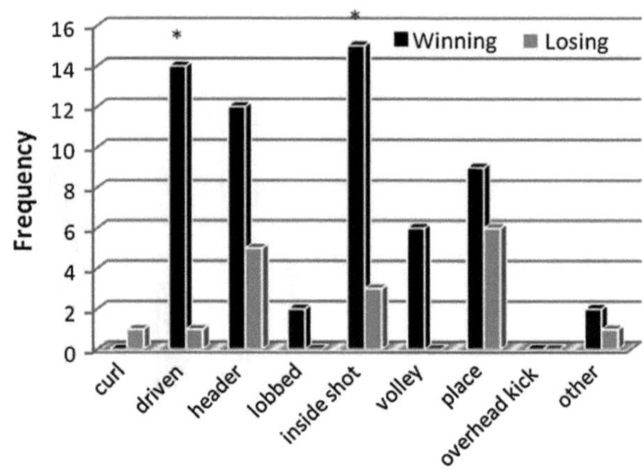

Both teams preferred to use driven shot, and inside shot in shooting at goal. For winning team, the driven shot accounts for 28.7 % of total shot at goal and 33.2 % for losing team. While the inside shot, the percentages were 23.8 % for the winning team and 16.9 % for the losing teams. However, when taking into consideration of types of shot used in scoring goal, significant differences were found for driven shots ($Z = -2.810$, $p < 0.05$) and inside shot ($Z = -2.489$, $p < 0.05$) for both winning and losing teams.

There are two sections in directions of goals, which are the lower 3 (low left, low middle, and low right) and the upper 3 (top left, top middle, and top right). Overall, both of the winning and losing teams shots most in the direction of lower left, lower middle, lower right and upper middle which represents 10.98, 8.87, 9.92, and 9.02 % of total number of shots, respectively. However, when looking at the direction of shots where the goals were scored, only the lower right ($Z = -2.930, p < 0.05$), and upper middle ($Z = -2.066$, $p < 0.05$) were found to be significant in scoring goal between both of winning and losing teams. While the rest of direction which were

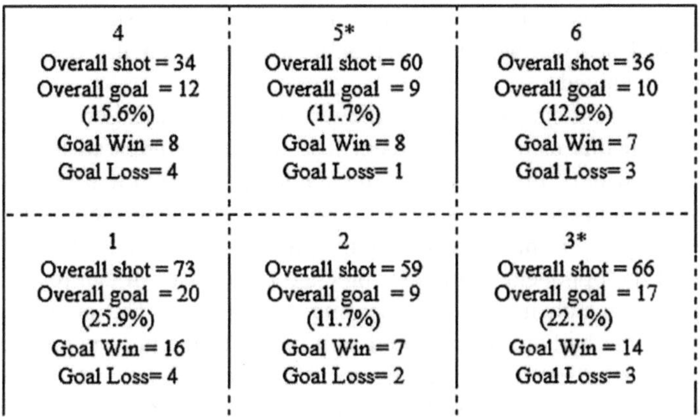

Fig. 6 Direction of goals scored (*$p \leq 0.05$)

the lower left ($Z = -1.774, p > 0.05$), lower middle ($Z = -1.475, p > 0.05$), top left ($Z = -0.551, p > 0.05$) and top right ($Z = -0.332, p > 0.05$) were found to be not significant. The goals scored in lower and upper sections of are equal with both accounted for 50 % of goal scored, respectively.

4 Discussion

The aim of the study was to differentiate the selected performance indicators between the winning teams and losing teams in UEFA-EURO 2012 Championship. The findings of current study will benefits the coaches and players and giving the suggestions to made on the tactical and technical development that most appropriate approach that was used during the UEFA-EURO 2012 Championship.

4.1 How Possession Was Gained

Five variables in how possession was gained were found to be significant different ($p < 0.05$) between the winning and losing teams which were intercept, corner, poor passing, poor control and goalkeeper (GK) distribution. For the winning teams, 25 % of the goals come from the intercept, 11 % from corner, both opponent's poor passing and opponent's poor control were 13 % and GK distribution was accounted for 7 %. While for the losing teams, only 20 % of the goals come from intercept, 5 % for corner, 10 % for opponent's poor passing, 5 % for opponent's poor control and 0 % for GK distribution. Contrast finding was found in one of the unpublished study by Falla [21] where the free kick (23.3 %) was the most frequent variable used in gained possession. This may be due to how the data was treated as current study investigates on how the possession was gained that resulted in goal scoring whereas [21] only investigate on the overall how the possession was gained.

In current study, it can be said that the winning teams was good in ball control and the losing teams using the intercept less frequent compared to the winning teams. Another way of the winning teams gained possession that resulted in goal scoring was through the corner shot. It can be suggested that the attackers should play the ball from the right or left side of the pitch and try to make as many crosses as possible. If the opponent's defender blocked the crosses, it can result in corner kick (set plays), which increase the scoring possibilities. The poor passing and poor control of the ball are inter-related as it indicates the players are weak in the controlling of the ball and poor passing accuracy. In this case, it can be suggested that the losing teams make more mistakes in controlling the ball and poor in the passing accuracy which enables the winning teams gained the possession and build up an attacking play and scoring more goals. Goal keeper also plays an important role in the possession gained as the winning teams' goal keeper contributed in goal scoring in winning teams and compared to losing teams which no goal keeper distribution contributed in goal scoring.

4.2 Where the Possession Was Gained

Current study has found that first quarter, second quarter third quarter and fourth quarter were the area that the winning teams gained possession that led to a goal compared to the losing teams. During the 2006 FIFA World Cup, all of the ethnic groups scored goals that resulted from the final attacking third of the pitch with 55.3 % of the goals from the area [21]. This was also supported by Reep and Benjamin [22] and Garganta et al. [23] which they have agreed that most of the goals scored were originated from the attacking area (third quarter and fourth quarter). Another unpublished study which analyzed the penalty area entrance of the South African Men's professional football team and found that the majority of the goals of the South African Men's team were resulted from the final attacking third of the field [24].

Interestingly, in current study, the researcher has also found that the winning team also gained possession at the first quarter and second quarter of the field which led them in scoring goals. There were no studies yet to be found that have significant difference in goal scoring resulted from gaining in the defensive third. This may be due to and can be related with how the possession was gained. The winning teams' defenders must be skilled players in intercepting the ball from the attackers of losing teams in the first and second quarter during the losing teams' attacking. Besides that, the poor control and poor passing of the ball as mentioned earlier may be the cause why the winning team can gain possession from the area and thus scoring more goals.

4.3 Touches Before Shooting

Current study found that touches ranged from 0 to 2 touches were found to be significant differences between the winning and losing teams. After excluding 9 goals from penalty kicks, the winning teams scored 31 goals (61 %) by 0 touch of the ball, 11 goals (22 %) by 1 touch of the ball, and 6 goals (12 %) from 2 touches of the ball. While for the 3 touches or more resulted in only 3 goals (5 %). For the losing teams, they scored 8 goals (73 %) by 0 touch, 2 goals (18 %) by 1 touch, and no goals at all by 2 touches and only 1 goal (9 %) by 3 touches or more.

Current study finding was supported by Kirkendal et al. [25] study where they studied the patterns of successful attacks between the men's and women's World Cup matches. They have found that the goals resulted from 1 touch were 58 and 47 % for women and men, respectively. While for the 2 touches, the women accounted for 26.3 % of the goal scored and men was 35.3 % and the goals resulted from 2 touches or more were 15.8 and 17.6 % for women and men. These findings were similar in present study as more goals were scored by less touches of the ball (<2 touches). It can be suggested that the players may do not have sufficient time to make more touches of the ball or dribble around the defenders. It can also be said that the attackers were stressed by the defenders of the opposing team thus make the attackers to make the shots. Thus, it can be suggested that during the training or game exercises, the coaches or tactical personnel should focus on the 1 or 2 touch shooting with the stress from the defense players to mimic the real situation of soccer match.

4.4 Types of Shot Used in Scoring

Overall, both of the winning and losing teams used driven shots (winning vs. losing = 29 % vs. 33 %) more often in shooting at goal followed by the inside shot (24 % vs. 17 %), header (14 % vs. 17 %) and volley shot (12 % vs. 13 %). For the rest of the variables, which are place shots (7 % vs. 6.7 %), others (6.5 % vs. 5.4 %), lobbed shot (3.5 % vs. 4.1 %), curl shot (3.3 % vs. 4.1 %) and overhead kick (0.8 % vs. 0.7 %). The driven shots finding were tallies with other studies by Hook and Hughes [26] and Hughes and petit [28] where both studies found that the most common method used in shooting were driven shots.

However, current study more focused on the types of shots used in scoring goal. The winning teams scored goals more often using the inside shots (25 %) followed by the driven shots (23.3 %), header (20 %), placed shots (15 %) volley shots (10 %), lobbed and other (3.3 %), while the curled shots and overhead kick shots produced no goal at all. For the losing teams, it was contrast with the winning teams finding as they produce more goals using the placed shots (35.3 %), followed by header (29.4 %), inside shot (17.6 %), curled, driven and other shots

(5.9 %). The other shots, which are volley, overhead kick, and lobbed shots, produced no goal.

Overall, only two types of shots in scoring goal that were found be significant difference between the winning and losing teams, which are the inside shot and driven shot. These findings was similar in study by Hughes and Snook [27] as they found that 14 % of the driven shots resulted in goals, despite the shots comes second most common method used in scoring goal in present study. Contrast finding were found in [28] study as they found the most common method used in scoring goals were place shots (16.7 %) followed by driven shots (9.8 %) as current study found that the inside shot was the most common method used in goal scoring. It was said that the winning team were able to feel more relax in front of the goal in placed shot as compared to the losing team who may be feel more pressure and strike for power instead of accuracy to score goals. It can be suggested that in current study, the driven shot of the winning teams are more accurate and more balance despite the power needed to strike the ball. The losing teams do not produce many goals using the driven shots may be due to the pressure by the winning teams' defenders thus making them to strike the ball with less accuracy and balance. Winning teams frequently used inside shot in scoring goal, even though the inside shot is not the most powerful shot in soccer, it is the most accurate. Besides that, players preferred to use the inside shot because it require less power in striking the ball which led to more shots and also the shots usually comes nearer to the goal either inside the penalty area or goal area. The driven shots are more suitable to use outside the penalty area, as the distance of from the goal was farther.

4.5 Directions of Shots in Scoring

Present study has found that in the direction of the shots, a total of 328 shots (49 %) were shots on target, while 337 shots (51 %) were off targets. When divided into winning and losing teams, the winning teams shooting 199 shots (53 %) on target and 170 shots off target (47 %), while the losing teams have 129 shots on target (43 %) and 167 shots off target (57 %). From the results, it can be said that the winning teams not only produce greater number of shots at goals than the losing teams, but they are also more accurate in shooting toward the goals (on target) compared to the losing teams. The losing team's shots off target were greater than the winning teams, and it indicates that the losing teams shooting skills are not accurate. There are many factors in contributing to these, for example, the winning team defenders put stressed on the attackers of losing teams thus making them lose control of the ball or not enough time to make a proper shooting at the goal.

In current study, the researcher focusing more on the shots on target (directions number 1 until direction number 6). The researcher divided the goal post into six areas (upper 3 and lower 3). The winning teams scored more goals at the lower 3 which accounted for 37 goals (62 %) compared to 23 goals (38 %) scored at the upper 3. However, the losing teams scored nearly even at the lower 3 and upper 3 with 9 goals (53 %) and 8 goals (47 %), respectively. When differentiating between the winning and losing teams, only direction lower right and middle upper section were found to be significant ($p < 0.05$). The lower right of the goalkeeper's only accounted for 22.1 % of the total goal scored, while the straight and high area of the goalkeeper's position accounted for 11.7 % of goals scored. These findings were similar with another study [28] as they found that most of the goals were scored low toward the post of the left or right of the goalkeeper's. Another similar finding was found by Bergier et al. [29] as they analyzed the shots at goals of the Women's European Football Championship and found that the goals were scored at the bottom sectors of the goals and [22] also found that most of the goals were scored at lower sector of the goal either left or right of the goalkeeper's.

5 Conclusion

Current study presents the game statistics that helps to identify which performance that differentiates between the winning and losing teams participated in the UEFA-EURO 2012 Championships. It can be conclude that:

(a) Player should train more on the ball control and passing so that they can retain the possession longer thus result in creating scoring opportunities.
(b) Training or game exercises should mimic the real situation of soccer match with focused training less than 2 touches before shooting to score.
(c) Training should be focus on shooting inside the penalty box using the inside shot and also outside the penalty box using the driven shot.
(d) Training should be conduct on shooting the ball toward the lower sector of the goal either the left or right.

These findings will helps the coaches and players in preparing possible training interventions in helping the team becoming a successful team. Further study should focus on the profiles of the teams that represent the continents, which will helps to differentiate between them. Focus on the playing profile of the ASEAN teams, especially the Malaysian football team with focusing on the winning and losing teams.

Acknowledgments The authors would like to thanks and express our gratitude to Research Management Institute (RMI) UiTM, Faculty of Sports Science and Recreation, Universiti Teknologi MARA (UiTM), Malaysia and also indebted to Sportstec SEA for the never ending feedback, information and support through the whole process of finishing the study.

References

1. Carling, C., Williams, A., & Reilly, T. (2005). *The Handbook of Soccer Match Analysis*. London: Routledge.
2. Hughes, M., & Franks, I. M. (2004). *Notational analysis of sport, System for better coaching and performance in sport*. London: Routledge.
3. Mohsen, S., Taylor, M., & Carlos, L. P. (2013). Performance consistency of international soccer teams in Euro 2012: A time series analysis. *Journal of human kinetics, 38*, 213–226.
4. Franks, I. M., Goodman, D., & Miller, G. (1983). Analysis of performance: Qualitative or quantitative?" In SPORTS, March 1983.
5. O'donoghue, P. (2006). The use of feedback videos in sport. *International Journal of Performance Analysis in Sport, 6*(2), 1–14.
6. Carling, C., Reilly, T., & Williams, A. (2009). *Performance assessment for field sports*. London: Routledge.
7. Ortega, E., Cardenas, D., Sainz De Baranda, P., & Palao, J. M. (2006). Differences between winning and losing teams in youth basketball games (14–16 years old). *International Journal of Applied Sport Sciences, 18*(2), 1–11.
8. T. Reilly (2003). *Science and soccer*. (2nd edn.) In: T. Reilly, & A. M. Williams (Eds.) London: Routledge.
9. O'donoghue, P. (2010). *Research methods for sports performance analysis*. London: Routledge.
10. Armatas, V., Yiannakos, A., & Sileloglou, P. (2007). Relationship between time and goal scored in soccer game: Analysis of three World Cup. *International Journal of Performance Analysis in Sport, 7*(11), 48–58.
11. Carlos, L. P., Joaquin, L. B., & Ezequiel, R. (2011). Differences in performance indicators between winning and losing teams in the UEFA Champions League. *Journal of Human Kinetics, 27*, 135–146.
12. Hughes, M., & Bartlett, R. M. (2002). The use of performance indicators in performance analysis. *Journal of Sport Sciences, 20*, 739–754.
13. Jones, N., Mellaleiu, S., & James, N. (2004). Team performance indicators as a function of winning and losing in rugby union. *International Journal of Performance Analysis in Sport, 4*(1), 61–71.
14. Carlos, L. P., Joaquin, L. B., Alexandre, D., & Maite, G. (2010). Game-related statistics that discriminated winning, drawing and losing teams from the Spanish soccer league. *Journal of sports science and medicine, 9*(2), 288–293.
15. Hughes, M., & Churchill, S., (2005). Attacking profiles of successful and unsuccessful team in Copa America 2001". In T. Reilly, J. Cabri & D. Araújo (Eds) *Science and football V* (pp. 219–224). London: Routledge.
16. Jones, P. D., James, N., & Mellalieu, S. D. (2004). Possession as a performance indicator in soccer. *International Journal Perform Analysis Sport, 4*(1), 98–102.
17. Hughes, M., & Franks, I. (2005). Analysis of passing sequences, shots and goals in soccer. *Journal of Sports Sciences, 23*(5), 509–5141.
18. Scoulding, A., James, N., & Taylor, J. (2004). Passing in the soccer World Cup 2002. *Int J Perform Analysis Sport, 4*(2), 36–41.
19. Hughes , M (1997). Notational analysis of football. In: T.Reilly, J. Clarys & A. Stribbe (Eds) *Science and football II* (pp 151–159). London: E & F.N. Spon.
20. Armatas, V., & Yiannakos, A. (2006). Evaluation of the goal scoring patterns in European Championship in Portugal 2004. *International Journal of Performance Analysis in Sport, 6*(1), 178–188.
21. Falla, A (2007). The different attacking styles and shooting patterns employed by different ethnic groups during the 2006 Fifa World Cup. Unpublished Masters' thesis. University of Wales Institute, Cardiff.

22. Reep, C., & Benjamin, B. (1968). Skill and chance in association football. *Journal of the Royal Statistical Society, 131*, 581–585.
23. Garganta, J., Maia, J., Basto, F (1997) Analysis of goal scoring patterns in European top level soccer teams. In: T. Reilly, J. Clarys & A. Stibbe E (Eds) *Science and Football II* (pp. 246–250). E. & F. Spon, London.
24. Bressan, E.S. & Engelbrecht W.P. (2010). Performance analysis of penalty area entrances of a South African men's professional football team, Unpublished masters' thesis. Stellenbosch University.
25. Kirkendal, D., Dowd, W. & DiCicco A. (2002). Patterns of successful attacks: a comparison of the men's and women's World Cup. Review Fútbol Cienc 1 (1).
26. Hook, C., & Hughes M.D. (2001). Patterns of play leading to shots in Euro 2000. In: Pass.com. Ed: CPA (Center for Performance Analysis). Cardiff: UWIC.pp. 295–302.
27. Hughes, M., & Snook, N. (2006). Effectiveness of attacking play in the 2004 European championships. In H. Dance, M. Hughes, & P. O'Donoghue (Eds.), *Notational analysis of sport—VII* (pp. 46–62). Cardiff: UWIC.
28. Hughes, M., & Petit, A. (2001). Patterns of play in the 1986 and 1998 World Cups for soccer. In: M.D. Hughes & I.M. Franks (Eds.) Pass.com: Computer science and sport III & performance of sport V (pp. 267–276). Cardiff: University of Wales Institute.
29. Bergier, J., Soroka, A., &Buraczewski, T. (2009). Analysis of actions ending with shots at goal in the Women's European Football Championship (England 2005). In: T. Relly & F. Korkusuz (Eds) Science and football VI: The Proceedings of the Sixth World Congress on Science and Football. pp. 197–201. London: E & F.N.
30. Cooper, S. M., Hughes, M., O'Donoghue, P., & Nevill, A. M. (2007). A simple statistical method for assessing the reliability of data entered into sport performance analysis systems. *International Journal of Performance Analysis in Sport, 7*(1), 87–109.

Assessment of Physical Fitness Performance Among Students with Cerebral Palsy on Selected Fitness Components

Nagoor Meera Abdullah, Wahidah Tumijan, Vincent Parnabas, Mazlan Ismail, Mohamad Nizam Mohamad Shapie, Muhammad Zulqarnain Mohd Nasir and Norlizah Abdul Hamid

Abstract Physical fitness and participation in sports are important to the students with cerebral palsy (CP). Persons with disabilities who lack of physical fitness can affect their overall performance. The objective of the study was to assess on physical fitness performance among students with CP based on the selected fitness components. Total of 25 students (male = 11; female = 14) were randomly selected range from classification classes of C2–C7. The sample of the study consists of 48 % Malay, 28 % Chinese, and 24 % Indian. The mean age of the subjects is 23.32 ± 2.07 years old (age range from 10 to 46 years old). Majority of the subjects are in the class C4 (16 %), followed by class C5 (36 %), from class C4 (16 %), from class C2 (8 %), and from class C3 (4 %). All the subjects were assessed through physical fitness tests such as sit-and-reach test, handgrip strength test, dumbbell press test, and arm ergometry (total pedal revolution, calorie, and time). All the data were analyzed using SPSS and presented as mean. The mean value for sit-and-reach test is 26.10 ± 1.65, while the mean value for handgrip strength test is 12.20 ± 1.25. The mean value for dumbbell press test is 27.76 ± 5.68, followed by arm ergometry; the total pedal revolution (170.80 ± 31.24), calorie (5.52 ± 0.98), and time (4.78 ± 0.67). The physical fitness overall was in satisfactory level, and the subjects tried their best to achieve the highest fitness status. Physical fitness should be conducted for students with disabilities so that it can help their activity of daily living (ADL) and enjoy their quality of life.

Keywords Component · Physical fitness · Students with cerebral palsy · Classification classes

N. M. Abdullah (✉) · W. Tumijan · V. Parnabas · M. Ismail · M. N. Mohamad Shapie · M. Z. Mohd Nasir · N. Abdul Hamid
Faculty of Sport Science and Recreation, Universiti Teknologi MARA, Shah Alam, Selangor, Malaysia
e-mail: nameera_ab@yahoo.com.my

R. Adnan et al. (eds.), *Proceedings of the International Colloquium on Sports Science, Exercise, Engineering and Technology 2014 (ICoSSEET 2014)*,
DOI: 10.1007/978-981-287-107-7_52,

1 Introduction

Physical fitness and sport participation are important for children and adults. Positive benefits can be seen in the following areas: (a) physical and psychological health; (b) self-concept; (c) body awareness; (d) motor development; (e) sportsmanship; (f) competition; and (g) social interaction [1]. Physical benefits include factors that will improve cardiovascular functions, muscular strength, and endurance as well as assisting in normal range of motion.

The physical, mental, and emotional needs of the disabled are being explored with a new awareness of the potential development of the physically disabled. In terms of physical capacity, current research is now investigating training techniques, fitness assessment, and the performance of disabled athletes.

Physical fitness is really important for any type of group of people. The term physical fitness should already be well known in a sport education among sport person or sport students. Physical fitness is defined as the capacity without excessive fatigue or as the capacity to perform everyday activities with reserve energy for emergency situations [2].

Cerebral palsy (CP) is a term that describes a variety of disorders of movement or posture. These movement problems are due to brain abnormalities that occur early in development. CP affects motion, muscle strength, balance, and coordination.

CP been described as "a group of permanent disorders of the development of movement and posture, causing activity limitation, that are attributed to nonprogressive disturbances that occurred in the developing fetal or infant brain" [3]. CP with a prevalence of 2–3 per 1,000 children is the most common motor disability in pediatric rehabilitation. Because of their motor problems, children and adolescents with CP experience participation restrictions and limitations in physical activities (PA). Based on [4], PA is defined as all body movements resulting in an increased energy output from the resting position. Children and adolescents with CP show lower levels of PA compared to their healthy peers. Lower levels of PA contribute to a reduced physical fitness, which may increase the risk of developing secondary health problems such as pain and fatigue, cardiovascular disease, and diabetes mellitus later in life. Moreover, PA is assumed to have a positive relation with health-related quality of life and psychosocial functioning [4].

CP is the commonest physical disability in childhood, yet in many cases, the cause remains unknown. The form "cerebral palsy" describes a group of disorders of movement and posture due to a defect or lesion of the immature brain [5]. CP is a symptom complex, with various types and degrees of motor impartment. These disorders become manifest early in life and are permanent and non-progressive conditions [6, 7].

Since physical fitness is important, sport participation by people with CP is increasing to improve their performance. Normal children in deed also try to enhance their motor development. Sport participation by people with CP is increasing worldwide. Since physical fitness is important in the performance of most PA and sports, it is necessary to better understand the needs of fitness of students with CP. Children who participate in physical fitness programs are reported to have higher self-concepts that those who are inactive [7].

According to Ferrari and Adriano [8], the disability in CP is also a problem of perception. Much perceptual information cannot be received or processed by the child with CP or else it may be analyzed in an excessive or distorted way. In certain children, depth and distance may be overestimated to the point of creating an overwhelming fear of falling or they may be underestimated to the point of creating an overwhelming fear of falling; or they may be underestimated to the point where the children cannot perceive how their balance is compromised, leading to continual loss of control of their postures. Phrases like "I'm falling" and "straighten up" soon become familiar to the CP child and the family. Rehabilitation in CP therefore cannot be limited to the motor aspect alone, but must also include promotion of perceptual functions.

Not too many years ago, if an individual with CP was involved with an elite sports team for athletes with disabilities, it was as manager, statistician, or other assistant to the athletes. A number of factors combined to prevent most people with CP from actually competing. Not the least of these was the commonly held belief that individuals with CP were particularly fragile and prone to injury [9].

In 1993, the US Department of Education funded project target, a research study primarily designed to develop a health-related, criterion-referenced physical fitness test for youngsters (aged 10–17) with disabilities. Specifically, the purpose was to recommend test items and standards for youngsters with mental retardation, spinal cord injuries, CP, blindness, congenital anomalies, and amputations. The project was centered at the State University of New Work, college at Brockport. A total of 1,542 youngsters with and without disabilities were tested and data associated with several projects analyzed as a part of the comprehensive project target. The Brockport physical fitness test (BPFT) includes a number of unique features. First, in an effort to personalize testing and assessment, the battery includes 27 different test items. However, a complete battery for one individual or category of disability generally includes four to six items. Second, it applies a health-related, criterion-referenced fitness approach to youngsters with disabilities. Third, it provides an approach based on health-related needs and a desired fitness profile. Finally, many of the test items are new (or at least non-traditional) and include a larger number of youngsters with disabilities in physical fitness testing program than previously reported in the literature [10].

Children with CP can participated in physical activity even though they are relatively weaker [11–16] have less endurance [17, 18] and exhibit reduced physical activity levels compared with children without CP. These past studies exposure are of concern because of the benefits of physical activity, and exercises to overall health are well known especially to people with disabilities who are less likely to engage in physically healthy lifestyles compared with people without disabilities. Additionally, inactive adults with disabilities exhibit increased severity of disease and reduced overall health and well-being [19]. Impairments such as weakness [20], muscle spasticity [21], and deficient balance [22] make it difficult for children with CP to participate in sport and play activities at a level of intensity sufficient to develop and maintain normal physical fitness levels. More studies are needed to identify safe and effective methods to improve physical fitness in this population.

There is strong scientific evidence that youths with low physical activity and fitness levels and high body fat levels are more likely to display additional risk factors for cardiovascular disease such as elevated blood pressure and serum cholesterol levels [23–25]. Additionally, an expert panel recently convened by the Centers for Disease Control and Prevention (CDC) concluded that daily participation in developmentally appropriate, moderate to vigorous physical activity lasting 60 min or longer can reduce body fat, encourage weight loss, and improve aerobic fitness in youth aged 6–18 years without disabilities [26]. The few studies available also suggest that some degree of positive association may exist between physical activity and various indexes of mental health, including anxiety, depressive symptoms, and physical self-concept [26].

Limited study has also indicated that children with CP display low levels of cardiorespiratory fitness, as evidenced by a reduced peak VO_2 max or a higher submaximal energy demand of walking [18, 27]. These findings are alarming because reduced cardiorespiratory fitness may contribute to poor general health. From a functional perspective, children with CP have difficulty performing purposeful and efficient physical movements for many reasons, including weakness, abnormal muscle coactivation, involuntary movement, poor selective voluntary motor control, spasticity, contractures, and decreased balance [18, 20, 28]. These impairments can limit a child's ability to play and exercise at certain intensity levels necessary to develop cardiorespiratory fitness. Fatigue, commonly reported by people with CP, is thought to be a result of using an abnormally high percentage of their peak energy resources during PA [18, 29]. For children with CP who are able to walk, the locomotor energy demands increase with age, making it difficult to sustain their walking endurance as they transition into adolescence and adulthood. Therefore, the aim of the study is to assess on physical fitness performance among students with CP based on the selected fitness components.

2 Methodology

2.1 Sample

A total of 25 subjects (male = 11; female = 14) were participated in the study. Their age is between 10 and 46 years old. The subjects are students with CP from a spastic center in Petaling Jaya, Selangor. For the selection, subjects are chosen randomly, but only 25 students will be selected according to the name list that was given by the teachers. The entire test was carried out at spastic center for CP in Petaling Jaya, Selangor. All participants were informed of the study procedure and purposes and all gave their inform consent.

For this study, all the subjects were chosen from classes 4–8. This is because the tests that are been selected for the study are suitable for them. People with CP have been selected according to their medical classification classes based on the Cerebral Palsy-International Sports and Recreation Association (CP-ISRA, 2000). People with CP have been divided into 8 medical classes. Classes 1–4 are classes for people with CP who are using the wheelchair, and for classes 5–8 is ambulatory CP. C1 includes individuals with the most severe involvement (e.g., those who depend on an electric wheelchair or assistance for mobility), while C8, the highest class, includes those who are minimally affected (e.g., those who can run and jump freely).

2.2 Instrumentation

Grip strength is an important prerequisite for good performance of the upper limb. In the study, handgrip strength was measured using a standard adjustable handgrip strength test (Takei model TKK5401). Maximum handgrip forces for dominant hand were recorded in kilograms as the highest of two trials. Before testing the participants individually, the researcher gave a brief orientation to the entire group. The dynamometer was adjusted to the size of the hand of participant. The arm, the hand, and the body position were standardized according to the suggestion of the American Society of Hand Therapists. Subjects were sitting with shoulder adducted and neutrally rotated, elbow flexed at 90° resting on the table surface and the forearm in neutral and wrist in 0–30° extension. The test was performed by squeezing calibrated hand dynamometer as forcefully as possible with the dominant hand. Static strength was assessed.

Sit-and-reach test is used to measure the flexibility of the hamstrings, buttocks, and lower back [30]. The participants were instructed to reach as far as possible from a sitting position (Acuflex model 1).

Dumbbell press test is used to measure arm and shoulder strength and endurance of the subjects. Subject is seated in a wheelchair or other sturdy chair. The tester serves as a spotter or has spotters available for safety. Subject grasps the dumbbell with the dominant hand, with the elbow flexed so that the weight is close to and in front of the dominant shoulder. One successful lift is counted each time the dumbbell is raised above the shoulder with complete elbow extension. The scoring ends when the subject is unable to lift the weight with complete elbow extension, rests for more than 4 s between repetitions, or completes 50 repetitions.

Arm ergometry is used to measure aerobic and endurance functioning. The ambulatory subjects were seated in standard on the chair with their feet placed on the floor while the wheelchair subject performed the test in their everyday wheelchair. For all subjects, the wheelchair was positioned at a comfortable distance example table from the ergometry allowing for a slight bend. The initial workload was 25 W. On the signal “Go” the participant start to grip the pedal and begin to cycle. Testing was terminated when the subject was unable to maintain the required cranking rate or upon the subject's volitional fatigue. The subject will cycle using their arm until the participant no longer to cycle. The subject's time, calorie, and total pedal revolution will be recorded.

2.3 Data Collections

All tests were administrated in group settings with up to 25 subjects. After the warm-up sessions, each of the subjects will follow the researcher to one of the station to begin their test. Each station was administrated by the tester and supervised by the researcher that follows the participants. The testers were responsible to explain and demonstrate the tests to participants. The skill demonstrated and its verbal description was standardized according to the listed in the Test Manual of Physical Fitness. After the demonstration, a practice trial was given for each participant. Then, participants were asked to perform only one test trials in turn, then the other one to complete each of the test. The participants will undergo the tests by following the test sequence: (a) the handgrip strength test, (b) sit-and-reach test, (c) dumbbell press test, and (d) the arm ergometry test. The test will be done in one day to ensure the good data collected.

2.4 Analysis of Data

The results are presented as means and the standard deviations. The SPSS (package) was used for the statistical analysis. All the data were presented as descriptive.

3 Results

The demographic profiles, physical fitness assessment, and classification of the subjects are show in Table 1. The present study consisted of n (25), 11 males with mean age of 24.18 ± 3.83, and 14 female with mean age 22.64 ± 2.28. Overall for demographic profiles males and female with mean was 23.32 ± 2.07.

Physical fitness assessment shown above is the result of all the 25 subjects including males and female. Based on the data, sit and reach presented study of 11 males and females for overall with mean 26.10 ± 2.07. Factors of the condition are the main reason why the subjects are not able to perform the sit-and-reach test very well. Specially, for those subjects who have type condition of diplegia. Meanwhile, hand grip test overall mean for males and female is 12.20 ± 1.23. Here, the subjects both male and female are weak and strength and cannot grip very well. Dumbbell press test for male and female subjects shows the mean was

Table 1 The subject characteristics and classification classes for male and female

Variables	Male Mean ± SEM	Female Mean ± SEM	Overall Mean ± SEM
Age	24.18 ± 3.83	22.64 ± 2.28	23.32 ± 2.07
Sit and reach	25.59 ± 1.88	26.50 ± 2.62	26.10 ± 1.65
Handgrip	12.45 ± 2.28	12.01 ± 1.37	12.20 ± 1.23
Dumbbell press	30.91 ± 9.14	25.29 ± 7.41	27.76 ± 5.68
Total pedal revolution	176.55 ± 55.47	166.29 ± 36.79	170.80 ± 31.24
Calorie	5.35 ± 1.64	5.67 ± 1.23	5.52 ± 0.98
Time	4.19 ± 1.06	5.26 ± 0.84	4.79 ± 0.67
Classification			
Classes	Male	Female	%
C2	2 (100 %)	–	25 (8 %)
C3	–	1 (100 %)	25 (4 %)
C4	2 (50 %)	2 (50 %)	25 (16 %)
C5	3 (33 %)	6 (66 %)	25 (36 %)
C7	4 (44 %)	5 (55 %)	25 (36 %)
Total	11	14	100 %

27.76 ± 5.68. Finally, for hand ergometry test, measure three variables such as total pedal revolution, calorie, and time. Total pedal revolution mean of male and female was 170.80 ± 31.24; the reason is lack in endurance and fatigue rapidly. It is occurring from female's subject. Based on the data, calorie overall mean of male and female was 5.52 ± 0.98. Finally, time overall mean of male and female was 4.79 ± 0.67, respectively.

The distribution for classification between the subject's class C2 (2) subjects that is 36 % followed by class C3 (1) subjects (4 %), class C4 (4) subjects (16 %), class C5 (9) subjects (36 %), and lastly class C7 (9) subjects makes up 36 %. Class C5 and class C7 were the most subject participated in the study. The conditional distribution between the subjects shows hemiplegia 9 subjects (36 %), quadriplegia 3 subjects (12 %), athetoid 3 subjects (12 %), ataxia 1 subjects (4 %), and diplegia 9 subjects (36 %) (refer Table 2).

4 Discussion

Physical fitness has been identified as one of the major role in managing performance and health risk. Lack on physical fitness may cause or may reflect the activity daily living (ADL) mostly the people with CP.

The physical disabilities in children include those, which impair physical functions. Examples are neuromuscular disease, including spina bifida, CP, and muscular dystrophy to name a few. The health status of children with physical disabilities is not remarkable. They may be at risk for hypertension, cardiovascular disease, diabetes, osteoporosis, and obesity [31].

Table 2 The classification of cerebral palsy classes and the type of cerebral palsy according to the number

	Total	%
Classification		
C2	2	8.0
C3	1	4.0
C4	4	16.0
C5	9	36.0
C7	9	36.0
Total	25	100.0
Types		
Hemiplegia	9	36.0
Quadriplegia	3	12.0
Athetoid	3	12.0
Ataxia	1	4.0
Diplegia	9	36.0
Total	25	100.0

All of the 4 parameters were suitable to predict the overall fitness and necessary to better understand the fitness of individuals with CP. The dumbbell press test is designed to measure arm, shoulder strength, and endurance. Previous reports indicate that children can increase their muscular strength above and beyond normal growth and maturation by participating in a progression strength training program [32]. Benefit of the test can build their strength and endurance to push their own wheelchair. According to Graves et al. [33], Hoffman et al. [34], and Hunter [35] said these finding are consistent with most of the studies involving adults that suggest more frequency training sessions cause greater increases in strength.

The handgrip test is also to measure strength and given 3 trials. The result shows that the CP male made significantly higher scores than CP female. According to previous studies by Montazer and Thomas [36] indicated a significant decrease in grip strength across trials. Their short intertrial rest may have been a major factor in reduction of grip strength across trial. This proposition is supported by Trossman and Li [37] conclude that an "intertrial rest period pf at least 1 min will be result in more stables score than will intertrial rest periods of 15–60 s."

The sit-and-reach test is to measure trunk flexibility. Flexibility is often underrated in importance as a component of physical fitness. Flexibility also is related to body size, gender, age, and physical activity. Mostly females are more flexible than males. Flexibility test are usually administrated to identify individuals with too little range of motion. According to present data, CP subject for male with mean 25.59 ± 1.88 and female with mean 26.50 ± 2.62, from the result we can see the female more flexible than male. Moreover, there can reach around 21–30 cm.

Hand ergometry test is to measure aerobic functioning. The higher result shows some of the subject can prolog around 13 min while others can reach around 4 min above. Their cardiorespiratory response during the maximal arm cranking exercise is lower. According to present study by Rodgers et al. [38] revealed that the force applied to the handrim of the wheelchairs was strongly correlated with the isokinetic torques of the anterior deltoid and triceps muscle groups. In addition, during fatiguing wheelchair propulsion, activation of the anterior deltoid increased as a function of percent cycle time. Their result, along common rehabilitation practices, suggests that specific overload of the triceps and anterior deltoid muscles would improve functional independence of the wheelchair user. For the wheelchair user, synchronous cranking is more task specific and perhaps aerobic conditioning programs should be designed that include synchronous movement patterns of the upper extremities similar to wheelchair propulsion, especially in the rehabilitation process [39].

The study shows that with the 4 performance items, there are significance differences on selected components physical fitness people with CP. To achieve the high level of performance in the physical fitness, people with CP should apparently need appropriate individual programs.

5 Conclusion and Recommendations

The objective of the study is to assess the performance of physical fitness on children with CP. This is because of that it may reflect the ADL mostly for the people with CP.

The study subjects comprise 25 students, (n = 11) male and (n = 14) females, which was selected from Spastic Children's Association of Selangor and Federal Territory at Petaling Jaya. All subjects completed doing the test and the result shows that the subjects can perform very well if they have individual programs of physical fitness.

The CP subjects have different scores in physical fitness test, and this will give them as a guideline and references to know their physical fitness level. However, the physical fitness tests selected are suitable to assess whether their physical finesses were valid and reliable indicator. Suggested for some CP subjects failure to performance in certain physical fitness test need more approaches in development of physical fitness. It is suggested that physical educators and coaches assume the responsibility of developing and implementing these programs.

In general, this study shows that all the subjects gave a full cooperation. All the test results are satisfactory. Gender did not have a similar influence or not a significant factor; however, different types of classification of CP subjects are made significant to reach at higher performance.

Both CP male and female showed a significant improvement with age on the dumbbell press. In addition, some types of CP, especially spasticity, are characterized by chronic muscular hypertonicity, which may create a higher-energy demand even at rest for certain CP individuals.

The following recommendations are made for further study in the area of physical fitness test and people with CP:

1. The present study should be getting more male subject, which is equal to female subject.
2. Additionally more and various standards of physical fitness test to examine their capability.
3. Educate parents about abilities of children with disability. Knowledge about the importance of appropriate physical fitness and a commitment to provide them are essential for improving the fitness and quality of life among people with disabilities.
4. Addition of motivation on physical fitness exercise is needed because the present study shows that these participants have interest in the physical fitness test.

Acknowledgments The authors wish to acknowledge the Research Management Institute (RMI), Universiti Teknologi MARA, to provide research project grant 600-RMI/DANA 5/3/RIF (774/2012) for the presentation and the publication.

References

1. Lai, A. M., Stanish, W. D., & Stanish, H. I. (2000). The young athlete with physical challenges. In M. D. Miller & L. J. Micheli (Eds.), *Pediatrics and adolescent sports injuries: Diagnosis, management and prevention* (pp. 793–822). Philadelphia: W. B. Saunders.
2. Miller, D. K. (2002). *Measurement by the physical educator: Why and how* (4th ed.). New York: McGraw-Hill.
3. Rosenbaum, P., Paneth, N., Leviton, A., et al. (2007). A report: The definition and classification of cerebral palsy April 2006. *Developmental Medicine and Child Neurology, 109*, 8–14.
4. Claassen, A. O. M., Gorter, J. W., Stewart, D., Verschuren, O., Galuppi, B. E., & Shimmell, L. J. (2011). Becoming and staying physically active in adolescents with cerebral palsy: Protocol of a qualitative study of facilitators and barriers to physical activity. *BMC Pediatrics, 11*(11), 1471–2431.
5. Bax, M. C. (1964). Terminology and classification of cerebral palsy. *Development Medicine and Child Neurology, 6*, 295–307.
6. Stanley, F. J., & Watson, L. (1985). Methodology of a cerebral palsy register: The Western Australia experience. *Neuroepidemiology, 4*, 146–160.
7. Sandstrom, K., Alinder, J., & Oberg, B. (2004). Descriptions of functioning and health and relations to a gross motor classification in adults with cerebral palsy. *Disability and Rehabilitation, 26*(17), 1023–1031. (September 2).
8. Ferrari, K., & Adriano, F. (1995). Cerebral palsy. *World Health, 48*, 22–23.
9. Little, J. (1996). Cerebral palsy and sports. *The Exceptional Parent* (On-line), *26*, 53–55. Available http://www.proquest.com
10. Winnick, J. P., & Short, F. X. (2000). *The Brockport physical fitness test manual.* United States: Human Kinetics.
11. Brown, J. K., Rodda, J., Walsh, E. G., & Wright, G. W. (1991). Neurophysiology of lower-limb function in hemiplegic children. *Developmental Medicine and Child Neurology, 33*, 1037–1047.

12. Elder, G. C., Kirk, J., Stewart, G., et al. (2003). Contributing factors to muscle weakness in children with cerebral palsy. *Developmental Medicine and Child Neurology, 45*, 542–550.
13. Engsberg, J. R., Ross, S. A., Olree, K. S., & Park, T. S. (2000). Ankle spasticity and strength in children with spastic diplegia cerebral palsy. *Developmental Medicine and Child Neurology, 42*, 42–47.
14. Rose, J., & McGill, K. C. (2005). Neuromuscular activation and motor-unit firing characteristics in cerebral palsy. *Developmental Medicine and Child Neurology, 47*, 329–336.
15. Wiley, M. E., & Damiano, D. L. (1998). Lower-extremity strength profiles in spastic cerebral palsy. *Developmental Medicine and Child Neurology, 1998*(40), 100–107.
16. Stackhouse, S. K., Binder-Macleod, S. A., & Lee, S. C. (2005). Voluntary muscle activation, contractile properties, and fatigability in children with and without cerebral palsy. *Muscle and Nerve, 31*, 594–601.
17. Campbell, J., & Ball, J. (1978). Energetics of walking in cerebral palsy. *Orthopedic Clinics of North America, 9*, 374–377.
18. Unnithan, V. B., Dowling, J. J., Frost, G., & Bar-Or, O. (1996). Role of co-contraction in the O2 Cost of walking in children with cerebral palsy. *Medicine and Science in Sports and Exercise, 28*, 1498–1504.
19. Rimmer, J. H. (2005). The conspicuous absence of people with disabilities in public fitness and recreation facilities: lack of interest or lack of access. *American Journal of Health Promotion, 19*, 327–329. (ii).
20. Damiano, D. L., Vaughan, C. L., & Abel, M. F. (1995). Muscle response to heavy resistance exercise in children with spastic cerebral palsy. *Developmental Medicine and Child Neurology, 37*, 731–739.
21. Sanger, T. D., Delgado, M. R., Gaebler-Spira, D., et al. (2003). Classification and definition of disorders causing hypertonia in childhood. *Pediatrics, 111*, e89–e97.
22. Nashner, L. M., Shumway-Cook, A., & Marin, O. (1983). Stance posture control in select groups of children with cerebral palsy: Deficits in sensory organization and muscular coordination. *Experimental Brain Research, 49*, 393–409.
23. DuRant, R. H., Baranowski, T., Rhodes, T., et al. (1993). Association among serum lipid and lipoprotein concentrations and physical activity, physical fitness, and body composition in young children. *Journal of Pediatrics, 123*, 185–192.
24. Tolfrey, K., Campbell, I. G., & Jones, A. M. (1999). Selected predictor variables and the lipidlipoprotein profile of prepubertal girls and boys. *Medicine and Science in Sports and Exercise, 31*, 1550–1557.
25. Williams, D. P., Going, S. B., Lohman, T. G., et al. (1992). Body fatness and risk for elevated blood pressure, total cholesterol, and serum lipoprotein ratios in children and adolescents. *American Journal of Public Health, 82*, 358–363.
26. Strong, W. B., Malina, R. M., Blimkie, C. J., et al. (2005). Evidence based physical activity for school-age youth. *Journal of Pediatrics, 146*, 732–737.
27. Hoofwijk, M., Unnithan, V., & Bar-Or, O. (1995). Maximal treadmill performance in children with cerebral palsy. *Pediatric Exercise Science, 7*, 305–313.
28. Sanger, T. D., Chen, D., Delgado, M. R., et al. (2006). Definition and classification of negative motor signs in childhood. *Pediatrics, 118*, 2159–2167.
29. Rose, J., Morgan, D. W., & Gamble, J. G. (2006). Energetics of walking. In J. Rose & J. G. Gamble (Eds.), *Human walking* (pp. 77–102). Philadelphia: Lippincott Williams.
30. Koutedakis, Y., & Bouziotas, C. (2003). National physical education curriculum: motor and cardiovascular health related fitness in Greek adolescents. *British Journal of Sports Medicine, 37*, 311–314.
31. Nelson, M. A., & Harris, S. S. (1995). The benefits and risks of sports and exercise for children with chronic health conditions. In B. Goldgerg (Ed.), *Sports and exercise for children with chronic health conditions* (pp. 13–30). Champaign: Human Kinetics.
32. Falk, B., & Tenenbaum, G. (1996). The effectiveness of resistance training in children. A meta analysis. *Sports Medicine, 22*, 176–186.

33. Graves, J., Pollock, M., Leggett, S., Braith, R., Carpenter, D., & Bishop, L. (1988). Effect of reduced frequency on muscular strength. *International Journal of Sports Medicine, 9*, 316–319.
34. Hoffman, J., Kraemer, W., Fry, A., Deschenes, M., & Kemp, M. (1990). The effects of self-selection for frequency of training in a winter conditioning program for football. *Journal of Applied Sports Science Research, 4*, 76–82.
35. Hunter, G. (1985). Changes in body composition, body build and performance associated with different weight training frequencies. *Strength and Conditioning, 7*, 26–28.
36. Montazer, M. A., & Thomas, J. G. (1992). Grip strength as a function of 200 repetitive trials. *Perceptual and Motor Skills, 75*, 1320–1322.
37. Trossman, P. B., & Li, P. W. (1989). The effect of the duration of intertrial rest periods on isometric grip strength performance in young adults. *Occupational Therapy Journal of Research, 9*, 362–378.
38. Rodgers, M. M., et al. (1994). Biomechanics of wheelchair propulsion during fatigue. *Archives of Physical Medicine and Rehabilitation, 75*, 85–93.
39. Mossberg, K., Willman, C., Topor, MA., Crook, H., & Patak S. (1999). *Comparison of asynchronous versus synchronous arm crank ergometry* (Vol. 37, pp. 569–574). Galveston, Texas, USA: Department of Physical Therapy, University of Texas Medical Branch.

Optimization Model of Instep Kick: The ABP-Based Mathematical Model

Zaifilla Farrina Zainuddin and Norasrudin Sulaiman

Abstract Our study is concerned with the development of a mathematical model of the biomechanical system dynamics, based on thc angular body posture (ABP), for generating high-impact instep kick. Instep kick is one of the most fundamental skills applied in many sports such as soccer, futsal, rugby or American football, swimming, martial art and sepaktakraw. For soccer and futsal, instep kick has been recognized for its role in goal scoring. This paper describes the 3D analysis of instep kicks based on data collected through controlled experiments at the National Sports Institute involving the Malaysian National Futsal team. Details of the method used are presented. This digitization of the images and the analysis are using Evart 5.0 software. Understanding the biomechanics system dynamics of the instep kicks will enable improvement of the kick performance, thus increasing the chance of scoring goals.

Keywords Instep kick · Biomechanical · Mathematical model · Futsal

1 Introduction

Kicking is one of the important aspects in games involving the lower limb segments. There are various types of kicks used in games like soccer, futsal, martial art and even in swimming. Each sport has a variety of kicking styles to suit

Z. F. Zainuddin (✉)
Institute of Product Design and Manufacturing UniKL, Kuala Lumpur, Malaysia
e-mail: filla_farin@yahoo.com

N. Sulaiman
Sports Science Center of Studies, Shah Alam, Malaysia

R. Adnan et al. (eds.), *Proceedings of the International Colloquium on Sports Science, Exercise, Engineering and Technology 2014 (ICoSSEET 2014)*,
DOI: 10.1007/978-981-287-107-7_53, © Springer Science+Business Media Singapore 2014

different game rules and types of ball being used. Side-foot kicks and instep kicks are two main types of kicks that have been subject to numerous studies from diverse aspects of kicking biomechanics. With the advance development of recording motion and computational analysis, the study of complex motor task of kicking was made possible. Researches on kicking motion range widely from the acceleration of the lower limb, kicking of the ball impact, electromyography characteristics of muscle activation during kicking, ground reaction force during kicking to the influence of soccer shoes on kicking velocity. All these studies were aimed to gain better insights on the characteristics and determinants of a successful ball kick. All the more, at stake are not just a club's win, but also the glory of winning in international events like the World Cup or Olympic games.

A successful kick is usually expressed in literature in terms of the velocity of the ball and the accuracy of direction of kick [1]. The velocity of the ball in turn is the result of various factors, including transfer of energy between segments [2]; approach speed and angle [3–5] and type of kick [2, 6] stated that powerful kicks were achieved not only through a high foot velocity but also coefficient of restitution (COR). COR describes the transfer of velocity during the impact phase and is the ratio between the relative velocity (velocity of the ball velocity of the foot) after impact and the relative velocity before impact [3]. The toe kicking was found to be more efficient to transfer the momentum to the ball at low velocities ($<15\ ms^{-1}$) than the instep kicking. However, no differences for the COR are observed between the toe kick and the instep kick at high velocity ($>15\ ms^{-1}$) [3]. Kicking accuracy in soccer also depends on how fast the player approaches the ball [6, 7], the angle of approach [1] and placement of the support foot behind and beside the ball [5]. Nevertheless, velocity of the ball is also considered as a determinant for a successful instep kick. The velocity of the ball in maximal instep kick for skilled soccer players had been reported by several researchers as between 17 and 28 m/s. Based on the 1990 World Cup in Italy, ball velocities produced by top players could reach the speed of 32–35 m/s. There is suggestion the ball should travel at about 1.2 times the velocity of the foot.

Instep kicking in soccer is one of the most fundamental and frequently used skills [8]. The technical description of the instep kick is as the following. An instep kick involves a complex interaction of angled approach to the ball and a subsequent support of foot contact with the ground accompanied by sequential transfer of momentum from proximal-to-distal body segments in the swing or kicking limb. Following the angled approach, the support foot is placed alongside and adjacent to the ball with the toe of the support foot pointed in the intended direction of ball movement. The kicking limb at support of foot contact is in a position of hip extension, knee flexion and ankle plantar flexion. In powerful instep kick, following preparation of the kicking limb, the hip is forcefully flexed and the knee is sequentially extended so that forces generated can be channeled into propelling the ball. Duško Bjelica [1] discussed on the different of non-preferred leg which shows that the accuracy of instep kicking between preferred and non-preferred leg is significantly different among the optimal and maximal intensity in a resting state and in a state of fatigue all in favor of preferred leg.

Fig. 1 Six-camera system

The biomechanics of instep kick in soccer has been widely studied. Kellis and Katis [2] provides a comprehensive review on the research findings on the instep kicks in soccer. Detail reviews on biomechanics of soccer movement can also be found in [6]. Barfield et al. [9] wrote that majorities of studies on kicking in soccer are reporting on the two-dimensional (2D) and the three-dimensional (3D) kinematics of the low maximum velocity instep kick [6, 8]. Meanwhile, according to Duško Bjelica [1], latest researches have identified new aspects of soccer kick performance. The authors have listed these aspects, which include the three-dimensional kinematics of the movements, joint moments that drive the movements, mechanism of soccer performance as well as various factors which affects the soccer kicks biomechanics such as age, gender, limb dominance and fatigue.

2 Experimental

The experimental study was conducted at the National Sports Institute Biomechanics Laboratory, Kuala Lumpur. Eight elite male national futsal players between the ages of 21 and 30 years old were recruited. Subjects were informed of the study procedures and total time needed for testing. All risks and possible harms were verbally explained and described in the consent form. All subjects signed a consent form approved by the Research Ethics Committee of Universiti Teknologi MARA. Their physical measurements for height (167 $\pm$ 4 cm), weight (60.3 $\pm$ 8.3) kg, waist to heel length (99.1 $\pm$ 3.9 cm), waist to knee length (51.3 $\pm$ 2.8 cm) and instep width (17.6 $\pm$ 1.5 cm) were recorded.

In the biomechanics laboratory of Institut Sukan Negara, a six-camera system (Motion analysis Corporation, Santa Rosa, CA) sampling at 200 Hz was used to collect motion analysis data (see Fig. 1). A hand held stalker Sport Speed Radar Gun was held by an independent observer behind the goal net to measure the velocity of the ball.

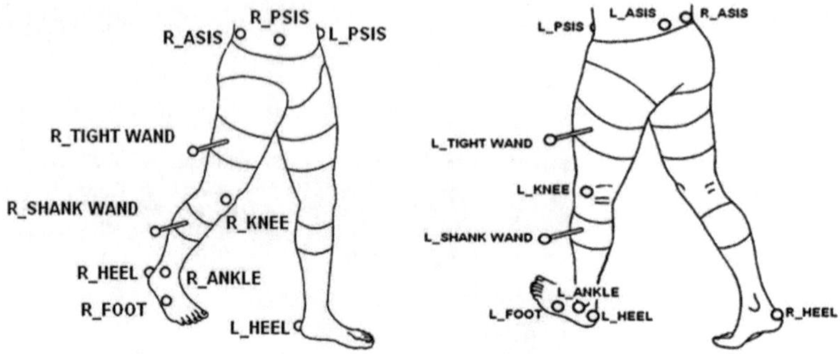

Fig. 2 Position markers modified from Helen Mayes system

Before the trials, 22 markers were placed on the participant's body based on the modified Helen Mayes system (Fig. 2). A T-pose of each participant was captured by the cameras from the front, left and right positions to be used as points of reference for the motion analysis. Six middle markers were removed leaving only 16 markers with 8 markers on each side of the lower extremities before the participant started to kick the ball. The positions of the reflective markers are as shown in Fig. 2. High-speed camera was employed to capture rapid movement of subjects. Marker position data were smoothed using Butterworth filter at a cutoff frequency of 6 Hz. The system was calibrated to the manufacturer's recommendations before each data collection.

Participants were instructed to kick a ball at a marked position into the goal post which was placed at the distance of 3, 5 and 10 m from the marked point. For each distance, three trials were performed.

3 Digitizing Data

Data were collected using the EVaRT (v.5.0, Motion Analysis Corporation, USA) and transformed into 3D coordinates via the direct linear transformation method and subsequently tracked using EvaRT 5.0. Missing data were rectified using solutions provided by EVaRT. Data were taken 40 frames before and 20 frames after the impact of the foot on the ball to ensure standardization. Figure 3 shows the stick figure of an instep kick movement.

4 Parameter

Kinetics analyses in biomechanics are based on forces that cause change in motion. These forces could be determined indirectly through the measurement of markers' positions from the captured motion. Velocities, speed, acceleration,

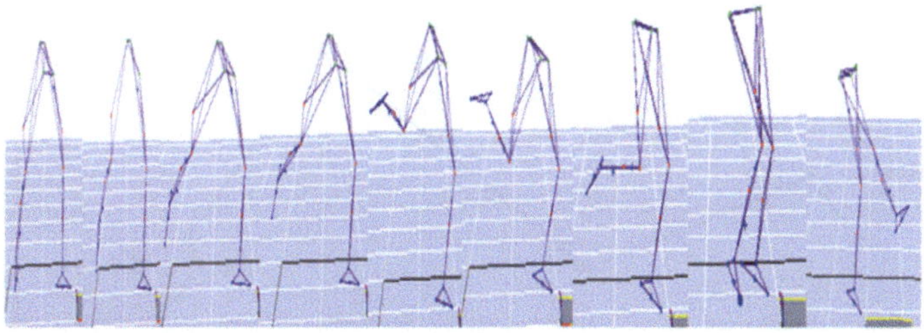

Fig. 3 Stick figure of an instep kick movement

force, torque, mass, body rotational, energy, impulse and posture of part of the body as parameters involved in biomechanics study. Parameters involved in 2D and 3D do not differ very much. Both 2D and 3D use similar characteristic, with the 3D analysis is capable of giving a more detailed analysis than 2D. Selecting true and thorough parameters realistically will always remain puzzling as there are numerous parameters. Therefore, this study concentrated on angles formed by various parts of the legs, angular velocities, velocities, center of mass and torques, which then selected as parameters of the study. Values obtained from these parameters were utilized in the formulation of the ABP-based mathematical model.

Steps implemented in the development of model used in the study are discussed. These include identification of parameters and variables, determination of the constraints, establishment of relationship between variables and formulation of model using mathematical programming concept. Discussion on selected parameters and variables to be used in the model is also presented. Constraints are formulated based on values obtained from data analysis, while the relationship between variables is established based on the biomechanics principle of the lower segmented body.

5 ABP-Based Mathematical Model

A mathematical model for generating high-impact instep kicks through maximization of torque (thus, ball velocity) and accuracy of kick (based on ball contact, run-up and motion of many body parts) subject to angular body posture constraints is expected to be the outcome of this study. This study focuses on lower part of the segmented body and assumed that the upper part of body gives the same body posture when deriving instep kick movement.

When performing the instep kick, one of the player's legs will be the supporting leg, while the other is kicking. Even though the supporting leg (stand leg) does not move much, the angles concerning both legs need to be considered in calculating the torque resulting from various angular body posture. There are five angles

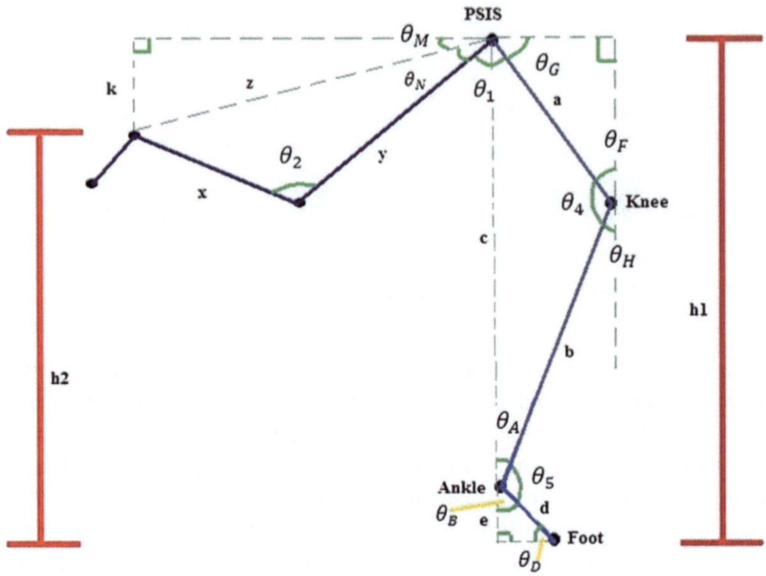

Fig. 4 Angles involved in ABP-based mathematical model

involved in formulating this equation. The first angle is θ_1, which is located at right PSIS, between the right thigh wand and left thigh wand. θ_2, positioned at right knee, which is the angle in between the right thigh wand and right shank wand. During the kicking motion, the impact is greatest at the foot where the ball will be placed during the motion.

Thus, considering this, third angle, θ_3, was taken to be between right shank wand and right foot. For non-dominant leg (not the kicking leg), two angles were defined. θ_4 represents the angle formed by the left thigh wand, left knee and left shank wand, while θ_5 was assigned as the angle between left shank wand, left ankle and left foot. The angles are as illustrated in Fig. 6. In calculating the torque involved in instep kick motion during shooting the ball, the leg bending positions during the motion were taken into consideration, as depicted in Fig. 4.

In analyzing angular body posture, it was noted that body bending during instep kick involves variation in height between the feet and the ground. As the kicking leg swing backward, it accumulates the torque in preparation for the downward motion toward the ball. Foot speed is governed by a combination of hip, shank and ankle rotational torques. Rates of charge of angles between the hip and shank, shank and ankle, right hip and left hip resulted in the angular velocity of the feet which in turn produced measureable torques during motion.

Biomechanical researches normally calculate the joint angles indirectly from linear coordinates taken from the motion capture of subjects in motion. For example, Fig. 6 shows a football player in a typical position of a kicking motion at a given time instant. The calculation in the ABP-based mathematical model will involve the parameter as:

Fig. 5 The *X*, *Y* and *Z* coordinate at a kicking position

5.1 Angular Displacement

The angular displacement between the thigh and shank is calculated from the *x*, *y* and *z* coordinates of markers at the corresponding position. Angular kinematics is usually derived using trigonometry.

The angle depicted in Fig. 5 is calculated based on the following equation:

$$\theta = \cos^{-1}\left(\frac{w^2 + s^2 - e^2}{2ws}\right). \tag{1}$$

5.2 Resultant Velocities

The resultant velocities of the movements of the thigh, knee, shank, ankle and foot for each subject were determined using the following formula.

$$v = \sqrt{\left(\frac{\Delta x}{\Delta t}\right)^2 + \left(\frac{\Delta y}{\Delta t}\right)^2 + \left(\frac{\Delta z}{\Delta t}\right)^2} \tag{2}$$

5.3 Angular Velocities

Angular velocities of joints are important in biomechanical studies because they represent the speed of flexion or extension and other anatomical rotations. Kicking motion has simultaneous rotation about *x*, *y* and *z* axes as:

$$\omega = \frac{\Delta\theta}{\Delta t} = \frac{\theta_n - \theta_{n-1}}{t_n - t_{n-1}} \tag{3}$$

5.4 Angular Acceleration

Angular acceleration, α, is the rate of change of angular velocity over time. Angular acceleration is also the second derivative of angular displacement with respect to the time. The unit for angular acceleration is rad/s^2.

$$\alpha = \frac{\mathrm{d}^2\theta}{\mathrm{d}t^2} \tag{4}$$

5.5 Center of Mass

The mass of an individual body segment represents a proportion of the total body mass. Each segment of body has its own COM and is usually expressed as a given fraction of the length of the segment from the proximal end. The terms COM and center of gravity are often used interchangeably. COM provides a measurement of the postural stability, which is frequently calculated based on markers placed at endpoints subset of body segment from the video captured. The idea is to calculate the moment attributable to the weight of the total moment about x and y axes. These values are parts of the anthropometric data defined by [10]. The formula for finding the coordinate of the COM is as follows:

$$x_{\mathrm{com}} = x_{\mathrm{proximal}} + \left[\mathrm{FL} \times \left(x_{\mathrm{distal}} - x_{\mathrm{proximal}}\right)\right] \tag{5}$$

Both kinetics and kinematics analysis require data on variables including mass distribution, mass centers, moment of inertia. Some of these measures can be determined directly. Human body consists of many types of tissues, muscles and bones, each with different masses.

5.6 Body Mass

Mass of body segment can be represented as a percentage of whole body mass (FM %). Based on the anthropometric data, the mass of the body segment can be described as a percentage from the total body mass. Mass of segmented body is calculated as:

$$\text{mass} = \text{segmented weight} \times \text{body weight} \tag{6}$$

5.7 Segmental Length

Most body segments do not rotate about their mass center but rather about the joint at either end. To calculate the moment of inertia of body segments using the proportion of segment length, the Radius of Gyration (ROG) must be computed first. The ROG is the distance that represents how far the mass of a rigid body would be from an axis of rotation if its mass was concentrated at a point. As compared to Center of Gravity (COG) of a rigid body, which is always at the same location on the body, the ROG varies depending on the axis about which the body is rotating. Using the anthropometric data of [10] in, the ROG at COG, at proximal end and at distal end can be represented as a proportion of the segmental length. For example, the ROG at COG relative to the segmental length for foot (SMF), thigh (SMT) and shank (SMS) are given as 0.475, 0.302 and 0.323, respectively. Segmental length, p_o, is formulated as:

$$p_o = (\text{segmented masses}) \times (\text{LOS}) \tag{7}$$

5.8 Moment

Moment of inertia or rotational inertia is the rotational analog of mass for linear motion, which depends on the mass of the object, its shape and its relative point of rotation. Moment of inertia is also known as polar moment of inertia of mass or the angular mass. It describes the tendency of a body to resist angular acceleration. It is imperative that the value of moment of inertia is known in order to further optimize sporting technique as oppose to what sport experts believed that it is relatively unimportant. Moment of inertia is expressed as the product of the mass, m, of each particle in the body and the square of its perpendicular distance from the axis of rotation, p_0, as follows:

$$I = m \times p_o^2 \tag{8}$$

5.9 Torque

Considering that the planar movement of a segment and reaction force at the distal end is known, the total torque, τ, during movement is given by:

$$\tau = \tau_1 + \tau_2 + \tau_3 + \cdots + \tau_n \tag{9}$$

Torques will be determined from the proximal-to-distal pattern of segmental angular velocities, which were identified by majority of studies as the key parameter in instep kicks [2]. Angular velocities in this study are derived from rate of change of angles during kicking motion. The moment (torque) at the proximal end cannot be calculated until the proximal reaction force has first been calculated. Therefore, the torque is written as:

$$\tau = I\alpha + \left((ma_x + R_{xd}) \times x_{\text{comp}}\right) + \left((ma_y + mg + R_{yd}) \times y_{\text{comp}}\right) \quad (10)$$

5.10 Optimization

The optimization of the total torque is subject to various angles identified, the height of the hip to the ground, h_1, and the height of the ankle to the ground, h_2. $\theta_i (i = 1, 2, \ldots, 5)$, in the constraints, are formulated to be between the range of lower bound (L_i) and upper bound (U_i), as described in Fig. 4. Note that, h_1 is a function of θ_4 and θ_5, where h_1 is the sum of c (length from the PSIS to ankle) and e (length from ankle to ground). Likewise, h_2 is function of h_1 and k, which is the distance between heights of hip to the highest raised right ankle.

Finally, the ABP-based mathematical model is defined as below with the parameter and constrains involved as illustrate in Fig. 4.

$$\begin{aligned}
\text{Maximize } \tau_t = {} & 0.0032715625W(\text{LOF})^2\left(\frac{\mathrm{d}^2\theta_1}{\mathrm{d}t^2}\right) \\
& + 0.004240986W(\text{LOS})^2\left(\frac{\mathrm{d}^2\theta_2}{\mathrm{d}t^2}\right) + 0.00104329W(\text{LOT})^2\left(\frac{\mathrm{d}^2\theta_3}{\mathrm{d}t^2}\right) \\
& + 0.0145Wa_{xf}(x_{\text{comf}} + x_{\text{coms}} + x_{\text{comt}}) - 0.0145Wa_{yf}(y_{\text{comf}} + y_{\text{coms}} + y_{\text{comt}}) \\
& + 0.0465Wa_{xs}(x_{\text{coms}} + x_{\text{comt}}) + 0.0465Wa_{ys}(y_{\text{coms}} + y_{\text{comt}}) \\
& + 0.100W\left(a_{xt}x_{\text{comt}} + a_{yt}y_{\text{comt}}\right) + 0.98Wy_{\text{comt}} + 0.4557W(y_{\text{comt}} + y_{\text{coms}}) \\
& + 0.1421W(y_{\text{comf}} + y_{\text{comt}} + y_{\text{coms}})
\end{aligned} \quad (11)$$

where

$$h_1 = a\cos\left(\frac{\pi}{2} - \theta_4 + \theta_5\right) + b\sin\theta_5 \quad (12)$$

and

$$\begin{aligned}
h_2 = {} & \left[a\cos\left(\frac{\pi}{2} - \theta_4 + \theta_5\right) + b\sin\theta_5\right] \\
& - \left[\left(\sqrt{y^2 + x^2 - 2xy\cos\theta_2}\right)\sin\left[\pi - \theta_1 - \theta_4 + \theta_5 - \left(\sin^{-1}\left(\frac{y}{\sqrt{y^2 + x^2 - 2xy\cos\theta_2}}\sin\theta_2\right)\right)\right]\right] \\
& - [z\sin(\pi - \theta_1 + \theta_2 - \theta_3 - \theta_4 + \theta_5)].
\end{aligned} \quad (13)$$

6 Summarize

The main focus of the ABP-based mathematics model is to determine torque optimization during instep kick. The variables involved were identified and elaborated in details. Various constraints were also considered. All these parameters and variables were thoughtfully deal with within the biomechanical scope. The general formulations mainly require Newton's law of motion, trigonometry, mathematical modeling and simulations. The quantitative biomechanical analyses used were limited to performances by the subject. The calculation involved was the angles made between the foot–shank–thigh of the kicking leg, the angles of the supporting leg during the instep kick motion. The anthropometric data and various mathematical theorems were used as a platform for the final inputs to the ABP-based mathematical model.

Acknowledgments The contribution, cooperation and support given by the personnel of the Biomechanical Department of the National Sports Institute (ISN), the National Sports Council (MSN) and the Malaysian Elite National Male Futsal Team.

References

1. Duško Bjelica, Stevo Popović and Jovica Petković. (2013). Comparison of instep kicking between preferred and non-preferred leg in young football players, monten. *Journal of Sports Science and Medicine*, 2(1), 5–10.
2. Kellis, E., & Katis, A. (2007). Biomechanical characteristics and determinants of instep soccer kick. *Journal of Sports Science and Medicine, 2007*, 154–165.
3. Plagenhoef, S. (1971). *Patterns of Human Motion: A Cinematographic Analysis*. Prentice–Hall, Inc: Englewood Cliffs.
4. Andersen, T. B., Kristensen, L. B., & Sorensen, H. (2008). Biomechanical differences between toe and instep kicking-influence of contact area on the coefficient of restitution. *Football Science, 5*, 45–50.
5. Nunome, H., Lake, M., Georgakis, A., & Stergioulas, L. K. (2006). Impact phase kinematics for instep kicking in soccer. *Journal of Sports Sciences, 24*, 11–22.
6. Petrone, J. (2006). Correlation between support foot placement and goal accuracy for instep kicks In The Soccer Field. *XXIV ISBS Symposium, Salzburg—Austria*.
7. Lees, A., & Nolan, L. (1998). Biomechanics of soccer—A review. *Journal of Sports Sciences,16*, 211–234.
8. Zhang, Y., Liu, G., & Xie, S. (2013). Movement sequences during instep rugby kick: a 3D biomechanical analysis. *International Journal of Sports Science and Engineering, 06*(2), 089–095.
9. Barfield, W., Kirkendall, D., & Yu, B. (2002). Kinematic instep kicking differences between elite female and male soccer players. *Journal of Sports Science and Medicine, 1*, 72–79.
10. Dempster, W. T. (1955). The anthropometry of body action. *Annals of the New York Academy of Sciences, 63*: 559–585. doi: 10.1111/j.1749-6632.1955.tb32112.x

The Level of Physiological Profile on Disability Track and Field Athletes on Selected Fitness Components

Nagoor Meera Abdullah, Wahidah Tumijan, Norlizah Abdul Hamid, Vincent Parnabas, Mohamad Rahizam Abdul Rahim, Sarimah Ismail and Mohamad Nizam Mohamad Shapie

Abstract The purpose of the study was to determine the physiological profile among disabled track and field athletes. 84 elite athletes ($n = 84$) from various types of disabilities such as visual impaired ($n = 32$; male = 19; female = 13), cerebral palsy ($n = 31$; male = 29; female = 2), intellectual disability (male = 18), and amputees (male = 3) were selected for this study. Four fitness tests were selected and conducted such as the handgrip strength test, vertical jump test, sit and reach test, and sit ups test. The result shows that the athletes such as visual impairment and intellectual disability have a higher muscle strength, power, and endurance. The mean score for athletes with intellectual disability in vertical jump test is 45.19 (±7.79), which shows the highest score among the athletes. While, the mean score for athletes with visual impaired in sit ups test is 39.53 (±13.42) which shows the highest score among the athletes. This shows that their disabilities was not affected the muscular strength and endurance. Athletes with amputees have a higher static strength and flexibility which shows the mean result in handgrip strength test is 46.80 (±5.57), and mean result in sit and reach test is 43.83 (±5.01). The athletes with cerebral palsy have the lowest score in all tests that shows lack of muscular strength, endurance, and flexibility because the disabilities such as spasticity do affect their physical such as balance, coordination, and direction. The significance value is ($p < 0.05$). The comparison for all types of disabilities in handgrip strength test shows significance difference which score is 0.00, while the score in vertical jump test is 0.00, and the score in sit ups test is also 0.00. However, the score for sit and reach test was not shown in the

N. M. Abdullah (✉) · W. Tumijan · N. Abdul Hamid · V. Parnabas · M. R. Abdul Rahim · S. Ismail · M. N. Mohamad Shapie
Faculty of Sport Science and Recreation, Universiti Teknologi MARA, Shah Alam, Selangor, Malaysia
e-mail: nameera_ab@yahoo.com.my

R. Adnan et al. (eds.), *Proceedings of the International Colloquium on Sports Science, Exercise, Engineering and Technology 2014 (ICoSSEET 2014)*,
DOI: 10.1007/978-981-287-107-7_54, © Springer Science+Business Media Singapore 2014

significance different where the score is 0.86. This shows that the test for measure the strength and endurance is suitable to determine the fitness profile for disabled track and field athletes and they are required the muscular strength and endurance.

Keywords Track and field · Disability · Physiological profile · Fitness test · Athletes

1 Introduction

The speed of a sprinter, the strength of a thrower, the power of a jumper, and the endurance of a distance runner is what draws the crowds to athletics. The track and field events at the Paralympic Games attract the largest number of athletes and spectators. Athletes, in general have a long history dating back to the times Greeks. For athletes with a disability, the first organized competition was held in 1952 when wheelchair racing was included at the Stoke Mandeville England within the framework of the games organized for World. They offer a wide range of competitions and the largest number of events. Athletics has been a part of the Paralympic Games since 1960 and events are open to male and female athletes in all disability groups. Advances in technology and athletes' dedication to the sport have made once unimaginable feats become realities. Some compete in wheelchairs, others with prostheses, and visually impaired athletes compete with the guidance of a sighted companion. Persons with disabilities (PWD) are a kind of person that has any loss or abnormality of the psychological, physiological, or anatomic structure of function. Disability is defined as a long-term reduction in a person's capacity to perform the average kind or amount of activities associated with his or her age group and typically resulting from chronic disease or impairment [1].

Persons with disabilities (PWDs) are a kind of person that has any loss or abnormality of the psychological, physiological, or anatomic structure of function. Disability is defined as a long-term reduction in a person's capacity to perform the average kind or amount of activities associated with his or her age group and typically resulting from chronic disease or impairment [2]. Since the World Health Organization (WHO) refers to disability as the loss or reduction of functional ability and/or activity and the American's With Disabilities Act (ADA 1990) defines it as a physical or mental impairment that substantially limits one or more of the major life activities, it would seem appropriate that functional ability be a goal of the assessment and fitness prescription used for PWD. Fitness can be defined as the undertaking of moderate to high intensity activity for the sole purpose of improving abilities in cardiovascular endurance, muscular strength and endurance, flexibility, morphology, motor ability, and metabolic function [1].

This definition includes not only activities undertaken for the intentional purpose of improving health and performance, but also daily living activities where energy expenditure is equivalent to that of fitness. PWD is very diverse group and their exercise prescription should be specific to their needs, goals, baselines level of fitness, and functional level of mobility [3]. The exercise prescription for PWD may be focuses more on health-related outcomes and include exercise for improvements in cardiorespiratory fitness, strength, and body composition. Rimmer et al. [3] states, through they must always be safe and within the participant's limitation, exercise prescriptions comprehensive exercise prescription would be design for the individual to provide the proper amount of physical activity to attain maximal benefit at the lowest risk and include factors that result in permanent lifestyle change and encourage a lifetime of physical activity [4].

Assuming that the decline in fitness among person with intellectual disabilities follow the same pattern as those PWD compounded with an already well-documented lower level of physical fitness, this population would be expected to experience difficulty in sustaining the energy to perform even the simplest of daily activities. Ponchillia et al. [5] believe that poor physical fitness is more likely due to lack of opportunity in planning and participation in physical activities. In addition, problems with processing information and lack of ability to problem solve may have some effect on physical functioning [6] suggested that many PWD have shown poor gross motor control which can affect performance on physical fitness tests, resulting in lower measures of fitness. As previously mentioned, differences in somatic growth and biological maturation between person with and without disabilities have been cited as having a strong influence on performance on physical fitness tests [1, 2]. In disability athletics (i.e., track and field), a number of classification systems are internationally recognized, including the systems of Cerebral Palsy-International Sport and Recreation Association (CP-ISRA), International Stoke Mandeville Wheelchair Sports Federation (ISMWSF), and the International Sport Organization for the Disabled (ISOD).

ISOD has one system for amputees and one for les autress. It has been widely suggested that these four systems of classification be replaced with a single system, an initiative similar to others undertaken in a range of disability sports, including swimming, table tennis, and equestrian. A major reason cited for such a move is that it would create an opportunity to reduce the absolute number of classes in the system, resulting in many advantages including increasing the number of athletes within classes in improving competitiveness within classes; increasing the number of competitive opportunities for athletes; and by inference, stimulating athlete participation and public interest [7].

Physical fitness illustrates to a physiologic state of well-being that allows individual to meet the demands in one or both physical fitness which is health-related fitness and skill- or performance-related fitness. However, skill-related physical fitness refers to components more closely related to sports and athletic performance such as balance, agility, coordination, and power; whereas health-related physical fitness involves components of everyday functional fitness including cardiorespiratory fitness, muscular strength or endurance, body composition and flexibility [8]. Levels of fitness reflect the outcomes of regular physical activity and are expect of health and functional abilities. Most of the researchers have investigated factor influence the physical fitness that have disabilities have lower physical fitness and are associated with participation in physical activity. The purpose of the study was to determine the physiological profile among disabled track and field athletes.

2 Methodology

2.1 Sample

84 elite disability tracks and field athletes ($n = 84$) from various types of disabilities such as visual impaired ($n = 32$; male = 19; female = 13), cerebral palsy ($n = 31$; male = 29; female = 2), intellectual disabilities (male = 18), and amputees (male = 3) was selected for this study. They are currently training for various competitions such as national and international track and field competitions. They are in the National Sport Council (NSC) Elite athletes with disability program. Their average age is between 18 and 25 years old. All subjects were informed of the study procedure, purposes, and all gave their inform consent by their coaches, teachers, parents, or the guardians.

2.2 Instrumentation

This study will use four (4) tests, which are handgrip strength test, vertical jump test, sit and reach test, and sit up test. The entire test has been conducted to the subjects. Before start the test, the permission has been granted by the team manager and the coaches. The subjects filled the personal detail form, the consent form and also been briefed about the tests. A demonstration session also been conducted by the test administrators. The entire test conducted for a day. The subjects had been given two (2) trials to complete the tests.

Standing height was recorded to the nearest half centimeter with the subject barefoot and with the back against a vertical wall. Body weight was measured to the nearest 0.5 kg with shoes, sweaters (SECA model 841). Body mass index (BMI) was defined as body mass (kg, measured using an electronic weighing scale to the nearest 0.1 kg) divided by height (m, measured to the nearest 0.1 cm) squared (kg/m^2).

Handgrip strength is an important prerequisite for good performance of the upper limb. In the study, handgrip strength was measured using a standard adjustable handgrip strength test (Takei model TKK5401). Maximum handgrip forces for dominant hand were recorded in kilograms as the highest of two trials. Before testing the subjects individually, the researcher gave a brief orientation to the entire group. The dynamometer was adjusted to the size of the hand of participant. The arm, the hand, and the body position were standardized according to the suggestion of the American Society of Hand Therapists. Subjects were sitting with shoulder adducted and neutrally rotated, elbow flexed at 90° resting on the table surface and the forearm in neutral and wrist in 0–30° extension. The test was performed by squeezing calibrated hand dynamometer as forcefully as possible with the dominant hand. Static strength was assessed.

Coaches often use vertical jump to measure athlete's lower body muscular power [9]. The vertical jump (counter movement jump with 90° knee flexion before the extension). The jump was performed on a hard and flat surface using the Vertec equipment adjusted to each of the participants. The subjects were asked to do a counter movement jump in which they began in a standing position, dropped into the semi-squat position, and immediately jumped as high as possible. The jump height was given automatically by the Vertec. Two tests were performed with five minutes of rest between them. The best jump was used for analysis.

Sit reach test is used to measure the flexibility of the hamstrings, buttocks, and lower back [10]. The subjects were instructed to reach as far as possible from a sitting position (Acuflex model 1).

Sit up test is used to measure muscular endurance of the abdomen. The subjects need to lie down on the exercise mat with both their leg bend at 90°. Upon receiving the signal Go, the subjects need to perform sit up with their chest touching their leg as many as they can within 1 min. The score will be recorded.

2.3 Test Administration and Data Collection

In this study, the sequence of the test follows as such (a) handgrip strength test (b) sit and reach test (c) sit up test and (d) vertical test been organized. The entire test has been conducted at the National Stadium, Bukit Jalil, Kuala Lumpur. Before conducting the test, the test administrators give the briefing for the subjects how the test was conducted. Besides that the weight and height of the subjects have been taken, and the test administrators conduct the warm up and stretching for 15–20 min to prepare the body from injury. The subjects have been divided into four groups accordingly. Then the subjects perform the test and follow the procedure and the best's score has been recorded in the scores sheet.

2.4 Analysis of Data

The results are presented as means and the standard deviations. The SPSS package (version 19.0) was used for the statistical analysis. One-way ANOVA has been used to compare between the four groups.

Table 1 Demographic data for the type of disability and gender

Disability	Data	Male		Female	
		Mean	SD	Mean	SD
Visually impaired	Age (yrs)	25.63	9.79	18.92	3.59
	Height (m)	164.12	7.43	153.00	7.55
	Weight (kg)	61.80	11.88	48.63	13.19
	BMI (kg m^{-2})	23.00	4.00	20.74	4.52
Cerebral palsy	Age (yrs)	23.93	6.07	22.50	6.36
	Height (m)	161.74	7.15	147.90	1.27
	Weight (kg)	54.51	10.68	35.85	3.04
	BMI (kg m^{-2})	20.96	3.92	16.49	1.56
Intellectual disabilities	Age (yrs)	19.83	3.99	–	–
	Height (m)	166.24	6.52	–	–
	Weight (kg)	56.44	12.27	–	–
	BMI (kg m^{-2})	20.37	3.33	–	–
Amputees	Age (yrs)	25.33	3.21	–	–
	Height (m)	163.23	3.43	–	–
	Weight (kg)	51.87	8.79	–	–
	BMI (kg m^{-2})	18.12	2.42	–	–

3 Results

Table 1 shows the demographic data for the subjects, types of disability and also gender. The mean for age among male subjects with visual impaired is 25.63 years (±9.79), while mean age among female subjects with visual impaired is 18.92 years (±3.59). The mean BMI among male subjects with visual impaired is 23.00 kg m^{-2} (±4.00), while mean BMI among female subjects with visual impaired is 20.74 kg m^{-2} (±4.52). The mean age among male subjects with cerebral palsy is 23.93 yr (±6.07), while the mean age among female subjects with cerebral palsy is 22.50 yr (±6.36). The mean BMI among male subjects with cerebral palsy is 20.96 kg m^{-2} (±3.92), while the mean BMI among female subjects with cerebral palsy is 16.49 kg m^{-2} (±1.56). The mean age among male subjects with intellectual disability is 19.83 yr (±3.99), and the mean BMI among male subjects with intellectual disability is 20.37 kg m^{-2} (±3.33). The mean age among male subjects with amputees is 25.33 yr (±3.21), while the mean BMI among male subjects with amputees is 18.12^{-2} (±2.42).

Table 2 shows the level of fitness among the subjects according to the fitness tests, type of disabilities, and gender. The mean score for handgrip strength test for male subjects with visual impaired is 39.88 (±5.71), while the female subjects with visual impaired is 24.88 (±5.98). The mean score for male subjects with cerebral palsy is 34.54 (±10.03), while the mean score for the female subjects with cerebral palsy is 23.65 (±1.63). The mean score for male subjects with intellectual disability is 39.72 (±8.34), while the mean score for the male subjects with amputees is 46.80 (±5.57). Amputees' subjects have a similar strength with

Table 2 Level of fitness among the subject according to fitness tests, disabilities, and gender

Fitness tests	Disability	Male		Female	
		Mean	SD	Mean	SD
Handgrip strength test	Visually impaired	39.88	5.71	24.88	5.99
	Cerebral palsy	34.54	10.03	23.65	1.63
	Intellectual disabilities	39.72	8.34	–	–
	Amputees	46.80	5.57	–	–
Vertical jump test	Visually impaired	43.73	9.33	25.77	3.22
	Cerebral palsy	32.69	14.43	20.50	7.78
	Intellectual disabilities	45.19	7.79	–	–
	Amputees	41.00	9.00	–	–
Sit and reach test	Visually impaired	33.56	8.64	33.50	9.02
	Cerebral palsy	29.24	10.84	29.35	1.06
	Intellectual disabilities	33.83	4.76	–	–
	Amputees	43.83	5.01	–	–
Sit up test	Visually impaired	39.53	13.42	21.85	9.70
	Cerebral palsy	22.76	9.73	11.00	0.00
	Intellectual disabilities	28.28	7.27	–	–
	Amputees	29.67	16.44	–	–

normal people and they just lost one of their limbs. Whereas, the cerebral palsy subjects expose less strength because of their unstable body condition, coordination, and balance due to their spasticity.

The mean score for vertical jump test for male subjects with visual impaired is 43.37 (±9.33), while the female subjects with visual impaired is 25.77 (±3.22). The mean score for the male subjects with cerebral palsy is 32.69 (±14.43), while the mean score for the female subjects with cerebral palsy is 20.50 (±7.78). The mean score for the male subjects with intellectual disability is 45.19 (±7.79), while the mean score for the male subjects with amputees is 41.00 (±9.00). This shows that the subjects with visual impaired and subjects with intellectual disabilities have a good leg strength and muscular power. The conditions/disabilities were not influence their muscular strength and power. While, the subjects with cerebral palsy and subjects with amputees have less leg power during execute the vertical jump test because their disabilities were affected their body condition such as spasticity and lost of limbs. For subjects with amputees, the usage of artificial limb such as prosthesis and orthosis was a limiting factor that their muscular strength and power.

The mean score for sit and reach test for male subjects with visual impaired is 33.56 (±8.64), while the female subjects with visual impaired is 33.50 (±9.02). The mean score for the male subjects with cerebral palsy is 29.24 (±10.84), while the mean score for female subjects with cerebral palsy is 29.25 (±1.06). The mean score for the male subjects with intellectual disabilities is 33.83 (±4.76), while the mean score for the male subjects with amputees is 43.83 (±5.01). This indicates that subjects with amputees have a high level of flexibility compare to other subjects. Even though the subjects with visual impaired and subjects with

Table 3 The comparison of fitness level among the groups

Fitness test		df	Mean square	f	P value
Handgrip strength test	Between groups	5	568.11	8.72	0.000
	Within groups	78	65.19		
	Total	83			
Vertical jump test	Between groups	5	948.73	8.43	0.000
	Within groups	78	112.49		
	Total	83			
Sit and reach test	Between groups	5	155.95	2.01	0.086
	Within groups	78	77.50		
	Total	83			
Sit up test	Between groups	5	876.94	8.08	0.000
	Within groups	78	108.51		
	Total	83			

$p < 0.05$

intellectual disabilities have a normal physical condition, they do not possess a good back flexibility.

The mean score for sit up test for male subjects with visual impaired is 39.53 (±13.42), while the female subjects with visual impaired is 21.85 (±9.70). The mean score for the male subjects with cerebral palsy is 22.76 (±9.73), while the mean score for female subjects with cerebral palsy is 11.00 (±0.00). The mean score for the male subjects with intellectual disabilities is 28.28 (±7.27), while the mean score for the male subjects with amputees is 29.67 (±16.44). The result exhibits subjects with visual impaired have a high level of muscular strength and muscular endurance that do not affect their visual acuity. Subjects with cerebral palsy are lack of muscular endurance because of the uncoordinated movement of the body such as coordination, balance, direction (Table 3).

The handgrip strength test score shows that this test is suitable to all these types of disabilities ($p < 0.05$; 0.000). There is no significant difference on all groups when performing the test. It means that all subjects capable of performing the test successfully and it also suitable to measure the static strength among the subjects from various types of disabilities.

The vertical jump test score shows that this test is suitable for all types of disabilities, even though subject with visual impairment need guide to perform the test ($p < 0.05$; 0.000). There is no significant difference for all the subjects doing the test successfully, and the test can be used to measure the muscular power among subjects from various types of disabilities. The muscular power includes the athletic power and work power.

The sit and reach test result shows that this test is not suitable to some subjects especially subjects with cerebral palsy who lacks of uncoordinated movement and spasticity. There is significance different of the hypothesis shows that the flexibility or the ability to move the body joints through a maximum range of motion which cannot be used for all the subjects ($p < 0.05$; 0.086).

The sit up test result shows that this test is suitable to measure muscular strength and endurance for all these kinds of athletes in order to determine the fitness level and profile for them. There is no significant different for the entire subject when performing the test ($p < 0.05$; 0.000).

4 Discussion

The handgrip strength is one of the components in order to determine the physiological level for an athlete. The handgrip strength test was used and conducted to measure the static strength among the athletes. Present study shows the significance difference in the level of handgrip strength test among the subjects. Strength is essential for high level performance in many sports, and, though not to the same degree, it also is essential for good health [11]. Past studies have consistently found men to have stronger grip strength than women [11]. Similar results were found by Rosandich [12]. Present study also shows that men have more grip strength than women. The subjects with amputees have a highest level of grip strength compared with others, and subjects with cerebral palsy have a lowest level of grip strength. According to [12], this method is considered one of the safest and most efficient especially for persons with disabilities in order to determine their fitness level and profile.

The need to acquire muscular power is very important in order to determine the physiological level for an athlete. O'Connor [13] presented information that muscles fiber types during explosive movements, such as vertical jump, play a big role in the height achieved while jumping. The vertical jump test was conducted to measure the muscular power among the subjects. The present study shows the subjects with visual impaired and intellectual disabilities achieve an acceptable level in vertical jump test because of their disabilities was not affected the body condition. While the subjects with cerebral palsy and amputees have a problem with their body condition that may affect the muscular strength. Tweedy [14] also states that if cross-bridge release is lengthened in time, force production and elastic energy can be wasted, therefore, the subjects is not able to create a maximal force production during a vertical jump performance. The comparison between all types of disability in this test shows significance different which means that it is suitable to determine the fitness profile for these kinds of subjects.

Flexibility is also one of the components in order to determine the physiological level for an athlete. The present study shows the subjects with amputees have a highest level in flexibility. Previous study exposes the data corroborate limited evidence from the literature that people with visual impairments score have lower than do sighted people in this area [15, 16]. The visually impaired women differed significantly from the sighted girls in the sit and reach, and the men's rated were 13 % below of the sighted boys (a statistically nonsignificant difference). Lower passing rates for this item may reflect the lack of participation in activities that demand lower back and hamstring flexibility [15]. This study shows the level of flexibility among male and female subjects are quite similar.

As a component in the physiological profile, muscular endurance, and strength are closely related; through weight-training methods for them are typically different. In general, endurance is improved through a low-resistance, high-repetition program [17]. The present study shows that subjects with visual impaired have a highest level in muscular strength. The subjects with visual impaired have a good body condition and it is similar with a normal people where they are no physical limitation. Several studies have reported that the person with visual impaired can involve in the physical activities with person without disabilities. Children who are visually impaired have high levels of health-related fitness than children with other disabilities [18].

5 Conclusion

The method to measure the fitness components in order to determine the physiological profile was also important to be considered. The suitable method will make sure the validity of the profile for athletes. O'Connor [13] has suggested that a more disability-specific approach is needed regarding testing and programing, this is challenging with the diverse population in a program where multiple types, conditions, and degrees of disability exist within individuals as well as within the group as a whole. Generally, there is lack of research on exercise physiology and behavior specific to persons with disabilities [2–4]. Without this study, it is difficult, if not impossible to implement valid interventions to alleviate the impact of physical activity toward the physiological profile. The finding of the lack of activities, combined with no one to do activities with, is similar to [5] finding that even elite athletes had a hard time finding opportunities for participation and peers with whom to engage in physical activity.

In general population, positive physical fitness changes have been positively related to psychological improvement. Past study does indicate that participation in physical activity is associated with greater satisfaction with one's appearance and one's level of perceived fitness. A review of the available literature indicates a great need for access to physical activity for person with disabilities if they continue relative healthy independence.

Acknowledgments The authors wish to acknowledge the Research Management Institute (RMI), Universiti Teknologi MARA to provide research project grant 600-RMI/DANA 5/3/RIF (774/2012) for the presentation and the publication.

References

1. Abolfotouh, M. A., & Telmesani, A. (1993). A study of some psychosocial characteristics of blind and deaf male students in Abha City, Asir Region Saudi Arabia. *Public Health, 107*, 261–269.
2. Fragala-Pinkham, M. A., Haley, S. M., & Goodgold, S. (2006). Evaluation of A Community-Based Group Fitness Program For Children With Disabilities. *Pediatric Physics Theory, 18*, 159–167.
3. Rimmer, J. H., Rubin, S. S., & David, B. (1999). Physical activity pattern of African American women with physical disability. *Medicine and Science in Sport Exercise, 31*(4), 613–618.
4. Robinson, J. (2005). The Comparison of Three Therapeutic Modalities on Hamstring Flexibility. *ProQuest Education Journal, 12*, 82–86.
5. Ponchillia, P. E., Strause, B., & Ponchillia, S. V. (2002). Athletes with visual impairment: Attributes and sports participation. *Journal of Visual Impairment and Blindness, 96*, 267–276.
6. Caroline, F., Neville, O., & Robert, P. (2001). Current injury of disabilities as a barrier to being more physically active. *Medicine and Science in Sports and Exercise, 33*(5), 778–782.
7. Houwen, S., Visscher, C., Hartman, E., & Lemmink, K. A. (2007). Gross motor skills and sports participation of children with visual impairment. *Research Quarterly for Exercise and Sport, 78*, 16–23.
8. Vincent, S. D., Barker, R., Clarke, M., & Harrison, J. (1999). A comparison heart rate elicited by 1 mile run walk and the progressive aerobic cardiovascular endurance. *Research Quarterly for Exercise and Sport, 70*(1), 75–78.
9. Huang, Y. C., & Malina, R. M. (2009). BMI and health-related physical fitness in Taiwanese youth 9–18 Years. *Medicine and Science in Sports and Exercise, 39*(4), 701–708.
10. Koutedakis, Y., & Bouziotas, C. (2003). National physical education curriculum: motor and cardiovascular health related fitness in Greek adolescents. *British Journal of Sports Medicine, 37*, 311–314.
11. Copper, R. A., Axelson, P. W., & Harlan, W. (1999). Research on physical activity and health among people with disabilities: A consensus statement. *Journal of Rehabilitation Research and Development, 36*(2), 1–13.
12. Rosandich, T. P. (2003). The international physical fitness test manual. *The Sport Journal, 6*(1), 1–4.
13. O'Connor, R. A. (2003). A comparison of selected parameters of physical fitness in adults with physical disabilities between periods of supervised and unsupervised physical activity. *ProQuest Education Journal, 11*, 28–32.
14. Tweedy, S. M. (2003). Biomechanical consequences of impairment: A taxonomically valid basis for classification in a unified disability athletics system. *ProQuest Education Journal, 74*(1), 9.
15. Laurence, E., Darren, G., & Delham, T. W. (1999). Modification of the standard sit and reach flexibility protocol. *Journal of Athletic Training., 34*(1), 43–47.
16. Winnick, J. P., & Short, F. X. (1999). *The Brockport Physical Fitness Test Manual*. United States: Human Kinetics.
17. Kahenisa, H., Funato, K., Abe, T., & Fukugama, T. (2003). Profiles of muscularity in junior Olympic weight lifters. *Journal of Sport Medicine and Physical Fitness., 45*, 77–83.
18. Oh, H. K., Ayvazoglu, N. R., & Kozub, F. M. (2006). Expalining physical activity in children with visual impairments: A family systems approach. *ProQuest Education Journals, 72*(2), 235.

Athletic Performance Comparison Across the Borders of Event, Gender, and Age

Sarel W. J. Bekker

Abstract The work in this document addressed the need of parents, school teachers, and athletic coaches to be able to assess the level of performance and to compare performances of individual athletes of either gender and in any age group. This was achieved by creating formulas for the following areas in track and field athletics: (1) Correctly set base standards for all events from the results of 5-year weighted average performances. (2) Use the base standards to determine comparative standards for all events in all age groups. (3) Define a function to compare male and female performances in all events. (4) For all events, generate tables with a range from 0 to 1,000 as a comparative measurement. These formulas were created using data from international and national meetings. Different case studies were used for each of the areas, and the results were evaluated using the data. In all four areas, it was possible to define hybrid functions with a confidence factor better than 99 %. The final performance tables can be updated during revision periods (every 2–4 years) by using the national results for the last 5 years. These revisions will be required as implements, tracks, and training methods that change with time and this will then in turn influence the performances. This is evident from the continuous improvement of records in all events. As this is a new and unique formulation for measuring performances at junior level in world athletics, it can be used as a basis for future improvements in the field of comparative measurement for athletics.

Keywords Athletics · Performance · Comparison · Measurement · Tables · Age groups · Events · Competitions

S. W. J. Bekker (✉)
Vaal University of Technology, Vanderbijlpark, South Africa
e-mail: sbekker@bsport.co.za

R. Adnan et al. (eds.), *Proceedings of the International Colloquium on Sports Science, Exercise, Engineering and Technology 2014 (ICoSSEET 2014)*,
DOI: 10.1007/978-981-287-107-7_55,

1 Introduction

In the competitive environment of track and field athletics, it is easy to determine the winner of any event. The problem arises to declare the performance of one better than the other with athletes participating in different events. Various arguments are then used. These arguments are generally based on emotional reasons and quite often biased toward the preferred event of the person voicing an opinion in the argument. When people argue about the merits of performances, logical reasoning is quite often ignored [1].

At school level, the same arguments become even more difficult to answer. Athletes may not only participate in different events but also participate in different age groups. Gender differences also contribute to the complexity of the comparison of performances. The standard list of age groups and the approved events for each age group, as approved by Athletics South Africa (ASA), is used.

This type of speculation requires an objective method, based on historical data, to give an answer to the question. The International Amateur Athletics Federation (IAAF) has attempted to solve this argument by creating a point scoring set of tables to compare performances of different events for senior men and women. These were originally only used for the multi-event competitions during the Olympic Games and were later expanded to include all the events supported by the IAAF for seniors [2]. Although the concept is based on sound principles, it is still not settling the cross-event arguments as it only sets standards within an event without appropriate interevent considerations.

The author's experience in athletics started in 1950 as an athlete participating at provincial level and thereafter as a fully qualified coach in sprints, hurdles, and long jump. During this time, he also qualified as an official in all events and participating as an official at all levels of meetings up to international meetings hosted by ASA.

2 Background

To compare and evaluate performances is part of human nature. In most cases, it is essential to obtain exact measurements, but because all measurements are taken using a tool, it implies that all measured results are approximations [3].

A different type of measurement is taken by using comparison. Some sport disciplines, such as gymnastics and skating, still rely on human judgment. The score given will depend on the impartiality, knowledge, and experience of the judge [4].

When applied to track athletics, the performance of individual athletes in a particular event can be determined with a varying degree of accuracy. The variance will depend on the measuring devices used and human error in handling these devices. To eliminate, or at least equalize these errors, athletes participate in groups, which implies that any errors due to instruments and/or the official will apply equally to all participating athletes.

All the work done by the IAAF is based on world performances and senior athletes. Very little work is done at junior level (19 years), and only a single attempt was made in England for the age groups 17 and 19 years.

In South Africa, various competitions based on combined events are available and sanctioned by Athletics South Africa (ASA). This requires a table-based system to ensure that the team with the best performances (not most first places) will be declared the winner. The first attempt at such a table was done in 1972, based on a linear approximation of the IAAF tables and using South African records in the age groups 15, 17, and 19 years.

3 Current Scoring Tables

The IAAF tables were created as an extension of the decathlon tables to cover all Olympic track and field events. These tables are only available for senior athletes and provide a point score in the range 1–1,400 points. The latest update was done in 2011, and it is available on the IAAF Web site. Most of the work done in developing these tables was through the IAAF working committee under the guidance of Dr Bojidar Spiriev (1932–2010). Dr Spiriev (from Bulgaria) passed away in 2010, and his work for the IAAF is carried on by his son, Attila Spiriev. These tables are currently the only tables with international acceptance [5].

The Purdy point system is calculated from a table of running performances. The table lists distance and velocity from 40 m to 100,000 m for senior events. These velocity measures are assumed to be maximum possible velocity in a straight line. These performances are arbitrarily given a Purdy point of 950. Times are calculated from the table (time = distance/velocity) by linear interpolation. Additionally, a time factor for start-up and running on a curve of a track is also added. These tables do not make provision for field events, hurdles, walk, or steeple chase [6].

Peter Riegel [7] also did research on the prediction of distance running. Riegel defined a value that he calls a "fatigue factor" which depends on the training, ability, and age of the athlete. By using this fatigue factor, the formula predicts the distance that an athlete can run at an average speed. This is not a comparative table but only a prediction of possible distance given the ability of the athlete.

David Cameron devised a formula that will predict an athlete's time at a different distance from the known time at another distance. These factors can be used to find comparable performances for different track events [8].

The Molvar Conversion tables were devised by John Molvar. It is also a prediction system that tries to convert the performance in one track event to a performance in another track event. These are perhaps the simplest way of equating performances. The conversion is simply done by multiplying the event time by the conversion factor [9].

Tables were created by Bill Rowan to help find the best athletes for events in the English high school championships. No formula was used but rather an intuitive allocation of performances. Bill never published these tables, and it was send to Tim Watts. The tables are extremely limited and only provide a scale from 1 to 30 points. It is also limited to the events for the age groups 17 and 19 years. Only events used during the championship are included.

The athletic performance evaluation (APE) tables are unique to South Africa and initially developed for a specific type of meeting. In 1985, ASA approached the author to modify the linear tables, based on the IAAF tables. The formula was then changed from linear to logarithmic. These tables have unique constants for each event but do provide smooth transitions between age groups. Statistical data and comments were obtained from Richard Stander (ASA). The tables include all events approved by ASA for all ages from 6 years to senior level. At present, these are the only tables available for school ages below 17 years, and the research is building on this work done to publish the existing tables. Although approved by ASA and recognized by the South African School Sport Association (SASSA), it is not achieving the requirement of correctly comparing events at all levels. In an effort to get event difference comparison, a complex function is used.

4 Performance Measurement

In spite of the existence of an official scoring table, this is seldom used to determine the best performances in athletics. This places doubt on the validity of the IAAF tables as the IAAF itself rarely refers to its own official tables. The IAAF only uses the tables for multi-event competitions and never publishes table scores with the results of individual performances. The IAAF also never refers to the tables in determining a series winner in individual events. They only use the medal-winning criteria for these awards.

The IAAF and most countries publish best performance lists by event. Enthusiasts then use these lists and start arguments by comparing different athletes who rank high in individual lists. They often refer to the consistency of performances and world records in their arguments to choose between athletes to determine the best performer. If the tables are used as a reference, people quite often reject the arguments by pointing out inconsistencies in the tables. These types of arguments have been observed during meetings at all levels over more than 30 years of experience.

Medal scoring is a method used at all Olympic Games: The team with the most gold medals is ranked first. If two or more teams have the same number of gold medals, then the team with the most silver medals will be ranked first. If there is still no result, the team with the most bronze medals will be placed first. After this, a tie is declared [7]. The argument against this method is that actual performance is not rewarded as the difference between a gold and a silver medal can sometimes be less than 0.01 of a second in a track event or a millimeter in a field event.

In a point scoring method, the athletes in each event earn points according to the placing in the event. Although this may appear to be well founded, it has the same negative effects as for medal scoring.

In 1972, Athletics South Africa (ASA) approved a unique type of meeting between schools. A boys' team consisted of 16 athletes, and a girls' team consisted of 14 athletes. For boys, the age groups were 15, 17, and 19 years, and for girls, age groups were 14, 16, and 18 years. Each athlete was allowed a maximum of two events. For boys, the best 16 results were combined as the team result and for girls the best 12 results. This required a table similar to the IAAF tables as athletes were participating in different events. The main challenge in this type of meeting is to ensure that any performance in any event, age, or gender can be correctly compared to any other event in a different age group or gender. The popularity of these meetings can be attributed to the smaller teams involved, which reduces transport cost and gives exposure to top athletes at more regular meetings against other top ranking athletes.

5 Generating Functions

Although the research is concentrated on South African school data, it is essential to start analyzing the available world statistics. It will ensure that the modeling results can be compared to existing models for validation. To ensure that the models proposed are reputable, they must be based on verifiable data. Various factors must be considered before data can be regarded as acceptable. The most reliable data available are from meetings directly controlled by the IAAF. This was therefore be used to formulate the models, and then, reputable South African data were used to validate the models.

The factors considered in the modelling are different for the different types of events, and separate case studies were formulated:

- For track events, the athlete relies on physical fitness, speed, speed endurance, reaction times, and blood oxygen content. The distance of the race will determine the prominent factor. The related events using obstacles such as hurdles and steeplechase races and race walking is considered as a subset of the track events. The formulation of tables for these events can be derived from the track events with corresponding track distances.
- In the case of long jump and triple jump, technique and the effect of gravity must be considered as well as the transfer of vertical speed to horizontal speed while maintaining the vertical component. A separate case study is used for these events.
- In the high jump, explosive power in converting vertical speed to horizontal speed while controlling the center of gravity during the jump is considered.

Table Normalized table scores

Points	100 m	200 m	400 m	800 m	1500 m	3000 m
1,400	9.46	9.46	9.46	9.46	9.46	9.46
1,200	10.02	10.08	10.12	10.08	10.11	10.14
1,000	*10.63*	*10.75*	*10.84*	*10.76*	*10.82*	*10.89*
800	11.30	11.49	11.63	11.50	11.61	11.71

- For all other field events, the transfer of body energy to kinetic energy in the implement at the point of release is the main technique to be considered. The distance achieved by the implement is then controlled by the laws of gravity.

The following four case study relations were considered in the development of the tables:

- A formulation for the relationship between different events within an age group.
- A formulation for the relation within an event for different age groups.
- A formulation for the difference between male and female athletes in similar events.
- A formulation for the performance rewards within an event.

The first three was not considered in any of the existing tables, and no reference to such relations was found in any existing references to tables.

In analysis of the IAAF tables, all track events are obviously progressive and performance reward increases correctly. This is not obvious for the field events. A slight progressiveness can be observed if the full range is considered. A bigger problem is observed if the track events are normalized. If the 100-m performance is projected to all other track events at the 1,400 point level, then the following table highlights the relation problem between events [10] (Table 1).

At the 1,000 point level, an 800 m athlete should run a faster time than a 400 m athlete for the same score. Similar problems exist and will be seen if all events are analyzed.

To determine the relation between events, the fifth-position average is shown in Fig. 1.

Examining the total distance and considering the spread of distances, and to get better relations, events can be grouped as sprints (50–200 m), intermediate distances (200–800 m), middle distances (800–2,000 m), and lastly long distances (2,000–10,000 m). In each of the sections, a trend can then be determined with continuance intercepts at the 200-, 800-, and 2,000-m events. This grouping proved to be correct, and second-order polynomials with a confidence factor of more than 99 % could be formulated.

The results obtained in each of the case studies were tested against the available data, and all results were supporting the individual case objectives. The original objective of finding a formula that will be able to correctly compare the

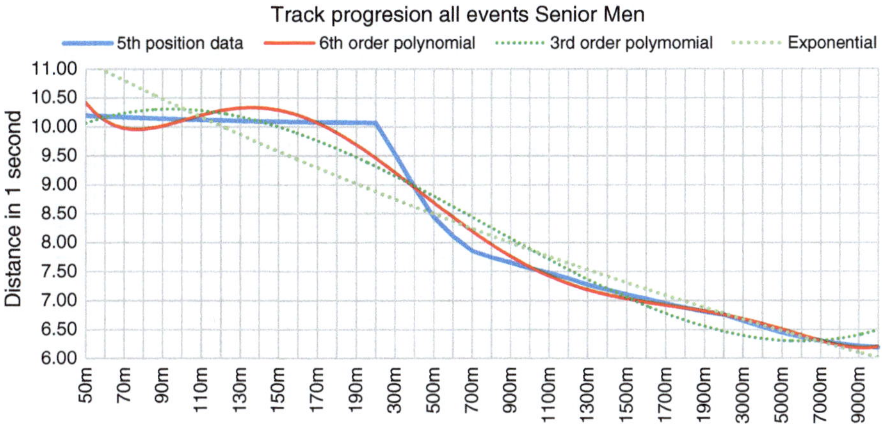

Fig. 1 Relation between track events

performances of different events, age groups, and gender was achieved. Using international and local data from a large number of meetings and from different venues and including a representative number of age groups from both genders, a consistent formula could be defined applicable to all tables.

The research shows that consistent functions could be developed to establish standards for different ages from the data of a single age with reliable data. It was also established that it is possible to generate standards for different events within an age group using a hybrid polynomial function.

Using fixed multipliers is also possible to determine the standards for all track events with obstacles and also all field events.

It was also possible to define a function for the determination of standards for girls using the standards obtained for boys.

This implies that it is possible to generate all standards and comparative tables from the measured standard of a single event. In practice, however, the freedom of variation is included to adjust the constants used in the functions. This is available if a new set of standards must be based on a number of measured values instead of a single value.

The base value(s) for a revision of the tables can be from any age or gender, and if multiple base values are used, these can be from different events, ages, and gender.

In practice, one or more weighted results will be used as confirmation of the generated standards and may also act as guidelines for minor adjustments which will be applied to all events.

The final function to generate all the tables has the form:

$$\text{Performance} = a \cdot b^{c \cdot \text{Point\_score}}$$

Fig. 2 Extract from boys 15 table

Boys 15 Field events						
Point	HJ	LJ	SP	HT	JT	Point
1000	2.11	7.29	20.50	73.61	72.29	1000
990	2.10	7.26	20.24	72.07	71.63	990
980		7.23	19.97	70.51	70.95	980
970	2.09	7.20	19.70	68.93	70.25	970
960	2.08	7.17	19.43	67.33	69.53	960
950	2.07	7.13	19.15	65.72	68.80	950
940	2.06	7.10	18.86	64.10	68.06	940
930	2.05	7.06	18.57	62.48	67.30	930
920	2.04	7.03	18.28	60.85	66.53	920
910	2.03	7.00	17.98	59.21	65.74	910
900	2.02	6.96	17.68	57.58	64.95	900

The table will reflect performances on a scale from 0 to 1,000. It is, however, not limited at the top mark and can be extended for individual performances exceeding the normally expected range for a specific event.

A portion of a typical table is shown in Fig. 2.

An additional requirement was the acceptability of table values between events which require event standards. As more reliable data are available at national level, the standards are set at the 5-year average of the fifth position in all events in the South African championship meetings. This will provide consistency in comparing other performances in different events. This average is defined as the standard performance and will be recalculated at periodic intervals to update the reliability of the tables. To ensure that the recent results carry more weight than older results, a weighted average of the fifth position is used. The weighted average is determined using the formula $average = (Y1 + Y2 + 2{\cdot}Y3 + 3{\cdot}Y4 + 3{\cdot}Y5)/10$ where $Y5$ is the last year of reference and $Y1$ is 5-year-older results.

For events not presented at the championship level and to accommodate slightly offset averages, the known event averages will be used to determine the constants of a set of second-order polynomial functions from which all standards are derived.

The same reasoning applies to field events where trending is used within a discipline. The relations between the weights of implements are used to determine the standards for related events.

This formulation was used to determine the constants of polynomial functions that will be used to set the relation between track events of different distances and for all events the relation between performances at different ages within a specific event.

For all events and age groups, the value of "c" in the performance function is fixed at a value of 0.0006. This value is also gender independent.

For all track events, the average speed to complete the event is used as the performance indicator of the athlete and not the time recorded for the event. This is to obtain consistency between the results of track and field events.

Table 2 Polynomial constants

Event group	*a*	*b*	*c*
50–200 m	0.00,050	−0.01640	10.20700
200–800 m	0.04890	−0.79060	10.84800
800–2000 m	0.00160	−0.10640	7.85510
2–20 km	0.00640	−0.13570	6.88530

The value of the base "*b*" in the performance function determines the spread of data, and this is different for performances below the standard performances and those above these values. For all events, ages, and genders, the base value for performances below the standard was found to be 3.3. This value gives a 0 point performance at 0.525 times the standard performance for all events. This implies that the zero point value on all tables will be at approximately half of the 900 point performances.

Above the standard performance, a base value is selected to reflect the current event records at approximately 1,000 points. For all track events, age, and gender independent, this value is set at 2.6, and for field events, it is 15.0. Testing these values, the current South African records range between 990 and 1,030 points on all tables.

The multiplier is event specific, but it was found that there is a consistent relation which can be used for age, gender, and event differences. This multiplier is also different for events above and below the standard performance of a specific event. The standard is set to produce 900 points, and the ratio between the multipliers in the two ranges must be such that continuity is maintained. This implies that if the standard performance is applied using either set of constants, it must return the 900 point value.

For all track events, the ratio between the top and bottom multipliers is consistent at 1.137396, and for all field events, it is 0.441477. Again, this is independent of the specific events, gender, and age. It is thus only required to determine one of these values using the event and age relations.

Track event age polynomials are used to generate the standards from the function:

$$\text{Standard} = a \cdot x^2 + b \cdot x + c$$

These will provide the standards for all the different track events at the base set by senior men data. The constants shown are for the data available at publication and obtained from international data. These constants will change on revision as in the generation procedure (Table 2).

The constants are obtained from the weighted averages, and they are continuous at the group intervals.

Event constant multipliers are automatically generated using the number of intervals between the available standard events. These constants are used to space the standards for different distances in the same age group.

Table 3 Polynomial constants

Event group	*a*	*b*	*c*
50–200 m	0.00050	−0.01640	*10.20700*
200–800 m	0.04890	−0.79060	10.84800
800–2000 m	0.00160	−0.10640	7.85510
2000–10000 m	0.00640	−0.13570	6.88530

Obstacle track event standards are adjusted using the corresponding track distance and a multiplier derived from the 5-year weighted averages for these events.

Field events use similar procedures as track events producing the second-order polynomials for each type of field event. This can be viewed in detail on the table generating spreadsheet.

With all standards available, any table can be generated using the hybrid exponential function.

Expanding the tables to other age groups is obtained by using the fourth-order polynomial

$$y = 4 \cdot 10^{-5} \cdot x^4 - 0.0016 \cdot x^3 + 0.00934 \cdot x^2 - 0.06852 \cdot x + k$$

where $x = 26$—*age* group value and k is an event-specific constant.

Some schools are using events during house meetings or interschool meetings which are not in the list of standard events. Now that the method of obtaining standards and the function for the table generation are available, it is a simple exercise to generate tables for nonstandard events.

Some of these standards can be directly derived from the current standards. For example, some schools have a "winners relay" consisting of 4 athletes from different age groups each running a 100-m leg. This standard can simply be determined by adding the four 100-m standards and multiplying it by the relay constant for the 4 × 100 m event.

Other events may need input from the school to provide their observed historical results to provide a suggested value that can be used as a standard. Then, it is simply a selection of the type of event to use the corresponding generation formula for the table. Only information in one age will be required to generate tables for a range of ages using the age polynomial function. Current events in this category used by some schools include turbo javelin throw and cricket ball throw.

In the case of track events (excluding events with obstacles), the standard values are used to determine the best-fit constants for the four groups of second-order polynomials. This method ensures that singe event exceptions are corrected and standards can be generated for the events not available in the age group used to establish the base standards. The current standards are shown in Table 3.

These constants are used in the fourth-order polynomial to generate age standards.

For hurdles, the standard is determined from the corresponding distance in the track multiplied by a factor determined using the function:

Table 4 Hybrid values

Event type	Above 900	Below 900
Track	2.6	3.3
Field	15	3.3

Standard = (Hurdle Height/106.7)$^{k}\cdot$ Distance-multiplier for short hurdles and Standard = (Hurdle Height/91.4)$^{k}\cdot$ Distance-multiplier for long hurdles

Similar multipliers are derived for steeplechase, race walking, and relay events.

The calculation of the standards for field events is similar to that of hurdles, but the lack of reliable data requires the calculation of deviation adjustments. This calculation is done automatically in the table generation spreadsheet and can be seen on the "Throw Standards" sheet.

It was determined that in all events, performances of female athletes follow the same reasoning as for male athletes with lower standards and a more pronounced deterioration as the performance position decreases. It is thus possible to use exactly the same functions for all events as those for male events to determine the standards. Because the tables use a progressive function, the change in standards will then automatically adjust the table performances to produce corresponding more pronounced adjustments in performance changes.

It was determined that a single function, $y = a{\cdot}b^{c.x}$, can be used for all events. To adjust the spread of data, different values of b are used for performances above and below the standard performances. Using the hybrid models a and b has different values for performances above and below the standard performances (Table 4).

The corresponding values of a are calculated for each age and event using the formula:

$$a = (\text{Event 900 point standard})/b^{0.54}$$

Using these values will ensure the following:

- A 0 point performance is about ½ of the standard.
- Records are at about 1,000 points.
- The function is continuous through the 900 point value.

6 Conclusion

It is conclusively shown that by setting statistical fifth-position averages for events in a reliable age group, standards for events in all other age groups and events can be generated.

The generated event standards can be used to generate table scores (from 0 to 1,000 points) for all events. The top end of tables is not limited to 1,000 points, but all published tables will be limited to this value.

Multi-event scores will be based on the actual performances of the individual events and not of the performances of athletes within the multi-event group. This will ensure that performances in multi-event competitions can be related to individual performances.

As standards are available for all events in all age groups and genders, comparison of performances based on the tables will be accurate and valid.

The standard adjustments of 0.14 and 0.24 used by the IAAF are used as conversion between hand and electronic timing for distances up to 400 m.

An approved set of adjustments is used to adjust the performances at altitude for all distances longer than 800 m.

Using this formulation, it is possible to extend the principles to include events for handicapped athletes by using existing multipliers used for these athletes.

It can also be applied on other sport types relying on the same principles such as swimming. This will then only require a new set of base values, and the same software can produce comparable results.

References

1. Rugg, G., & Petre, M. (2007). *A Gentle Guide to Research Methods*. Maidenhead: Open University Press.
2. de Coubertin P. (1997). Olympic memoirs (reprinted 1997; originally published in French, 1931) (online). Available at http://www.britannica.com. Accessed 12/06/2012.
3. Krebs Robert, F. (1999). *Scientific development and misconceptions through the ages: A reference guide*. Westport: Greenwood Press.
4. McFEE, G. (2004). *Sport, Rules, and Values: Philosophical Investigations into the Nature of Sport*. New York: Routledge.
5. IOC website (online). Available at http://www.olympic.org. Accessed from 01/06/2012 to 30/09/2012
6. Purdy, J. G. (1970). Computer generated track scoring tables. *Medicine and Science in Sports, 2*(3), 111–115.
7. Riegel, P. S. (1981). Athletic records and human endurance. *American Scientist, 69*, 285–290.
8. Kaplan D. (2010). Explaining the performance predictors article in track and field news. Winter 2010 Issue.
9. Handbook: IAAF rules (2012–2013). (online). Available at http//www.iaaf.org. Accessed 01/02/2012.
10. IAAF Scoring tables (online). Available at http://www.iaaf.org. Accessed 01/02/2012.

Part VI
Sports Industry and Management

An Overview of Sport Facilities Management in Malaysia

Milton Garaat, Abdul Hakim Mohammed and Mat Naim Abdullah

Abstract Sport facilities are integral part for physical activity. Improper design and unavailability of these facilities restrict participation. The government of Malaysia has taken steps to construct and provide sport facilities throughout the country at district and state levels, while the private sector is focusing in the main cities. However, sport facilities require an excellent management team to successfully manage all operations involved, meet the demands of the public, and counter the problems that arise in sport facilities. Therefore, a review of literature was carried out to identify the problems related to sport facilities. Among of the problems that are highlighted in this paper include globalization that requires the manager to be more competitive, the grand vision of the Malaysia Prime Minister Dato' Sri Mohd Najib Tun Haji Abdul Razak to construct world-class sports stadiums, financial problems, and lack of skilled staff and administrative professionals. This paper provides an overview of the current situation in the management of Malaysia's sports facilities.

Keywords Malaysia · Sport facilities · Competencies · Sport facility manager

1 Introduction

Physical activities are gaining popularity and more people are participating in it to enhance their health and well-being as well as life skills development [14]. In addition, by involving in physical activity, a wide range benefits could be gained such as physical, social, and mental health benefits, and moreover, people who do not meet the minimum requirements of physical activity of at least 30 min per day

M. Garaat (✉) · A. H. Mohammed · M. N. Abdullah
Faculty of Geoinformation and Real Estate, Universiti Teknologi Malaysia (UTM), Johor Bahru, Malaysia
e-mail: miltongaraat@gmail.com

R. Adnan et al. (eds.), *Proceedings of the International Colloquium on Sports Science, Exercise, Engineering and Technology 2014 (ICoSSEET 2014)*,
DOI: 10.1007/978-981-287-107-7_56,

would likely to be at risk of getting cardiovascular disease 1.5 times higher compared to those that are active [33]. According to WHO [34], two-third [or 71 %] of deaths in Malaysia was associated with the chronic disease, which encompassed 85,000 deaths in 2002. The best solution to tackle this problem was by involving oneself in physical activity and healthy lifestyles [33].

However, insufficient or unavailability of facilities restrict people from participating in sports, and different sports require various types of facilities. For example, the swimming pool is designed for swimmers, while the running track is designed for track and field's events, and football fields have their intended purposes. Thus, proper facilities are needed for specific sport activities inclusive of age range, to successfully implement the government policy to foster sport participation from among community members [8].

Foreseeing the demands and awareness of this situation, the Government of Malaysia has taken several steps to promote healthy life styles and physical activity, and this is carried out by the Ministry of Youth and Sports by advocating and organizing sports programs and providing sports facilities at state and district levels. Based on the Malaysian Plan (1966–2015) with respect to expenditure on sports facilities, it can be seen that millions of Ringgit Malaysia were allocated and invested for the construction of stadiums, sports complexes, sports centers, training centers, and public parks. Sport facilities have been increasing from time to time. According to the Malaysian Plan, every period of planning contains allocations on expenditure for the construction of new sports facilities [23–31]. As the sport facilities were increased, managers were needed for planning, organizing, leading, and evaluating. Megat Ahmad Kamaluddin [17] argued that the management of sports institutions in Malaysia still has plenty of room for improvement. However, in order to be effective and efficient, managers would require essential competencies. According to Ko et al. [13], competency is a combination of four essential elements, knowledge, skills, abilities, and personal traits. These basic elements are required to successfully perform the duties as sport facilities managers [11].

Hurd and McLean [10] had highlighted that the competencies serve a number of purposes to the organization, like employee evaluation, hiring criteria, professional development, mentoring others, benchmarking for determining skills and knowledge needed, developing new skills, develop job descriptions, creating interview questions and exercises, and describing the person type that is suitable for the organization. If competences are important in affecting the entire organization by selecting the right person based on the assessment of employees' performance, then necessary professional development programmes are needed to enhance employee accountability and credibility. Thus, it is required that sport facilities have a manager who possesses the necessary skills, or the manager should be equipped with sufficient training to successfully manage sports facilities. The managers should have the essential competencies based on the responsibilities and tasks to integrate the technological, human, organizational, and cultural factors. For example, the government of Malaysia is planning on expanding the sports facilities into a World Class Stadium [21]. With the construction of a World Class Stadium, there will be a shift in the work and production process changes, job responsibilities may expand, technology use will increase, and managers are expected to become more self-managing and team-oriented. Thus, the pressure is more on managers to produce good results in a shorter period of time, with limited resources and less formal authority [16]. For these reasons, managers with high-specific skills possessing essential competencies are needed. Thus, this research is timely and is important for the successful management of new sports facilities in Malaysia.

2 Background of Sports Facilities in Malaysia

A sport facility is a place where the sporting activities carried out and viewed by spectators [6]. Hallman et al. [8] defined it as the physical form to facilitate sports participation. The new forms of sports facilities can revive decaying areas of old cities into 'new cities' where the sports facilities could bring together the people into one place either as sports participants or sports spectators. In addition, sports facilities define the quality of life for a city to be successful in the twenty first century [4]. Barghchi et al. [4, 5] had also stated that sports facilities play a number of roles in the city, including urban development, economic growth, and as a catalyst for redevelopment, community generation and so forth.

In Malaysia, sport facilities development started to change and improve when the 16th Commonwealth Games was held in 1998. This was the biggest event ever held in Malaysia [5]. The first person who had the grand vision of promoting sporting excellence among Malaysians was the second Prime Minister, Tun Abdul Razak Hussein. He had devoted his vision of building up the National Sports Complex, but due to various circumstances his vision could not be achieved till the fourth Prime Minister Tun Dr. Mahathir Mohamad who officially authorized the office. Under the premiership of Tun Dr. Mahathir Mohamad, the agricultural-based economy evolved to an industrial nation, and it was then that Tun Abdul Razak's grand vision became a reality. Many other stadiums were constructed and sport programs multiplied in numbers, like the Sports for All program, Rakan Muda Program, Mt Everest Challenge, Commonwealth Games, Le Tour de Langkawi, and so forth [7].

To date, a massive number of multi use sports facilities had been constructed all over the country (Table 1). According to the report by Azman [3], 1,233 sport facilities had been constructed throughout Malaysia, and in comparison with the report retrieved from the Malaysia Plans [23, 28–31], there were thousands of sports facilities that had been constructed. It was clear that the government had executed most of the plans successfully and millions of Ringgit was invested for the purposes of construction, renovation, maintenance, improvement, and expansion of sports facilities [23–31].

The main sports facilities provider in Malaysia is the government itself. However, the government alone cannot cater for the publics' demand for sporting facilities. Thus, the government encourages the private sector to work together

Table 1 Number of constructed sport facilities

Agency	Sport facilities	Reference(s)
Ministry of youth and sports	• Futsal court: 1,100 • 1 Malaysia gym: 90 • Rakan Muda complex: 16 • Youth and Sport Complex: 14 • Sport Community Complex: 10 • Air and Water Sport Complex: 2 • Disable sport Complex: 1	[3]
The economic planning unit	• 12 multipurpose centers in state level will be constructed	[23]
The economic planning unit	• Training centers were added from 115 (1995) to 916 (2000) for the purposes of archery, track and field, badminton, bowling, and golf	[28]
The economic planning unit	• 24 public parks constructed and the existing 109 public park were improved	[29]
The economic planning unit	• Another 13 multipurpose sport complex will be built at district level • A total of 564 sport association was registered in line with the sport development act, 1997. However, 29 associations were deregistered for various reasons	[30]
The economic planning unit	• A total of 1,282 multipurpose sports courts and field was built	[31]

with them to provide sports facilities in accessible place [31]. However, despite the vast numbers of sports facilities constructed throughout the country, there seems to be loop holes, in particular problems related to management of these facilities. The purpose of this study was to discuss the problems currently faced by sports facility managers. Good management could effectively maximize their potential as sports providers at state and district levels and thus meet public satisfaction.

3 Issues of Malaysian Sports Facilities

Globalization is defined as an increased flow of information, goods, capital, and people across political and economic boundaries [12]. Walt [32] described that globalization has a significant impact on economic, social, political, and cultural aspects of countries. Reflecting on globalization, Hashim [9] argued that globalization is a movement of business, industrial, and professional activities into a global market place.

According to Amminudin and Perilah [2], globalization affects sports in several ways. The internationalization of sports competitions and events, and the international diffusion of sports and mega sports events have strong cultural-political and economic significance to the hosting nation. Globalization seems to provide an opportunity for the local sports industry with globalization strategies as suggested by Amminudin and Perilah [2]; however, this globalization can also be a threat if not carefully planned. For instance in Malaysia, sports organizations are unable to penetrate the international market as globalization pressures the sports managers internally. In European countries, Jacoby and Meunier [12] argued that globalization poses more threats than opportunities.

Sport in Malaysia has been considered as an industry only in the last 10 years, and there is significant pressure posed by globalization. The globalization pressures sports facility managers to be more competitive in business in order to be able to compete with the outside forces. For instance, the small local companies have to compete with the foreign brands which are barriers for them to penetrate the domestic and also the international market [2]. Poor business administration and business knowledge also leads to difficulties in penetrating the market and adapt to environmental changes.

To cope with the global movement and with the increase in the demand as sports portrays a good reputation of the country in the eyes of the world, and together with the vision to increase sports participation for health benefits, the Malaysia Prime Minister Dato' Sri Mohd Najib Tun Haji Abdul Razak, allocated funds amounting to RM80 million for building world-class sports facilities, such as the cycling Velodrome in Seremban and the Badminton Academy in Bukit Kiara [21]. However, Barghchi et al. [4] argued that the world-class sports facilities are having difficulties to generate funds to cover the annual operating costs, especially in term of maintenance activities. A world-class level and high capacity stadiums can be constructed, but excellent trained staff teams are required to manage the entire stadium operations effectively, and to efficiently generate funds for the relative maintenance costs involved. Thus, a competent manager is required.

Nonetheless, it is well known that strong financial support is the key for organizations to continuously operate. It is undeniable that a financial crisis will freeze the activities planned by the sports organization, including maintenance work and sporting events. According to Hurd and McLean [10], sport facilities often times face budget constraints, with public demand and pressure to provide services without sufficient allocation and resources. The budget constraints are not only in terms of services that have been highlighted by Hurd and McLean [10], but financial support is also needed for sportsmen, coaches, and trainers to have good training [20], to pay salaries of employees [18], and also for maintenance purposes [1]. Discussions about maintenance of sports facilities did not arise till the 7th Malaysia Plan (1966–2000). Prior to this, the focus was more on the construction of new sports facilities in every state. Even the management of sport facilities was not given much attention. The maintenance practices of sports facilities in Malaysia have been discussed by Mohd Taib et al. [19]. Though the sports facilities officers at the Malaysian Stadium Corporation rated the effectiveness of maintenance at a satisfactory level, but how well the maintenance activities were actually implemented are arguable [19]. The seriousness of the government in implementing maintenance works in sports facilities was indicated in the 8th Malaysia Plan, where RM113.2 million was allocated, and another RM280.9

million was allocated in the 9th Malaysian Plan for upgrading and maintenance purposes.

As a developing country, the sports industries are also going through a maturation process. Hence, the lack of skilled staff and professional administrative personnel is unavoidable. The issues related to the lack of skilled staff were highlighted by Hurd and McLean [10]. People with knowledge and skills is an important asset in the management of sports facilities, because they are the ones responsible for activities run within the facilities Ratten [22]. It was argued that the people were the most important asset in an organization as they affect the organizations operations, performance, and success, where wrong selection could results in loss of time and money. However, this could be prevented if the management personnel could be provided with the essential skills and knowledge to manage the sports facility professionally.

In Malaysia, for instance, the involvement of professional administrators is considered as new phenomena and some of the sport governing bodies lack basic administrative knowledge and professional support will be necessary for the sport facilities management [15]. Megat Ahmad Kamaluddin [17] argued that the sports industry needs extensive knowledge and skills, and as such there was a lack qualified of human resources, under utilization of sports facilities, and poor media skills and relations. In addition, Mohd Taib et al. [19] had noted that there was still a low number of experienced personnel in sport facilities maintenance, and hence they suggested that existing personnel be equipped with training and necessary skills. However, Malaysia was not the only country which faced this problem. Japan was also faced with a similar problem. Mastuoka [18] commented that sport management officers in Japan needed to have adequate knowledge and skills in sport management. Furthermore, Mastuoka [18] also argued that the problem in attracting participation in sports was also due to marketing inefficiency. Meanwhile, Akinsolo et al. [1] highlighted that poor maintenance management in sports complexes in South West Nigeria resulted in poor performance of sports facilities, delays in the response to repairs on complaints and unavailability of skillful personnel for maintenance activities.

Thus, professional administrators are needed due to sport facilities becoming more complex and difficult to manage, and it is deemed important to position skillful staff to adequately address this issue [4]. In addition, the manager should be extra vigilant and competent in many areas to excel with the job requirements to ensure the facility operates normally under good, reliable, and safe conditions for the user.

4 Conclusion and Recommendations

The purpose of this study was to discuss the current issues in sport facilities management in Malaysia. The study showed that Malaysian sport facilities had improved a lot and had multiplied in numbers. However, although this improvement is vividly seen, the management teams have to be given extra attention and be equipped with the necessary skills and knowledge in accordance with global events. Thus, this study contributes to the sport facility world in performing their duties, responsibilities, and function by identifying the issues and problems in sports organizations.

Based on the discussion, there were four main problems that have been identified. These include the fact that globalization has penetrated the sports industry in Malaysia, the grand vision of the Malaysia Prime Minister Dato' Sri Mohd Najib Tun Haji Abdul Razak to construct world-class sporting stadiums, financial problems, and the lack of skilled staff and professional administrators. In discussing these issues, it was clear that the major problem was actually the lack of competencies among sport facility managers.

Future research should integrate the competency variables in detail. For instance, the globalization as an issue discussed is actually related to skills in business administration. Hence, being able to identify the meaning of each problem will be useful for the study. It would also be interesting to study the perceptions of sport facility managers about the competencies that will be needed by them in performing their task.

References

1. Akinsola, O. E., Fapohunda, J. A., Ogunsanmi, O. E., & Fatokun, A. O. (2012). Evaluation of the scenario of facilities maintenance management of sport complexes in South West Nigeria. *Journal of Sustainable Development, 5*(4), 99–115.
2. Aminuddin, Y., & Perilah, M. S. (2008). Globalization and the Malaysian sports industry. *Research Journal of International Studies, Issues, 8*, 112–116.
3. Azman, M. (2011). Opportunity, recognition and future of Malaysian youth via sports for all program. *International Conference of Sports Science and Recreation 2011*. Holiday Villa Hotel and Suites Subang, Subang Jaya, Selangor, Darul Ehsan. 4–6 Oct.
4. Barghchi, M., Omar, D., & Aman, M. O. (2009). Sports facilities development and urban generation. *Journal of Social Sciences, 5*(5), 460–465.
5. Barghchi, M., Omar, D., & Aman, M. O. (2009). Cities, sport facilities development, and hosting events. *European Journal of Social Sciences, 10*(2), 185–195.
6. Fried, G. (2005). *Managing sport facilities*. Human Kinetics: Champaign.
7. Gold sports Magazine, (2003). Dr. Mahathir Mohamad father of sports development Malaysia. Special edition. Ramcel communications.
8. Hallman, K., Wicker, P., Breuer, C., & Schonherr, L. (2012). Understanding the importance of sport infrastructure for participation in different sports—findings from multilevel modeling. *European Sport Management Quarterly, 12*(5), 525–544.
9. Hashim, J. (2008). Competencies acquisition through self-directed learning among Malaysia mangers. *Journal of Workplace Learning, 20*(4), 259–271.
10. Hurd, A. R., & McLean, D. D. (2004). An analysis of the perceived competencies of CEOs in public park and recreation agencies. *Managing Leisure, 9*, 96–110.
11. Hurd, A. R., & Buschbom, T. (2010). Competency development for chief executive officers in YMCA. *Managing Leisure, 15*, 96–110.
12. Jacoby, W., & Meunier, S. (2010). Europe and the management of globalization. *Journal of European Public Policy., 17*(3), 299–317.

13. Ko, L. M., Henry, I., & Kao, J. C. H. (2011). The perceived importance of sport management competencies by academics and practitioners in cultural/industrial context in Taiwan. *Managing Leisure, 16*, 302–317.
14. Kruk, J. (2007). Physical activity in the prevention of the most frequent chronic diseases: an analysis of the recent evidence. *Asian Pacific Journal of Cancer Prevention, 8*, 325–338.
15. Li, M., Macintosh, E. W., & Bravo, G. A. (2012). *International sport management*. Human Kinetics: Champaign.
16. Longenecker, C. O., & Fink, L. S. (2001). Improving management performance in rapidly changing organizations. *Journal of Management Development, 20*(1), 7–18.
17. Megat Ahmad Kamaluddin, M. D. (2007). Sport management in Malaysia. *Asian Sport Management Review, 1*(1), 21–31.
18. Matsuoka, H. (2007). Progression of sport management in Japan. *Asian Sport Management Review, 1*(1), 42–47.
19. Mohd Taib, H., Noorlena, S., & Hasnul Faizal, H. (2013). Appraisal of the spot facilities maintenance management practices of Malaysian stadium corporations. *Asian Social Science, 9*(12), 93–98.
20. Oyunbat, N. (2007). Current situation of the development sport management in Mongolia. *Asian Sport Management Review, 1*(1), 15–20.
21. Prime Minister's Office of Malaysia (2013). Budget speech. Retrieved on 25 Nov 2012 from http://www.pmo.gov.my/.
22. Ratten, V. (2011). International sports management: current trends and future developments. *Thunderbird International Business Review, 53*(6), 679–686.
23. The First Malaysia plan (1966–1970). The economic planning unit. Prime Minister's Department, Putrajaya, Malaysia.
24. The Second Malaysia Plan (1971–1980). The economic planning unit, Prime Minister's Department, Putrajaya, Malaysia.
25. Third Malaysia Plan (1981–1985). The economic planning unit, Prime Minister's Department, Putrajaya, Malaysia.
26. The Fourth Malaysia Plan (1986–1990). The economic planning unit, Prime Minister's Department, Putrajaya, Malaysia.
27. The Sixth Malaysia Plan (1991–1995). The economic planning unit, Prime Minister's Department, Putrajaya, Malaysia.
28. The Seventh Malaysia Plan (1996–2000). The economic planning unit, Prime Minister's Department, Putrajaya, Malaysia.
29. The Eighth Malaysia Plan (2001–2005). The economic planning unit, Prime Minister's Department, Putrajaya, Malaysia.
30. The Ninth Malaysia Plan (2006–2010). The economic planning unit, Prime Minister's Department, Putrajaya, Malaysia.
31. The Tenth Malaysia plan (2011–2015). The economic planning unit, Prime Minister's Department, Putrajaya, Malaysia.
32. Walt, G. (2001). Globalization and health. *Medicine, Conflict and Survival, 17*, 63–70.
33. World Health Organization (2003). Health and development through physical activity and sport. Retrieved on 10 May 2013 from http://apps.who.int/iris/.
34. World Health Organization (2010). The impact of chronic disease in Malaysia. Retrieved on 10 May 2013 from www.who.int.

Examination of Golf Resort Service Attributes and Customer Satisfaction: An Application of Importance Performance Analysis

Tah Fatt Ong and Abdul Hadi Muhamad

Abstract Over the past two decades, an overwhelming increase in the popularity and participation of golfers has spurred studies on the importance of customer satisfaction with services received while playing golf. However, few studies have focused on the essential nature of golf resort services and the satisfaction of the golfers. Thus, the present study seeks to examine the relationship between golf resort service attributes and customer satisfaction at a local Golf and Country Club in Selangor. This study was designed to explore the relationship of the services involved in the golf industry and customer satisfaction from the perspective of the customers utilizing the Importance Performance Analysis (IPA) technique. Convenience sampling technique was employed to collect data from 149 golfers visiting a local Golf and Country Club. Self-administered and adapted questionnaires were distributed in the process of data collection. Factor analysis was performed on the collected data, and five underlying factors of the golf resort service attributes were identified, which were named "lodging," "food," "environment," "accessibility" and "activity". Using Pearson's correlation analysis, golf resort service attribute was found to possess a significant relationship with customer satisfaction ($R = 0.305$, $p = 0.001$). The result of the present study provides important information for golf resort operators in the development of personnel training and marketing strategies so as to enhance customer satisfaction in relation to golf resort services.

Keywords Golf resort service attributes · Customer satisfaction · Importance performance analysis

---

T. F. Ong (✉) · A. H. Muhamad
Faculty of Sports Science and Recreation, Universiti Teknologi MARA, Shah Alam, Malaysia
e-mail: ong278@salam.uitm.edu.my

R. Adnan et al. (eds.), *Proceedings of the International Colloquium on Sports Science, Exercise, Engineering and Technology 2014 (ICoSSEET 2014)*,
DOI: 10.1007/978-981-287-107-7_57,

1 Introduction

Golf has been a popular sport in around the world for a long time. According to [1], golf tourism is a profit business which attracts high-income tourists who generate significantly above average per capita revenues for a destination. Golf industry will be expected as the fastest growing segment of the tourism industry in Malaysia. For the last year, golf industry has contributed revenue of RM286 million, exceeding the original forecast of RM249.5 million to the Malaysian economy [1]. Though golf is becoming a fast-growing sport tourism product in the country, there is limited knowledge concerning golf tourist preference in relation to golf course attributes [2]. There are many possible reasons for golf tourists to visit a destination, often impersonal and driven by numbers of internal and external forces. The relationship between the variables involved that influence golfer's behavior needs to be investigated in order to fully understand current and potential visitors to a destination.

Due to the abundant choices of golf resorts available, golfers are becoming more demanding and selective with their choice of golfing destinations. Golf destination service providers need to recognize this trend and cater to the particular needs of the customer. In order to stay competitive in the golfing industry, golfing service providers are constantly seeking for ways to gain a competitive advantage, attracting and retaining their customers. Increased demands of consumers and competition have caused the management teams to place more interest in quality service practices [3]. Hence, quality of service has been recognized to be one of the important features that golfing organizations could differentiate themselves.

In relation to this, concern in the measurement of service quality is elevating. Measuring the quality of service encounter has become an essential part of most managers' responsibilities [4]. In order to better understand customers' expectation of service quality, service managers could begin with evaluating customer satisfaction. Cline [5] highlighted that it is necessary for service organizations in the hospitality industry to become conversant with customer satisfaction. One of the best ways to measure customer satisfaction is to identify whether you are meeting customer needs. Getting feedback information from customers is an imperative way to identify expectations, determine if customers are satisfied, and if the agency is putting their resources in the right areas [6].

In the hospitality industry, among some of the popular tools used to gather feedback information from the customers include comment cards and guest satisfaction surveys. Guadagnolo [7] noted that most evaluation instruments assess features of services that have been prescribed in advance by the agency that are thought to be important to the customers. A thorough customer understanding and focus are missing in these types of technique. Evaluation that relies on agency predictions, based on past observations, is unlikely to be as accurate and useful as one based on current customer disclosures.

Customer satisfaction comprises of two aspects: an evaluation by a customer regarding certain attributes and the customer's expectations of these attributes.

Due to the existence of different desires in each customer, the evaluation of a customer's satisfaction must include both an assessment of the attributes the customer feels are important and an assessment of each attribute's performance [8]. The Importance Performance Analysis (IPA), developed by Martilla and James [9], is a technique that accesses importance and performance information related to a customer's experience. It can translate this information into "easy to understand" management suggestions. The IPA technique has been popularly used in the hospitality industry to comprehend the importance placed on attributes as well as how the agency performed on these same attributes.

Although many studies have examined the relationships between services attribute and customer satisfaction in the leisure industry, there is a paucity of study conducted specifically on the golf industry. Few local researches have examined in detail the attributes of golf resort and courses or their operations which might be expected to influence golfer satisfaction. The relationship between golf service attribute importance and customer satisfaction should be studied to fully understand the existing and potential users to the destination.

Based on review of relevant literature, the present study aims to investigate: What are the dimensions of golf resort service attributes that are considered important among golfers? And how does golf resort service attribute importance relate to overall customer satisfaction? How is the applicability of Importance Performance Analysis (IPA) technique in assessing the performance of a golf resort?

The purpose of this study was to investigate the application of Importance Performance Analysis (IPA) technique in an outdoor sport activity, namely golf. It aimed to introduce a new approach to explore the relationship between the dimensions of golf resort service attributes and customer satisfaction, from the perspective of the customers.

2 Methodology

2.1 Design of the Study

The present study depicts a quantitative research design utilizing survey methods in collecting data, to examine the relationship of golf service attributes and customer satisfaction. Questionnaires were distributed to respondents who were met on-site and voluntarily participated in the survey. To avoid any confusion regarding the questionnaire, the researchers were present throughout the session to answer any question from the respondents. There was no time limit for the respondents to complete the survey.

2.2 Selection of Subjects

Respondents for this study were golfers playing at a local Golf and Country Club in Shah Alam, Selangor. Convenience sampling technique was used, and the data were collected using self-administered questionnaires. Questionnaires were distributed personally by the researcher and allowing respondents to fill up the questionnaire by themselves. A total of 149 questionnaires were distributed to golfers who have played at the particular Golf and Country Club. The age of golfers involved in this study was from 18 years old and above.

2.3 Instrumentation

The questionnaire consisted three sections. The first section focuses on socio-demographic questions, in which respondents were asked to provide information regarding their gender, age, marital status, occupation and country of origin. Section 2 was designed to gather information related to the golfers' rating concerning golf service attributes importance. Five-point Likert-type scale ranging from 1 (very unimportant) to 5 (very important) was used. This section consists of 14 items for golf resort attributes. Section 3 was designed to collect data related to the golfers' satisfaction concerning golf service attributes. This section also used five-point Likert-type scale ranging from 1 (very dissatisfied) to 5 (very satisfied). This section consists of the same 14 items for golf resort attribute. These items were developed and adapted from past research by Ong et al. [2]. A list of service attributes of golf resort was developed and determined based on expert interviews, and with reference to previous research. A pilot study was conducted to test the validity and reliability of the survey instrument by conducting a pilot study that involved 50 golfers at a local golf resort. Cronbach's alpha reliability coefficient for the golf resort service attributes was found to be $\alpha = 0.73$. Thus, the instrument is considered to have acceptable reliability, with alpha value >0.70 [10].

2.4 Data Analysis

The Statistical Package for Social Sciences (SPSS Ver.16) was utilized for the data analysis process. Both descriptive statistics and inferential statistics were engaged. The demographic variables: gender, age, marital status, occupation and country of origin were summarized using descriptive statistics. Exploratory factor analysis was performed to identify the underlying dimensions associated with golf resort service attributes. To determine the relationship between attribute importance and satisfaction, the Pearson's Correlation statistical analysis was employed. Lastly, IPA was conducted to further examine the relationship between golf resort service attributes importance judged by the participants and their satisfaction with the related service attributes.

3 Research Findings

3.1 Demographic Profile of Respondents

Table 1 summarized the demographic backgrounds of respondents. Majority of the respondents were males, married, aged from 35–54 years old, with occupation as businessman, professional and company employee. Most of them were domestic golfers.

3.2 Golf Resort Service Attributes

The golf resort service attributes were measured by 14 items. To identify the underlying dimensions of golf resort service attributes, the data were factor-analyzed using principal components analysis with varimax rotation method. The results are shown in Table 2. The resulting five-factor solution, with eigen values >1, explained 62.77 % of the variance. The five identified dimensions of golf resort service attributes were named "lodging," "food," "environment," "accessibility" and "activity."

3.3 Importance and Satisfaction of Golf Resort Service Attributes

The mean score of attribute importance of each service attribute was calculated and ranked accordingly as shown in Table 3.

The mean scores of the 14 golf club attributes importance were ranged from 4.10 to 4.50. The table showed that quality of food ($M = 4.50$, $SD = 0.64$), considered the most important, convenience of meal ($M = 4.46$, $SD = 0.61$) and reasonable price accommodation ($M = 4.38$, $SD = 0.61$) were the three most important service attributes, while the three least important service attributes were peaceful and restful atmosphere ($M = 4.10$, $SD = 0.72$), ease of access ($M = 4.10$, $SD = 0.69$) and availability of travel information ($M = 4.17$, $SD = 0.80$).

The mean score of satisfaction (which represents perceived performance) of each service attribute was calculated and ranked, as shown in Table 4. The mean scores of the 14 golf club attributes satisfaction were ranged from 4.01 to 4.32. The result showed that quality of food ($M = 4.32$, $SD = 0.77$) and convenience of meal ($M = 4.32$, $SD = 0.78$) were considered the most satisfied and follow by service in restaurants ($M = 4.25$, $SD = 0.77$). The availability of travel information was considered the less satisfied ($M = 4.01$, $SD = 0.78$). The attribute of convenient and ample parking ($M = 4.07$, $SD = 0.80$) and variety of outdoor recreation (attribute 14; $M = 4.08$, $SD = 0.78$) was the three less satisfied service attributes.

Table 1 Demographic profile of respondents

Demographic characteristics	Frequency ($N = 149$)	Percentage (%)
Gender		
Male	108	72.5
Female	41	27.5
Age		
18–24	18	12.1
25–34	21	14.1
35–44	66	44.3
45–54	35	23.5
55 and above	9	6.0
Marital status		
Married	113	75.8
Single	25	16.8
Other	11	7.4
Occupation		
Company employee	31	20.8
Businessman	47	31.5
Civil servant	23	15.4
Professional	34	22.8
Others	14	9.4
Country		
Malaysia	129	86.6
Japan	10	6.7
Singapore	7	4.7
Indonesia	3	2.0

3.4 Relationship Between Golf Resort Services Attribute Importance and Customer Satisfaction Based on IPA

The relationship between golf resort services attribute importance and customer satisfaction is illustrated using an importance–performance grid as shown in Fig. 1. The result shows that five attributes were included into the "Keep up the Good Work" quadrant which include, "quality of food," "convenience of meal," "service in restaurants," "reasonable price of accommodation" and "service in lodging facilities."

The "Concentrate Here" quadrant indicates a possible risk to customer satisfaction and can be an opportunity for the resort to improve customer satisfaction. The only attribute involved was "convenient and ample parking."

There were six attributes that fall in the "Low Priority" quadrant. These include attributes such as "quality and cleanliness of lodging facilities," "peaceful and restful atmosphere," "ease of access," "convenience of local transportation," "availability travel information" and "variety of outdoor recreation."

Table 2 Factor analysis of service attribute

Factor and variables	Factor loading	Variance explained (%)	Cronbach alpha (α)
Factor 1 : lodging		15.2	0.72
Service in lodging	0.81		
Quality and cleanliness of lodging facilities	0.78		
Reasonable price accommodation	0.65		
Factor 2: food		13.27	0.68
Convenience of meal	0.82		
Quality of food	0.76		
Service in restaurants	0.73		
Factor 3: environment		12.85	0.61
Cleanliness environment	78		
Safety and security	0.76		
Factor 4: accessibility		11.77	0.67
Convenience of local transportation	0.80		
Ease of access	0.77		
Peaceful and restful	0.55		
Factor 5: activity		9.67	0.66
Convenient and ample parking	0.86		
Availability of travel information	0.76		
Variety of outdoor recreation	0.66		
Overall		62.77	0.73

Table 3 Mean score of service attribute importance of golf resort

Service attribute	Mean	SD	Rating
Quality of food	4.50	0.64	1
Convenience of meal	4.46	0.61	2
Reasonable price accommodation	4.38	0.61	3
Service in restaurants	4.37	0.62	4
Service in lodging facilities	4.30	0.60	5
Convenient and ample parking	4.29	0.72	6
Convenience of local transportation	4.26	0.87	7
Safety and security	4.23	0.65	8
Cleanliness environment	4.22	0.61	9
Variety of outdoor recreation	4.19	0.73	10
Quality and cleanliness of lodging facilities	4.18	0.64	11
Availability of travel information	4.17	0.80	12
Ease of access	4.16	0.69	13
Peaceful and restful atmosphere	4.10	0.72	14

Table 4 Mean score of satisfaction (perceived performance) of service attribute

Service attribute	Mean	SD	Rating
Quality of food	4.32	0.77	1
Convenience of meal	4.32	0.78	1
Service in restaurants	4.25	0.78	3
Safety and security	4.23	0.67	4
Service in lodging facilities	4.21	0.71	5
Cleanliness environment	4.21	0.68	5
Reasonable price accommodation	4.19	0.71	7
Peaceful and restful atmosphere	4.14	0.79	8
Ease of access	4.13	0.79	9
Convenience of local transport	4.10	0.89	10
Quality and cleanliness of lodging facilities	4.09	0.87	11
Variety of outdoor recreation	4.08	0.78	12
Convenient and ample parking	4.07	0.80	13
Availability of travel information	4.01	0.85	14

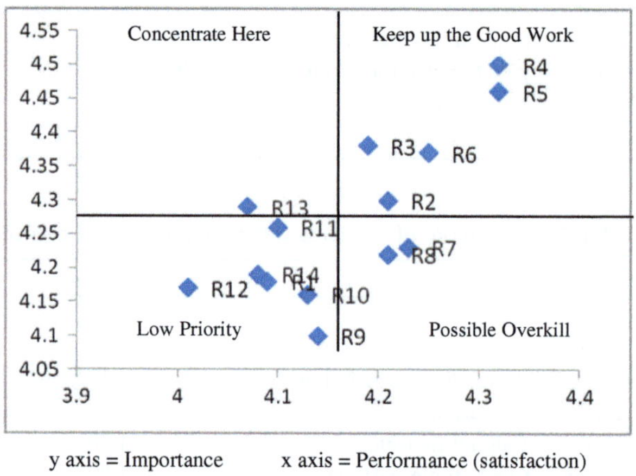

Fig. 1 Importance-performance grid of golf resort service attribute

Finally, the "Possible Overkill" quadrant comprises the comparatively less important service attributes, but those of relative high customer satisfaction. These attributes were "safety and security" and "cleanliness environment."

4 Conclusion

The finding of the current study revealed that the golfers visiting this golf resort are mostly married male, aged from 35 to 54 years old, with occupation as businessman, professional and company employee. Most of them were domestic golfers. In general, the demographic profile of the golfers identified in this study is rather consistent with previous studies conducted locally [2] and abroad such as in Canada [11] and Korea [12].

For the golf resort customers, quality of food was considered the most important service attribute. This result is consistent with past studies from several researchers. Clark and Wood [13] found that quality of food was the most influential predictor of consumer loyalty. Mattila [14] also indicated that quality of food, service and atmosphere are the top three reasons for customers to go to their target restaurants. So, resort managers should ensure the quality of food provided in the restaurant of the resort is always at high quality and meets the expectation of the customers. Meanwhile for golf club satisfaction, quality of food and convenience of meal were considered the most satisfied. This finding is consistent with the study of Ryu and Han [15], stating that quality of food is a significant predictor and has potential to influence customer satisfaction toward service experienced.

Though the mean scores for all attributes (importance and satisfaction) are averagely high (more than 4), it was noticed that most mean scores of performance do in fact fall short against importance. It should be interpreted that there is still room for improvement in many aspects of the service provided by the golf resort.

The "Keep up the Good Work" quadrant comprises five aspects including: quality of food, convenience of meal, service in restaurants, reasonable price of accommodation and service in lodging facilities. This sends a meaningful message to managers that they should concentrate to maintain this aspect, from the customer point of view. The current result revealed that customers at this resort have high level of satisfaction with service attributes in this quadrant. Quality and convenience of food served at the club are important to ensure customer can have their meal as expected. [16] suggested when the customers compare their perceptions of services performance with the expectations, the feelings of satisfaction will arise. Employees that are responsible for services in restaurant and lodging facilities need to constantly improve their skills in responding to different needs and wants of the customers, in order to provide quality services. Reasonable price in accommodation should be continued by the resort manager, and ensure services delivered at the golf resort is worth the value of the price charged. Basically, service attributes in this quadrant need to be reinforced to keep up the quality services that meet customer satisfaction.

"Concentrate here" quadrant indicates a possible risk to the resort as attributes in this quadrant is low in meeting customer satisfaction. The only attribute that falls in this quadrant is "this club has convenient and ample parking." Attribute in this quadrant is perceived to be very important, but customer's satisfaction level is fairly low due to weak service performance related to this particular service attributes. Therefore, resort manager should concentrate on this attribute, and immediate action should be taken to provide more parking areas for their customers. With better facility performance on this aspect, it should be able to improve and upgrade the level of customer satisfaction.

Service attributes in the "Low Priority" quadrant include quality and cleanliness of lodging facilities, peaceful and restful atmosphere, ease of access, convenience of local transportation, availability travel information and variety of outdoor recreation. As most respondents are locals, it is understandable that some of these attributes are rated as less important. Although these attributes had a lower importance, resort managers will still need to maintain and improve upon them to enhance customer satisfaction. Marketing approach could be made by promoting those attributes that can influence customer perception, such as the beautiful landscape of the resort, possibly attracting more foreign golfers. Variety of events, competitions and outdoor recreation activities could be interchangeably held and promoted to maintain the image and services of the resort.

Finally, the "possible overkill" quadrant comprises the less important service attribute, but those of relative high satisfaction. These attributes were safety and security and cleanliness environment. It is suggested that club manager should consider reallocating these resources and efforts elsewhere as the present efforts related to these attributes are regarded as being overdone.

The current study has demonstrated the applicability of the Importance Performance Analysis (IPA), as an evaluation technique that can be used to provide information and improve golf resort practices to better meet or exceed the needs of customers. This study also provided a prioritized list for quality service improvements that emphasized the need for golf organizations to recognize customer expectations and levels of satisfaction.

As the golfing industry is becoming more competitive, it is essential for golf resort managers to uncover tools and techniques for addressing quality service practices. In this study, IPA has shown to be a practical tool to be used by resorts management as it is comprehensive and easy to use. It does not involve difficult statistics to interpret. In addition, the IPA provided a means of addressing the need to exceed customer expectations. IPA offers the way to understand both agency performance and customer expectations in a meaningful way.

For future research, it is recommended that larger numbers of golf clubs and resorts, involving bigger sample size of respondents should be used to reflect the expectation of golf service attributes among golfers in the country. Comparison study of local and foreign golfers could be performed to identify the characteristics, preference and expectation on golf service attributes and satisfaction among golfers from different countries. Perhaps, this could present more information for different marketing strategies in the golf tourism industry. Finally, repetitive studies in the same industry should be conducted to ensure the consistency of result among golfers in the country.

References

1. Malaysia Tourism, (2012). *Tourism Malaysia welcomes tiger woods and top golfers to cimb classic 2012*. Tourism Malaysia: Communications & Publicity Division.
2. Ong, T. F., Wee, E. H., Hazali, M. G., & Kee, K. M. (2009). *Destination preferences and travel behavior among golf tourists in Malaysia.* Shah Alam: University of Technology MARA. (FRGS Report: 5/3/2097).

3. Kandampully, J., & Suhartanto, D. (2000). Customer loyalty in the hotel industry: The role of customer satisfaction and image. *International Journal of Contemporary Hospitality Management, 12*(6), 346–351.
4. O'Neill, M. (2001). *Service quality management in hospitality, tourism, and leisure, measuring service quality and customer satisfaction*. New York: The Haworth Hospitality Press.
5. Cline, R. S. (1997). The value of human capital. *Lodging Hospitality, 53*, 20–22.
6. Evenson, R. (2001). What is the best way to measure customer service? *Supervision, 62*(3), 14–16.
7. Guadagnolo, F. (1985). The importance-performance analysis: An evaluation and marketing tool. *Journal of Park and Recreation Administration, 3*(2), 13–22.
8. Ennew, C. T., Reed, G. V., & Binks, M. R. (1993). Importance-performance analysis and the measurement of service quality. *European Journal of Marketing, 27*(2), 61–77.
9. Martilla, J. A., & James, J. C. (1977). Importance-performance analysis. *Journal of Marketing, 41*(1), 77–79.
10. Nunnally, J. C. (1978). *Psychometric theory* (2nd ed.). New York: McGraw-Hill.
11. Hennessey, S., MacDonald, R., MacEachern, M., & Yun, D. (2007). The effects of perceived quality of performance on price-value, satisfaction, and behavioural intentions by golfers' resident types. In *2007 TTRA Annual Conference Proceedings.*
12. Kim, S. S., Chun, H., & Petrick, J. F. (2005). Positioning analysis of overseas golf tour destinations by Korean golf tourists. *Tourism Management, 26*, 905–917.
13. Clark, A. C., & Wood, R. C. (1998). Consumer loyalty in the restaurant industry—a preliminary exploration of the issues. *International Journal of Contemporary Hospitality Management, 10*(4), 139–144.
14. Mattila, A. S. (2001). Emotional bonding and restaurant loyalty. *The Cornell Hotel and Restaurant Administration Quarterly, 42*(6), 73–79.
15. Ryu, K., & Han, H. (2010) Influence of the quality of food, service, and physical environment on customer satisfaction and behavioral intention in quick-casual restaurants: moderating role of perceived price. *Journal of Hospitality and Tourism Research, 34*(3), 310–329.
16. Oliver, R.L. (1980). A cognitive model of the antecedents and consequences of satisfaction decisions. *Journal of Marketing Research, 17*, 460–469.

Travel Motivation and Points of Attachment Among Golf Spectators

Tah Fatt Ong and Siti Hannariah Mansor

Abstract Golf has emerged as a popular spectator sport since the twentieth century. This trend is becoming more revealing in Malaysia, as shown by the rise of International Golf Tournaments organized across the nation, in the recent decade. However, there has been little research examining the consumption behavior of those who attend golf tournaments across the country. The purpose of this study was to examine the travel motivation and points of attachment among golf spectators. Convenience sampling technique was used. Data were collected using self-administered questionnaires, from 132 spectators attending three golf tournaments across the Peninsula Malaysia. The Motivation Scale for Sport Consumption was adapted to measure seven different motives of golf spectators namely achievement, aesthetics, drama, physical skills, knowledge, social and escape. The points of attachment index (PAI) was utilized to measure four dimensions of 'identification of sport', 'specific golfer', 'level of sport' and 'the tour.' 'Aesthetics' was found to be the most important factor that motivates spectators to attend golf tournaments. 'Identification with sport' was the most important point of attachment that influences spectators in attending golf tournaments. The outcome of this study provided information that could assist golf tournament organizers/managers to design more effective planning, management and promotion strategies to achieve higher spectator rate to the organized tournaments.

Keywords Travel motivation · Points of attachment · Golf spectators

---

This paper is a result of research funded by Fundamental Research Grant Scheme (FRGS) awarded by Universiti Teknologi MARA, Malaysia.

T. F. Ong (✉) · S. H. Mansor
Faculty of Sports Science and Recreation, Universiti Teknologi MARA,
Shah Alam, Malaysia
e-mail: ong278@salam.uitm.edu.my

R. Adnan et al. (eds.), *Proceedings of the International Colloquium on Sports Science, Exercise, Engineering and Technology 2014 (ICoSSEET 2014)*,
DOI: 10.1007/978-981-287-107-7_58,

1 Introduction

Sport spectating is fast becoming a popular leisure activity all over the world. In fact, revenue or profits for spectator sport have steadily grown up in the USA for past few decades. In 2005, it has been estimated that the profit from spectator sport was approximately $24 billion, which is 3.5 % higher than the previous year and 26.6 % more than in 2000 [1]. It has also been estimated that people spent approximately $15.9 billion in 2005 for admission to professional and for amateur athletic events. Particularly, Robinson and Carpenter [2] stated that golf has emerged as a popular spectator sport during twentieth century in North America. Three major professional golf tours: The Professional Golf Association tour, The Ladies Professional Golf Association tour and the PGA Senior tour have drew over 10 million spectators and awarded over US$263 million in prize money to competitors in year of 2002.

Malaysia is rapidly emerging as Asia's golf heaven for many golfers in the region. Golf is a year round activity and a game that can be enjoyed at any time in Malaysia. Golf has been related to tourism and plays an important role in generating profit or revenue to the country's economy. Statistic from Malaysia Golf Association (MGA) [3] in calendar year of 2012 shows that 36 local golf tournaments and seven international golf events were being organized in Malaysia. Golfers from different countries have known to travel to Malaysia for golfing activities. Thus, there is a need to better understand the golf spectators market. The profile of golf spectators and the reasons for their travel are important information that can help in segmenting the market, so that appropriate services or travel packages can be provided.

Motivation is commonly described as those processes that arouse and instigate behavior, that give direction to behavior. Travel motivation can be looked as reasons that influence spectators to travel to watch sport tournaments. In relation to sport spectating motivation, researchers have identified numerous motives for spectator attendance, which include vicarious achievement, aesthetics, social interaction, drama or excitement, escape and role models. Several researchers have noted that the relationship between motivation performance and resultant satisfaction invites consideration of a number of variables [4, 5]. Some of the researchers perceived importance of an activity in terms of self-development, self-enhancement, ego, role fulfillment and responding to perceived requirements of significance others as important variables that determining motivation, behavior and desired satisfaction.

In examining the motives of sport spectators attending sport events, numerous researches have focused on the team sport environment. Researchers have examined the differences in motives of sport spectators by gender [6, 7], types of sport attended [7, 8], stadium size and location [9], race [10], age and economic factors [11], among others. However, there is a paucity of research specifically on examining the motives of attendees in golf tournaments, which is an individual sport event. As suggested by Robinson et al. [12], more attention and focus are needed to explore potential variables with which to segments consumer in the areas of spectators' motives and points of attachment.

Golfers are known to have different levels of skills and behavior. One of the ways to better understanding the differences among golfers is by the specialization concept. When a person participate in a given recreation activity, they are likely to become more specialized in that activity over time. Most researcher identified that more specialized user differ from less specialized users on numerous attributes. According to Ditton et al. [13], recreational specialization is an area of study that attempts to describe this variation through segmentation of participant into meaningful and identifiable subgroups.

Though, some researchers have investigated variety motives as important dimensions that influence people's decisions to spectating at sports events [14–16], other researchers have suggested that there may be other factors that influence sport consumer spectating behaviors and decision making. Kwon et al. [17] stated that some people watch sports events because they have strong social-psychological connection with a team, coach, player, university, community, level of sport and/or type of sport. All these connections were recognized as points of attachment. In relation to this, researchers found that certain spectators' motives are associated with certain points of attachment [6, 8] in some team sports. However, study on the different motives and points of attachment among spectators is still lacking, specifically in the context of individual sport such as golf. Thus, the present study aims to investigate: What characteristics do golf spectators possess? What are the motivation factors and point of attachment that influence the attendance of golf spectators to golf tournaments? And whether there are any differences between different segments of golf spectators identified by golfing specialization, in terms of motivation factors and point of attachment?

Thus, the present study aims to apply the recreation specialization concept to categorize the golf spectator with different specialization levels. It also intend to examine the different travel motivation and points of attachment exist among golf spectators who attended local and international tournaments.

2 Methodology

2.1 Design of the Study

This study involved a survey research design to identify different segments of golf spectators in the golf tourism market. Self-administered questionnaires were distributed and collected from spectators attending three golf tournaments. Respondents were met on-site and were kindly asked to participate in the survey voluntarily. The researchers were present throughout the session to answer any question from the respondents concerning the questionnaires. The time taken for each respondent to complete the inventory was about 10 min.

2.2 Selection of Subjects

Convenience sampling technique was used to collect data for this study. Data were collected using self-administered questionnaires, from spectators attending three different tournaments at different places in the country. A total of 150 questionnaires were distributed, 18 were spoiled. Thus, the total usable questionnaires were only 132. The age of respondents involved in this study were golf spectators above 18 years old.

2.3 Instrumentation

The questionnaire consisted four sections. The first section was designed to gather information related to travel motivation. The items were adapted from motivation scale for sport consumption (MSSC), developed by Trail and James [14]. There were seven subscales and each subscale comprises three items. Respondents were asked to clarify the importance of each item related to travel motivation on a 7-point Likert-type scale, which range from 1 (strongly disagree) to 7 (strongly agree). This section consists of 21 items. The second section, related to golfer specialization, comprises 5 items (3 items—behavioral dimension, and 2 items—cognitive dimension) adapted from previous studies by Ong et al. [18]. Section 3 was designed to collect data related to the points of attachment. The items were adapted from points of attachment index (PAI), developed by Robinson and Trail [8]. For this study, PAI was modified to include only 4 subscales, which are (1) the golfers, (2) the tour, (3) the sport and (4) the level of sport. Each subscale consists of three items and the total numbers of item were twelve (12). Each item uses the 7-point Likert-type response format ranging from 1 (strongly disagree) to 7 (strongly agree). The fourth section focuses on socio-demographic questions, in which respondents were asked to provide information regarding their gender, place of origin, marital status, age, education level, occupation and monthly income. A pilot study was conducted to test the validity and reliability of the survey instrument by distributing the questionnaire to 40 golfers at a local golf resort in the Klang Valley. Cronbach's alpha reliability coefficient for the three measurements, i.e., travel motivation ($\alpha = 0.80$), golf specialization ($\alpha = 0.81$) and points of attachment ($\alpha = 0.86$) were considered as acceptable, as they were better than 0.70.

2.4 Data Analysis

Data analysis was conducted utilizing the statistical package for social sciences (SPSS Ver. 17). For categorical and demographic variables of respondents, descriptive statistic consists of percentage, frequencies, mean and ranking was utilized to explain and describe the findings. The items in demographics variables (i.e., gender, age group, place of origin, marital status, occupation, income and education level) were summarized using the descriptive statistics. Meanwhile for travel motivation and points of attachment, cross-tabulations were used to profile the different segments of golf spectators.

3 Research Findings

3.1 Demographic Profile of Respondents

The demographic backgrounds of respondents were summarized in Table 1. The respondents were majority males, married, age from 18 to 50 years old and relatively affluent with monthly income of RM 3,001 and above. The respondents are mostly company employee, professional and businessman.

3.2 Golf Specialization Level of Golf Spectators

The respondents were categorized into three specialization levels. To classify golf tourists into their respective level of specialization (low, moderate and high), total mean scores in cognitive and behavior characteristics were calculated. The different level of golf specialization was classified according to the resulting range of mean score, such as low specialization (1.00–2.00), medium specialization (2.01–3.00) and high specialization (3.01–4.00). The result indicated that 40.9 % (54) of the respondents were categorized in the medium level of golfer specialization, followed by 31.1 % (41) of them at the low golfer specialization level, and 28.0 % (37) of the respondents were at the high golfer specialization level.

3.3 Travel Motivation Among Golf Spectators of Different Specialization Level

Travel motivation has seven dimensions which were named achievement, aesthetics, drama, escape, knowledge, physical skills and social. The mean score related to each dimensions were calculated and was ranked with reference to each specialization group of golf spectators. The results were tabulated as shown in Table 2.

In the overall ranking, 'aesthetics' dimension with mean, $m = 5.95$, was ranked the highest travel motivation among the golf spectators. Meanwhile 'knowledge'

Table 1 Demographic profile of respondents

Demographic characteristics	Frequency ($N = 132$)	Percentage
Gender		
Male	103	78.0
Female	29	22.0
Marital status		
Single	55	41.7
Married	77	58.3
Others	0	0
Place of origin		
Local	117	88.6
Foreign	15	11.4
Age		
18–30 years	22	16.7
31–40 years	47	35.6
41–50 years	33	25.0
51–60 years	24	18.2
>60 years	6	4.5
Education level		
Primary	3	2.3
Secondary	27	20.5
Diploma	22	16.7
Degree	80	60.6
Occupation		
Company employee	38	28.8
Businessman	29	22.0
Civil servant	2	1.5
Professional	32	24.2
Housewife	7	5.3
Student	16	12.1
Retired	8	6.1
Income		
RM 1,500 or less	9	6.8
RM 1,501–RM 3,000	32	24.2
RM 3,001–RM 4,500	37	28.0
RM 4,501–RM 6,000	16	12.1
Over RM 6,001	38	28.8

dimension was ranked second ($m = 5.92$) and was followed by dimension of 'physicals skills' ($m = 5.82$), 'drama' ($m = 5.66$), 'achievement' ($m = 5.44$), 'social' ($m = 5.43$) and the lowest ranked was the dimension of 'escape' ($m = 4.87$).

Both the medium and high specialization group ranked dimension of 'aesthetics' as the most important travel motivation to attend golf tournaments. For the low specialization group, the dimension of 'knowledge' was ranked the most important travel motivation to be considered, while dimension of 'escape' was ranked the least important for all the three specialization groups.

Table 2 Mean of travel motivation among different specialization level of golf spectators

Travel motivation	Low level		Medium level		High level		Overall	
	M	*R*	*M*	*R*	*M*	*R*	*M*	*R*
Achievement	5.27	5	5.49	5	5.56	6	5.44	5
Aesthetics	5.80	2	6.02	1	6.04	1	5.95	1
Drama	5.56	3	5.76	3	5.64	4	5.66	4
Escape	5.07	7	4.73	7	4.87	7	4.87	7
Knowledge	6.17	1	5.66	4	5.86	3	5.92	2
Physical skills	5.51	4	5.96	2	5.99	2	5.82	3
Social	5.20	6	5.48	6	5.62	5	5.43	6

Table 3 Mean of points of attachment among different specialization level of golf spectators

Point of attachment	Low level		Medium level		High level		Overall	
	M	*R*	*M*	*R*	*M*	*R*	*M*	*R*
Specific golfer	4.73	2	5.27	2	5.24	2	5.09	2
The tour	4.40	4	4.96	4	4.95	4	4.78	4
Identification with sport	4.98	1	5.53	1	5.65	1	5.35	1
Level of sport	4.72	3	5.09	3	4.98	3	4.94	3

M mean; *R* rank

3.4 Points of Attachment Among Golf Spectators of Different Specialization Level

In terms of points of attachment, there were four dimensions namely: 'specific golfer', 'the tour', 'identification with sport' and 'level of sport'. The mean score for each dimension was calculated and was ranked with reference to each specialization group of golf spectators as presented in Table 3.

As shown in Table 3, the highest ranked points of attachment was the dimension of 'identification with sport' ($m = 5.35$). Second highest rank was the dimension of 'specific golfer' ($m = 5.09$), followed by dimension of 'level of sport' ($m = 4.94$), and last ranked was dimension of 'the tour' ($m = 4.78$). All the three specializations groups have the same ranking for the four dimensions.

4 Conclusion

The findings concluded that golf spectators in the present study are mostly male, married, age from 18–50 years old and relatively affluent (monthly income of RM 3,001 and above). The golf spectators are mostly businessman, company employees and professionals. The demographic profile of golf spectators identified in the present study is rather consistence with other studies, in terms of golfer's characteristics [18, 19].

In terms of motivation to attend golf tournaments, 'aesthetics' dimension ($m = 5.95$) was ranked the highest factor. The present finding is consistent with Funk and colleagues [20], who noted in their study of soccer players in the Women's World Cup, that the motives of 'aesthetics' and 'excitement' were significantly correlated with interest in soccer. Although the sports are comparatively different, the present study revealed consistent characteristics among sport participants. This means that most of the golf spectators very much appreciate the exquisiteness strokes and enjoy the graceful movements associated with golf. 'Knowledge' dimension ($m = 5.92$) was ranked second. This showed that majority of golf spectators are interested in gaining more knowledge about golf. This would help to enhance and increase their understanding on various aspects of golf such as skills, the technical aspects and game strategies. In order to attract more spectators, tournament organizers can take the opportunity to offer golf seminars or courses to attendees, with the purpose to enhance their knowledge about golf during the tournament. This is one of the promotional strategies that organizer can capitalized on to motivate the golf spectators who are interested to increase their knowledge while attending golf tournaments.

With regards to points of attachment, 'identification with sport' ($m = 5.35$) was ranked the highest factor of point of attachment. Study on professional basketball by Gencer et al. [21] also highlighted 'identification with sport' is a significant factor considered by spectators in attending basketball tournaments. Thus, every sport has its own sport fans as supporters or spectators. The current study presented the common characteristics found in sport spectators that is most of the spectators are golfers who are fans of golf tournaments.

The second ranked points of attachment was 'specific golfer' ($m = 5.09$). The finding showed that majority of golf spectators has their own favorite players. They will attend the golf tournament when their favorite player participated in the tournament. For example, AT&T National's yearly attendance figure was increasing from 2007 until 2009. However, the number of spectators experience a drastic drop in year of 2010 (192,633) until 2011 (150,060). The main reason for the decline in spectator attendance was the absence of Tiger Wood in the tournaments. This reflected the important roles of 'specific golfer' to the tournament organizer. Thus, it is vital for tournament organizers to invite or include famous golf celebrity to compete in their tournaments, in order to boost spectator's attendance. In addition, it is interesting to note that all the three specialization groups have the same ranking for all the points of attachment factor. This means that spectators in all specialization levels have same sense of attachment and belongingness to the sport of golf. It was observed that the mean score for each point of attachment also increase as the specialization level increased.

From the findings of this study, several recommendations can be suggested. Since the sport spectators are likely to be male, married and age 18–50 years old, special family package ticket price or discount could be offered for spectators to bring their spouse or family members along so as to increase spectator's attendance. Realizing the importance of 'specific golfer', organizer of tournament also should highlight the well-known golfers during their event promotion. These would help in attracting more spectators to the tournament. In addition, organizing golf seminars or courses conducted by well-known players would be able to attract golf spectators to increase their knowledge about golf. This can be one of the promotion strategies that organizer can compose to motivate the golf spectators who are interested to increase their knowledge in the duration of golf tournament.

In conclusion, the current study presented valuable information regarding the demographic, travel motivation and point of attachment of golf spectators. This information may assist golf tournament organizers to better segment the golf spectator market and identifying prominent factors that can attract more spectators to future tournaments.

For future research, it is recommended that wider areas and more golf tournaments should be included to determine whether the finding of the present study is applicable to other areas in Malaysia such as Sabah and Sarawak. Other than the variables in the present study, other variable such as sponsorship should be included as variable that could influence the attendance of spectators. It is also recommended future research should use multi-language to involve more foreign spectators in order to get a better understanding of the study.

References

1. U.S. Census Bureau. (2008). The 2008 statistical abstract. Retrieved February 10, 2012, from http://www.census.gov/compendia/statab/2008edition.html
2. Robinson, M. J., & Carpenter, J. R. (2003). The day of the week's impact on selected socio-demographic and consumption pattern of spectators at a LPGA event. *Sport Marketing Quarterly, 11*, 242–247.
3. Malaysia Golf Association. Statistic of local and international golf tournaments. Retrieved July 13, from www.mgaonline.com.my
4. Dimanche, F., Havitz, M., & Howard, D. R. (1991). Testing the involvement profile (IP) scale in the context of selected recreational and touristic activities. *Journal of Leisure Research, 23*, 51–66.
5. Kopferer, J. N., & Laurent, G. (1993). Further evidence on the consumer involvement profile: Five antecedents of involvement. Psychology and Marketing, *10*, 347–355.
6. Fink, J. S., Trail, G. T., & Anderson, D. F. (2002). Environmental factors associated with spectator attendance and sport consumption behavior: Gender and team differences. *Sport Marketing Quarterly, 11*, 8–19.
7. James, J. D., & Ridinger, L. L. (2002). Female and male sports fan: A comparison of sport consumption motives. *Journal of Sport Behavior, 25*, 260–278.
8. Robinson, M. J., & Trail, G. T. (2003). Motives and points of attachment: Differences between college football, men's basketball, and women's basketball spectators. Manuscript submitted for publication.
9. Nakazawa, M., Mahony, D. F., Moorman, A. M., & Hirakawa, S. (2000). The relationship between stadium size and location and spectator characteristics: Implications for marketing strategies. *International Sport Journal, 4*, 9–25.

10. Wan, D. L., Bilyeu, J. K., Brennan, K., Osborn, H., & Gambouras, A. F. (1999). An exploratory investigation of the relationship between sport fans' motivation and race. *Perceptual and Motor Skills, 88*, 1081–1084.
11. Pan, D. W., Gabert, T. E., McGaugh, E. C., & Branvold, S. E. (1997). Factors and differential demographic effects on purchases of season tickets for intercollegiate basketball games. *Journal of Sport Behavior, 20*, 447–463.
12. Robinson, M. J., Trail, G. T., & Kwon, H. (2004). Motives and points of attachment of professional golf spectators. *Sport Management Review, 7*, 167–192.
13. Ditton, R. B., Loomis, D. K., & Choi, S. (1992). Recreation specialization: reconceptualization from a social worlds perspective. *Journal of Leisure Research, 24*, 33–51.
14. Trail, G. T., & James, J. D. (2001). The motivation scale for sport consumption: Assessment of the scale's psychometric properties. *Journal of Sport Behavior, 24*(1), 108–127.
15. Trail, G. T., Fink, J. S., & Anderson, D. F. (2003). Sport spectator consumption behavior. *Sport Marketing Quarterly, 12*, 8–16.
16. Won, J. U., & Kitamura, K. (2007). Comparative analysis of sport consumer motivations between South Korea and Japan. *Sport Marketing Quarterly, 16*(2), 93–105.
17. Kwon, H. H., Trail, G. T., & Anderson, D. S. (2005). Are multiple points of attachments necessary to predict cognitive, affective, conative, or behavioral loyalty? *Sport Management Review, 8*, 25–270.
18. Ong, T. F., Wee, E. H., Ghazali, M., & Kee, K. M. (2009). *Destination preferences and travel behavior among golf tourists in Malaysia*. Shah Alam: University Teknologi MARA. (FRGS Report: 5/3/2097).
19. Scarborough research: "Men's golf demographics," Street and Smith's Sports Business Daily. Retrieved April 14, 2008 from http://www.sportsbusinesadaily.com/article/108393
20. Funk, D. C., Mahony, D. F., & Ridinger, L. L. (2002). Characterizing consumer motivation as individual difference factors: Augmenting the sport interest inventory (SII) to explain level of sport. Sport Marketing Quarterly, *11*, 33–43.
21. Gencer, T. R., Kiremitci, O., & Boyacioglu, H. (2011). Spectator motive and point of attachment: an investigation on professional basketball. *Journal of Human Kinetics, 30*, 189–196.

Inclusive Outdoor Recreation: Transformation of the Social Acceptance and Outdoor Experience of Person with Disabilities

Rezian-na Muhammed Kassim, Hisyam Che Mat, Norasrudin Sulaiman, Nagoor Meera Abdullah, Rozita Abdul Latif and Mohamad Rahizam Abdul Rahim

Abstract This paper aims to give persons with and without disabilities an opportunity to share their experiences in an inclusive outdoor recreation (IOR) program, thus to expand the knowledge of understanding their emotion and identifying the outcomes. This study employed a qualitative design to respective sample n: 20 with disabled and n: 10 without disabled people in inclusionary outdoor programs. People without disabilities transform their social communications and developed better perceptive toward IOR program. Persons with disabilities have been positively accepted by the society and this increased their personal growth.

Keywords Social acceptance · Persons with disabilities · Inclusive outdoor recreation

1 Introduction

Although outdoor recreation has been in existence for many years, programming for persons with disabilities is new in Malaysia. This is due to the awareness of creating recreation program and facilities. People with disabilities are frequently excluded in many of community recreation programs. People always assume that persons with disabilities reluctant to take part in outdoor or adventure recreation program because these activities are considered not safe, risky, and out their domain.

Furthermore, the perception of natural outdoor environment is too inaccessible and impossible to be achieved [1]. Nevertheless, studies show that people with

R. Muhammed Kassim (✉) · H. Che Mat
Sports Tourism and Recreation Lab, Universiti Teknologi MARA, Shah Alam, Malaysia
e-mail: rezian@salam.uitm.edu.my

R. Muhammed Kassim · H. Che Mat · N. Sulaiman · N. M. Abdullah · R. Abdul Latif · M. R. Abdul Rahim
Faculty of Sport Science and Recreation, Universiti Teknologi MARA, Shah Alam, Malaysia

R. Adnan et al. (eds.), *Proceedings of the International Colloquium on Sports Science, Exercise, Engineering and Technology 2014 (ICoSSEET 2014)*,
DOI: 10.1007/978-981-287-107-7_59,

disability may participate and perform just like people without disabilities if the society accepts them as normal people [1, 2]. Most of them are likely to participate in leisure activities such as sightseeing, picnicking, bird watching, camping, and hunting. The sad back is only few of the disabled people likely participate in more challengers' water activities such as kayaking or canoeing due to risk and challenging [3].

The normal inclusion programs for disabled person served the general population and occurred in areas such as Public Park, school, and working environment. Conversely, they are lacking areas of organized camp and outdoor program for the disabilities person [4]. The outdoor program normally offers a bundle of benefits to everybody involved including people with disabilities. Some researchers had indicated that persons with disabilities are attracted to outdoor environment but reveals that they do not significantly influence the participation rates in outdoor recreation [4, 5].

Still, person with disabilities favor the same benefits of outdoor recreation activities that the person without disabilities enjoyed [4]. Benefits include enhancing self-concept, self-esteem, and increased personal growth [4–6]. Besides that, social contacts between individual with disabilities and the individual without disabilities will improve attitudes, increase empathetic, develop acceptance, and improve social development [7]. Of late, the social acceptance has been acknowledged as the foundation for friendship development [8].

Some studies [6] noted that the lack of social acceptance of people with disabilities has limited their inclusion in society. This is because person with disabilities mostly are those with visibly obvious impairments perceived as not been accepted by the public and tend to avoid inclusive leisure services [5].

Empirical inclusion study found that lack of social acceptance is due to unequal treatment toward the persons with disabilities. In other study, which contradicts with pervious study, it reveals that persons with disabilities have a significant relationship of social acceptance by persons without disabilities [8]. Which conclude that, the more or frequent a person with disabilities participated in an inclusive leisure programs with their non-disabled peers may be connected to levels of perceived social acceptance [8].

Thus, the main aim of this study is to describe and observe the transformation of social acceptance and outdoor experience of persons with disabilities and without disabilities in an inclusive outdoor recreation (IOR) program setting.

2 Methodology

2.1 Setting

This study focused on an inclusive outdoor adventure program organized by Faculty of Sport Science and Recreation, Universiti Teknologi MARA, Malaysia. The program is located at Shah Alam Lakes, Selangor, Malaysia. This program was aimed to expose people with disabilities about inclusive outdoor program mainly looking into the social acceptance factors. With guidance from outdoor recreation lecturer and n: 10 students without disabilities, the participants n: 20 (with disabilities) and a group of numerators were exposed to activities such as canoeing and kayaking for a day program.

2.2 Sample

Twenty persons with disabilities voluntarily participated in the study. There were males ($n = 10$) and females ($n = 10$) who had physical impairment such as being wheelchair bound, amputees, and les autres (locomotor conditions) that was visibly obvious to the researcher upon observation during the outdoor recreation program, and this is a technique used successfully in leisure programs by Devine and Dattilo [5]. Their age ranges from 20 to 40 years old.

2.3 Data Collection Procedure

This study used qualitative data collection procedures (19). This method seemed to be the most appropriate way in understanding the actual experience of the persons with disabilities. A convenience sampling and the snowball approach to sampling were adopted (Patton, 199). The characteristics of central importance of this included (a) respondents who have a physical impairment such as being wheelchair bound, amputees, and les autres (locomotor conditions); (b) individuals who enrolled and participated in the IOR programs; (c) individuals between the age of 18 and 40 as individuals in this age range are most likely to participate in inclusive outdoor programs and able to articulate their experience; and (d) individuals who could express perceptions and discuss their experience relative to IOR program.

Data sources were primarily from the responses of the respondents. Face-to-face interview and structure questions were employed. To accomplish this study, all the transcripts and observation data concerning their experience were systematically recorded and analyzed. For this reason, no statistical comparisons were analyzed. Unstructured questionnaires that addressed the facets of outdoor education experience such as perception of social acceptance, attitudes; social acceptance, participations, satisfaction, and meaning were given to the numerators. The major research questions posed for this study were delineated as below.

RQ1 What are the observations of social acceptance of persons with disabilities joining in the inclusive outdoor programs?
RQ2 Is there any interactive process of accepting before and after the program?
RQ3 How does social acceptance affect the integrated inclusive outdoor program experiences of persons with disabilities?

The primary instrument in this study relies on unstructured interviews for data collection. Most importantly, the success in collecting this data relies on the skill of the numerators. Therefore, the numerators have been trained beforehand. The face-to-face interviews took around 15–45 min. To comprehend fully about the phenomena, dialect observation of interest was as established. The numerators then spend time with the participants and record observation in the forms of detailed field notes.

The analysis of data comprises of three link of subprocess, data reduction, conclude data display, and conclusion picture; these procedures took place before data collection, during the planning of design and during data collection. Phenomena of this study were identified, classified, and compared across categories. Therefore, it is important to generate the categories of themes and analyzed the transcripts. Then, data or information were grouped into general categories in nature and informed the research questions. The categories have then been organized into a central theme, which was of greater generality and reflect common phenomena. All of the interview transcripts were read by the numerators and coded by the researchers in the style of a phenomenology approach [8].

3 Results

The results were analyzed in synchrony with the research questions posed for the study.

3.1 Research Question 1

The observation of social acceptance, use of social statement, means of communicating with friends and publics build friendship bonds either intragroup or intergroup. A number of respondents found that the learning and coping with the IOR program positively influenced their perceptions of social acceptance, and other participants feel that the social acceptance is still lacking.

3.2 Lack of Social Acceptance

Respondent 1, a handicapped (21 years old) female, suffers from moderate mental retardation. She did not involve herself much with other. She is very quiet and very reserved and behaved rather judgmental toward social acceptance with others. Meanwhile, respondent 2 (age 38) felt that the social acceptance is still lacking due to her thinking of anxiety and lack of confidence toward outdoor activities. She felt that she had to struggle when doing the activity compare to normal people therefore she preferred to be by herself. There are at least 4 respondents who felt that there is a gap in terms of social acceptance unrelated to physical skills. Respondent 3 (age 32) who is a house wife with physical impairments stated that she is so shy because of her disability. “I had tried this activity once but lack of confidence with my disability to pursuit this and felt that normal people won’t accept me.” Later, her perception changed when she did the second time with the help of four volunteers.

Some of the verbal communication that we heard between respondent 4 and respondent 5 were: “That they don’t feel good when a group of youngsters (public) around the lake was watching them and make a joke about their disability.” It really hurts their emotions. The significance of lacking social acceptance among them was represented by the delay in communication and building the social relationship. A perceived lack of acceptance by persons with disabilities is consistent with previous studies on social acceptance. Found that negative meanings of disability were particularly prevalent in environments that required physically active leisure skills.

3.3 Social Acceptance

Social acceptance in this study was conceptualized by the interaction with others and felt of acceptance and group belonging in certain environment or situation influenced the person. A majority of participants with and without disabilities perceived themselves to be physically, socially, and psychologically accepted by others in the integrated outdoor education program.

Respondent 4 (age 34), with physical impairments that used wheelchair placed himself in contexts where they can develop social relationships around shared interests.

He felt that without these opportunities for social inclusion in the recreation activity, the range of sport choice is limited. He wrote: “I was very lucky because I have a group of friends with or without disability who are actively involved in the recreation program especially kayaking. To me the social acceptance is not a big issue as long as we could communicate with other.”

During the interview, data were transcripted into the reflective journal, which was obtained from both formal and informal conversation with participants.

Through the researcher’s reflective journal, respondent 5, a 24-year-old male (Chinese) with a locomotor disability (cerebral palsy) felt very happy and enjoyed the program. He can get along with the participants and the volunteers. He has no problem communicating and expressing his feeling with others.

The only Indian participation, respondent 6, a 24-year-old male with physical impairments (neurological dysfunction) felt that he was accepted by the group because of his attitude that like to help others with his incapability by motivating others to join the water-confident activities where some of the participations were not confidents. He stated that “we can do it, show to others.”

Respondent 8, the younger aged 20: "I'm extrovert and I can talk with others and confident to do this activity. My parent told me that I have hands but not fingers; therefore I am as normal people that can do what others can. I felt accepted by others. Thanks to Allah (god) for the blessing."

Study reported that disabled persons who participate in outdoor recreation settings have increased social contact with people with and without disabilities [9]. In other study [10], found out a significant degree of social acceptance and positive perception of disability when teens involved inclusive activities that featured common interests and built social relationships.

3.4 Research Question 2

The construction of acceptance is an interactive process of accepting the outdoor program before and after. The focus of this investigation was related to question "yes" and "no" before and after the program. Few observation crucial factor of interaction has been observed in terms of language and behavior. Most of the participant perceived that the outdoor recreation program said "No" (70 %) before the program starts (due to the feeling of not welcome). When the program ends, the actual feeling of fun and happy and actively involved has changed their mindset. Most of them wanted to join again (93 %).

3.5 Research Question 3

Part of the neutral of this study was to observe the participants' roles that they had played which directly influenced their outdoor activities experiences. In this study, the model was adapted from Che Mat [11] that was conceptualized as individuals connected participation in integrated outdoor recreation environments to social acceptance (Fig. 1).

This category well-versed a relationship between social acceptance by persons with and without disabilities and outdoor frequency, friendship development, acceptance of differences, intentions, social interaction, and spirituality for participation in other integrated.

In this study, a relationship appeared between construction of social acceptance and acceptance of differences in the way in which outdoor recreation skills were carried out. This takes place when those participants without disabilities viewed adaptations as just another way to engage the activities and exposed a high level of social acceptance. Both groups positively viewed their integrated experiences, while participants without disabilities had improved value of social acceptance and bonding relationship.

Respondent 9, a 24-year-old Universiti Teknologi MARA who pursuits her study in the outdoor recreation areas, had no disability. She so astonishes with the

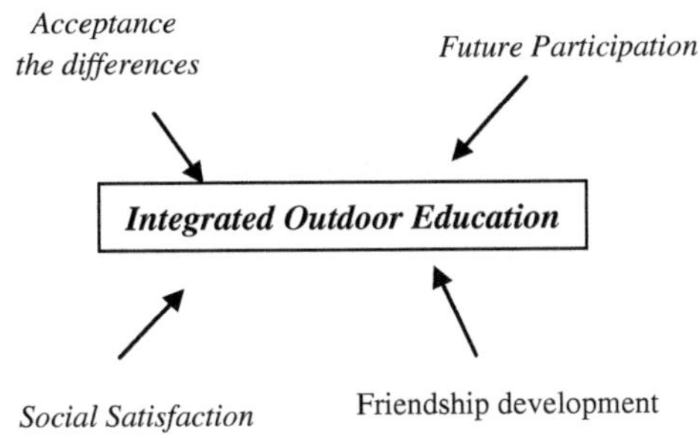

Fig. 1 Social acceptance model (adopted from Che Mat [11])

capabilities of the disabled participants and her opinion was: "They shared outdoor recreation experience; it is not only for us normal people but also for them to have the right to enjoy and socialize. I learned to communicate and not afraid because they can do lots of things. They are amazing and the most importantly friendship was developed." Our first impression of the observation (before the program started) is most of the disabled participant's acceptances of difference were very high not because of their physical skills but more of the social integration. There were few participants who wrote about the lack of social acceptance.

Respondent 10, 34-year-old male with physical impairments stated: "At first, I'm too timid and I couldn't discuss about the similar issues (canoe) with normal people did. I'm also not sure what to talk in front of other people but my perception was wrong and they accepted me with good hospitality." Respondent 11, a 23-year-old female UiTM student without disabilities also felt the same. "At first, I thought that they were very sensitive and do not want to accept us and when I started to talk with them and the acceptance changed and we created good bonding."

In the indication of respondent 12, person challenged with physical and developmental disabilities with the strong desire, Annie without disables can create bonding relationship the obstacles can be seemly whelming.

Friendship had different meanings between those with and without disabilities. Intrinsic worth that they considered significant in sharing similar interests, caring, helpful, kind, and most important accepting of individual differences. One of the participants that catch our attention was respondent 13, 23 years old, with disability. He was quite worried and kept wondering but after the water-confident activities has changed his perception and to tolerate with others. Respondent 14, a 23-year-old male participant without disabilities, shares his new experience.

He said: "This event gave me a new paradigm of looking at them—I learn a lot in handling their emotions and physical—I appreciated them a lot."

Respondent 15 (25 years old with disability) described that friendship build during the program as, "they like me and care. They also do not tease me about my sight; a number of participants spoke of the importance of friends development and been accepted. This study revealed that inclusion recreation activity is more than a feel-good experience supported by positive friendship development and supported by research [11] stated that inclusion enhances people with disabilities' capability to formulate and uphold friendships with normal people."

The social satisfaction been measured by the formation of both groups sharing their sense of enjoyment, increased sense of personal confidence, lowered level of stress, and decreased social isolation. Mostly, this finding showed that most of the participants with and without disabilities noticed that they were socially accepted and satisfied with the program.

For instance, Respondent 16, a 21-year-old participant with disability wrote: "I am happy and enjoyed every moment. I will share my experience with my family. They must be proud of me. I love kayaking."

Another female (respondent 17) participant without disabilities, who was attached to respondent 16, stated: I'm very happy because respondents 16 participates well in the event and successfully manages to help in doing outdoor activities. Respondent 17 supported the statement by saying "Yes, I successfully persuade her too (respondent 16) until she has the confidence to participate the activities." In particular, some studies have speculated that disability is most likely to take on negative meanings of social satisfaction, more socially contexts [4], but this study revealed that inclusive outdoor program changed the perception. The social acceptance of people with disabilities reported significant relationship of social satisfaction [5].

3.6 Question Such as: "Will You Like to Join Again?"

Precisely, the greater their sense of social acceptance, the greater their intention to participate in future integrated IOR programs is. When the participants were asked whether they intended to participate in other integrated outdoor education programs, most of them agreed to participate in future programs. "We really like outdoor activities especially the water confidence and kayak session."

Most of them felt that they can do things as normal people and enormously will participate in more adventure recreation program. They also nodded their heads agreed with the statement.

4 Conclusion and Recommendation

This study expands an understanding of the relationship between social acceptance and outdoor education experiences of persons with and without disabilities in the inclusive recreation program. Findings provide insight into integrated outdoor education experiences of persons with and without disabilities by focusing on their role in constructing social components of their participation. Closing this discussion, it was very rewarding in realizing that both groups had experienced their "first-ever" IOR program. Most of them had a positive experience that lead to future intention.

It is important to increase attention in building trust, personal growth, and confidence that inclusive recreation outdoor recreation program have positive value of sharing, caring, and social acceptance between them. It is also important for recreation providers to assure more program, support, and commitment involving the nation and decrease social isolation experienced by many persons with disabilities.

In the future, it will be valuable to propose and execute a longitudinal study to see whether and how inclusive outdoor program of acceptance with wider setting of the nation involvement is. In addition, future study should address the following questions: "What is the support needed for successful participation between family who have disabled children in outdoor recreation setting and why it is important?" Limitation of these results may not represent the general population of people with disabilities. In addition, the qualitative method may create threat to the validity of the study.

Acknowledgments The authors would like to thank the Ministry of Education of Malaysia and Research Management Institute (RMI) for providing fund. Our gratitude to Faculty of Sport Science and Recreation, Universiti Teknologi MARA (UiTM), Shah Alam, Selangor, for providing support and facilitating this research.

References

1. Allen, B., & Allen, S. (1995). The process of a social construction of mental retardation: Toward value-based interaction. *The Journal of the Association for Persons with Severe Handicaps, 20*, 158–160.
2. Cordell, H. K., Tarrant, M. A., & Green, G. T. (2003). Is the public viewpoint of wilderness shifting? *International Journal of Wilderness, 9*(2), 27–32.
3. Robb, G. M., & Ewert, A. (1987). Risk recreation and persons with disabilities. *Therapeutic Recreation Journal, 21*(1), 58–69.
4. Brown, T. J., Kaplan, R., & Quaderer, G. (1999). Beyond accessibility: Preference for natural areas. *Therapeutic Recreation Journal, 33*(3), 209–221.
5. Devine, M. A., & Dattilo, J. (2000). The relationship between social acceptance and leisure lifestyles of people with disabilities. *Therapeutic Recreation Journal, 34*, 306–322.
6. McAvoy, L. H., & Schleien, S. J. (1988). Effects of integrated interpretive programs on persons with and without disabilities. In L. A. Beck (Ed.), Research in interpretation: Proceedings of the 1988 National Association of Interpretation Research Symposium (pp. 13–26). San Diego: Institute for Leisure Behavior, San Diego State University.
7. Dattilo, J., & Murphy, W. D. (1987). Facilitating the challenge in adventure recreation for persons with disabilities. *Therapeutic Recreation Journal, 21*(3), 14–21.
8. Schleien, S. J., & Ray, M. T. (1988). *Community recreation and persons with disabilities: Strategies for integration*. Baltimore: Paul H. Brookes.
9. Sugerman, D. (1996). Diversity: Including people with disabilities in outdoor adventure programs. *Parks and Recreation, 31*(8), 44–47.
10. Sugerman, D. (1993). The impact of the Americans with Disabilities Act on adventure education programs. *The Journal of Experiential Education, 16*(1), 28–30.
11. Che Mat, H. (2010). Integrated outdoor education in Malaysia: Reflections of persons with and without disabilities. Malaysian Journal of Sport Science and Recreation, *9*(2), 25–54.

The Construction of Women Position in Sport: A Textual Analysis of the Articles and Images on Female Athletes in Malaysia Toward Two National Dailies Newspapers During 26th Sea Games 2011

Sarimah Ismail, Siti Amirah Amiruddin, Vincent Parnabas, Norlizah Abdul Hamid and Nagoor Meera Abdullah

Abstract Sport media is seldom covering women's athletes by excluding them from sports news. Yet, female athletes often to compare between male athlete's ability and hegemonic powers during sport media coverage. This study identifies how national printed media in Malaysia (New Straits Times and Berita Harian) provide coverage to female athlete's performance and also analyze construction of empowering positions and disempowering positions for women especially in newspapers. Methods: a content analysis is used to determine the descriptor in the newspaper's content of both newspapers based on articles and images in sport sections during 26th Sea Games 2011 at Indonesia. Kian [2] (Newspaper Sport Journal 29(3), 2008) coding system was used to analyze the articles and Sport Illustrated Content (1997–2000) coding system for images. Result: both of newspapers show the different angle of view for women athletes where the highest numbers of articles published by Berita Harian focusing on psychological strengths and emotional strengths while New Straits Times is focusing on the positive skills level and accomplishment. However, both newspapers give a lowest coverage of articles, which was focusing on athletic weaknesses and limitations. For images, NST and Berita Harian focused on the same code of portrayal for nonsport setting images rather than athletic action.

Keywords Women position · Sport · Printed media · Articles · Images

S. Ismail (✉) · S. A. Amiruddin · V. Parnabas · N. Abdul Hamid · N. M. Abdullah
Faculty of Sport Science and Recreation, Universiti Teknologi MARA,
Shah Alam, Selangor, Malaysia
e-mail: sarim375@salam.uitm.edu.my

R. Adnan et al. (eds.), *Proceedings of the International Colloquium on Sports Science, Exercise, Engineering and Technology 2014 (ICoSSEET 2014)*,
DOI: 10.1007/978-981-287-107-7_60,

1 Introduction

Media play an important role in spreading out of information. The effect of media in our lives is highly influenced where people are easily trusted to the dissemination of news. Media can also be an effective perfect medium by applying the modes of communications to reach out information using the channels (radio, television, films and printed words and image). Yet, media influence the process of socialization and forming ideology and thinking by shaping public opinion, personal beliefs and even people's self-perceptions. Eagleman and Pederson [1] described the sport and the mass media have a strong and mutually beneficial in helping the media to sell newspapers, magazines and build the perception of the public as well be used as a medium to create awareness and promote the sport through coverage of stories and pictures published. Neither printed media nor electronic media have their own works to describe and give publicity to athletes during tournaments and competitions by disseminating news through their masculinity, femininity, athlete's achievement and failures as well. In a certain situation, female athletes often to compare between male athlete's ability and hegemonic powers during sport media coverage. Sports reporter gives a sport coverage of athlete by trivialized, minimized their accomplishment and achievement as well as described them as a sex object.

Most of the studies found gender bias and reinforced gender-specific stereotypes were happening in printed media since two decades of centuries. Usually, male athletes were often to get complimented for their power and athleticism. Instead, sport media keep on downgrading skill level and the accomplishments of women athletes Kian [2]. Yet, Fondazione et al. [3] explained gender stereotyping promotes an asymmetrical vision of women and men in society and the unequal distribution of power and resources among them in all spheres and at all stages of life.

So far in Malaysia, there is a limited specific coverage for women's sports competitions provided by the media because the organizations of women's sports are not fully developed compared with other countries, which extremely organizes and develops their young and talented athletes based on implementation of speciality competition for women. It supported by coverage during women's college basketball in the USA where gets highly implication during media give full coverage to the tournament Kian [2]. Others, Lumpkin [4] from the research which reviewed Sport Illustrated also agreed, mass media in women's sport coverage do not reflect the development of women's popularity and achievements. They tend to "symbolically annihilate" the female athletes.

Media in Malaysia basically work closely connected to government objectives and were binding with the policies as well. The roles of mass media are not only required to inform, educate and motivate the masses toward development goals stipulated by the government, but are also expected to go along hand in hand with the government's policies in order to remain viable and survive in the journalism business [5].

Malaysia is a diversity of ethnics, multiracial and unique with several cultures. The mass media are an essential part of today's social life because the media create and transmit important cultural information. It influences how mass media especially printed media communicated to the citizens and be equal in disseminating news when a women be a priority in their stories because the uniqueness of Malaysia also includes many traditional and norms where mass media have to considered. So, the news can reflect the society and government operations by touching the legislative, economic and sociocultural methods in order to the sake of national development, stability and identity.

Basically, in action of male athlete's pictures are differently from women sport media articles and images. Most of female's portrayal and the articles were described by readers as a private matter, focusing on physical appearance rather than spot their career development in sport position.

By identifying the newspaper coverage of the women's sport in Malaysia, researcher will be able to examine whether the sports media has accepted the popularity of women's sports by empowering women athlete's position or disempowering women's athlete's position with using femininity athletes as in place for their prior women stories. This study investigates and identifying problems that are facing by women athletes from their description of news, which published by these national daily newspapers (New Straits Times and Berita Harian).

2 Methodology

2.1 Setting

The textual and images from Berita Harian and New Straits Times through the sport coverage along 6-week period toward female athletes participation in the games, the differences of textual and images from Berita Harian and New Straits Times through the sport coverage and the construction of empowering positions and disempowering position for female athletes by textual analysis of articles and images along 6-week period during 26th Sea Games 2011 at Indonesia were analyzed.

2.1.1 Data Collection Procedure

The major data collections were based on the objective of this study which were as follows:

1. To measure the textual and images from Berita Harian and New Straits Times through the sports coverage along 6-week period toward female athletes' participation during 26th Sea Games 2011 at Indonesia.

2. To investigate the differences of textual and images from Berita Harian and New Straits Times through the sport coverage toward women athletes' participation of 26th Sea Games 2011 at Indonesia.
3. To measure the construction of empowering positions and disempowering position for female athletes by textual analysis of articles and images along 6-week period during 26th Sea Games 2011 at Indonesia.

A content analysis is used to determine the descriptor in newspaper's content where analyzing was based on textual analysis of articles and images in sport sections. Two most influential Malaysia's newspapers under NSTP Group Press which were Berita Harian and News Strait Times were monitored for six-week period to investigate the coverage toward female athlete's news during Sea Games 2011 at Indonesia.

To identify the empowering position and disempowering position in sport by media coverage, Kian [3] coding system was used to analyze the articles which includes nine categories of codes which are as follows:(i) physical appearances, sexuality and attire; (ii) athletic prowess, strengths; (iii) athletic weaknesses, limitations; (iv) positive skill level, accomplishments; (v) negatives skill level, failures; (vi) family roles, personal relationships; (vii) psychological strengths, emotional strengths; (viii) psychological weaknesses, emotional weaknesses; and (ix) humor. For image descriptions, data adapted from Sport Illustrated content from 1997 to 2000. The coding of photograph divided into four categories. The categories are (i) athletic action, (ii) dressed but poised and pretty, (iii) nonsport setting and (iv) pornographic/sex-inappropriate. Analysis based on the total amount and percentage of articles and images published by newspapers.

2.2 *Sample*

The two national daily newspapers, namely Berita Harian and News Strait Times which were published for 6 weeks during 26th Sea Games, were choose for the sample of this research.

3 Results

The results were analyzed in synchrony with the research questions posed for the study.

The first objective of this study was analyzed by measuring a textual and images from both newspapers on women athletes' participation during 26th Sea Games 2011 at Indonesia. It indicates that total articles published by both newspapers are 51.1 % ($n = 114$) compared with images 48.9 % ($n = 109$). During the competition, these newspapers more published a stories rather than pictures (Table 1).

Table 1 Total of articles and images

Newspapers	Categories	*n*	Percentage
NST and Berita Harian	Articles	114	51.1
	Images	109	48.9

Table 2 Focus for women's articles and images

Categories	Newspapers				Differences	
	New straits times		Berita Harian		*n*	Percentage
	n	Percentage	*n*	Percentage		
Articles	58	50.9	56	49.1	2	1.8
Images	36	33.1	73	66.9	37	33.8

The result in Table 2 answered the second objectives by identified the differences of total articles and images between New Straits Times (NST) and Berita Harian. NST was published 50.9 % ($n = 58$) articles that are higher than Berita Harian which is 49.1 % ($n = 56$) articles. It has a distinguish views when Berita Harian more portrayed images than a story, which is 66.9 % ($n = 73$) compared with 33.1 % ($n = 36$).

The analysis also shown only 2 articles that set a number of stories published by NST than Berita Harian about 1.8 % compared with a large number of images occurred by 33.8 %.

The result in Table 3 answered the third objectives in this studied by measuring the construction of empowering position and disempowering position for female athletes by textual analysis of articles. The coding was used to analyze the constructing women athletes in sport by printed media coverage. The highest numbers of articles published by Berita Harian focusing on psychological strengths and emotional strengths 57.7 % ($n = 15$) while New Straits Times is focusing on positive skills level and accomplishment 66.7 % (24). However, both newspapers give a lowest coverage of articles which was focusing on athletic weaknesses and limitations, which NST is 66.7 % ($n = 2$) and Berita Harian 33.3 % ($n = 1$). Other, both of newspapers do not allocate any coverage for articles published about physical appearances, sexuality, attire and humor. The result also shown psychological weaknesses, emotional weaknesses and family roles; personal relationships get the balanced coverage by both newspapers.

The result in Table 4 shows the code used for images. The result indicates that most of the images portrayed by photographer for NST and Berita Harian have a same code of portrayal for nonsetting images, which is each of newspapers presented about 28.8 % ($n = 19$) for NST and 71.2 % ($n = 47$) for Berita Harian rather than athletic action where NST 39.5 % ($n = 17$) and Berita Harian 60.5 % ($n = 26$). Both of the results for dressed but poised and pretty and pornographic/sex-inappropriate do not portray any pictures by NST and Berita Harian.

Table 3 Total coded articles

Category/code	Newspaper			
	News straits times		Berita Harian	
	n	Percentage	*n*	Percentage
Physical appearances, sexuality, attire	0	0	0	0
Athletic prowess	5	31.3	11	68.7
Athletic weaknesses, limitation	2	66.7	1	33.3
Positive skills level, accomplishments	24	66.7	12	33.3
Negative skill level, failures	7	46.7	8	53.3
Family roles, personal, relationships	4	50.0	4	50.0
Psychological strengths, emotional strengths	11	42.3	15	57.7
Psychological weaknesses, emotional weaknesses	5	50.0	5	50.0
Humor	0	0	0	0

Table 4 Total coded for images

Category/code	Newspaper			
	News straits times		Berita Harian	
	n	Percentage	*n*	Percentage
Athletic action	17	39.5	26	60.5
Dressed but poised and pretty	0	0	0	0
Nonsport setting	19	28.8	47	71.2
Pornographic/sex-inappropriate	0	0	0	0

4 Discussion

The purpose of this study was discussed the finding based on three objectives stated where it were used to seek and explore the coverage toward women athletes by printed media in Malaysia. Even though both of newspapers under the same roof and publication, but they had difference angel of views to focus on and the way they allocating some coverage. It happened because each newspaper has a difference of management, editor, editorial section and also manpower (journalists and photographers). It proved when both newspapers published a vary stories and images in a same date of competition. For example, by a date of November 16, 2011, NST was focused a story about fencing's athlete who won bronze medal rather than Berita Harian focused on athletic athletes who received a new record for her pole vault.

For images, by a date of November 8, 2011, NST was published a picture of female self-defense art's contingent arrived at Jakarta Airport and welcomed by Indonesia ushered compared with Berita Harian, which portrayed a picture of female athlete holding a rifle. It indicates each newspaper in Malaysia has their own priorities and view to publish a stories and pictures in order to grab reader's interest and attention whereby to increase the sales itself. It was supported by Kian [6] who said the structure and foundation of these attitudes must be altered among editors and writers before women's sports start receiving substantially more coverage and better quality coverage in daily newspapers.

Most of the stories and pictures published at the sport section in both newspapers are helpful in constructing empowering woman position and career. Even though both newspapers had difference angle of views but it still give a positive storyline and captured the good images as well. It contradict to previous researcher studied by Shaller [7] found print media has a lack portrayal of female athletes, and when a woman is depicted, she is more likely to be in a socially acceptable or nonactive situation.

Yet, media in Malaysia give a motivation to athletes in developing their career by published and portrayed the positive images and stories where most of the articles publish regarding athlete's psychological and emotional strength as well as positive skills level and accomplishment. It contradicts by Allison and Shirley [8] from their studied revealed women are significantly under-represented in athlete performance, coaching positions, management of sport teams and as referees and umpired. Moreover, women athletes are always under-represented by sport media where they are often trivialized in received coverage and also by amount of coverage that they receive Eagleman and Pedersen [1]. Media often and more focused to describe women personal life rather than come out with their achievement and using a sense of humor to elaborate female athletes' career Kian [2] but in Malaysia, humor and physical appearance, sexuality and attire do not get any coverage toward printed media in Malaysia. It is because editor, writers and photographers in Malaysia still bind with journalistic ethics. It proved through this study when the news published by journalist mostly wrote stories positively.

In Malaysia situation, media members are looked to the prescribed norms and upheld as a Malaysian. It also must maintain the decency to respect people cultures and religions. It proved that when the photographers not even portrayed and published images related to pornographic/sex-inappropriate or personal life. The studied shown women athletes in Malaysia are assisting by media to develop their career by not trivialized in receiving coverage and also underrepresented them.

Hence, media have the ability and power to change the coverage provided for women's sports. Whether as commentators or sports writers/editors, more women in the field could potentially decrease the inequitable coverage leading to more interest. It depends when the media have been an importance site for reinforcing dominant ideologies and power structured, Kane [9]. Despite, any news released in newspaper media have an importantly quality reflection for portrays ideas to the public's. It supported from Grappendorf et al. [10] found that media have an immediate and future impact on society's view of female athletes, but also, it can alter viewers' and readers' perceptions of women's coverage.

5 Conclusion

Media in Malaysia still give a priority in women athletes' stories especially about achieving their victory. It necessarily to help women athletes to be more well known by empowering their position in sport by provide them with a positive value of stories and pictures.

Acknowledgments We would like to thank the Research Management Institute (RMI), UiTM for funding this study through the Excellence Fund.

References

1. Eagleman, A. N., & Pederson, P. M. (2008). An Analysis of the Coverage (Promotion) of the Females and Males in ESPN Magazine.
2. Kian, E. M. (2008). Study examines stereotypes in two national newspapers. A study of the New York Times and USA Today articles on College Basketball Counters the Gender-specific stereotypes found in previous studies on television commentary of March Madness. Newspaper Sport Journal, *29*(3).
3. Fondazione, G. B., Elisa, G., Silvia, S., & Anna, L. T. (2013). Policy Department C: Citizens' Rights and Constitutional Affairs, Gender equality. *Women and girls as subjects of media's attention and advertisement campaigns: The situation in Europe, best practices and legislations.*
4. Lumpkin, A. (2009). Female representation in feature articles published by Sports Illustrated in the 1990s.
5. Faridah, I., Normah, M., Chang, P. K., & Fauziah, A. (2011). Images and issues of superpowers: An analysis of international news coverage by The Government-owned News Agency, Bernama in Four National Dailies in Malaysia. Innovation Journal, *16*(3), 6.
6. Kian, E. M. (2009). Gender in sports writing by the print media: An exploratory examination of writers' experiences and attitudes. The SMART Journal, *4*(1).
7. Shaller, J. (2006). Female athletes in the media: Under representation and inadequacy. Undergraduate Review: a Journal of Undergraduate Student Research, *8*(10).
8. Huggins, A., & Randell, S. (2007). The contribution of sports to gender equality and women's empowerment. *International Conference on Gender Equality on Sports for Social Change.*
9. Kane, M. J. (2008). Media coverage of the post title ix female athlete: A feminist analysis of sport, gender and power. *Duke Journal of Gender Law and Policy Journal, 3*, 95.
10. Grappendorf, et al. (2009). Is it equitable online? The media coverage of the 2007 NCAA division I basketball tournament on Foxsports.com. *The SMART Journal: Sport Management and Related Topics, 5*(1), 30–42.

Attitude and Perceived Constraints Towards Physical Activity Among Gender in Alor Setar, Kedah

Rozita Abdul Latif, Nora Idura Othman, Nagoor Meera Abdullah, Norlizah Abdul Hamid and Chee Hian Tan

Abstract This study aims to identify attitudes among adults on physical activity and also identify key constraints in physical activity among gender in Alor Setar, Kedah. The study was conducted in Pekan Rabu and Star Parade Shopping Mall on 151 adult male and 237 female adults between the ages of 25–59 years. Data were collected using administered questionnaire, which adapted from Kenyon Attitude Towards Physical Activity Questionnaire [13] and the Perceived Constraints to Leisure Time Physical Activity Questionnaire [14] . Results showed respondents had high attitude towards physical activity ($M = 3.68$, $SD = 0.35$), and they also perceived high constraints in participating physical activity ($M = 3.67$, $SD = 0.87$). Data also showed significant differences for gender on attitude for catharsis domain ($t = 1.11$, $p < 0.02$) and ascetic experience domain ($t = 2.23$, $p < 0.03$), as well as partners domain in the perceived constraints ($t = 2.31$, $p < 0.02$). Most respondents have positive attitude and believe that physical activity helps to improve their health and fitness. They also reported that they are busy with their family and friends; therefore, they do not have enough time to do physical activity sufficiently. In conclusion, even if they have a good attitude towards physical activity, but because they still perceived a lot of constraints, then, participation in physical activity became low. To encourage continuing participation, certain aspects need to be addressed by the parties through providing activities, programmes, places and guideline that can be used either in the home or workplace in connection to physical activity participation.

Keywords Attitudes · Constraints · Physical activity · Gender

R. Abdul Latif (✉) · N. I. Othman · N. M. Abdullah · N. Abdul Hamid · C. H. Tan
Faculty of Sport Science and Recreation, Universiti Teknologi MARA,
Shah Alam, Malaysia
e-mail: rozita.abdlatif@salam.uitm.edu.my

R. Adnan et al. (eds.), *Proceedings of the International Colloquium on Sports Science, Exercise, Engineering and Technology 2014 (ICoSSEET 2014)*,
DOI: 10.1007/978-981-287-107-7_61,

1 Introduction

Physical activity helps to enhance physically psychological and emotional health [1], as well as helps to emphasize positive physical, functional, psychological effects related to feelings of well-being [2]. Besides, physical activity helps reduce the risk of coronary heart disease [3], and physical activity helps against depressive symptoms in young women even with low level of activity [4].

However, rates of physically active people in Malaysia are declining. Only fourteen per cents (14 %) of Malaysians had exercised sufficiently and generally sedentary [5]. World Health Organization (WHO) [6] revealed statistics report that 43.7 % of those above 18 years of age are not exercising, and Malaysians are the fattest in Southeast Asia. The rate of physical inactivity is getting higher because of urbanization; material wealth is increasing; and active commuting to work is declining. Decreasing in physical activity may be affected by the attitude of people towards physical activity. Positive attitude is contributing to a large extent towards active participation in physical activity. Frequent participation in physical activity is because of positive and higher perception and attitude towards physical activity [7]. Apparently, positive attitude alone is not enough to increase physical activity because of the constraints and other demands in life.

Failure to increase physical activity may be the result of constraints in doing physical activity that hinder people from being more active. Some studies showed that the greatest constraints to physical activity participation are time, family obligation and lack of energy. This is supported by Kamaria and Sofian [8] that stated the main barriers to physical activity are time constraints followed by costly to exercise, lack of encouragement from family and embarrassment. There are three main constraints among working adults from being physically active, which are structural (time, facilities, knowledge, accessibility), interpersonal (partners) and followed by intrapersonal (psychological) [9]. People knew and understood the importance of physical activity for their health, but unable to take action to increase their physical activities [10]. Time constraints, family obligation and lack of energy are several constraints that hinder people being more active. If the attitude and constraints to physical activity are not addressed, the level of physical inactivity will keep increasing and generally sedentary regardless of their age and gender. So, this study is to determine the attitudes towards physical activity participation among gender in Alor Setar, Kedah and to identify the perceived constraints to physical activity participation among genders in Alor Setar, Kedah.

2 Method

The design of this study can be categorized as quantitative research because it is objective, systematic process and information obtained by numerical data. Quantitative researches determine the relationships among variables and identify cause-and-effect interactions between them. This study used closed-ended questionnaire to obtain information from groups of people and concerned with the relationships between variables within these groups. Closed-ended question are relatively quick, easy to standardize and more objective. It is suitable for large group of people with minimal cost and greater anonymity [11, 12].

2.1 Sampling

The survey questionnaire was completed by three hundred and eighty-eight respondents ($n = 388$) where male respondents were 151 and 237 (61.1 %) female respondents, and they were young to middle-aged adults (between 38 and 56 years old). This study was carried out in Pekan Rabu and Star Parade Shopping Mall in Alor Setar, Kedah,

2.2 Instrumentation

The instrumentations consists three sections; Socio-demographic questionnaire to gather respondents demographic data, 'Kenyon Attitude towards Physical Activity Questionnaire' (Malays version) [13] to identify attitude towards physical activity, and the perceived constraints to physical activity was assessed by Perceived Constraints to Leisure Time Physical Activity Questionnaire [14].

3 Result

Table 1 shows that the highest attitude score towards physical activity is 'as Health and Fitness' ($M = 4.35$, $SD = 0.66$) followed by 'as a Social Experience' ($M = 4.18$, $SD = 0.77$), and the least attitude towards physical activity is 'as an ascetic experience' ($M = 3.25$, $SD = 0.81$). The data indicated that all participants showed that they still have higher attitude towards physical activity ($M = 3.68$, $SD = 0.35$).

Table 2 shows the attitudes score for males and females. The data indicated that the highest attitude score towards physical activity is 'as health and fitness': male ($M = 4.33$, $SD = 0.66$), female ($M = 4.36$, $SD = 0.66$). The least attitude score towards physical activity is 'as an ascetic experience': male ($M = 3.36$, $SD = 0.80$), female ($M = 3.18$, $SD = 0.80$). There was a statistically significant difference in catharsis ($t = 2.34$; $p < 0.02$) and ascetic experience ($t = 2.23$; $p < 0.03$) between gender. No significant difference for the social, health, catharsis and vertigo sub-domains of attitude between the genders.

Table 1 Attitude towards physical activity

Attitude	Mean ($N = 388$)	SD
As health and fitness	4.35	0.66
As a social experience	4.18	0.77
As the pursuit of vertigo	3.51	0.56
As catharsis	3.47	0.56
As an aesthetic experience	3.34	0.78
As an ascetic experience	3.25	0.81
Total	3.68	0.35

Table 2 Attitude towards physical activity by gender

Domain	Male (*M*, *SD*)	Female (*M*, *SD*)	*t* test	*P*
Health and fitness	4.33	4.36	0.18	0.85
	−0.66	−0.66		
Social experience	4.19	4.18	0.4	0.69
	−0.79	−0.85		
Pursuit of Vertigo	3.55	3.49	0.96	0.34
	−0.55	−0.56		
Catharsis	3.5	3.45	2.34	0.02[a]
	−0.55	−0.57		
Aesthetic experience	3.46	3.26	1.11	0.27
	−0.76	−0.78		
Ascetic experience	3.36	3.18	2.23	0.03[a]
	−0.8	−0.8		

[a] Significant at $p < 0.05$

Table 3 shows the score for perceived constraint among gender in Alor Setar, Kedah. It is indicated that all adults still perceived higher constraints for them to be involved in physical activity ($M = 3.67$, $SD = 0.87$). The highest mean score was 'I am too busy with my family and friends' item ($M = 4.74$, $SD = 0.60$), and the lowest mean score was 'I do not want to interrupt my daily schedule' item ($M = 4.06$, $SD = 0.81$). For facilities constraints, the highest score was 'the parks are crowded' ($M = 3.99$, $SD = 0.78$) and lowest score was 'the quality of the parks is not good enough for me' ($M = 3.75$, $SD = 0.77$).

For knowledge constraints, 'I do not know anyone who can teach me' was the highest score ($M = 3.34$, $SD = 1.07$), and lowest score was 'I do not have information on where I can go' ($M = 2.90$, $SD = 0.76$). For accessibility constraints, 'I cannot afford the money to do many physical activities/sport opportunities' item ($M = 4.14$, $SD = 0.97$), and the lowest mean score was 'I do not have transportation to get to places to be physically active' item ($M = 3.04$, $SD = 1.07$). For partners constraints, the highest mean score was 'My friends do not like to be physically active or play sports' item ($M = 4.46$, $SD = 0.66$), and the lowest mean score was 'It is difficult to find others to carry out physical activities with me' item ($M = 3.65$, $SD = 1.05$).

Table 3 Constraint towards physical activity

Constraints	Mean ($N = 388$)	SD
Psychology	3.07	0.73
I am not fit enough	3.16	1.02
I am not skilled for most activities	3.06	0.87
I do not feel confident enough	3.00	1.05
Time	4.33	0.41
I am too busy with my family and friends	4.74	0.60
The schedules do not fit in with my life	4.19	0.60
I do not want to interrupt my daily schedule	4.06	0.81
Facilities	3.88	0.47
The parks are crowded	3.99	0.78
The quality of the parks is not good near my home	3.90	0.77
The quality of the parks is not good enough for me	3.75	0.77
Knowledge	3.13	0.58
I do not know anyone who can teach me	3.34	1.07
I do not know where to participate	3.15	0.97
I do not have information on where I can go	2.90	0.76
Accessibility	3.45	0.54
I cannot afford the money to do many physical activities /sport opportunities	4.14	0.97
There are no opportunities near my home	3.16	0.96
I do not have transportation to get to places to be physically active	3.04	1.07
Partners	4.15	0.52
My friends do not like to be physically active or play sports	4.46	0.66
I do not have anybody to play sports with	4.35	0.72
It is difficult to find others to carry out physical activities with me	3.65	1.05
Total	3.67	0.87

Table 4 indicated that the highest perceived constraints score to physical activity was time: male ($M = 4.37$, $SD = 0.36$), female ($M = 4.30$, $SD = 0.44$). The least perceived constraints score towards physical activity was psychology constraints: male ($M = 3.08$, $SD = 0.72$), female ($M = 3.07$, $SD = 0.75$). Table 4 also indicated the analysis of perceived constraints to physical activity, which shows a statistically significant difference between the genders for the partners constraints ($t = 2.31$; $p < 0.02$). No significant difference for the psychology, time, facilities, knowledge and accessibility constraints between genders was found.

4 Discussion

According to data in the Table 1, overall, the highest attitude score towards physical activity is attitude 'as health and fitness' ($M = 4.35$, $SD = 0.66$). Most of the respondents believe that physical activity helps to improve their health and

Table 4 Constraints towards physical activity by gender

Domains	Male (*M*, *SD*)	Female (*M*, *SD*)	*t* test	P
Time	4.37 −0.36	4.3 −0.44	1.67	0.09
Partners	4.23 −0.46	4.1 −0.55	2.31	0.02[a]
Facilities	3.93 −0.47	3.85 −0.46	1.62	0.1
Accessibility	3.48 −0.57	3.43 −0.52	0.86	0.39
Knowledge	3.14 −0.59	3.13 −0.57	0.12	0.9
Psychology	3.08 −0.72	3.07 −0.75	0.11	0.91

[a] $P < 0.05$

fitness. This positive attitude may occur from health awareness campaign by government and sport organization regarding the importance of physical activity. The result of the current study demonstrates some similarities with the previous studies by Jussila et al. [15], which reported that Badi women had positive attitude and perceptions towards physical activity, and they believe that physical activity is essential to improve health and mental health.

Eun et al. [10] supported in feminine perspective where white midlife women in the USA understand the importance of physical activity for their physical and mental health. Even though physical activity is prior as a medium of socialization, the female students prefer to perceive physical activity for improved health and fitness [16], which show enjoyment and regularly doing physical activity will give benefit for physical and mental health benefit.

Table 2 indicated that the highest attitude score towards physical activity is 'as Health and Fitness' among both gender, while the least attitude score towards physical activity is 'as an ascetic experience' among both gender. There were no significant differences between genders in attitude towards physical activity. The results are consistent with findings by Tan [13] among teachers' trainee who shows that both gender perceived physical activity as a social experience. However, this result contrasts with study by Kee et al. [17] among university students who indicated that effect size in attitude 'Health and Fitness' between the genders was large enough and educationally meaningful. There are similarities in results of the present study with Kee et al. [17] about the least attitude towards physical activity for both genders, which was 'as an ascetic experience'. Possible explanations maybe the respondents may still be unfamiliar with the term 'ascetic' and the physical activity for high-level achievement.

According to data in the Table 3, overall, the highest perceived constraints score towards physical activity is time constraints. Most respondent reported that they are busy with their family and friends; therefore, they do not have enough time to do physical activity sufficiently. This constraint maybe resulted from urbanization and working nature. People spend more time on working and managing children. Previous studies point out that time constraint is factor that favoring people to being inactive. Study by Sonja et. al. supported that structural constraint such as lack of time, affordability, family obligations and lack of energy were the most constraints and reason people being inactive. Chinese female international students (CFIS) [16] were identified as the least physically active group in the USA, and the results indicated that time constraints were the reason being physically inactive.

Table 4 also indicated that the highest perceived constraints score towards physical activity is time among both gender, while the least perceived constraints score towards physical activity was psychology constraints among both gender. Overall, there are no differences between genders in terms of perceived constraints to physical activity. The t test result shows that there was statistically significant difference in the overall perceived constraints to physical activity based on genders (t (386) $= 2.07$, $p = 0.039$). There was a statistically significant difference between the genders for partner domain of constraint. The result is quite similar with Wee [18] findings that females faced more interpersonal constraints, which are followed by structure constraints as compared to males. No significant difference for the psychology, time, facilities, knowledge and accessibility subdomains of perceived constraints between genders was found.

5 Conclusion

Based on the finding, most respondents perceived that physical activity help to improve health and fitness followed by providing opportunities to meet new people. However, they less believe that physical activity is a medium to release hostility, tension and aggression and exposed to danger.

Time constraints are the main reason why respondents were not doing physical activity sufficiently. Most respondents stated that they are busy with work, family responsibilities and friends. Besides, less opportunities, financial affordability and transportation may hinder respondent's accessibility to physical activity places even though they have strong and positive attitude towards physical activity.

The best strategy to increase their physical activity is providing advice and guidelines about how to do physical activity around daily routines and simple exercise at home and workplace. Government should encourage private and government office to build more physical activity places in workplace and develop more physical activity programme in their schedule, and more non-competitive activities or games should be developing to gain their interest towards physical activity workplace.

Future study should include societal and cultural factors that affect attitude towards physical activity.

Acknowledgments We would like to express our gratitude to the Faculty of Sports Science and Recreation Universiti Teknologi MARA (UiTM) Shah Alam, Selangor, Malaysia, for the support on the feedback and information throughout the whole process of this study.

References

1. Gladys, S. L., Frank, J. H., & Amy, H. W. (2009). Exploring the relationships of physical activity, emotional intelligence and health in Taiwan college students. *Journal of Exercise Science Fitness, 7*(1).
2. Nuria, G., Olga, M., Raquel, M., Rodrigo, J., Javier, G., & Sara, M. (2007). Feelings of well being in elderly people: Relationship to physical activity and physical function. *Archives of Gerontology and Geriatrics, 48.*
3. Kristina, S., Jan, Q., Sven-Erik, J., Jan, S. (2005). The long-term effect of physical activity on incidence of coronary heart disease: A 12-year follow-up study. *Preventive Medicine, 41*(1).
4. Denise, A., Kylie, B., Jo, S., & Verity, C. (2008). The association between physical activity and depressive symptoms in young women: A review. *Mental Health and Physical Activity I,* 82–88.
5. Poh, B. K., Safiah, M. Y., Tahir, A., Siti Haslinda, M. D., Siti Norazlin, N. et al. (2010). Physical activity pattern and energy expenditure of malaysian adults: findings from the malaysian adult nutrition survey (MANS). *Malaysian Journal of Nutrition, 16*(1).
6. World Health Organization (WHO). (2010). *Global Strategy on Diet, Physical Activity, and Health.* Geneva, Switzerland: World Health Organization.
7. Lim, K. (2009). University students' attitude, self-efficacy and motivation regarding leisure time physical participation. *Jurnal Pendidik dan Pendidikan, 24.*
8. Kamaria, K., & Sofian, O. F. (2007) Attitudes towards physical activities among college students. *Pakistan Journal of Psychological Research, 22*(1).
9. Casper, J., Michelle, G., & Inés, P. (2011). *Perceived constraints to leisure time physical activity among latinos.* Ninth Active Living Research Annual Conference, San Diego, CA.
10. Eun, O. I., Bokim, L., Wonshik, C., & Stuifbergen, A. (2011). Attitudes towards physical activity of white midlife women. *Journal of Obstet Gynecol Neonatal Nursing, 40*(3).
11. Burns, N., & Grove, S.K. (2005). *The practice of nursing research: Conduct, critique, and utilization* (5th ed.). St. Louis: Elsevier Saunders.
12. Constantinos, N. P., Alice, B., Clive, S. (2011). *Structured methods: Interviews, questionnaires and observation.* Retrieved August 10, 2013, from http://www.sagepub.com/upm-data/47370_Seale_Chapter_11.pd.
13. Tan, S. E. (1991). *The attitude of the University of Malaya Diploma in Education teacher trainees towards physical activity.* M.Ed. Thesis, University of Malaya.
14. Jonathan, C., Michelle, G. H., Inés, P. (2011). Perceived Constraints to Leisure Time Physical Activity among Latinos. *American Journal of Health Promotion, 27*(3), *139–42.* doi: 10.4278/ajhp.110401-ARB-145.
15. Jussila, J. (2013). Experience and perception of physical activity among sexually exploited Badi women in Nepal.
16. Yana, Z., & Bradley, J. C. (2013). Perception of physical activity participation of chinese female graduate students: A case study. *Research Quarterly for Exercise and Sport, 84*(3).
17. Kee, K. M., & Wee, E. H. (2005) Attitude toward physical activity: An investigation of sports science students in Malaysian public universities. *Research Journal, 2*(1).
18. Wee, E. H. (2007). Motives and constraints to physical activities participation among undergraduates. http://www.isdy.netwww.isdy.net/pdf/eng/2007_21.pdf.

Factors Influencing Spectators' Attendance of Malaysian Super League Using Bootstrap Linear Model

Rumaizah Che Mohd Nor, Norazan Mohamed Ramli, Nik Arni Nik Mohamad and Nor Hayati Abdul Hamid

Abstract Football is the world's most popular sports. One of the indicators that show the football popularity is the large number of stadium spectators. It is crucial for the event management to determine the factors that significantly affect the stadium attendance as this will help them in scheduling the football match locations. This study aims to determine the factors that might influence the attendance of spectators to the Malaysian Super League (MSL) matches. Spectators' attendance for 2008/2009 seasons were used in the analysis and bootstrap linear model was preferred to ordinary linear model to model the spectators' attendance as it can improve the model efficiency for MSL data with outlier problem. Results of the analysis show that bootstrap linear model produces the most reliable result with a model producing higher adjusted R-Square value. From the selected model, it can be concluded that performance and percentage of matches won by the home team are the significant factors which can influence the attendance of spectators to the matches.

Keywords Spectators' attendance · Malaysian super league · Bootstrap · Linear model

1 Introduction

The game of football can be traced back to the ancient Greek and Romans, around 388–311 BC. There is also evidence that showed football to be a popular sport in the Chinese military between the third and the first century BC. A reference to a ball game played in Britain can be found from the ninth century Historia Brittonum,

R. Che Mohd Nor · N. Mohamed Ramli (✉) · N. A. Nik Mohamad
Centre for Statistical and Decision Science Studies, Universiti Teknologi MARA, Shah Alam, Malaysia
e-mail: norazan@tmsk.uitm.edu.my

N. H. Abdul Hamid
Centre for Information Technology, Universiti Teknologi MARA, Shah Alam, Malaysia

R. Adnan et al. (eds.), *Proceedings of the International Colloquium on Sports Science, Exercise, Engineering and Technology 2014 (ICoSSEET 2014)*,
DOI: 10.1007/978-981-287-107-7_62,

which describes “a party of boys… playing at ball”. The rules and codes of football can be traced back to the English public schools in the eighteenth century. The first known codes were those of Eton in 1815 and Andelham in 1825 [1].

Football was adopted as a way of encouraging competitiveness and keeping youths fit. The influence and power of the British Empire allowed football to spread to areas under its control. Sports clubs dedicated to playing football began in the eighteenth century [1]. Since then, it has become the world’s most popular sports, with 4 % of the world’s population, or an estimated 260 million people are involved in the sport [2]. Today, professional football seems to focus the attention of entire nations and club football has become a massive industry that attracts millions of spectators on a yearly basis, especially in Europe [3].

One of the factors that made football to become popular is the large number of spectators’ attendance at the stadium to watch the football match. According to Jeffrey et al. [4], a sport event is interesting due to the attendance of spectators in the stadium. In Malaysia, the first modern set of rules and codes for football were established in 1921. The sport of football in the country is run by the Football Association of Malaysia (FAM). The association administers all the leagues and competitions. There are four national leagues, and the MSL is considered to be the most prestigious. It was introduced in 2004, with eight teams. To encourage more competition, FAM has increased the number to fourteen [5].

Since 1991, stadiums have been equipped with better facilities to attract more spectators’ attendance. Even so, the attendance was poor, only averaging about 2,000–6,000 per football game, an indication of poor quality matches. The drop in the number of spectators’ attendance implied a fall in FIFA rankings for the Malaysian football team [6]. The number of spectator in attendance for the matches for the 2006/2007 season was more than 600,000. For the 2007/2008 season, the number increased to more than 700,000 but decreased for the 2008/2009 season. According to Hamid and Kendall [5], this decrease is partly due to the fact that some matches are televised. Since then FAM is focusing to increase the spectators’ attendance, for the game’s popularity and survival and for its enormous potential for social and economic development. So the knowledge concerning factors that influence the attendance of spectators for MSL football matches would be useful, especially for FAM and also the other football associations in Malaysia.

In the UK, with the Barclay’s Premier League (BPL), for the 2013/2014 season, the spectators’ attendance sometimes exceed stadium capacity. For the Spanish Primera Liga, for 2013/2014 season, attendance has been almost full, although for some matches, less than half of the stadium is filled (http://stats.football365.com). Unfortunately, recent data for the MLS are not available for comparison.

2 Factors Affecting Spectators’ Attendance

Borland and MacDonald [7] reported that spectators’ attendance in England was higher on weekends. Team success and their recent performance have been shown to affect spectators’ attendance. It has also been found that home-team performance consistently has a positive effect on attendance [7]. The report also noted that when the league ranking is fairly even and thus a large number of matches are significant; attendance is generally higher. There is also some indication that attendance is higher when the matches are spread across a longer time period [7, 8].

According to Javanmardi and Noghondarian [9], the ranking of the home team and the number of the championships of the guest team are found to be significant. They also proved that to spectators, the ranking of the team is the most important factor. The estimated population size within the city where the match is held and the distance between the cities also significantly influenced the number of spectators to watch the match at the stadium [9]. Forest and Simmons [8] found that the distance between the cities can give effect to the spectators to attend to the stadium [8]. According to Kringstad and Solberg [10], the higher the rank for a team, the more attendance the match attracted. This applied to both the home team and away team. They also found that whether the geographical distance between the two teams was less than 120 km, this had a positive effect. Matches between historical rivals also attracted more attendance [10].

Unfortunately, in this study, we did not include all the variables or factors as discussed in the above paragraphs since the data were not available. The data used in this study were obtained from the website www.soccerway.com and from the officers of FAM. It consists of the number of attendance for the 2008/2009 seasons. The response data represent the number of spectators' attendance of the MSL football matches. From the website, information about the date and day of matches, teams that were playing for the matches and results of all the matches were extracted. Table 1 summarizes the variables identified as factors that probably influence spectators' attendance.

3 Modeling Spectators' Attendance Data

Previous studies used linear models to determine the factors that affect attendance at stadiums. For example, Javanmardi and Noghondarian [9] and Kringstad and Solberg [10] used regression models. Aminuddin and Lee [6] also used a multiple linear regression model to study the relationship between stadium factors and team quality on spectators' satisfaction and their intentions to attend football matches. Thus, in this study, multiple linear regression analysis will be used to achieve the best model for determining the significant factors that influence the spectators' attendance to MSL. The following multiple linear model in (1) is considered as appropriate to model the spectators' attendance:

$$y_i = \beta_0 + \beta_1 x_{i1} + \beta_2 x_{i2} + \ldots + \beta_9 x_{i9} + \varepsilon_i \quad (1)$$

Table 1 Data description

Label	Variable	Description
y	Spectators' attendance	Total of spectators' attendance to a particular match
x_1	Weekend	1 = weekend match 0 = weekday match
x_2	First	1 = first match 0 = not first match
x_3	Importance	1 = access important league positions 0 = not access important league positions
x_4	Derby	(neighboring team) 1 = local derby 0 = not local derby
x_5	Holiday	1 = school holiday match 0 = not school holiday match
x_6	Performance	The average points for the last five matches
x_7	Home Length	The length of time (in days) for the particular match to the next match for the home team
x_8	Away length	The length of time (in days) for the particular match to the next match for the away team
x_9	Home win	The percentage of matches that was won by the home team

where y_i represents the spectators' attendance, $\beta_0,\ldots, \beta_9$ are the coefficient parameters and ε_i is the error term of the ith match. The model is chosen with the following assumptions:

1. The response variable y is from a normal distribution.
2. The relationship between y and x is linear.
3. All x's are uncorrelated.
4. The error term, ε is normally distributed.

Diagnostic plot in Fig. 1 indicates that the spectators' attendance data, represented by y, is not normally distributed. The presence of extreme outliers as shown in the plot causes data skewness and violates the normality assumption as required in the linear model. Cook and Weisberg [11] proposed that appropriate transformations on the response variable would reduce the skewness problem. After taking a log transformation, a great improvement on the normality is indicated in Fig. 2.

Several methods can be used to estimate the value of β in Eq. (1). The ordinary least squares (OLS) method is the most common regression estimator and has become the most popular estimator of β. Its objective function is simply to minimize the sum of the squared residuals. In addition, the OLS method is normally preferred as it is very attractive for the simplicity in its computation and it produces the minimum variance if the normal assumption of the error term is not violated. In this study, fixed-resampling bootstrap method is proposed to be used as it is proven to be more efficient if the response data do not meet normality

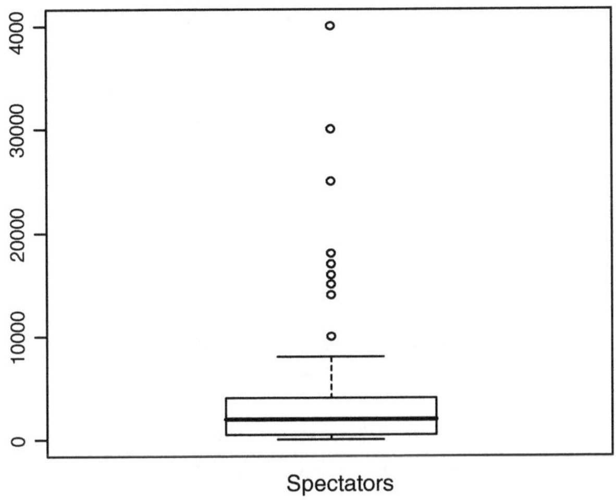

Fig. 1 Box-plot of spectators' attendance

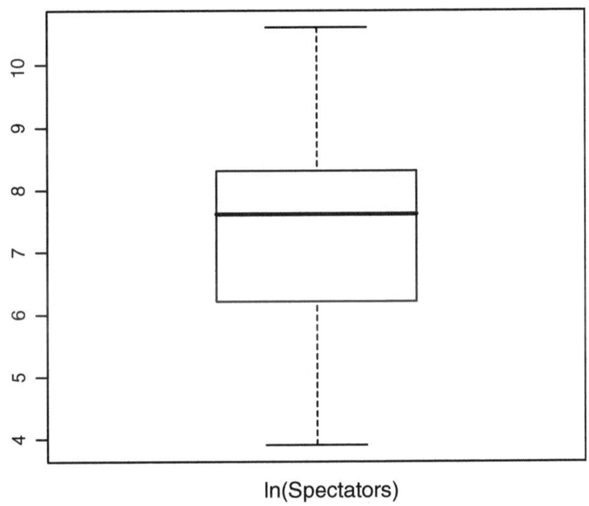

Fig. 2 Box-plot of spectators' attendance after taking a log transformation

assumption [12]. The procedure for bootstrapping the least square estimates in the linear regression model can be summarized in the following steps:

Step 1: Fit a model to the original sample of observations using OLS method to get $\hat{\beta}_0, \hat{\beta}_1, \ldots, \hat{\beta}_9$ and the fitted values

$$\hat{y}_i = \hat{\beta}_0 + \hat{\beta}_1 x_{i1} + \cdots + \hat{\beta}_{i9}$$

Step 2: Get the residuals $\varepsilon_i = y_i - \hat{y}_i$.

Table 2 R-square, adjusted R-square, mean square error and F-test with critical value (in bracket)

	OLS model	Bootstrap-OLS model
R^2	0.3064	0.3876
Adjusted R^2	0.2743	**0.3592**
MSE	1.2500	1.2800
F-test (critical value)	9.5500	13.6800
	(1.8500)	(1.8500)

The adjusted R^2 value (in bold) from the model that based on the bootstrap OLS method is higher compared to the model of OLS, indicating that the bootstrap OLS model produces higher percentage of variability in the response variable Y

The critical value for the F-test was obtained at $\alpha = 0.05$ level of significance with sample size $n = 188$

Step 3: Form a bootstrap sample. By means of simple random sampling, ε_i^* is drawn from ε_i and attached to $\hat{y}_i$ to get a fixed-x bootstrap values y_i^* where

$$y_i^* = \hat{\beta}_0 + \hat{\beta}_1 x_{i1} + \cdots + \hat{\beta}_{i9} + \varepsilon_i^*$$

Step 4: Regress the bootstrapped values y_i^* on the fixed x to obtain $\hat{\beta}^*$.

Step 5: Repeat steps 3 and 4 for B times to get $\hat{\beta}^{*1}, \ldots, \hat{\beta}^{*B}$.

The bootstrapped OLS estimates for $\beta_0, \beta_1, \ldots, \beta_9$ are based on 1,000 replications of bootstrap samples, i.e., $B = 1{,}000$, and computed as follows:

$$\hat{\beta}_j^* = \frac{\sum_{k=1}^{B} \beta_k^*}{B} \quad (2)$$

where $j = 0, 1, 2, \ldots, 9$, and the standard error for $\hat{\beta}_j^*$ is

$$se\left(\hat{\beta}_j^*\right) = \sqrt{\frac{\sum_{k=1}^{B} \left(\beta_k^* - \hat{\beta}_j^*\right)^2}{B - 1}} \quad (3)$$

See Imon and Ali [12] for details of the method. The values of R^2, Adjusted R^2, Mean Square Error (MSE) and F-test were used to assess the model performance. A good model is the one with high values of R^2 and Adjusted R^2 and low value of MSE.

4 Results and Discussion

The adjusted R^2 value from the model that based on the bootstrap-OLS method is higher compared to the model of OLS as shown in Table 2, indicating that the bootstrap-OLS method is more suitable to be used for modeling spectators' attendance data.

Table 3 *t* test statistic and its respective *P*-value for each factor

Factor	Estimator	OLS model	Bootstrap-OLS model
Intercept	$\widehat{\beta}_0$	19.4739 (0.0000)	20.5160 (0.0000)
Weekend	$\widehat{\beta}_1$	−1.6523 (0.1003)	−1.6330 (0.1043)
First	$\widehat{\beta}_2$	2.1821 **(0.0305)**	3.6230 **(0.0004)**
Importance	$\widehat{\beta}_3$	0.5760 (0.5653)	1.6380 (0.1033)
Derby	$\widehat{\beta}_4$	0.2600 (0.7951)	−0.3580 (0.7210)
Holiday	$\widehat{\beta}_5$	0.1775 (0.8593)	−1.3380 (0.1826)
Performance	$\widehat{\beta}_6$	4.1605 **(0.0000)**	5.4870 **(0.0000)**
Home length	$\widehat{\beta}_7$	−2.1174 **(0.0357)**	−3.0140 **(0.0030)**
Away length	$\widehat{\beta}_8$	5.3622 **(0.0000)**	5.2510 **(0.0000)**
Home win	$\widehat{\beta}_9$	6.6614 **(0.0000)**	3.9560 **(0.0001)**

The factors with p-values (in bold) are statistically significant to the model at $\alpha = 0.05$ level of significance (p-value < 0.05)

Table 3 presents the *t* test statistics and *p*-values of the regression coefficients. Looking at the performance of the estimates, variables 'First', 'Performance', 'Home length' and 'Home win' are significant factors for attracting the spectators to the MSL football matches. These variables are significant in both methods. A study done by Iho and Heikkilä [13] showed that the first match of the season draws 9.7 % higher attendance than the rest of the matches. They also showed that current dismal performance of the team could lower the attendance by as much as 31.2 %.

The number of days between a current match with the next match for a particular home team also influences spectators' attendance. The longer the gap between the match, the higher the number of spectators will come to the stadium to watch the game [13]. Simmons and Forrest [8] and Borland and MacDonald [7] indicate that attendance is higher when the matches are spread across a longer time period.

Figure 3 shows the plot of the residuals versus fitted values for the OLS while Fig. 4 gives the plot for the bootstrap-OLS model. The plots confirm there is no clear relationship between the residuals and fitted values. The regression analysis was repeated by including only the significant variables. Tables 4 and 5 present the results of the analysis. We notice that the value of the R^2 for the bootstrap-OLS model is greatly improved and the standard errors for the bootstrap-OLS model are lower than the OLS model.

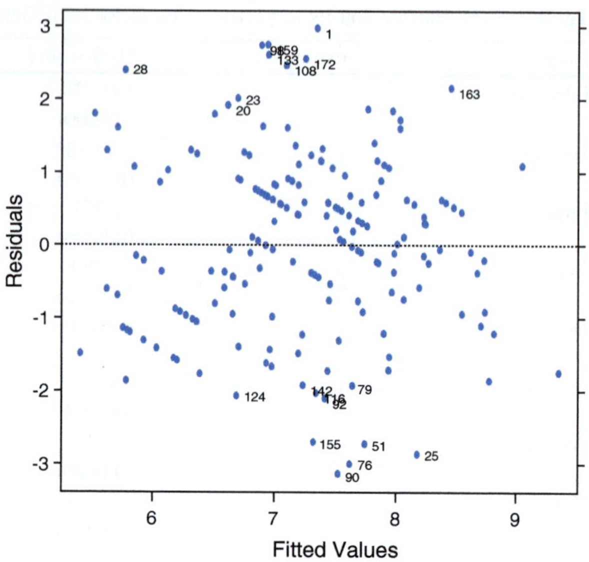

Fig. 3 Residuals versus fitted values of OLS model

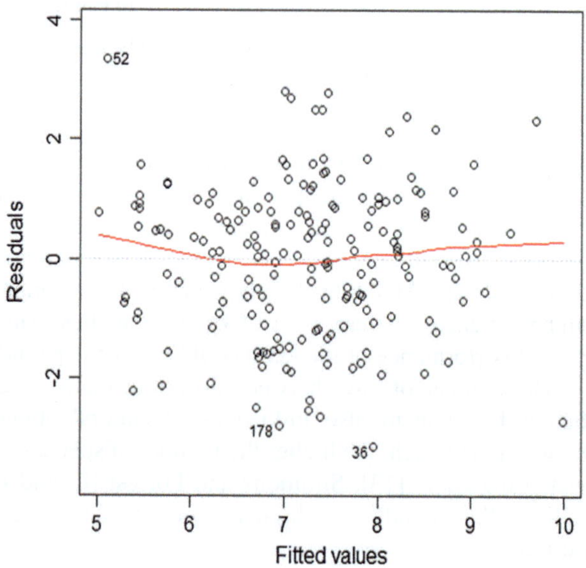

Fig. 4 Residuals versus fitted values of bootstrap-OLS model

Table 4 R-square, adjusted R-square, mean square error and F-test with critical values (in bracket) after removing the insignificant factors

	OLS model	Bootstrap-OLS model
R^2	0.2935	**0.4080**
Adjusted R^2	0.2775	**0.3947**
MSE	1.2500	**1.1300**
F-test statistic (critical value)	18.3800 (2.4200)	30.5000 (2.4200)

The R^2 and adjusted R^2 values (in bold) from the bootstrap OLS method are higher compared to the ones from the OLS model, indicating the goodness-of-fit of bootstrap OLS model improves without the insignificant factors together with smaller error measures of MSE (in bold)

Table 5 Estimates for beta and its respective standard error after removing the insignificant factors

Factor	Estimator	OLS model	Bootstrap-OLS model
Intercept	$\widehat{\beta}_0$, se($\widehat{\beta}_0$)	6.0117 (0.2129)	5.6126 (0.2020)
First	$\widehat{\beta}_2$ se($\widehat{\beta}_2$)	0.2050 (0.0986)	0.4306 (0.1783)
Performance	$\widehat{\beta}_6$ se($\widehat{\beta}_6$)	0.4697 (0.1108)	0.5219 (0.1001)
Home length	$\widehat{\beta}_7$ se($\widehat{\beta}_7$)	−0.0211 (0.0116)	−0.0391 (0.0105)
Home win	$\widehat{\beta}_9$ se($\widehat{\beta}_9$)	1.5237 (0.2892)	1.7524 (0.2614)

5 Conclusion

The study has identified that the teams' performance, the first match and the percentage of matches won by the home and away team as the significant factors for attracting spectators to the MSL football matches. The study also identifies that the Bootstrap-OLS model is better than the initial OLS model in determining the significant factors influencing the spectators' attendance. The Bootstrap-OLS model can be applied and extended to other leagues and should also consider of including other factors such as ticket prices and stadium facilities in the model.

The findings for this study are important to the organizers of Malaysian football tournaments, FAM and other football associations in Malaysia. They can re-schedule the matches based on the factors above so as to attract higher spectators' attendance to the stadiums so that high profit can be generated. Furthermore, higher number of spectators may attract interested football sponsors to finance the teams for future football tournament seasons.

Acknowledgments The authors gratefully acknowledge the financial support received from the Ministry of Higher Education of Malaysia and the Universiti Teknologi MARA under the Fundamental Research Grant Scheme (FRGS) 600-RMI/FRGS 5/3 (7/2012).

References

1. Football, http://en.wikipedia.org/wiki/Football
2. Federation International the Football Association (FIFA), 2006.
3. Stolt, M., & Waldenor, C. (2010). *The demand for football attendance in Sweden*. Stockholm School of Economics. Bachelor's thesis.
4. James, J. D., & Ridinger, L. L. (2002). Creating a kind of loyalty and a high level of dependence female and male sport fans: A comparison of sport on team is important for managers and marketers. Among consumption motives. *Journal of Sport Behavior*, 25.
5. Hamid, N., & Kendall, G. (2008). Maximizing stadium attendance: Case study of Malaysian Football.
6. Aminuddin, Y., & Lee, H. S. (2008). Spectators perceptions of physical facility and team quality: A study of a Malaysian Super League Soccer Match. *Research Journal of International Studies, 8*, 132–140.
7. Borland, J., & MacDonald, R. (2003). Demand for sport. *Oxford Review of Economic Policy, 54*(6), 633–650.
8. Simmons, R., & Forrest D. (2005). *New issues in attendance demand: The case of the English Football League*. Lancaster University Management School working Paper 2005/004.
9. Javanmardi, E., & Noghondarian, K. (2011). A study of factors affecting the demand for watching football in stadius. *Sport Management International Journal, 7*(2).
10. Kringstad, M., & Solberg, H. A. (2012). www.easm2012.com
11. Cook, R. D., & Weisberg, S. (1999). *Applied Regression Including Computing and Graphics*. New York: Wiley.
12. Imon, R., & Ali, M. M. (2005). Bootstrapping regression residuals. *Journal of Korean Data Information Science and Society, 16*(3), 665–682.
13. Iho, A., & Heikkilä, J. (2009). Impact of advance tickets sales on attendance in the Finnish Football League. *Journal of Sport Economics, 11*(2), 214–226.

Sport Science Graduates' Employability in the Job Market

Chee Hian Tan, Abbylolita Sullah and Tham Yin Choong

Abstract Present study was designed to investigate the sport science graduates' employability, which consisted of graduates' employment circumstances and learning, self-attitude toward responsibility, and soft skills in work life in the job market. In order to justify the objectives in elaborating these dimensions in the overall graduate's employability level, the mailed survey method using the Graduates Employability Questionnaire (Abdul Razak Ahmad in Opinion: the unemployable Malaysian, 2005) [1] was adopted. The results showed that every dimension of the GEQ with (a) level of employment circumstances/learning had a mean score of 3.35; (b) level of self-attitude toward responsibility which stated self-willingness to take up responsibility without interference of third part as well as other characteristic, had a mean score of 3.89; and (c) level of soft skills in work life had a significant mean score of 3.99. As far as determination of selected dimensions of graduates employability of sport science is concerned, the descriptive statistics profiled respondents according to gender, ethnicity, university attended, monthly income, job search mechanism, status and sectors of employment as well as length of time before they gaining their first job with inferential results of personal attributes and qualities and showed contributively statistics ($r = 0.608$, $N = 184$, $p < 0.01$; and $r = 0.428$, $N = 184$, $p < 0.01$), respectively. These results indicate that there was a moderate correlation or influence on graduate employability at the 0.05 level of significance.

Keywords Graduates employability · Personal readiness (personal attribute and personal quality) · Program attended

C. H. Tan (✉) · A. Sullah · T. Y. Choong
Faculty of Sport Science and Recreation, Universiti Teknologi MARA, Shah Alam, Malaysia
e-mail: tanchee@salam.edu.my

R. Adnan et al. (eds.), *Proceedings of the International Colloquium on Sports Science, Exercise, Engineering and Technology 2014 (ICoSSEET 2014)*,
DOI: 10.1007/978-981-287-107-7_63, © Springer Science+Business Media Singapore 2014

1 Introduction

The graduates of Public Higher Education Institutions (PHEI) in general would establish their delivery of knowledge and skills to their clients. However, it was necessary to investigate further on the social and psychological perspective as far as scientific and systematic research study is concerned.

As far as extension is concerned, it is a process which links subjects on a continuous basis with evolving research-based and tested knowledge, technologies, procedures and perspective that may serve its own purposes [2–4]. The extension mission eventually is to help people to develop their capacity and potential to manage and to cope with changes in their environment, and especially so among university graduates in this study and their contribution to their clients after graduation. On top of that, this study even ventures into factors that influence employability level with a social psychological perspective.

Products of PHEI are treated as human resources with outcomes of planned programs.

In planned programs, a point to bear in mind would have been the avoidance of human resource wastage, and this is a very critical issue and is fundamental in extension education, which focuses on Community Human Resource Development (CHRD). This study focuses particularly on sports science human resource graduates produced by PHEI concerned.

In reality, the conceived program may not achieve the planned objectives, and this could be due to many factors without exclusive extension in practice, especially when doubtful of the graduate's capabilities in the work place. They indirectly became an unemployable human resource or the low employability among sports science graduates as end product to practice in the actual work place, as at present there are many jobless diploma and degree holders of PHEI. For instance, in 2012 there were comments that local graduates were not up to the mark and were out in the work place with their respective documents, but lacking in readiness or were low in employability and were unable to obtain a job [5]. Why this is happening constitutes the main concern of this study?

Problems arise especially regarding the level of employability of the graduates. It significantly contributes to the fundamental human resource problem that has been major concern for a long time. More to the point is the question about important indicators that could address shortcomings in the present system of education. How successful are "tailored" programs of particular PHEI? What are the other factors such as graduates' self-readiness that could influence the employability of the graduates? However, from the human resource point of view, a self-equipped (graduates of Sports Science) disposition should develop in the university setting. In the context of a wider range of global experiences, in terms of people and future economies, it would be necessary for an employee to be sensitive to cultural sensibilities and local politics. This social science perspective is strictly becoming an important reference point and a source of human resource, especially in the areas of training subjects or implementation of human resource development. On top of that philosophically HRD has been defined as helping individuals to develop their full potential and readiness, although the question remains as to what end [6].

Factors influencing graduates' employability, a survey, based on related literature from the 1950s until the present, has identified macro and micro factors that influence graduates' employability. The macro factors would include changes in science and technology, the globalization process, structural changes, geographical changes, economic changes, wars, riots and others, which could give rise to changes in the social economy or changes in the social culture [7]. As far as this study is concerned, these macro factors cannot be controlled and are part of its delimitations.

However, in related literature, what comes firmly across are that these micro factors had significantly attracted many researchers from the 1950s until 2012, to venture into exploration of attributes that influence graduates employability. These factors cover personal readiness as the main factor and also the perception of graduates toward their own program. The interpretation of the relationships with graduates' employability would enhance the total growth of a nation. Thus, there was a significant need to further investigate these micro aspects, especially independent factors, which would give an impact of added value to graduates.

2 Methodology

2.1 Research Design

This study was a descriptive cum correlation research design, which was undertaken by means of a mail survey method [8]. A correlation design study determines and describes the relationship between independent and dependent variables. In terms of scientific research, this study had two research objectives to be examined, i.e., the differences and relationships between the selected independent factors and the employability of graduates.

Non-experimental in nature, the correlation design research utilizes a wide variety of methodologies to collect data such as survey, interview, direct measurement and observations being prevalent techniques [8]. This study used the mail survey method to collect data from graduates of sports science in the workplace. They had graduated from their respective PHEI and had been working between 6 months to a year. They were in the age group of 19–28 years old.

2.2 Sampling and Population

The term population is referred to an entire group or aggregate of people or elements having one or more common characteristics. In this study, the population

Table 1 Proportionate-stratified random samples from the population of Malaysian graduates who graduated from the selected PHEI

University	Population	GEQ distributed	Responses	Percentage
UTM	350	96	50	7.2
UM	277	72	38	20.6
UiTM	218	60	60	32.6
UKM	150	41	18	9.8
UPSI	113	31	18	9.8
Total	1,108	300	184	61.33

was graduates who have had 6 months to 1 year experience, involved in the workplace, and aged between 19 and 28 years old.

The population of graduates, who had been selected according to the criteria listed above, represented a total sample size of 300 graduates. This number was calculated proportionately according to the subtotal population produced by the selected respective PHEI. It came to appropriately around 27 % of the whole population. It was considered an appropriate sample size that was accepted based on the applications from the statistical package G-Power [9].

Based on the principle of mathematically proportionate sampling method, the subtotal of each PHEI were drawn. As a result, after the application of the procedures in calculating with electronic measuring tools, sample sizes of 300 respondents were selected. The sample of 300, which was appropriately proportionate to the population, were randomly selected from among samples from a name list that was obtained from the respective alumni units, student welfare departments or faculty, and samples were chosen according to the table of random numbers [10]. Hence, an equal chance was given for every potential graduate throughout the nation. This protocol represented appropriateness for this study, and hence, the results of this study would be valid and reliable to generate the overall cohort regarding the employability of graduates of sports science within PHEI.

The adopted questionnaire [1] was distributed to the 300 samples according to the number of graduates produced in each selected sports science graduates of PHEI. This protocol resulted as an appropriate measure of the cohort employability with regard to the selected sample size with percentage of subtotals as follows: UTM with a total 50 respondents (27.2 %), UM with 38 respondents (20.6 %), UiTM with 60 respondents (32.6 %), UKM with 18 respondents (9.8 %), and UPSI with 18 respondents (9.8 %). In the process of collecting data with the mail survey method, an overall total of 184 respondents (61.33 %) from the selected sample responded to the final GEQ (Table 1).

3 Results

In this study, the samples were formulated to generate representation of the population of graduates with a significant alpha level of 0.05, with three independent factors and one entity as the dependent variable, which was the graduates' employability. As far as the G-power statistical tool was concerned, the statistical power was 0.95. This was considered with the F-test on means in the ANOVA test. It comprises of nine demographic profiles to measure a graduate's employability. A priori analysis with various demographic profiles such as monthly income, length of time to obtain first job and age of respondents on employability were used to determine accuracy to generate the results of this study. Demographic profiles were collected as part of this study and the items included: age, gender, ethnicity, status of employment, sector of employment, university attended, monthly income, job searching mechanism, and length of time to obtain first job.

3.1 Personal Readiness

The dimension on personal attributes and qualities consisted of 27 factors, which were divided into 15 statements or factors of personal attributes and 12 statements or factors on personal qualities, respectively; and they were highly reliable with Cronbachs' alpha of 0.882 and 0.831 compiled in Part B of the adopted questionnaire. Moreover, the overall personal readiness was with the Cronbachs' alpha of 0.856, and this implied that this factor was highly reliable and valid for consideration as the independent factor in this study.

3.2 Demographic Findings

A total of 184 respondents (sport science graduates) from five PHEI responded to the formulated questionnaire in this mailed survey. The demographic profiles consisted of nine factors that were included in the final questionnaire in order to gather demographic data as stated in the research questions. The demographic data are summarized in Table 2.

Graduates of UiTM comprised of about one-third (32.6 %) of the respondents and followed by 27.1 % of UTM graduates, 20.6 % of UM graduates and 9.8 % each from UKM and UPSI.

Males and Malay graduates formed the highest percentage of respondents with 61.4 and 77.2 %, respectively. The majority of respondents (94.6 %) were employed in the government sector, while the remainder were employed in private companies, and 45.2 % had pursued and acquired a Masters' degree.

Table 2 Demographic profiles of respondents ($N = 184$)

Variables	Frequencies	Percent
University		
UiTM	60	32.6
UTM	50	27.2
UM	38	20.6
UKM	18	9.8
UPSI	18	9.8
Gender		
Males	113	61.4
Females	71	38.6
Ethnicity		
Malay	142	77.2
Chinese	17	9.2
Indian	14	7.6
Others (Bumiputera)	11	6.0
Status of employment		
Employed	174	94.6
Unemployed	10	5.4
Sector of employment		
Government	117	63.6
Private company	57	31.0
Freelance	10	5.4
Lifelong learning		
Master degree	83	45.1
Others	1	0.5
None	100	54.4
Job search mechanism		
Advertisement	44	23.9
Internet	31	16.8
Friends/family	25	13.6
Offered during internship	22	12.0
HR department	13	7.1
Univ career unit	8	4.3
Career agency	5	2.7
Others	36	19.6

In terms of job search mechanism used by graduates, advertisements were the most popular way to enhance job (23.9 %), followed by internet listing (16 %), and through friends/family (13.6 %). This implied that newspaper advertisements or other written forms of information were preferred by the respondents.

3.3 Employability Among Graduates of PHEI

The first objective of this study was to determine the employability among graduates.

Table 3 Overall the employability of sports science graduates ($N = 184$)

Level	Frequency	Percent	Mean	S.D
Graduates employability			3.84	.33
Low (1–2.33)	–	–		
Moderate (2.34–3.66)	55	29.9		
High (3.67–5.00)	129	70.1		
Employment circumstance/learning			3.53	.37
Low (1–2.33)	–	–		
Moderate (2.34–3.66)	132	71.7		
High (3.67–5.00)	52	28.3		
Self-attitude toward responsibility			3.89	.36
Low (1–2.33)	–	–		
Moderate (2.34–3.66)	49	26.6		
High (3.67–5.00)	135	73.4		
Soft skills in work life			3.99	.36
Low (1–2.33)	–	–		
Moderate (2.34–3.66)	32	17.4		
High (3.67–5.00)	152	82.6		

Measuring the employability among graduates of PHEI was a useful and meaningful study. It had been widely researched, especially from different perspectives within different time frames and in different countries since the 1950 s until 2012. However, this study was considered as the most recent research with the initiative to determine the employability of graduates among local PHEI.

The data on the employability of graduates are summarized in Table 3. Overall, the respondents indicated the choice of 'very strongly agreed' on each statement or element for items D1 to D45. The overall mean score was 3.84 with a standard deviation of 0.33, and this indicated that overall the graduates were at a high level of employability, where 70.1 % of respondents showed a high level of above 3.67, while only 29.9 % of the respondents showed a moderate level of employability.

In conclusion, these graduates showed high employability in the workplace, and they were able to obtain employment in whatsoever environment. They were definitely not the group that contributed to unemployment or were unemployable youth.

In other words, the self-assessment of these graduates gave positive results as well as the capability for self-sufficiency into and within the workplace. They managed to fulfill self potential through sustainable and accessible employment, which depended on the soft skills they possessed and their self-attitude. They were self-equipped to be presented to employers, and this showed that the human resources were highly value-added and could perform their employability wherever there went and in whichever way they were involved in the extension of human resources.

4 Conclusion

The adopted GEQ could identify the employability of PHEI graduates holistically through the self-assessment questionnaire within 38 min. It was relevant for graduates of PHEI, and this was the early initially empirically constructed questionnaire. It was adaptable and adoptable for the use of the whole nation's public higher education institution graduates. It was contributed a significant body of knowledge, especially to the field of community human resource in general.

At the same time, this study measured the different levels of employability of graduates with personal readiness as well as in determining the relationship factors between them, in order to profile the employability level of graduates among PHEI. The added value extension will contribute to the nation's gross domestic product.

In conclusion, the findings of this study showed that local PHEI graduates were fully equipped with personal readiness to react and present themselves to employers as well as to obtain employment in the workplace within a short duration. Hence, they would not be categorized as unemployable. This would not increase the unemployment rate throughout the nation. It is recommended that there was a need to highly emphasize on constructional proficiency programs in the university delivery system, especially for undergraduate programs.

References

1. Abdul Razak Ahmad. (2005). *Opinion: The unemployable Malaysian graduates*. March 20, 2005. http://pgoh.free.fr/unemployable200305.html ACER (2001). Acer Review. p. 38.
2. Allport, G. W. (1967). Attitudes in attitude. In M. Fishbein (Ed.), *Theory and measurement*. New York: Wiley.
3. Alreck, P. L., & Settle, R. B. (1985). *The survey research handbook*. Homewood: IRWIN.
4. American Psychological Association. (2001). *Publication manual of American psychological association* (5th ed.). Washington, D.C.: American Psychological Association.
5. Aneshensel, C. S. (2002). *Theory based data analysis for the social sciences*. Thousand Oaks: Sage Publication.
6. Arulampalam, W., & Steward, M. B. (1995). The determinants of individual unemployment duration in an area of high unemployment. *Economic Journal, 105*, 321–332.
7. Ashton, D. N., & Field, D. (1976). *Young workers: From school to work*. London: Hutchinson.
8. Samah, B. A., & Suandi, T. (1999). *Statistics for social research-with computer application* (1st ed.). Malaysia: UPM Publisher.
9. Bailey, T. (1990). Jobs of the future and skills they will require: New thinking on an old debate. *American Educator, 14/1*(10–15), 40–44.
10. Bandura, A. (1986). *Social foundations of thought and action: A social—cognitive theory*. Englewood Cliffs, NJ: Prentice Hall.

Preferred Coaches' Leadership Styles of Malaysian Football Teams

Abbylolita Sullah, Chee Hian Tan and Sarimah Ismail

Abstract Coach's leadership style has a great effect on performance and coach–athlete relationship in a team. Therefore, the purpose of this study was to assess players' preferences for their coaches' leadership styles as well as to evaluate the congruence between players' preferences and the respective coaches' actual behaviour in line with the multidimensional model of leadership. Successful and less successful football coaches ($n = 16$) and players ($n = 200$) of the Malaysia Super League and Malaysia Premier League were identified to participate in the refined questionnaire survey. The five dimensions measured were training and instruction (TI), democratic behaviour (DB), autocratic behaviour (AB), social support (SS) and positive feedback (PF). Descriptive statistics of the 5 dimensions showed high means for successful ($M = 4.25$; 4.15; 2.51; 4.10; and 4.13) and less successful ($M = 4.30$; 4.21; 2.51; 4.20; and 4.22) teams, respectively. Multiple regression analyses were applied to evaluate the congruency between players and coaches. The results indicated that there was a congruence in dimension between successful and less successful football teams with statistical readings of DB of $t(108) = 2.009$ ($p < 0.05$) and $t(108) = 2.544$ ($p < 0.05$). This study revealed an important finding, especially in the congruence of coach–athlete relationships, which demonstrate equal influence towards DB in order to perform. This study has contributed a significant body of knowledge to coaching science.

Keywords Leadership scale for sport (LSS) · Personality traits · Leadership behaviours

A. Sullah (✉) · C. H. Tan · S. Ismail
Faculty of Sport Science and Recreation, UiTM, Shah Alam, Malaysia
e-mail: abbylolita_12@yahoo.com.my

R. Adnan et al. (eds.), *Proceedings of the International Colloquium on Sports Science, Exercise, Engineering and Technology 2014 (ICoSSEET 2014)*,
DOI: 10.1007/978-981-287-107-7_64,

1 Introduction

Leadership in athletics refers to the process of inspiring or influencing athletes of a team to perform their tasks enthusiastically and competently to meet the team's goals [1], and it has been often considered as the main reason for the success or failure of an athlete or a sports team [3]. Within the context of sports coaching, the theoretical multidimensional model of leadership behaviour developed by suggests that actual coach behaviour is influenced not only by the characteristics of the coach (e.g. personality, experience and attitude), but also by the coaching behaviour (required and preferred) that is directly influenced by antecedent situational characteristics (e.g. team sport, individual sport, home or away location, success and failure) and member characteristics (e.g. gender, achievement, motivation and skill level of the athlete) [5]. The model therefore suggests that in order to maximize performance (and satisfaction), there should be congruence between the actual coaching behaviour and preferred and required behaviours.

1.1 Multidimensional Model of Leadership

Although contingency and situational theories are focused on behavioural and situational factors, research has not provided conclusive support for these theories in the sport setting. Chelladurai and Carron [7], Chelladurai [4] recognized in the sport domain that situational theories were unable to explain fully the components of effective leadership. The literature suggested that investigations of leadership in the sport environment required a multiple factor approach. On response, a multidimensional model of leadership [5–7] was synthesized and extended to the athletic context. This model is based on the contingency theory [12], path-goal theory [13, 14], the adaptive reactive theory [15] and the discrepancy model of leadership [20].

According to the MML, situational characteristics, leader characteristics and member characteristics are antecedents of leader behaviour. Furthermore, leader behaviour can be classified as required, preferred or perceived, and the congruence of these factors determines the levels of performance and satisfaction. For the measurement of the three different leader behaviours in sport, Chelladurai and Saleh [8] developed the leadership scale for sport (LSS).

1.2 Coach–Athlete Relationship

Considering the importance of the coach in determining the quality and success of player's sport experience, surprisingly little research exists that identifies most favourable coaching leadership behaviour and factors that influence the efficiency of particular behaviours. Obviously, coaching leadership behaviours are perceived and given meaning by each player resulting in an attitude towards both the coach and the sport experience. Similarly, Shaver [17] has suggested that an individual's behaviour itself determines one's feelings or actions towards the other person. Moreover, Carron and Bennet [2] suggested that in determining coach–athlete relationships, it is necessary to assess not only the coach's personality and leadership styles, but also the player's preferences for such leadership behaviour in their coach.

1.3 Purpose of Study

The purpose of this study was to assess players' preferences for their coaches' leadership styles as well as to indicate the relationship of personality traits (personal attributes and personal qualities) between successful and less successful Malaysian football teams. This insight study from the perception of the players could result in improved experience for all parties as well as develop player potential and could result in a deeper connection between the coach and the players and strongly relate to performance and contribute to better achievement of the football team.

2 Methodology

2.1 Participants

The respondents were 16 football coaches (2 coaches from each state football association) and 200 football players (100 players from the Malaysia Super League and 100 from the Malaysia Premier League) based on the classification of less successful and successful football teams [9, 19].

2.2 Instrumentation Constructiveness

The coaches' self-perception version of the LSS developed by Chelladurai and Saleh [8] was used to measure the leadership behaviours applied by the coaches, and athlete's preferences version of the LSS developed by Chelladurai and Saleh [8] was used to evaluate players preferred leadership behaviours of their coaches. The LSS contains five dimensions of leader behaviour with 40 items: training and instruction (TI) (thirteen items); democratic (nine items); autocratic (five items);

social support (SS) (eight items); and positive feedback (PF) (five items). In the perception version of the LSS, the items were preceded by the statement "my coach…" and were measured by a five-point Likert scale, anchored with never (1), seldom (2), occasionally (3), often (4) and always (5).

In addition, the questionnaire included a section to measure coaches and players' demographic characteristics and the personal attributes and personal qualities consisted of 27 items and were divided into 15 statements of factors of personal attributes and 12 statements of factors of personal qualities, respectively [18].

2.3 Procedures

The football associations identified were contacted via mail, email and phone. They were brief on the purpose of the study and were encouraged to participate in the study. Permission was granted by Football Association of Malaysia (FAM) and each state football association that were involved, and the appointment for onsite data collection was confirmed at least 1 week in advance. On the day of data collection, informed consent was obtained from the coaches and players. The respondents were briefed again on the purpose of the study. Prior to questionnaire administration, the respondents were assured that their participation in the study was completely voluntary and that they may withdraw at any time without penalty. Data were collected using self-administered questionnaires and monitored by helpers. All the respondents completed the questionnaires within half an hour.

2.4 Data Analysis

Descriptive analysis was used to analyse respondent's background information. Reliability analyses used the Cronbach's alpha value to assess the reliability of instruments. Means and standard deviations were calculated to determine the importance of each variable. In addition, multiple regression analyses were conducted to predict the effect of congruence between preferred and actual leadership behaviour variables (TI; democratic behaviour (DB); autocratic behaviour (AB); SS; and PF).

3 Results

The first sample consisted of data from sixteen coaches. Eight of the coaches were from successful teams, and the other eight were from less successful teams which were categorized based on the percentage. This would indicate that the sample, even though small, was representative of the population being studied. The results

Table 1 Frequencies and percentage of coaches' demographics ($N = 16$)

Demographic profiles	Frequency	Percent (%)
Age		
43–50	12	75
50 and above	4	25
Race		
Malay	13	81.25
Chinese	2	12.5
India	1	6.25
Marital status		
Married	16	100
Education		
SPM	16	100
Participation in coaching football		
More than 15 years	16	100
Coaching certification		
A licence	16	100

showed that twelve (75 %) of the coaches were in the 43–50-year-old age group, while the other four (25 % of the coaches) were above 50 years old (Table 1). All sixteen of them were married. In terms of racial origin, most of them (13) were Malays (81.25 %), while 3 were Chinese coaches (12.5 %), whereas Indian coaches comprised of only 6.25 %.

With respect to educational qualifications, sixteen (100 %) of the coaches had at least SPM (Sijil Pelajaran Malaysia) qualification. Overall, all the 16 coaches reported that they were previously involved in coaching football for more than 15 years. This may suggest that the longer the participation in coaching, the more experience they would have acquired in providing knowledge to the team players. Among these 16 coaches, none of them had accomplished the highest coaching certification, which is a pro-diploma licence. The majority of them obtained an A licence coaching certificate. This indicated that not all coaches could simply hold the pro-diploma licence, even though they had long participated in coaching football, in some cases for more than 15 years.

The samples in this study comprised of a total of two hundred football players from four different teams in the Malaysia Super League ($n = 100$) and four different teams in the Malaysia Premier League ($n = 100$). Therefore, the sample in the present study was a close representative of the population of Malaysia football players being investigated. The frequencies and percentages of overall players' demographics are presented in Table 2. More than half of the respondents, or 109 players (54.5 %), were aged between 24 to 29 years, with less than 25 % above 30 years of age. Moreover, it comprised of about forty-eight young players (24 %) aged in the range of 18–23 years. Most of the players were married (122 or 61 %), and seventy-seven players (38.5 %) reported that they were still single. With regard to racial composition, the majority of players were Malay (149 or 74.5 %) with barely four Chinese players (2 %) involved in this study.

Table 2 Frequencies and percentage of players' demographics ($N = 200$)

Demographic profiles	Frequency	Percent
Age		
18–23	48	24
24–29	109	54.5
30–35	43	21.5
Race		
Malay	149	74.5
Chinese	4	2
India	28	14
Others	19	9.5
Marital status		
Single	77	38.5
Married	122	61
Divorced	1	0.5
Education		
SPM	188	94
STPM	1	0.5
Diploma/diploma in education	3	0.5
First degree	2	1
Others	6	3
Participation in football		
11–15 years	55	27.5
More than 15 years	145	72.5
Level of competition		
Super league	125	62.5
Premier league	75	37.5

As for the educational qualifications, about 188 (94 %) of the players had SPM (Sijil Pelajaran Malaysia) qualification, while one player (.5 %) had achieved STPM (Sijil Tinggi Pelajaran Malaysia) qualification, and three players (1.5 %) had attained a diploma/in education. The other two players (1 %) managed to obtain their first degree. Fifty-five players (27.5 %) had between 11–15 years of participation in football, while 145 (72.5 %) players had engaged in football for over 15 years. This may suggest that most of the players had started their football profession as very young adults. There were 125 (62.5 %) players reported playing for the Malaysia Super League, while seventy-five players (37.5 %) were reported as playing for the Malaysia Premier League.

This study attempted to answer which coaching leadership style (TI, PF, AB, DB and SS) applied by the coaches was preferred by players of Malaysian successful and less successful football teams. Comparison between player's successful football teams' preferences and less successful football teams' preferences for coaching leadership behaviour is presented using descriptive analyses.

The ratings for leadership preferences were almost similar for both players from successful football teams and less successful football teams. The ranking in the mean value for players from successful teams was higher than the less successful teams. Overall, the players consistently identified TI and DB as the most preferred

Table 3 Descriptive statistic for players by team

Dimension	Successful teams (n = 100)		Less successful teams (n = 100)	
	Mean	Standard deviation	Mean	Standard deviation
TI	4.25	0.248	4.30	0.322
DB	4.15	0.304	4.21	0.342
AB	2.51	0.655	2.51	1.067
SS	4.10	0.231	4.20	0.351
PF	4.13	0.412	4.22	0.386

Table 4 Predicted factors on preferred coaching leadership by coaches and players of successful football teams ($N = 216$)

Variables	b	Beta	t	p
TI	0.222	0.206	2.009	0.047
DB	0.182	0.209	2.009	0.047
AB	0.042	0.102	1.053	0.295
SS	0.050	0.044	0.455	0.650
PF	0.039	0.060	0.582	0.562

leadership style employed by the coaches, while AB was displayed as the least preferred by the participating football players. Moreover, means for the TI and DB were all significantly higher in the successful football teams as well. The PF and SS were considerably higher in the less successful football teams.

The descriptive statistics for each of the five leadership dimensions and the different groups are presented in Table 3.

This study also answered to what extent was the preferred leadership reported by players congruent with the actual leadership behaviours of their coaches. Multiple regression analysis, a form of general linear modelling, was the multivariate statistical technique used to examine the relationship between the single dependant variable and the set of independent factors. The research problem in this study involved the fundamental purpose of multiple regression analysis to verify the predicted influencing independent factors such as TI, DB, AB, SS and PF towards coaches and players of successful and less successful Malaysian football teams.

As far as coaches and players were concerned, the identifying factors that influence the actual coaches' behaviour and players' preferences could be predicted by applying multiple linear regressions to select independent variables.

Factors of leadership behaviour shown by successful Malaysian football teams were TI as well as DB (Tables 4, 5). This was predicted to influence coaches' actual behaviour and players' perception towards overall coaching leadership, supported with statistical results as follows: $t(108) = 2.009$; $p < 0.05$ and $t(108) = 2.009$; $p < 0.05$, respectively (Table 4). However, AB, SS and PF did not significantly contribute to the predicted preferred coaching leadership as follows: $t(108) = 1.053$; $p > 0.05$, $t(108) = 455$; $p > 0.05$ and $t(108) = -0.582$; $p > 0.05$, respectively.

Table 5 Predicted factors on preferred coaching leadership by coaches and players of less successful football teams ($N = 216$)

Variables	*b*	Beta	*t*	*P*
TI	0.196	0.236	1.654	0.101
DB	0.264	0.339	2.544	0.012
AB	0.022	0.087	0.827	0.410
SS	0.027	0.036	0.300	0.765
PF	0.070	0.103	0.912	0.364

At the same time, the factor of leadership behaviour that was confirmed by less successful Malaysian football teams was merely DB as explained by the predicted influence on coaches' actual behaviour and players' perception towards overall coaching leadership, supported with statistical results as follows: $t(108) = 2.544$; $p < 0.05$ (Table 5). Conversely, TI, AB, SS and PF did not significantly contribute to the predictors of preferred coaching leadership as follows: $t(108) = 1.654$; $p > 0.05$, $t(108) = 0.827$; $p > 0.05$, $t(108) = 0.300$; $p > 0.05$ and $t(108) = 0.912$; $p > 0.05$, respectively (Table 5).

The raw and standardized regression coefficients of the predictors for coaching leadership behaviour are presented in Tables 4 and 5. As for successful football teams, TI received the strongest weight in the model, followed by DB, SS, AB and PF. On the contrary, in less successful football teams, DB received the strongest weight in the model, followed by TI, PF, SS and last of all AB.

The multiple regression results showed that the congruency of those players' coaching leadership preferences with coaches' actual coaching leadership towards Malaysian successful and less successful football teams was measured to have merely DB as predictive influence by both Malaysian successful and less successful football teams. The results have significantly contributed to the body of knowledge on coaching science.

The current study has extended past leadership research by applying the MML [6] into the Malaysian context. The study utilized a well-developed scale of the LSS [8]. In addition, considering the importance of interaction between individuals (coaches and players) and situational characteristics associated with leadership behaviour [12, 13], the study perceived the congruency between the actual coaching behaviour and behaviours preferred by players. Understanding player's preferences on leadership behaviours could help coaches and sport administrators to increase athletes' satisfaction and performance [6] and eventually achieve sporting excellence in any sports.

4 Conclusion

In summary, the LLS consists of one direct task factor (TI), two decision-style factors (democratic and AB) and two motivational factors (SS and PF). The LLS provided the researcher with a valuable tool that has advantages over other proposed structures [10]. These dimensions were consistent with the path-goal theory of leadership [14]. They were conceptually distinct categories of coaching behaviour, and each of the dimensions was relatively reliable. This study believed that there were several situational factors that could sensibly validate personality in predicting coaching leadership behaviours.

The goal of this study was to resolve issues regarding the preference and congruence of leadership. A review of the literature suggests that there is a considerable gap between the importance assigned to coaching leadership behaviour and efforts to understand it [16]. This study on hand has focused on only two aspects of leadership as the important leader behaviour: preferred and actual. This is in line with MML which like the present study focused on the leader, athlete and situation [11]. This research looked at the MML which involves the coach, players and the situation in training or competition. This approach in the study of coaching leadership behaviour of Malaysian football teams allows for a more complete picture.

The unique aspects of the sports environment may demand leader behaviours other than consideration and initiating structure. If in fact the sport context is unique, it becomes necessary to identify the dimensions of leader behaviour that are relevant to football. Utilizing the MML investigations of successful and less successful football players' preferences of leadership behaviour of their coaches appears to be an important variable in the understanding of coach leadership behaviour. The data from this study reinforce research findings and identify TI and DB as the most preferred coach leadership behaviours. Previous research recognizes PF, TI and DB as most preferred by athletes. Hence, the factor that influences success is the training process, which is the coach' ability to transfer knowledge and operate by leading the individuals and the group. Moreover, mutual communication between coach and athlete must be appropriate and reach decisions towards achieving goals.

In spite of these results, when the congruence of coach–athlete relationship was analysed by multiple regressions, the results demonstrate equal influence towards DB. So the players' preferences of coaching leadership behaviour are matching with their actual respective coaches' behaviour. Leadership is always something of a two-way street in which coaches influence players and players, in turn, hold some degree of influence over coaches. Through an analysis of coaching leadership behaviour and events in which they are demonstrated, the coach–athlete relationship can be implied as a complex phenomenon called leadership. As far as coaching leadership behaviour and coach–athlete relationship of Malaysian football teams are concerned, the findings of the present study have contributed a significant body of knowledge to coaching science, from the perspective of multidimensional leadership nationwide.

References

1. Bridges, F. J., & Roquemore, L. L. (1996). Management for athletic/sport administration: Theory and practice. Calgary.
2. Carron, A. V., & Bennett, B. B. (1977). Compatibility in the coach-athlete dyad. *Research Quarterly, 48*, 671–679.
3. Ch'ng, A., & Koh-Tan, A. (2006). *Managing sport: Concepts and issues of non-profit organizations*. Singapore: Prentice Hall.
4. Chelladurai, P. (1990). Leadership in sport: A review. *International Journal of Sport Psychology, 21*, 328–354.
5. Chelladurai, P. (1993). Leadership. In R. N. Singer., M. Murphy & L. K. Tennant (Eds.). New York: Macmillan: Handbook of research on sport psychology.
6. Chelladurai, P., & Carron, A. (1978). Leadership. Ottawa: Sociology of sport monograph series. *Canadian Association for Health, Physical Education and Recreation: University of Calgary.*
7. Chelladurai, P., & Carron, A. (1978). Leadership. Ottawa: Sociology of sport monograph series. *Canadian Association for Health, Physical Education and Recreation: University of Calgary.*
8. Chelladurai, P., & Saleh, S. D. (1980). Dimensions of leader behavior in sports: Development of a leadership scale. *Journal of Sport Psychology, 2*, 34–45.
9. Dana, B. (1974). *The relationship between personality, leadership style and team success* (pp. 1–12).
10. Danielson, R. R., Zelhart, P. F., & Drake, D. J. (1975). Multidimensional scaling and factor analysis of coaching behaviour as perceived by high school hockey players. *Research Quarterly, 46*, 323–334.
11. Duda, J. L. (1998). *Fitness information technology: Advances in sport and exercise psychology measurement*. Morgantown: W. Va. Basic books.
12. Fielder, F. E. (1967). A theory of leadership effectiveness. New York: McGraw Hill. Higher Education Institutions. UPM. PhD thesis (Unpublished).
13. House, R. J. (1971). A path-goal theory of leader effectiveness. *Administrative Quarterly, 16*, 321–338.
14. House, R. J., & Dessler, G. (1974). *A path-goal theory of leadership* (contingency approaches to leadership ed.). Carbondal, IL: Southern Illinois University Press.
15. Osborne, R. N., & Hunt, J. G. (1995). *An adaptive-reactive theory of leadership: The role of macro variables in leadership research*. In J. G. Hunt & L. L. Larson (Eds.). Kent, OH: Kent State University: Leadership frontiers.
16. Riemer, H. A., & Chelladurai, P. (1995). Leadership and satisfaction in athletics. *Journal of Sport and Exercise Psychology, 17*, 276–293.
17. Shaver, K. J. (1975). *An introduction to attribution processes*. Cambridge, MA: Winthrop.
18. Tan, C. H. (2009). *Employability of sport science graduate of Malaysian public*.
19. Turman, P. D. (2001). Situational coaching styles: The impact of success and athlete maturity level on coaches' leadership styles over time. *Small Group Research, 3*, 576–594.
20. Yukl, G. (1971). Toward a behavioral theory of leadership. *Organizational Behavior and Human Performance, 6*, 414–440.

Author Index

R. Adnan et al. (eds.), *Proceedings of the International Colloquium on Sports Science, Exercise, Engineering and Technology 2014 (ICoSSEET 2014)*,
DOI: 10.1007/978-981-287-107-7, © Springer Science+Business Media Singapore 2014

MIX
Papier aus verantwortungsvollen Quellen
Paper from responsible sources
FSC® C105338

If you have any concerns about our products,
you can contact us on
ProductSafety@springernature.com

In case Publisher is established outside the EU,
the EU authorized representative is:
Springer Nature Customer Service Center GmbH
Europaplatz 3, 69115 Heidelberg, Germany

Printed by Libri Plureos GmbH
in Hamburg, Germany